家蚕功能基因组研究

上册

夏庆友 主编

西南师範大學出版社
国家一级出版社 全国百佳图书出版单位

图书在版编目(CIP)数据

家蚕功能基因组研究 / 夏庆友主编. — 重庆 : 西南师范大学出版社, 2021.6
ISBN 978-7-5621-9789-8

Ⅰ. ①家… Ⅱ. ①夏… Ⅲ. ①蚕—基因组—生物学—研究 Ⅳ. ①S881.2

中国版本图书馆CIP数据核字(2019)第084269号

家蚕功能基因组研究 JIACAN GONGNENG JIYINZU YANJIU

夏庆友 主编

责任编辑: 杜珍辉
责任校对: 魏烨昕
装帧设计: 观止堂_未 氓 朱 璇
排　　版: 黄金红
出版发行: 西南师范大学出版社
地址:重庆市北碚区天生路2号
邮编:400715
市场营销部电话:023-68868624
经　　销: 全国新华书店
印　　刷: 重庆升光电力印务有限公司
幅面尺寸: 210 mm×285 mm
印　　张: 63.75
插　　页: 121
字　　数: 1500千字
版　　次: 2021年6月 第1版
印　　次: 2021年6月 第1次印刷
书　　号: ISBN 978-7-5621-9789-8
定　　价: 598.00元(上下册)

夏庆友

男，1965 年生，教授，博士生导师，家蚕基因组生物学专家。牵头完成了家蚕基因组研究标志性的“三部曲”工作，两次作为第一作者在 *Science* 杂志发表论文。主持完成了家蚕全基因组基因芯片、生物信息学数据库、蛋白质组学、定位克隆、转基因、基因干涉以及素材创新等研究技术体系建立，主持开展以丝蛋白合成、免疫防卫、性别决定和发育变态为主的家蚕功能基因组学研究，并取得重要成果。在 *Science*、*Nat Biotech*、*Genome Res*、*PNAS*、*Genome Biol*、*Nucleic Acids Res*、*PLoS Pathog*、*BMC Dev Biol* 和 *J Biol Chem* 等重要杂志发表论文 500 余篇。为国家“973”首席科学家（两届），教育部高级专家，首批“万人计划”领军人才。获国家自然科学奖二等奖、重庆市自然科学奖一等奖、日本蚕丝学会特别奖、第八届光华工程科技奖、第九届中国青年科技奖等奖励和荣誉 30 余项。

向仲怀

男，1937年生，教授、博士生导师、中国工程院院士。长期从事家蚕基因资源研究，建成国际最大的家蚕基因资源库。研究发现新基因20余个，主持育成推广家蚕新品种三对，倡导桑树多倍体育种。领导完成家蚕、微孢子、桑树基因组计划。获国家自然科学奖二等奖、四等奖，四川省科技进步奖一等奖、重庆市自然科学奖一等奖、日本蚕丝学会特别奖、香港何梁何利基金会科学与技术进步奖、香港桑麻基金会纺织科技奖，香港柏宁顿（中国）教育基金会首届、第三届"孺子牛金球奖"，中华农业科教奖等多项成果奖励，荣获全国农业科技先进工作者、四川省有重大贡献科技工作者、全国优秀教师，四川省先进工作者、重庆直辖10周年建设功臣、全国先进工作者等多项荣誉称号。

钟伯雄

男，1958年生，教授，博士生导师，家蚕生物学与家蚕转基因专家。采用数量遗传学、基因组学、蛋白质组学、转基因家蚕等分析方法和技术，从事家蚕茧丝产质量性状基因表达调控、家蚕丝腺生物反应器、杂种优势机理等的研究工作，获得了能够生产黑寡妇蜘蛛牵引丝、人血清白蛋白、溶菌酶、狂犬病病毒糖蛋白等转基因家蚕。主持国家重点基础研究（"973"）、国家自然科学基金、浙江省自然科学基金等科研项目和课题50多项。在 *Journal of Proteome Research*、*BMC Genomics*、*Insect Biochemistry and Molecular Biology* 等国内外杂志发表论文160多篇，获国家发明专利授权18项。

冯启理

男，1958年生，教授，博士生导师，昆虫变态发育分子生物学专家。主持“973”计划项目课题1项，国家自然科学基金重点项目2项、国家自然科学基金国际（地区）合作与交流重点项目1项、国家自然科学基金面上项目4项、广东省自然科学基金重点项目和面上项目各1项。主持完成了昆虫激素信号转导途径对家蚕变态发育作用机制的研究，并取得重要进展。在*PNAS*、*Annu Rev Entomol*、*Nat Ecol Evol*、*Nucleic Acids Res*、*Epigenetics & Chromatin*、*Insect Biochem Mol Biol*、*Autophagy*、*Apoptosis*、*Genome Res*和*BMC Genomics*等国际学术刊物上发表研究论文100多篇。曾获加拿大森林服务部优秀成就奖，广东省“优秀教育工作者”称号。

崔红娟

女，1968年生，教授，博士生导师，研究方向为干细胞生物学与转化；主要以家蚕为研究对象，深度开展生物体表型和细胞功能的研究，构建了适合家蚕等昆虫类的干细胞生物学研究体系和技术平台；将家蚕拓展延伸为医学模式生物，开展人类癌症、退行性、代谢性疾病模型的研究，探索疾病的发生发展的机制，实现医学转化。以第一或通讯作者在*Nat Cell Biol*, *Cel Rep*, *Oncogene*, *Neuro-Oncology*, *J Biol Chem*等国际期刊发表SCI论文50余篇，影响因子超过300分；并申请专利多项。主持“973”课题、国家重点研发计划课题、国家自然科学基金、国家留学基金、教育部基金等多个项目。入选重庆市首批“百人计划”、重庆市首批“特支计划”、重庆市“巴渝学者”等人才计划；获得重庆市自然科学奖二等奖和国务院政府特殊津贴等荣誉奖励。

周泽扬

男，1958年生，教授，博士生导师。首批“新世纪百千万人才工程”国家级人选，享受国务院政府特殊津贴专家，农业部有特殊贡献的中青年专家。原西南农业大学常务副校长、重庆师范大学原校长、家蚕基因组生物学国家重点实验室学术委员会副主任。推动开展了家蚕基因组计划，组织完成了家蚕病原性微孢子虫基因组的测序分析，主持开展了蚕桑病原微生物抗病、检测、功能基因组学、寄生虫与宿主互作关系以及资源微生物开发利用等研究。近年来，先后主持“863”项目、“973”项目子课题、国家转基因植物专项、国家自然科学基金重点项目等多项重大课题，先后在 *Science*、*PLoS Pathogens*、*Proteomics*、*Infection and Immunity* 等国内外学术杂志发表研究论文130余篇。编写专著《家蚕微孢子虫基因组生物学》。获国家自然科学奖二等奖、重庆市自然科学奖一等奖、教育部高等学校自然科学奖二等奖、第四届中国科协西部开发突出贡献奖、“全国五一劳动奖章”等奖励。

李　胜

男，1971年生，教授、博士生导师，教育部高级专家、国家杰青、中科院“百人计划”入选者、全国优秀科技工作者。华南师范大学昆虫科学与技术研究所所长，广东省昆虫发育生物学与应用技术重点实验室主任，广东省广梅园昆虫新型研发机构主任。任 *Journal of Insect Physiology* 等5个昆虫学SCI期刊编委，任中国昆虫学会理事、广东省昆虫学会副理事长，中国昆虫学会昆虫发育与遗传专业委员会主任等学术职务，是我国现代昆虫科学研究领头人之一。从事昆虫科学和技术研究20余年，在动物发育、代谢和进化领域，尤其是“昆虫变态发育的调控机制与进化规律”方向取得了一系列国际领先和原创性的研究成果，在资源昆虫和再生动物的资源开发利用方面做出了开拓性的贡献。发表SCI论文104篇，其中 *Annu Rev Entomol*、*Autophagy*、*Nat Commun*、*PNAS* 等通讯作者和第一作者论文64篇，多篇论文遴选为封面论文或者作为专题评论；申请专利15项，授权专利10项。组织和承担了国家自然科学基金重点项目和重点国际合作项目、“973”课题、“863”项目、转基因专项等国家级研究任务。多次应邀在重大国际会议做主题和大会报告，多次担任国际大会主席和组委员成员，享有重要的国内外学术声誉。

黄勇平

男，1963年生，博士生导师，昆虫分子遗传学专家。中国科学院上海生命科学院植物生理生态研究所研究员。在昆虫分子遗传、昆虫功能基因组和害虫遗传防治方面开展基础与应用基础研究。国务院政府特殊津贴和国家杰出青年基金获得者。曾任上海昆虫学会理事长、亚太化学生态学会主席。现任中国昆虫学会副理事长。在*PNAS*、*Development*、*Cell Research*、*Journal of Proteome Research*、*BMC Genomics*、*Insect Molecular Biology*等国际学术杂志上发表研究论文100余篇。已申请国家发明专利8项，获得授权6项(正在申请国际PCT)1项。

鲁　成

男，1957 年生，教授，博士生导师，长期从事家蚕资源、遗传育种、细胞生物学和功能基因研究。建成世界最大的家蚕遗传资源中心；作为中国家蚕基因组计划主要研究人员，共同完成世界首张家蚕基因组框架图；主持创建家蚕卵巢、胚胎等细胞系 7 个。主持承担“973”、“863”、“948”、国家科技基础调查专项、国家自然科学基金等科研课题 20 余项。获国家自然科学奖 2 项、省级一等奖 4 项等 10 余项科技成果奖；在 *Science*、*PNAS* 等国内外刊物上发表论文 200 余篇；主编、参编专著 10 余部。被评为国家“百千万人才工程”一、二层次人才，国家级有突出贡献的中青年专家、全国优秀科技工作者（2016）等。国家蚕桑体系首席科学家、中国蚕学会名誉理事长、重庆市遗传学会理事长。

沈卫德

男，1950年生，教授，博士生导师，国家蚕桑产业技术体系执行专家，苏州大学蚕桑研究所原所长、蚕桑学科带头人。用8年时间，在专科层次的蚕桑学科上建立了硕士点、博士点、博士后流动站，书写了学科建设的佳话。培养博士研究生15人，硕士研究生57人，获全国优秀硕士论文指导教师称号。主持“973”课题，“863”课题，“948”课题和国家公益性行业专项，蚕桑体系岗位科学家等重大项目多项。在家蚕生理生态，分子毒理，以及功能基因组领域取得重要进展，发表SCI收录研究论文45篇，授权发明专利15件，成果获得省部级科技进步奖二等奖5项，三等奖2项，市厅级科技进步奖一等奖3项。

张耀洲

1964年生，教授，天津大学人类基因组研究中心原主任，天津滨海新区科协原主席，获国务院政府特殊津贴。教育部跨世纪人才，国家百千万人才第一二梯队人才。主要研究基因大数据、中药现代化与生物工程制药，获得省部级以上科研奖项8项，其中家蚕生物反应器制备生物制品的方法获2004年国家技术发明奖二等奖（第一完成人）。在*Nature Structural Biology*、*Proteomics*、*RNA*、*BMC Genomics*等杂志上发表论文400余篇，其中SCI 90余篇，专著6部，申请1 000余项专利，已授权200余项，1994年首次报道家蚕*Bm*NPV的*P10*基因构建了双启动子表达载体，完成家蚕蛹全转录组，首次实现口服hGM-CSF获准进入临床试验。培养研究生400余名，其中博士、博士后40余名。

王　俊

男，1976年生，研究员、博士生导师。主要从事基因组学、生物信息学大数据分析，机器学习和人工智能等方面的研究。在家蚕基因组计划中主要负责了测序、组装、注释、比较和进化在内的各种数据分析及工具、数据库开发等。在专业期刊上共发表300篇论文，第一作者或通讯作者超过100余篇，多篇论文发表在*Science*、*Nature*、*Cell*及*NEJM*等杂志，获评英国《自然》杂志2012年科学界年度十大人物。973首席科学家，国家杰出青年基金获得者，曾任华大基因CEO，华大基因研究院院长；丹麦哥本哈根大学、香港大学客座教授。获国家自然科学奖二等奖、陈嘉庚青年科学奖等奖励。

曹　阳

男，1958年生，华南农业大学动物科学学院教授，博士生导师，家蚕遗传发育与免疫学专家。“广东省农业动物基因组学与分子育种重点实验室”和“广东省蚕桑工程技术研究中心”PI，学校二级学科带头人。在遗传毒理、昆虫（家蚕）免疫与细胞自噬和凋亡（PCD）等方向有深入研究。先后承担国家“973”、“863”计划、国家自然科学基金和省部级课题30余项，在*Autophagy*、*Apoptosis*、*IBMB*、*BMC Genomics*、*BMC Mol Biol*等国际期刊和国内核心期刊发表论文60余篇，其中SCI期刊收录论文20余篇。获得发明专利4项，获得中国人民解放军科学技术进步奖二等奖和广东省科学技术奖三等奖等4项奖励。已培养硕士38名、博士10名和出站博士后2名。《昆虫学报》和《蚕业科学》编委，《中国农业科学》栏目主审专家。

郭锡杰

男，1959年生，研究员，博士生导师，工作单位江苏科技大学、中国农业科学院蚕业研究所。主要从事家蚕分子病理学与蚕病防治研究，近年来在家蚕应对病毒、真菌感染的免疫应答和抗病分子机制研究方面取得重要进展。在国内外发表学术论文100余篇，获得教育部科技进步奖二等奖等科技成果奖励。现任农业部兽药审评委员会委员，农业部全国动物防疫专家委员会委员，中国蚕学会主办《蚕业科学》主编。

何宁佳

女，1971年生，1998年获农学博士学位，教授，博士生导师，“万人计划”领军人才，“巴渝学者”特聘教授，重庆市首届十佳科技青年奖获得者，中国桑树基因组计划首席科学家。曾留学日本信州大学、九州大学和美国佐治亚大学。是我国第一批从事蚕桑分子生物学研究的科研人员，在家蚕作为鳞翅目害虫防治的模式昆虫方面开辟了新的研究领域。先后主持承担了“973”子课题、“863”重点项目、国家自然科学基金、重庆市杰出青年基金等多项研究课题。带领西南大学桑树研究团队开展了桑树基因组的测序研究工作，解析获得了接近3万个桑树基因，建立了桑树基因组数据库、蛋白质组学平台、次生物质代谢组学平台，开创了“家蚕-桑树”在研究植食性昆虫与植物关系的新领域，全面确立我国在桑树功能基因组研究上的核心优势。现担任重庆市蚕学会常务理事，重庆市青科联常务理事、重庆市蚕桑品种审定委员会副主任委员，教育部蚕桑资源及分子改良工程研究中心主任和农业部桑树品种改良中心重庆分中心主任。在国际杂志上发表高水平SCI研究论文40多篇，并担任重要学术杂志审稿人。参编专著一部，翻译译著一本，为学科发展产生了积极的带动作用。

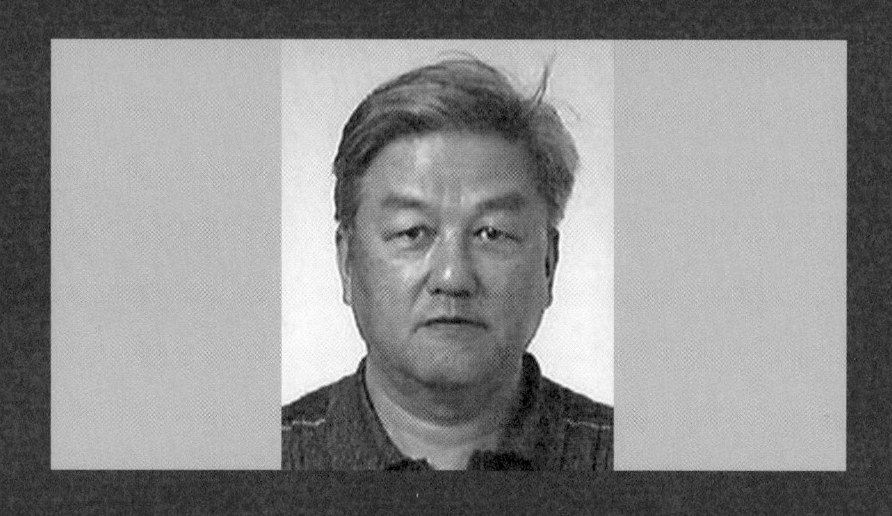

张芝利

男，1938年生，研究员，原北京市农林科学院植保所所长，享受国务院政府特殊津贴。多年来从事农业害虫综合治理研究，获多项省、部级科技奖。1991年主持“无公害菜果生产技术研究”获北京市科技进步奖一等奖，该项成果对6种菜果提出了11套实用的无公害生产技术规程和检测技术，在国内外产生重要影响。任北京昆虫学会名誉理事长，中国植物保护学会副理事长，兼任科技部农业领域“973”专家咨询组成员及多个国家重点实验室学术委员会委员。作为秘书长组织筹备了在中国召开的第19届国际昆虫学大会，大大地促进了我国昆虫学研究的国际化进程。荣获全国优秀科技工作者、北京市先进工作者等荣誉称号。获中国昆虫学会首届终身成就奖、全国五一劳动奖章、首都劳动奖章。

陈　杰

男，1938年出生，南京农业大学动物生理学教授、博士生导师。从事营养生理学，泌乳、生长、生殖生理及调控，反刍动物微生物内分泌学等研究近60年。曾在瑞士和德国进修，建立了长期的国际科研合作关系。曾任南京农业大学农业部动物生理生化重点开放实验室主任，全国动物生理生化学会理事长，国务院学位委员会第三届学科评议组成员、第四届召集人，国务院学位委员会兽医专业学位指导委员会主任，国家自然科学基金委员会第六、九届评审组成员、召集人，农业部教育指导委员会成员，科技部“973”计划第三届咨询专家组成员，享国务院特殊津贴。先后荣获部、省级科技进步奖8项，国家发明专利3项。编写专著两部、主编家畜生理学全国统一教材（第四版）、发表论文160余篇。

喻子牛

男，1941年出生，华中农业大学教授、博士生导师、国家级专家。科技部“973”农业领域咨询专家、湖北省科技精英，享受国务院政府特殊津贴。他是20世纪60年代陈华癸院士培养的研究生，曾赴美国农业部、英国剑桥大学等访学，进行合作研究。先后承担国家重点基金，“863”、“948”、世界卫生组织、美国杜邦公司等资助的课题50多项。获国家和省部级奖励14项、国家发明专利40多项，发表学术论文500多篇，其中SCI收录240篇。主编了专著和教材14部。培养了硕士、博士研究生和博士后180多人。作为学科负责人创建了国家重点实验室、国家工程研究中心和农业农村部质检中心。研发了微生物农药和废弃物高效生物转化多种产品，科技成果转化经济效益显著。被联合国世界卫生组织、美国杜邦公司等40多个单位聘为兼职教授或高级顾问。主持召开国内外学术会议20多次。出访了40多个国家，在专业领域颇具影响力。

张家骅

男，1946年出生，1990年留学美国，获美国伊利诺伊大学兽医学院兽医基础科学博士学位，西南大学二级教授，博士生导师，原西南大学动物科技学院院长、西南农业大学副校长、重庆市高校牧草与草食家畜重点实验室主任、重庆市临床兽医学重点学科学术带头人。社会工作包括：国家"973"项目农业领域咨询专家组组长、中国畜牧兽医学会理事、兽医产科学分会名誉理事长等。长期以来一直从事兽医产科学、动物生殖内分泌学和动物遗传育种与繁殖的教学与科研工作。先后主持10余项国家和省部级课题，获省部级二等奖3项，三等奖5项，在国内外刊物公开发表论文100余篇，编写教材和专著10部，培养博士生10名、硕士生20余名。主持选育的高繁殖率大足黑山羊已通过国家遗传资源认定。享受国务院政府特殊津贴，获得省部级有突出贡献专家、重庆市优秀共产党员、重庆市优秀农业科技工作者、两部四省市武陵山区扶贫先进个人、科技部"973计划重要贡献者"、新中国60年畜牧兽医科技贡献奖（杰出人物）等称号和荣誉。

杨公社

男，1959年生，陕西富平人，西北农林科技大学二级教授、博士生导师、校学术委员会委员，动物遗传育种与繁殖国家重点学科带头人。第十一届全国政协委员，第五、六届国务院学位委员会学科评议组成员、国家“973”计划领域咨询组专家、国家转基因猪新品种培育重大专项执行专家、国家生猪遗传改良计划专家组成员、国家畜禽遗传资源委员会猪专业委员会委员和国家生猪产业技术体系岗位科学家。主要从事肌肉生物学与猪遗传改良教学和科研工作。先后主持国家和省部级课题20余项，发表论文170余篇，其中SCI论文136篇，主编《猪生产学》和《肉类学》等多部学术著作，获省部级科技奖4项。培养博硕士研究生130余人，多人入选国家级人才。

徐卫华

男，1961年生，教授，博士生导师，从事昆虫滞育的研究。以农业害虫棉铃虫为材料，围绕滞育的诱导、准备、起始、维持和打破开展工作，鉴定了许多滞育相关重要基因，如促前胸腺激素、滞育激素、蜕皮激素、胰岛素等，阐明了这些激素如何作为上游信号控制滞育的起始；发现了激素通过调节脂肪体的代谢活性，反馈抑制脑的活性从而诱导滞育进入的机制；找到了昆虫脑控制代谢活性的关键点——三羧酸循环；鉴定出一条调节脂肪体发育分化的信号路径：蜕皮激素-NF·κB-半胱氨酸蛋白酶；发现光照通过表观遗传因子调节促前胸腺激素基因上的甲基化修饰控制滞育表型。在*PNAS*、*PLoS Genetics*、*J Biol Chem*、*Biochemical J*、*ACTA*、*Journal of Proteome Research*等杂志发表论文50余篇。担任*Journal of Insect Physiology*编委，《昆虫学报》副主编。

周丛照

男，1968年出生于湖北麻城。1987年至2000年就读于中国科学技术大学生物系并先后获学士、硕士和博士学位。2000年至2003年在法国国家科研中心及巴黎南大学结构基因组学实验室从事博士后研究。2004年入选中国科学院“百人计划”，2009年至2014年任中国科学技术大学生命科学学院副院长，2014年起任中国科学技术大学教务处处长，2019年起任中国科学技术大学校长助理、教务长。在*Nature*子刊、*PNAS*、*PLoS Pathogens*、*EMBO Reports*、*Nucleic Acids Res*、*J Biol Chem*和*J Mol Biol*等杂志发表研究论文100余篇，总引用近5 000多次。曾获得1998年度安徽省自然科学二等奖和2000年中法科技交流协会生物技术奖。2005年10月当选中国生物化学与分子生物学学会理事，2014年起任安徽省生物化学与分子生物学学会理事长，2015年当选亚洲及大洋洲生物化学家与分子生物学家联盟(FAOBMB)中国代表。2006年至2010年任科技部重大研究计划首席科学家，2015年起为科技部重点研发计划指南编制成员和总体专家组成员。

李 龙

男，1960年生，教授，蚕学专家。中国农业科学院蚕业研究所副所长，农业部蚕桑产业产品质量监督检验测试中心主任，农业部蚕桑产品及食用昆虫质量安全风险评估实验室主任，全国桑蚕业标准化技术委员会副主任。国家蚕桑产业技术体系岗位专家，中国蚕学会常务副理事长，中国丝绸协会常务理事，国家商务部市场运行调控专家，江苏省原子能农学会副理事长。先后受聘担任国际昆虫生理生态中心（ICIPE）、瑞士开发署（SDC）、世界银行（World Bank）、联合国国际农业发展基金会（IFAD）蚕桑专家，中国-古巴蚕桑科技合作中心中方负责人。先后主持商务部、科技部、农业部、江苏省等国家级和省部级项目20余项。获2016年古巴国家科学奖（国际合作部分）。

鲁兴萌

男，1962年生，教授，博士生导师，家蚕病理学与病害控制技术专家。以家蚕微粒子病为主要研究领域，从病原生物学、病原与家蚕及中肠微生态的相互作用关系、病原扩散和病害流行规律、检测与检疫技术以及防控技术体系等维度，开展了较为系统的研究，发表家蚕微粒子病相关论文70余篇。此外，也涉足家蚕病毒性软化病、家蚕血液型脓病、污染物和农药对家蚕的毒性、养蚕用药的评价技术，以及其他养蚕病害防控相关技术的研究与产品开发。主编蚕学相关图书4部、担任副主编1部、参编6部，参与项目获省部级科技进步奖4项，参与获国家级教学改革成果奖2项，获专利5项。农业部现代农业产业（蚕桑）技术体系蚕病害控制岗位科学家。

编写说明

《家蚕功能基因组研究》收录了发表于国际知名期刊，有关家蚕研究的各类文献的摘要。本书的编写，在充分尊重原文献与作者的前提下，对书中文字的大小写、正斜体、文献出处、作者单位编写格式等做了适当的统一。

下面对本书中的一些特殊情况做相关说明：

1. 由于各类期刊，特别是国外期刊对卷和期的标准并不统一，原文献中有卷或期信息的则如实写明。原文献中没有此类信息的，为确保信息准确，则书中未擅自添加。格式上尽量保持原文刊发时的原貌，故可能不同篇幅间有不统一之处。

2. 本书中的人名、单位、地址等以作者提供的信息为准。经作者要求，外文人名并未翻译为中文。因各文献发表时间的不同，有些信息可能已经过时，比如电子邮箱的信息，仍以当时发表时的信息为准。

3. 由于篇幅有限，书中的插图并未将所有文献中的图片一一列出，只选取了部分文献中有代表性的插图。

4. 全书的中文译文由作者提供，个别语句（包括个别标题）采用了意译的方式，并未做一字一句的对译，只列出了主要的信息。

本书编委会

前言

《家蚕功能基因组研究》一书的成稿历时数年，终得付梓，实为不易。细细回想，本书的出版问世，多亏了三件事情。

家蚕（*Bombyx mori*）具有比较特殊的地位。一方面其作为人类驯化最为成功的昆虫之一（另外一个当数蜜蜂），支撑了整个蚕丝产业，并传承数千年至今；另一方面，因其具有诸多研究的便利，成为重要的生物模式，为遗传学的发展做出了积极贡献。2000年前后，随着基因组规模化测序技术的建立和明星生物模式果蝇的全基因组序列发表，人们很容易就想到了应尽快启动家蚕的基因组测序工作。正因为如此，到底谁能率先完成该项工作，还经历了一场较为激烈的国际竞争。2003年初，我所在的团队联合华大基因（BGI）启动了家蚕全基因组测序计划，并克服了诸多困难，于2004年正式向世界同行公布了家蚕430M基因组序列组装和分析结果，使家蚕成为继果蝇之后完成全基因组序列的第二个昆虫。这是第一件十分幸运的事情。

随着基因组数据的公布和发表，家蚕一时成为同行讨论的话题，吸粉无数。但热闹总是短暂的，留给课题组的压力却如石如山，如影相随。当时我们面临的主要困境有三：一是家蚕的关键科学问题尚未凝练清楚，二是规模化的研究技术与平台严重缺乏，三是基础研究队伍十分弱小。这使得家蚕基因组信息的解读和功能基因组的研究困难重重，举步维艰。在这样一个关键的时期，科技部批准了家蚕"973"计划立项，犹如及时雨一般，恰逢其时，意义重大。彼时国家"973"项目数量十分有限，家蚕项目三上香山参加终轮答辩，最终我们的态度和决心打动了专家们，2005年"家蚕主要经济性状功能基因组与分子改良研究"项目获得科技部正式批准立项，开启了家蚕功能基因组研究一个崭新的时代。这是第二件十分幸运的事情。

2011年，家蚕"973"项目一期顺利结题，因完成情况较好，2012年得到滚动支持，"家蚕关键品质性状分子解析及分子育种基础研究"得以启动实施，至2016年正式结题，前后共计10年。有了"973"计划的强大支撑，我们才得以凝练科学问题，建立研究技术平台，并组织了精干队伍，积极推动了家蚕功能基因组学和分子生物学的发展。尤其让人感动，并足以载入蚕学发展史册的是，由西南大学牵头组织，不但有来自浙江大学、苏州大学、华南农业大学等单位一批国内优秀的蚕学家，而且更有来自中国科学院上海植生所、中国科技大学、华南师范大学等单位的众多杰出昆虫学和生化学家，加入了家蚕"973"项目，大家共同探讨，密切合作，为家蚕的基础研究奉献了一批精彩的成果。从2006年到2016年的这10年，可以说是中国蚕业科学发展最快，成果最丰，影响最大的一个时期，也正是得益于这个时期的奋斗，中国在世界家蚕基础研究领域的地位才得以超越。此为第三件幸运的事情。

2015年国家实施科技体制改革，"973"计划正式并入国家重点研发计划。尽管两期家蚕项目结题都被评为优秀，但终因国家体制变动，家蚕研究未获得连续资助，家蚕"973"计划也就成为了离我们最近的一个经典记录。正因为如此，2016年项目结题的时候，不少课题组成员建议汇集这10年的研究成果，以专题著作的形式出版，以资纪念，并方便后继青年学子们查询检索，学习借鉴。此即为本书的由来。

全书共分13章，整理收录了部分家蚕“973”项目资助发表论文484篇，所涉及的内容十分丰富，关键科学问题方面有家蚕丝蛋白合成调控、变态发育机制，免疫与疾病抵抗性、性别决定机制、比较基因组与进化等；关键技术方面有遗传操作、基因编辑和基因定位等；关键平台包括生物信息、干细胞、代谢组、表观组和miRNA等；关键应用领域包括生物反应器、素材创制和品种培育等。几乎涵盖了家蚕从基础到应用所有主要领域和方向。为了最大限度反映原创成果的真实风貌，除每个章节附上了一个简短的介绍外，全书采用了以论文为基本叙述单元的形式。该形式尽管系统性和整体性偏弱，但我们相信细心的读者自会从中获得家蚕功能基因组研究的整体构架，洞悉家蚕分子生物学的基本逻辑和特点。倘若读者诸君能够从中感悟一二，并对未来的研究工作有所裨益的话，那就是另外一件让我们十分高兴的事情了。

在此，我们要满怀崇敬，衷心感谢为家蚕“973”计划做出了巨大贡献的向仲怀院士、张芝利先生、陈杰先生、喻子牛先生和杨公社先生。向院士既是家蚕基因组计划的领导者，又是家蚕“973”项目的策划者，自始至终主导了家蚕的基础科学研究。张芝利先生、陈杰先生和喻子牛先生作为“973”项目的咨询专家，自始至终陪伴着家蚕项目一路前行。老一辈科学家们不但学术造诣深厚，富有远见卓识，而且对我们后辈学人的关心、关爱和指导也无微不至，感人至深。他们的科学修养和人文精神，是我们永远学习的榜样。

特别感谢参与家蚕“973”项目的所有学者，他们不但是本书所记录成果的直接完成人，同时也对本书出版给予了热情的支持。特别感谢为本书付出了艰辛工作的侯勇、董照明、程道军、王菲、查幸福、马三垣、王峰、李志清、杨丽群、刘春、王鑫、程廷才、蒋亮、钱文良等老师，以及刘红玲、董浩南、郭凯雨、刘青松、彭章川、秦道远、田弛、王若琳、孙强、何雪川、刘丽婧、张晓璐、王伟、秦利均、常珈菘、陈恩祥、李懿、陆卫、张天镭、施润、孙乐等诸多研究生们，他们在百忙的科研工作中抽出了宝贵的时间和精力，并完成了如此繁重的整理、翻译和校对工作。但因为时间、水平和篇幅有限，未能包含所有涉及论文，尚有遗漏甚至错误之处，还请作者和读者批评指正。

最后，借用伟大诗人的一句诗，为这本迟到的但倾注了我们心血的专著做一个广告语：

仍然拥有的彷佛从眼前远遁，已经逝去的又变得栩栩如生。——歌德《浮士德》

夏庆友

2021年1月22日于重庆

目 录

第一章 家蚕丝蛋白合成机制与蚕丝生物材料

第二章 家蚕变态发育机制研究

第三章 家蚕免疫机制研究

第四章 家蚕性别调控机制研究

第五章 蚕的遗传操作与基因编辑

第六章 家蚕生物反应器研究

512 An optimized sericin-1 expression system for mass-producing recombinant proteins in the middle silk glands of transgenic silkworms

513 转基因家蚕中部丝腺中丝胶1表达系统高量表达重组蛋白的优化

514 Immobilization of foreign protein in *Bm*NPV polyhedra by fusion expression with partial polyhedrin fragments

515 通过与部分多角体片段的融合表达将外源蛋白固定在*Bm*NPV多角体中

516 Advanced silk material spun by a transgenic silkworm promotes cell proliferation for biomedical application

517 通过转基因家蚕获得的先进蚕丝材料可以促进细胞增殖从而应用于生物医学领域

518 Reducing blood glucose levels in TIDM mice with an orally administered extract of sericin from *hIGF-I*-transgenic silkworm cocoons

519 口服*hIGF-I*转基因家蚕蚕茧丝胶蛋白的提取物降低TIDM小鼠血糖水平

520 Nonvirus encoded proteins could be embedded into *Bombyx mori* cypovirus polyhedra

521 非病毒编码的蛋白质能被包被进家蚕质型多角体

522 Overexpression of recombinant infectious bursal disease virus (IBDV) capsid protein VP2 in the middle silk gland of transgenic silkworm

523 利用转基因家蚕中部丝腺过表达重组传染性法氏囊病毒衣壳蛋白VP2

524 Large-scale production of bioactive recombinant human acidic fibroblast growth factor in transgenic silkworm cocoons

525 具有生物活性的重组人酸性成纤维细胞生长因子在转基因家蚕茧中的大规模生产

526 Development of a VLP-based vaccine in silkworm pupae against rabbit hemorrhagic disease virus

527 基于VLP进行家蚕蛹抗兔出血症病毒疫苗开发

528 Transgenic silkworms secrete the recombinant glycosylated MRJP1 protein of Chinese honeybee, *Apis cerana cerana*

529 转基因家蚕分泌中华蜜蜂的重组糖基化MRJP1蛋白

530 Analysis of the *sericin1* promoter and assisted detection of exogenous gene expression efficiency in the silkworm, *Bombyx mori* L.

531 家蚕*sericin1*启动子的活性分析及外源基因表达的辅助检测

532 Expression of hIGF-I in the silk glands of transgenic silkworms and in transformed silkworm cells

533 转基因家蚕丝腺组织和转化家蚕细胞表达hIGF-I

534 The promoter of *Bmlp3* gene can direct fat body-specific expression in the transgenic silkworm, *Bombyx mori*

535 *Bmlp3*启动子可以启动基因在转基因家蚕的脂肪体中特异性表达

第七章 表观遗传与miRNA

562 Involvement of microRNAs in infection of silkworm with *Bombyx mori* cytoplasmic polyhedrosis virus (*Bm*CPV)

563 家蚕感染质型多角体病毒（*Bm*CPV）相关的microRNAs

564 Characterization and expression patterns of *let-7* microRNA in the silkworm (*Bombyx mori*)

565 家蚕*let-7* microRNA基因的特征和表达模式

566 Identification and characteristics of microRNAs from *Bombyx mori*

567 家蚕microRNA的鉴定与特征

568 Insect-specific microRNA involved in the development of the silkworm *Bombyx mori*

569 昆虫特异的microRNA调控家蚕的生长发育

570 Computational identification and characteristics of novel microRNAs from the silkworm (*Bombyx mori* L.)

571 家蚕新型microRNA的计算机鉴定及其特征

572 MicroRNAs show diverse and dynamic expression patterns in multiple tissues of *Bombyx mori*

573 MicroRNAs在家蚕多种组织中呈现多样性及动态表达模式

574 MicroRNAs of *Bombyx mori* identified by solexa sequencing

575 利用高通量测序鉴定家蚕中的MicroRNAs

576 Bioinformatics analysis on structural features of microRNA precursors in insects

577 昆虫中microRNA前体结构特征的生物信息学分析

578 Expression analysis of miRNAs in *BmN* cells

579 *BmN*细胞中miRNAs的表达分析

580 MicroRNA expression profiling of the fifth-instar posterior silk gland of *Bombyx mori*

581 五龄家蚕后部丝腺microRNA表达谱分析

582 Characterization and profiling of microRNAs in posterior silk gland of the silkworm (*Bombyx mori*)

583 家蚕幼虫后部丝腺microRNAs的鉴定与表达分析

584 *BmNPV-miR-415* up-regulates the expression of *TOR2* via *Bmo-miR-5738*

585 *BmNPV-miR-415*通过*Bmo-miR-5738*上调表达*TOR2*基因

586 *Bmo-miR-2758* targets *BmFMBP-1* (Lepidoptera: Bombycidae) and suppresses its expression in *BmN* cells

587 家蚕*miR-2758*体外下调丝腺转录因子*FMBP-1*的表达

第八章 蚕的细胞生物学

第九章 家蚕突变体分子机理研究

第十章 家蚕蛋白质组研究

714 Comparative proteomic analysis reveals the suppressive effects of dietary high glucose on the midgut growth of silkworm
715 比较蛋白质组学分析揭示摄食高葡萄糖对蚕中肠生长的抑制作用

716 Development of an effective sample preparation approach for proteomic analysis of silkworm eggs using two-dimensional gel electrophoresis and mass spectrometry
717 用于双向电泳和质谱检测的蚕卵蛋白质高效样品制备方法开发

718 Shotgun analysis on the peritrophic membrane of the silkworm *Bombyx mori*
719 家蚕围食膜“鸟枪法”蛋白质组学分析

720 Comparative proteomic analysis between the domesticated silkworm (*Bombyx mori*)reared on fresh mulberry leaves and on artificial diet
721 新鲜桑叶和人工饲料饲养的家蚕间的比较蛋白质组学研究

722 Shotgun proteomic analysis of *Bombyx mori* brain: emphasis on regulation of behavior and development of the nervous system
723 鸟枪法蛋白质组分析家蚕脑组织:着重于行为的调控以及神经系统发育

724 Proteomic analysis of the silkworm (*Bombyx mori* L.) hemolymph during developmental stage
725 家蚕不同发育阶段血淋巴蛋白质组学分析

726 Proteomics of larval hemolymph in *Bombyx mori* reveals various nutrient-storage and immunity-related proteins
727 家蚕幼虫血淋巴蛋白质组学揭示各种营养储存和免疫相关蛋白

728 Determination of protein composition and host-derived proteins of *Bombyx mori* nucleopolyhedrovirus by 2-dimensional electrophoresis and mass spectrometry
729 通过二维电泳和质谱法检测家蚕核型多角体病毒的蛋白质组成及宿主来源的蛋白质的分析

730 Comparative proteomic analysis reveals that caspase-1 and serine protease may be involved in silkworm resistance to *Bombyx mori* nuclear polyhedrosis virus
731 比较蛋白质组学分析显示半胱天冬酶-1和丝氨酸蛋白酶可能参与了家蚕对家蚕核型多角体病毒的抗性

732 Proteomics analysis of digestive juice from silkworm during *Bombyx mori* nucleopolyhedrovirus infection
733 感染核型多角体病毒后家蚕消化液的蛋白质组学分析

734 Analysis of protein expression patterns of silkworm *Jinqiu* and its cross parents
735 家蚕“金秋”及其亲本之间的蛋白表达模式分析

736 A novel droplet-tap sample-loading method for two-dimensional gel electrophoresis
737 一种新的二维凝胶电泳液滴取样加载方法

816 Genome-wide comparison of genes involved in the biosynthesis, metabolism, and signaling of juvenile hormone between silkworm and other insects

817 全基因组比较分析家蚕和其他昆虫中参与保幼激素生物合成、代谢和信号传导的基因

818 *In silico* identification of *BESS-DC* genes and expression analysis in the silkworm, *Bombyx mori*

819 家蚕*BESS-DC*基因的生物信息学鉴定与表达分析

820 Identification and characterization of the *cyclin-dependent kinases* gene family in silkworm, *Bombyx mori*

821 鉴定并分析家蚕细胞周期蛋白依赖性蛋白激酶基因家族

822 Identification and analysis of YELLOW protein family genes in the silkworm, *Bombyx mori*

823 家蚕YELLOW蛋白家族基因的鉴定与分析

824 Identification and expression pattern of the chemosensory protein gene family in the silkworm, *Bombyx mori*

825 家蚕化学感受蛋白基因家族的鉴定和表达特征分析

826 Analysis of the structure and expression of the 30K protein genes in silkworm, *Bombyx mori*

827 家蚕30K蛋白基因的结构和表达分析

828 Species-specific expansion of C2H2 zinc-finger genes and their expression profiles in silkworm, *Bombyx mori*

829 家蚕C2H2型锌指蛋白基因物种特异性扩增及表达谱分析

830 Characterization of multiple *CYP9A* genes in the silkworm, *Bombyx mori*

831 家蚕多个*CYP9A*基因的特征分析

832 Identification and analysis of Toll-related genes in the domesticated silkworm, *Bombyx mori*

833 家蚕Toll相关基因的鉴定和分析

834 Identification of *MBF2* family genes in *Bombyx mori* and their expression in different tissues and stages and in response to *Bacillus bombysepticus* infection and starvation

835 家蚕*MBF2*家族基因的鉴定及其在不同组织和时期的表达以及对家蚕饥饿和黑胸败血芽孢杆菌的应答反应

836 Identification and characterization of *piggy*Bac-like elements in the genome of domesticated silkworm, *Bombyx mori*

837 家蚕基因组中*piggy*Bac类似转座元件的鉴定和描述

838 The UDP-glucosyltransferase multigene family in *Bombyx mori*

839 家蚕尿苷二磷酸-糖基转移酶多基因家族的分析

第十二章 素材创制与品种培育

第十三章 其他研究

960 Molecular characterization of a peritrophic membrane protein from the silkworm, *Bombyx mori*
961 一种家蚕围食膜蛋白的分子表征

962 A novel method of silkworm embryo preparation for immunohistochemistry
963 一种制备家蚕胚胎用于免疫组织化学实验的新方法

964 Expression analysis of chlorophyllid α binding protein, a secretory, red fluorescence protein in the midgut of silkworm, *Bombyx mori*
965 家蚕中肠红色荧光分泌蛋白叶绿酸α结合蛋白CHBP的表达分析

966 Molecular cloning and characterization of a *Bombyx mori* gene encoding the transcription factor Atonal
967 家蚕基因编码转录因子Atonal的分子克隆和特征分析

968 *Bm*HSP20.8 is localized in the mitochondria and has a molecular chaperone function *in vitro*
969 *Bm*HSP20.8蛋白定位于线粒体并具有分子伴侣功能

970 Expression analysis and tissue distribution of two 14-3-3 proteins in silkworm (*Bombyx mori*)
971 家蚕中两种14-3-3蛋白的表达分析和组织分布

972 Expression analysis and characteristics of *profilin* gene from silkworm, *Bombyx mori*
973 家蚕*profilin*基因的表达和特征分析

974 Expression analysis and characteristics of *Bm-LOC778477* gene, a hypothetical protein from silkworm pupae, *Bombyx mori*
975 家蚕蚕蛹中假定蛋白基因*Bm-LOC778477*的表达分析和特征

976 Cloning and partial characterization of a gene in *Bombyx mori* homologous to a human adiponectin receptor
977 家蚕脂联素受体同源基因的克隆和功能初探

978 Identification and characterization of a novel 1-Cys peroxiredoxin from silkworm, *Bombyx mori*
979 一个新的家蚕1-Cys过氧化物氧还蛋白的鉴定和功能分析

980 Cloning and characterization of the gene encoding an ubiquitin-activating enzyme E1 domain-containing protein of silkworm, *Bombyx mori*
981 编码家蚕含E1蛋白结构域的泛素活化酶基因的克隆与鉴定

982 Cloning, expression, and cell localization of a novel small heat shock protein gene: *BmHSP25.4*
983 一个新的小热激蛋白基因*BmHSP25.4*的克隆、表达和细胞定位

984 Molecular cloning and expression analysis of *Bmrbp1*, the *Bombyx mori* homologue of the *Drosophila* gene *rbp1*
985 家蚕中果蝇的同源基因*Bmrbp1*的分子克隆和表达分析

1008 A novel method for isolation of membrane proteins: a baculovirus surface display system

1009 一种分离膜蛋白的新方法：杆状病毒表面展示系统

1010 Analysis of similarity / dissimilarity of DNA sequences based on a class of 2D graphical representation

1011 基于一类二维图形表示的DNA序列相似性/相异性分析

1012 Analysis of similarity/dissimilarity of protein sequences

1013 蛋白质序列相似性/相异性分析

1014 Effect of extracts of mulberry leaves processed differently on the activity of α-glucosidase

1015 不同加工方法的桑叶提取物对α-糖苷酶活性的影响

1016 Bactrocerin-1: a novel inducible antimicrobial peptide from pupae of oriental fruit fly *Bactrocera dorsalis* Hendel

1017 一种来自桔小实蝇蛹的新诱导型抗菌肽

1018 Molecular and functional characterization of a c-type lysozyme from the Asian corn borer, *Ostrinia furnacalis*

1019 一种亚洲玉米螟（*Ostrinia furnacalis*）溶菌酶的分子特征及其功能表征

第一章

家蚕丝蛋白合成机制与蚕丝生物材料

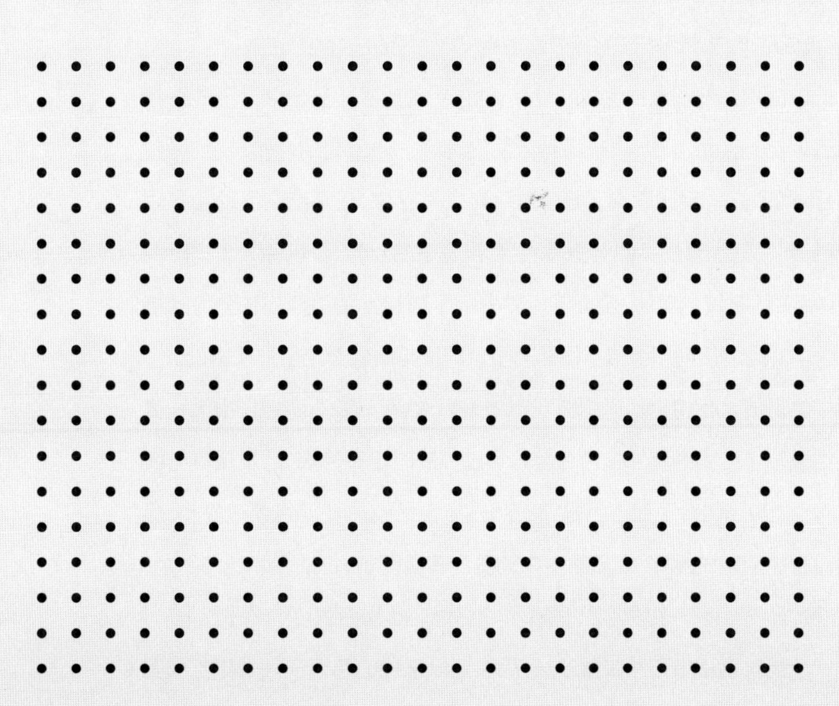

Chapter 1

蚕之所以重要，在于其分泌蚕丝。蚕丝是包含了众多成分的蛋白质复合物，由不同的丝腺区段合成并分泌。丝的产量受制于基因和代谢物等的调控，而丝的质量则取决于丝蛋白的成分及结构。近年来，项目组对家蚕的丝腺和丝蛋白进行了重点关注，综合应用组学、遗传学、分子生物学和材料学的手段，在丝蛋白的合成调控机制研究、丝蛋白成分和活性的发掘、丝蛋白的遗传改造和应用等方面取得了一系列突破性的进展。

丝蛋白在家蚕的中部和后部丝腺中合成并贮存，通过前部丝腺分泌出体外形成蚕丝纤维，然而丝蛋白合成与分泌的机制一直未得到全面的阐释。项目组结合转录组、蛋白质组和代谢组等多组学的方法，对比研究了野蚕丝腺和家蚕丝腺、正常结茧品种和裸蛹品种的丝腺、高丝量品系和低丝量品系的丝腺、高温和常温下的丝腺、桑叶和人工饲料饲养的家蚕丝腺、前中后不同区段的丝腺，全面鉴定并分析了影响丝蛋白合成和分泌的相关基因、蛋白质和代谢物，为今后利用遗传手段提高蚕丝产量与质量奠定了分子基础。

丝蛋白在丝腺中的合成是在特定的时间被开启和关闭，然而其表达调控在很大程度上是未知的。项目组揭示了丝腺细胞的核内有丝分裂过程的时期调控规律，发现其与细胞周期相关蛋白的表达相关；调查了参与丝腺细胞凋亡与自噬相关基因的表达模式，揭示了其在丝腺退化过程中的发生规律；调查丝腺中的多种转录因子，包括 *Dimm*、*Sage*、*MBF2*、*POU-M2*、*FMBP-1* 等的表达特征，阐释了它们调控丝蛋白表达的分子机制。

丝蛋白的主要成分是丝素蛋白和丝胶蛋白，然而长期以来我们忽视了丝素与丝胶蛋白的结构，更忽略了蚕丝中潜藏的其他蛋白。项目组通过对丝素蛋白 N 端的晶体结构进行解析，发现其能够响应 pH，从而启动丝蛋白的自组装；通过鉴定家蚕幼虫不同时期纺出的蚕丝蛋白成分，新发现了众多蛋白酶抑制剂和未知功能的蛋白质也是蚕丝的主要成分，通过调查不同茧层的丝蛋白成分，发现蚕茧外层比内层拥有更多的蛋白酶抑制剂，从蚕茧中提取的蛋白酶抑制剂对真菌蛋白酶表现出高效的抑制活性，从而为蚕茧提供更好的保护。

丝蛋白具有良好的生物相容性与生物降解性，被广泛制成各种生物材料，然而蚕丝在力学性能和离子结合性能等方面还存在一些缺陷。项目组利用前部丝腺特异的启动子使得离子转运蛋白在前部丝腺过量表达，获得了韧性和延展性都显著增强的蚕丝；利用丝素轻链启动子使得*Fib L-GFP-Cec*融合基因在后部丝腺表达，获得了具有抗菌特性的蚕丝；利用家蚕丝腺作为生物反应器生产了含有珍珠母贝基质蛋白序列的丝蛋白，获得了钙结合活性明显提高的蚕丝，作为矿化材料具有潜在的应用价值。项目组将蚕丝中的丝素蛋白、丝胶蛋白分别与胰岛素交联结合，发现结合物可以显著提高这种多肽药物的理化稳定性和生物学稳定性；应用有机溶剂制备获得了丝素纳米颗粒，将L-天冬酰胺酶和胰岛素分别共价结合在丝素纳米颗粒上，极大地提高了其稳定性和抗酶水解能力，表明丝素纳米颗粒是一种良好的固定化载体，在药物缓释系统方面具有潜在的研究和开发价值。项目组还在丝胶凝胶表面设计了聚电解质膜并在原位合成了纳米银颗粒，赋予了丝胶凝胶抗菌活性并提高了其热稳定性，有望应用于生物医学等领域。

董照明

The gene expression profile of *Bombyx mori* silk gland

Tang J[1,2] Li WF[1,2] Zhang X[1,2] Zhou CZ[1,2*]

Abstract: During prepupal stage, the genes expression in silk gland is considered as a model for gene expression and regulation of eukaryotes. Aiming to comprehensively interpret gene expression profile in the silk gland, we collected all currently available EST, complete cDNA and protein expression information and other gene expression testing data published before, and explored their roles in their function pathways level. With the analysis of interaction between the known proteins and putative bio-macromolecules partners *in silico*, we list our prediction results in the form of pathway classification and test some of their expressions by experiments.

Published On: Gene, 2007, 396(2), 369-372.

1 Hefei National Laboratory for Physical Sciences at the Microscale, University of Science and Technology of China, Hefei 230027, China.

2 School of Life Sciences, University of Science and Technology of China, Hefei 230027, China.

*Corresponding author E-mail: zcz@ustc.edu.cn.

家蚕丝腺基因表达谱的研究

唐 杰[1,2] 李卫芳[1,2] 张 璇[1,2] 周丛照[1,2*]

摘要：在家蚕预蛹阶段，丝腺中的基因表达被认为是研究真核生物基因表达和调控的最佳模型之一。为了更全面地理解丝腺基因的表达谱情况，我们收集了目前公共数据库的EST、cDNA和蛋白质表达的数据以及已出版的其他的基因表达数据，对家蚕不同组织和不同发育阶段的表达信息系统进行了详尽的分析，鉴定出一些在丝腺中特异表达的基因，研究了它们的功能和可能参与的代谢途径。综合之前报道的蛋白质与生物大分子互作的大量数据，将我们的预测按照途径分类的方式列出，并通过实验证实了其中的一些预测。

1 合肥微尺度物质科学国家研究中心，中国科学技术大学，合肥

2 生命科学学院，中国科学技术大学，合肥

Studies on middle and posterior silk glands of silkworm (*Bombyx mori*) using two-dimensional electrophoresis and mass spectrometry

Hou Y Xia QY* Zhao P Zou Y
Liu HL Guan J Gong J Xiang ZH

Abstract: Silk glands, present in the larval stage of the silkworm, produce threads of silky material to form the cocoon and are mainly composed of three parts: the anterior, the middle, and the posterior silk glands, each playing different roles in silk secretion. High-resolution two-dimensional polyacrylamide gel electrophoresis and computer-assisted analysis were used to investigate quantitative and qualitative differences between the middle and posterior silk glands. Silver staining revealed over 600 spots for each sample, mostly distributed from 15 to 100 kDa with pH 4-7. Computer-assisted image analysis, matrix-assisted laser desorption ionization time-of-flight mass spectrometry, and post-source decay technology suggested that there were significant differences in spot distribution and expression between the middle and posterior silk glands. In addition, 98 spots from the posterior silk gland were excised and further investigated following trypsin digestion. The results suggested that more than 20% of the 88 proteins identified were related to heat-shock proteins and chaperones. Redox system and DNA replication proteins involved in silk protein synthesis were also detected in the posterior silk gland. Interestingly, two novel serpin proteins were identified in the middle silk gland, and to a lesser extent in the posterior gland, which were presumed to be involved in regulation of proteolytic activity and protection of silk proteins from degradation.

Published On: Insect Biochemistry and Molecular Biology, 2007, 37(5), 486-496.

The key Sericultural Laboratory of Agricultural Ministry, Southwest University, Chongqing 400716, China.
*Corresponding author E-mail: xiaqy@swu.edu.cn.

利用双向电泳和质谱分析技术鉴定家蚕中部丝腺与后部丝腺蛋白质组学差异

侯 勇 夏庆友* 赵 萍 邹 勇
刘鸿丽 官 建 龚 竞 向仲怀

摘要:丝腺是家蚕幼虫阶段的重要组织器官,可分泌丝物质以形成茧壳。人们根据丝腺功能的差异将其分为前部丝腺、中部丝腺和后部丝腺。我们利用高精度双向电泳和计算机图像分析技术对家蚕中部丝腺和后部丝腺的蛋白质进行了调查分析。结果表明,在每一个样品中都可以检测到超过600个蛋白点,其中大部分蛋白点的分子质量为15—100 kDa,pH为4—7。计算机辅助图像分析、基质辅助激光解析电离飞行时间质谱和源后裂解分析表明在中部丝腺和后部丝腺的蛋白质图谱当中存在一些明显的差异蛋白点。此外,我们对后部丝腺中表达量较高的98个蛋白点进行胶内酶解和质谱分析,结果表明,在成功鉴定出的88个蛋白点当中,20%以上的蛋白与热激蛋白和分子伴侣蛋白相关,同时与氧化还原、复制、转录等生物学过程相关的大量蛋白质也在这次实验中被鉴定出来。有趣的是,两个新的丝氨酸蛋白酶抑制剂蛋白也被发现,它们在中部丝腺的表达水平明显比在后部丝腺高,推断它们可能与蛋白水解活性的调节及丝蛋白的保护功能有关。

农业部蚕桑学重点实验室,西南大学,重庆

Expression profiling and regulation of genes related to silkworm posterior silk gland development and fibroin synthesis

Li JY[1,3] Yang HJ[1] Lan TY[1] Wei H[1] Zhang HR[2]
Chen M[4] Fan W[1] Ma YY[2] Zhong BX[1*]

Abstract: The posterior silk gland (PSG) is the most important suborgan responsible for the synthesis and secretion of silk core fibroin proteins in silkworm. Here, we performed genome-scale expression profiling analysis of silkworm PSG at the fourth molting (M4) and at day 1 (V1), day 3 (V3), day 5 (V5), and wandering stage (W) of the fifth instar by microarray analysis with 22 987 probes. We found that the five genes of silk proteins secreted from PSG including *fibroin heavy (H)* and *light (L) chains*, *P25*, *seroin 1*, and *seroin 2* basically showed obvious up-regulation at V3 which lasted to V5, while slight down-regulation at W. The expression of translation-related genes including ribosomal proteins and translation initiation factors generally remained stable from M4 to V5, whereas it showed clear down-regulation at W. Clustering analysis of the 643 significantly differentially expressed transcripts revealed that 43 of the important genes including *seroin 1* and sugar transporter protein had co-expression patterns which were consistent with the rate changes of fibroin synthesis and PSG growth. Pathway analysis disclosed that the genes in different clusters might have coregulations and direct interactions. These genes were supposed to be involved in the fibroin synthesis and secretion. The differential expression of several hormone-related genes also suggested their functions on the regulation of PSG development and fibroin synthesis. 2D gel-based proteomics and phosphoproteomics profiling revealed that the phosphorylated proteins accounted for no more than one-sixth of the total proteins at each stage, which was much lower than the level in normal eukaryotic cells. Changes in the phosphorylation status and levels of several proteins such as actin-depolymerizing factor 1 and enolase might be deeply involved in fibroin secretion and tissue development. Shotgun proteomic profiling combined with label-free quantification analysis on the PSG at V3, V5, and W revealed that many small heat shock proteins (sHSP) were specially expressed at W, which was substantially consistent with the results from 2-DE analysis, and implied the close correlations of sHSP with the physiological states of PSG at W. A majority of significantly up-regulated proteins at V5 were related to ribosome pathway, which was different from the microarray results, implying that the translation-level regulation of ribosomal proteins might be critical for fibroin synthesis. In contrast, the ubiquitin-proteasome pathway related proteins appeared obviously up-regulated at W, suggesting that the programmed cell death process of PSG cells might be started before cocooning.

Published On: Journal of Proteome Research, 2011, 10(8), 3551-3564.

1 College of Animal Sciences, Zhejiang University, Hangzhou 310029, China.
2 Zhejiang California International Nanosystems Institute(ZCNI), Zhejiang University, Hangzhou 310029, China.
3 Institute of Developmental and Regenerative Biology, Hangzhou Normal University, Hangzhou 310036, China.
4 College of Life Sciences, Zhejiang University, Hangzhou 310058, China.
*Corresponding author E-mail: bxzhong@zju.edu.cn.

家蚕后部丝腺发育和丝素蛋白合成相关基因的表达及调控

李建营[1,3] 杨惠娟[1] 兰天云[1] 危 浩[1] 张华蓉[2]
陈 铭[4] 范 伟[1] 马莹莹[2] 钟伯雄[1*]

摘要：家蚕的后部丝腺是负责丝素蛋白合成和分泌的重要器官。在研究中，我们利用具有22 987个探针的芯片在基因组水平上对家蚕后部丝腺在四眠期，五龄第1，3，5天以及熟蚕期中的表达谱进行了分析。我们发现编码丝蛋白的5个基因，包括丝素重链、轻链、*P25*、*seroin 1*和*seroin 2*，从五龄第3天开始出现明显上调，并持续到五龄第5天，而在熟蚕期略微下调。翻译相关基因，包括核糖体蛋白和翻译起始因子的表达从四眠期到五龄期第5天大体上维持稳定，而在熟蚕期出现明显的下调。对643个显著差异表达的转录本进行聚类分析发现，43个重要基因包括*seroin 1*和糖运输蛋白具有相似的表达谱，而这与丝素合成和后部丝腺生长的速率变化趋势是吻合的。通路分析揭示了在不同的聚类里面的基因可能存在共调控和相互作用，这些基因被认为可能与丝素的合成和分泌有关。而若干个激素相关基因的差异表达也暗示它们具有调节后部丝腺发育和丝素合成的功能。基于双向凝胶电泳的蛋白质组和磷酸化蛋白质组分析发现磷酸化蛋白的表达在每个时期都不超过总蛋白的1/6，而这远低于正常真核细胞中的水平。有几个蛋白质比如肌动蛋白-解聚因子1和烯醇酶磷酸化状态、水平的变化可能与丝素分泌和组织发育有密切关系。鸟枪蛋白质组分析结合非标定量分析发现在熟蚕期有许多小热激蛋白特异表达，这与双向电泳的结果基本一致，暗示了小热激蛋白可能与该时期后部丝腺的生理状态有关。在五龄期第5天显著上调的蛋白质绝大多数都与核糖体通路有关，而这不同于微阵列结果，暗示了核糖体蛋白翻译水平的调控可能对于丝素的合成非常关键。相比之下，泛素-蛋白酶体通路相关蛋白在熟蚕期明显上调，表明后部丝腺细胞的程序性死亡可能在结茧之前就开始了。

1 动物科学学院，浙江大学，杭州
2 浙江加州国际纳米技术研究院，浙江大学，杭州
3 发育与再生研究所，杭州师范大学，杭州
4 生命科学学院，浙江大学，杭州

Comparative proteomic and phosphoproteomic analysis of the silkworm (*Bombyx mori*) posterior silk gland under high temperature treatment

Li JS[1,2] Ye LP[1] Lan TY[1] Yu ML[3] Liang JS[4] Zhong BX[1*]

Abstract: The proteins from the posterior silk gland of silkworm hybrids and their parents reared under high temperatures were studied by using comparative proteomic and phosphoproteomic analysis. A total of 82.07%, 6.17% and 11.76 % protein spots showed additivity, overdominance and underdominance patterns, respectively. Fifteen differentially expressed protein spots were identified by peptide mass fingerprinting. Among these, four spots, including sHSPs and prohibitin protein that were directly relevant to heat response, were identified. Eleven protein spots were found to play an important role in silk synthesis, and nine protein spots expressed phosphorylation states. According to Gene ontology and KEGG pathway analysis, these nine spots played an important role in stress-induced signal transduction. Expression of most silk synthesis-related proteins was reduced, whereas stress-responsive proteins increased with heat exposure time in three breeds. Furthermore, most proteins showed under- or over-dominance in the hybrids compared to the parents. The results suggested that high temperature could alter the expression of proteins related to silk synthesis and heat response in silkworm. Moreover, differentially expressed proteins occurring in the hybrid and its parents may be the main explanation of the observed heterosis.

Published On: Molecular Biology Reports, 2012, 39(8), 8447-8456.

1 College of Animal Sciences, Zhejiang University, Hangzhou 310029, China.

2 Institute of Sericulture, Chengde Medical University, Chengde 067000, China.

3 College of Life Sciences, Zhejiang Sci-Tech University, Hangzhou 310029, China.

4 College of Environmental and Resource Sciences, Zhejiang University, Hangzhou 310029, China.

*Corresponding author E-mail: bxzhong@zju.edu.cn.

高温胁迫下家蚕后部丝腺差异蛋白质组和磷酸化蛋白质组学分析

李季生[1,2]　叶露鹏[1]　兰天云[1]　俞梅兰[3]　梁建设[4]　钟伯雄[1*]

摘要：研究采用差异蛋白质组和磷酸化蛋白质组学技术研究了高温胁迫下家蚕杂交种及其亲本后部丝腺的蛋白表达差异。通过比较杂交种及其亲本相应蛋白点的表达差异，结果发现有82.07%的蛋白点表现出加性效应（additivity），6.17%的蛋白点呈超显性（overdominance），而11.76%的蛋白点为显性不足（underdominance）。对显著差异的蛋白质进行肽质量指纹谱分析，结果鉴定出了15个蛋白点，其中4个蛋白点（包括一些热激和抑制素类蛋白）显示与胁迫有关，其余11个蛋白点（其中有9个发生了磷酸化）则参与丝物质的合成。GO和KEGG进一步分析显示，9个磷酸化蛋白与胁迫信号的传导有关。另外，从蛋白质的变化趋势分析，大部分与丝物质合成相关的蛋白质表达量下降，而与胁迫相关的蛋白质则表达量上升。同时，杂交种中大部分蛋白质呈现超显性或显性不足的现象。研究结果表明高温胁迫引起了后部丝腺胁迫应答，同时影响了丝物质的合成。结合杂交种及其亲本差异蛋白点的不同表现，我们的研究从另一个侧面反映了杂种优势的存在。

1　动物科学学院，浙江大学，杭州
2　蚕业研究所，承德医学院，承德
3　生命科学学院，浙江理工大学，杭州
4　环境与资源学院，浙江大学，杭州

Shotgun proteomic analysis of the *Bombyx mori* anterior silk gland: An insight into the biosynthetic fiber spinning process

Yi QY[#] Zhao P[#] Wang X Zou Y Zhong XW
Wang C Xiang ZH Xia QY[*]

Abstract: The *Bombyx mori* anterior silk gland (ASG) is a natural fiber manipulator for the material provided by the middle and posterior silk glands. In view of the significant role of the ASG in the liquid-crystal spinning process, a shotgun proteomics approach was taken to study the relationship between the function of proteins in the silkworm ASG and the spinning mechanism. A total of 1 132 proteins with 7 647 unique peptides were identified in the ASG dataset including some involved in the cuticle, ion transportation, energy metabolism, and apoptosis. Two putative cuticle-specific proteins were highly and specifically expressed in the ASG; therefore, the ASG dataset could provide clues for comprehensive understanding of the natural silk spinning mechanism in the silkworm.

Published On: Proteomics, 2013, 13(17), 2657-2663.

State Key Laboratory of Silkworm Genome Biology, Southwest University, Chongqing 400716, China.
[#]These authors contributed equally.
[*]Corresponding author E-mail: xiaqy@swu.edu.cn.

鸟枪法分析家蚕前部丝腺蛋白组学：深入了解生物合成纤维纺丝过程

衣启营[#] 赵 萍[#] 王 鑫 邹 勇 钟晓武
王 晨 向仲怀 夏庆友[*]

摘要：家蚕的中部丝腺与后部丝腺合成丝蛋白，而前部丝腺负责加工丝蛋白。鉴于前部丝腺在液晶纺丝过程中的重要作用，我们采用鸟枪法蛋白质组学研究了家蚕前部丝腺中蛋白质的功能与纺丝机理之间的关系，共鉴定到1 132个蛋白质且有7 647个蛋白特异的肽，其中包括表皮、离子转运、能量代谢和细胞凋亡等相关蛋白。两种假定的特异性表皮蛋白均在前部丝腺中高量且特异表达。因此，家蚕前部丝腺蛋白组学数据集可以为全面了解家蚕自然纺丝过程提供线索。

家蚕基因组生物学国家重点实验室，西南大学，重庆

Quantitative proteomic and transcriptomic analyses of molecular mechanisms associated with low silk production in silkworm *Bombyx mori*

Wang SH[1] You ZY[1] Ye LP[1] Che JQ[1]
Qian QJ[1] Yohei Nanjo[2] Setsuko Komatsu[2*] Zhong BX[1*]

Abstract: To investigate the molecular mechanisms underlying the low fibroin production of the *ZB* silkworm strain, we used both SDS-PAGE-based and gel-free-based proteomic techniques and transcriptomic sequencing technique. Combining the data from two different proteomic techniques was preferable in the characterization of the differences between the *ZB* silkworm strain and the original *Lan10* silkworm strain. The correlation analysis showed that the individual protein and transcript were not corresponded well, however, the differentially changed proteins and transcripts showed similar regulated direction in function at the pathway level. In the *ZB* strain, numerous ribosomal proteins and transcripts were down-regulated, along with the transcripts of translational related elongation factors and genes of important components of fibroin. The proteasome pathway was significantly enhanced in the *ZB* strain, indicating that protein degradation began on the third day of fifth instar when fibroin would have been produced in the *Lan10* strain normally and plentifully. From proteome and transcriptome levels of the *ZB* strain, the energy-metabolism related pathways, oxidative phosphorylation, glycolysis/gluconeogenesis, and citrate cycle were enhanced, suggesting that the energy metabolism was vigorous in the *ZB* strain, while the silk production was low. This may due to the inefficient energy employment in fibroin synthesis in the *ZB* strain. These results suggest that the reason for the decreasing of the silk production might be related to the decreased ability of fibroin synthesis, the degradation of proteins, and the inefficiency of the energy exploiting.

Published On: Journal of Proteome Research, 2014, 13(2), 735-751.

1 College of Animal Sciences, Zhejiang University, Hangzhou 310058, China.
2 National Institute of Crop Science, NARO, Tsukuba 305-8518, Japan.
*Corresponding author E-mail: skomatsu@affrc.go.jp; bxzhong@zju.edu.cn.

定量蛋白质组学和转录组学分析家蚕低产丝量的分子机制

王少华[1] 尤征英[1] 叶露鹏[1] 车家倩[1]
钱秋杰[1] Yohei Nanjo[2] Setsuko Komatsu[2*] 钟伯雄[1*]

摘要：在研究中，我们利用基于SDS-PAGE凝胶和无胶的蛋白质组学技术以及转录组测序技术来研究家蚕突变体品系*ZB*低丝产量的分子机制。通过结合两种不同组学技术获得的数据，更有利于比较*ZB*和原始的*Lan10*品系间的差异。关联性分析表明单个的蛋白质和转录本之间的相关性并不明显，但是在通路水平上差异蛋白与差异转录本表现出了明显的相关性。研究结果显示，相对于转基因家蚕供体*Lan10*品系，*ZB*品系中大量核糖体蛋白在蛋白质和转录水平上的表达量降低，翻译相关延伸因子转录水平降低以及重要的丝蛋白成分表达量降低。另外，在五龄第3天丝蛋白合成与分泌达到顶峰的时期，*ZB*品系中蛋白酶体通路增强，无疑对丝蛋白的合成造成了损害。*ZB*品系丝产量较低，但是其能量代谢通路包括氧化磷酸化、糖酵解/糖异生和三羧酸循环增强，说明在*ZB*品系中存在着能量的空耗。这些结果说明，*ZB*品系丝产量的降低可能是由丝蛋白合成能力降低，蛋白质降解提前以及能量的非有效利用这几方面造成的。这为今后培育高产丝量家蚕品种奠定了分子生物学的理论基础。

1 动物科学学院，浙江大学，杭州
2 国立作物科学研究所，国立农业与食品研究组织，茨城，日本

Transcriptomic analysis of differentially expressed genes in the *Ras1*^{CA}-overexpressed and wildtype posterior silk glands

Ma L[1] Ma Q[1,2] Li X[1] Cheng LL[3*] Li K[2*] Li S[1*]

Abstract:

BACKGROUND: Using the *piggy*Bac-mediated *GAL4 / UAS* transgenic system established in the silkworm, *Bombyx mori*, we have previously reported that overexpression of the *Ras1*^{CA} oncogene specifically in the posterior silk gland (PSG) improved cell growth, fibroin synthesis, and thus silk yield. However, the detailed molecular mechanism remains to be fully elucidated. To achieve this goal, Illumina sequencing was used in the present study to compare the transcriptomes of the *Ras1*^{CA}-overexpressed and wildtype PSGs.

RESULTS: The transcriptomic sequencing results in 56 million reads following filtering steps. Most of the reads (~70%) are successfully mapped to the *Bombyx* genome. The mapped reads are situated within at least 9 133 predicted genes, covering 62.46% genes of the *Bombyx* genome. GO annotation shows that 2 512 of the 2 636 differentially expressed genes (DEGs) are mostly distributed in metabolic process, cell and cell part, and binding, and KEGG annotation shows that 1 941 DEGs are mapped into 277 pathways. Importantly, *Ras1*^{CA} overexpression in the PSG upregulated many DEGs distributed in "pathway in cancer", "insulin signaling pathway", and "MAPK signaling pathway" as well as "purine metabolism" and "pyrimidine metabolism". Transcriptional regulation of these DEGs was verified by quantitative real-time PCR. Moreover, injection of small-molecule chemical inhibitors of the Ras1 downstream effectors into the *Ras1*^{CA}-overexpressed silkworms revealed that both Raf-MAPK and PI3K-TORC1 pathways are required for the Ras1-induced DEG expression.

CONCLUSION: The transcriptomic analysis illustrates that, apart from phosphorylational regulation, Ras1 activates its downstream Raf-MAPK and PI3K-TORC1 pathways at the transcriptional level. Meanwhile, Ras1 increases DNA content and induces endoreplication, at least in part, by upregulating genes in "nucleotide metabolism" and "cell cycle". This study provides further insights into the molecular mechanism of how *Ras1*^{CA} overexpression in the PSG improves silk yield.

Published On: BMC Genomics, 2014, 15.

1 Key Laboratory of Developmental and Evolutionary Biology, Institute of Plant Physiology and Ecology, Shanghai Institues for Biological Science, Chinese Academy of Sciences, Shanghai 200032, China.

2 School of Life Science, East China Normal University, Shanghai 200062, China.

3 Shanghai Institute of Cardiovascular Diseases, Zhongshan Hospital, Fudan University, Shanghai 20032, China.

*Corresponding author E-mail: cheng.leilei@zs-hospital.sh.cn; kaili@admin.ecnu.edu.cn; lisheng01@sibs.ac.cn.

超表达*Ras1*^{CA}家蚕后部丝腺转录组变化研究

马　俐[1]　马　倩[1,2]　李　轩[1]　程蕾蕾[3*]　李　恺[2*]　李　胜[1*]

摘要:

背景——利用*Piggy*Bac转座子介导的*GAL4/UAS*转基因系统,我们之前的研究发现在家蚕的后部丝腺特异地过表达致癌基因*Ras1*^{CA}后,促进了细胞的生长和丝素蛋白的合成,提高了丝产量,然而相关分子机制并不清楚。因此,在本研究中我们应用了Illumina测序比较研究了野生型和超表达*Ras1*^{CA}基因的家蚕后部丝腺的转录组。结果——转录组测序结果获得了5 600万个reads,这些reads中的约70%可以成功地与家蚕的基因组相匹配,这些匹配上的reads属于9 133个预测的基因,覆盖了家蚕整个基因组的62.46%。GO注释显示2 636个差异表达的基因中2 512个基因属于代谢途径、细胞与细胞组分及结合,KEGG注释显示1 941个差异表达的基因分布在277个信号通路中。更重要的是,*Ras1*^{CA}过表达后部丝腺上调的差异基因基本都存在于癌症相关信号通路、胰岛素信号通路、MAKP信号通路、嘌呤代谢和嘧啶代谢中。定量PCR进一步验证了这些差异表达基因,此外对*Ras1*^{CA}超表达的家蚕注射Ras下游效应因子的小分子抑制剂显示Raf-MAPK和PI3K-TORC1信号是Ras1诱导的基因差异表达所必需的。结论——转录组分析表明,除了磷酸化调控外,Ras1还在转录水平上激活下游的Raf-MAPK和PI3K-TORC1途径的基因表达;Ras1通过上调"核酸代谢"和"细胞周期"中的基因,增加了DNA含量并诱导核内复制。我们的研究为将来深入研究后部丝腺*Ras1*^{CA}超表达提高家蚕蚕丝产量的机制提供了参考。

1　昆虫发育与进化生物学重点实验室,上海生命科学研究院植物生理生态研究所,中国科学院,上海

2　生命科学学院,华东师范大学,上海

3　上海心血管疾病研究所,复旦大学附属中山医院,上海

Transcriptomic analysis of the anterior silk gland in the domestic silkworm (*Bombyx mori*): Insight into the mechanism of silk formation and spinning

Chang HP[1,2] Cheng TC[1] Wu YQ[1] Hu WB[1] Long RW[1] Liu C[1] Zhao P[1] Xia QY[1*]

Abstract: Silk proteins are synthesized in the middle and posterior silk glands of silkworms, then transit into the anterior of the silk gland, where the silk fibers are produced, stored and processed. The mechanism of formation and spinning of the silk fibers has not been fully elucidated, and transcriptome analyses specific to the anterior silk gland have not been reported. In the present study, we explored gene expression profiles in five regions of silk gland samples by using the RNA-Seq method. As a result, there were 959 979 570 raw reads obtained, of which 583 068 172 reads were mapped to the silkworm genome. A total of 7 419 genes were found to be expressed in terms of reads per kilobase of exon model per million mapped reads ≥5 in at least one sample. The gene numbers and expression levels of the expressed genes differed between these regions. The differentially expressed genes were analyzed, and 282 genes were detected as up-regulated in the anterior silk gland, compared with the other parts. Functions of these genes were addressed using the Gene Ontology and Kyoto Encyclopedia of Genes and Genomes databases, and seven key pathways were enriched. It suggested that the ion transportation, energy metabolism, protease inhibitors and cuticle proteins played essential roles in the process of silk formation and spinning in the anterior silk gland. In addition, 210 genes were found differently expressed between males and females, which should help to elucidate the mechanism of the quality difference in silk fibers from male and female silkworms.

Published On: PLoS ONE, 2015,10(9).

1 State Key Laboratory of Silkworm Genome Biology, Southwest University, Chongqing 400716, China.

2 College of Biotechnology, Southwest University, Chongqing 400716, China.

*Corresponding author E-mail: xiaqy@swu.edu.cn.

家蚕前部丝腺的转录组分析——蚕丝形成与吐丝机理的探讨

常怀普[1,2] 程廷才[1] 吴玉乾[1] 胡文波[1] 龙仁文[1]
刘 春[1] 赵 萍[1] 夏庆友[1*]

摘要：丝蛋白在家蚕的中部和后部丝腺中合成，然后转运到前部丝腺并被生产、储存和加工。丝纤维的形成和吐丝机理尚未得到充分的阐述，前部丝腺的转录组分析也没有报道。在研究中，我们使用RNA-Seq方法探究了丝腺样品5个不同区段中的基因表达谱。结果共获得了959 979 570个原始reads，其中583 068 172个reads能匹配家蚕基因组。在至少一个样本中，共发现7 419个基因的RPKM≥5，表达基因的基因数量和表达水平在这些区域存在差异。分析差异表达基因，与其他部位相比，282个基因在前部丝腺中上调表达。对这些基因的功能，使用GO、KEGG数据库进行了研究，其中有7个关键通路被富集。研究表明离子运输、能量代谢、蛋白酶抑制剂和表皮蛋白在前部丝腺丝的形成和吐丝过程中发挥了重要作用。此外，雄、雌之间有210个基因表达不同，这有助于阐明雄蚕和雌蚕丝纤维质量差异的机制。

1 家蚕基因组生物学国家重点实验室，西南大学，重庆
2 生物技术学院，西南大学，重庆

Comparative proteomic analysis of the silkworm middle silk gland reveals the importance of ribosome biogenesis in silk protein production

Li JY[1,2#] Ye LP[2#] Che JQ[2] Song J[2] You ZY[2] Ki-chan Yun[2]
Wang SH[2] Zhong BX[2*]

Abstract: The silkworm middle silk gland (MSG) is the sericin synthesis and secretion unique sub-organan. The molecular mechanisms of regulating MSG protein synthesis are largely unknown. Here, we performed shotgun proteomic analysis on the three MSG subsections: the anterior (MSG-A), middle (MSG-M), and posterior (MSG-P) regions. The results showed that more strongly expressed proteins in the MSG-A were involved in multiple processes, such as silk gland development and silk protein protection. The proteins that were highly expressed in the MSG-M were enriched in the ribosome pathway. MSG-P proteins with stronger expression were mainly involved in the oxidative phosphorylation and citrate cycle pathways. These results suggest that the MSG-M is the most active region in the sericin synthesis. Furthermore, comparing the proteome of the MSG with the posterior silk gland (PSG) revealed that the specifically and highly expressed proteins in the MSG were primarily involved in the ribosomes and aminoacyl-tRNA biosynthesis pathways. These results indicate that silk protein synthesis is much more active as a result of the enhancement of translation-related pathways in the MSG. These results also suggest that enhancing ribosome biogenesis is important to the efficient synthesis of silk proteins.

Published On: Journal of Proteomics, 2015, 126, 109-120.

1 Institute of Developmental and Regenerative Biology, Hangzhou Normal University, Hangzhou 310036, China.
2 College of Animal Sciences, Zhejiang University, Hangzhou 310029, China.
#These authors contributed equally.
*Corresponding author E-mail: bxzhong@zju.edu.cn.

比较蛋白质组学分析家蚕中部丝腺揭示核糖体生成对丝蛋白生产的重要性

李建营[1,2#]　叶露鹏[2#]　车家倩[2]　宋　佳[2]　尤征英[2]
Ki-chan Yun[2]　王少华[2]　钟伯雄[2*]

摘要:家蚕中部丝腺是丝胶合成和分泌的唯一器官,而中部丝腺调控蛋白合成的分子机制还不清楚。在研究中,我们利用鸟枪法蛋白组学分析了中部丝腺的3个区段,包括前段、中段和后段。结果显示在前段高表达的蛋白质参与了多个生物过程,包括丝腺的发育和丝蛋白保护。中段高表达的蛋白质富集于核糖体通路,后段高表达的蛋白质主要涉及氧化磷酸化和三羧酸循环通路。这些结果表明,中段是丝胶合成最活跃的区域。而且,比较中部丝腺和后部丝腺发现在中部丝腺特异和高表达的蛋白质主要涉及核糖体通路以及氨基酰-tRNA合成通路。这些结果表明,由于中部丝腺翻译相关蛋白的表达增强,造成了其丝蛋白合成比较活跃,这也暗示着加强核糖体生物合成对于提高丝蛋白的合成效率非常重要。

1　发育与再生研究所,杭州师范大学,杭州
2　动物科学学院,浙江大学,杭州

Analyses of the molecular mechanisms associated with silk production in silkworm by iTRAQ-based proteomics and RNA-sequencing-based transcriptomics

Wang SH[1] You ZY[1] Feng M[2] Che JQ[1] Zhang YY[1]
Qian QJ[1] Setsuko Komatsu[3] Zhong BX[1*]

Abstract: Silkworm is used as a model organism to analyze two standard complex traits, which are high and low silk yields. To understand the molecular mechanisms of silk production, the posterior silk glands aged to the third day of the fifth instar were analyzed from the *ZB* strain with low silk production and from the control strain *Lan10*. Using isobaric tags for relative and absolute quantification (iTRAQ) quantitative shotgun proteomics and RNA-sequencing-based transcriptomics, 139 proteins and 630 transcripts were identified as novel in the *ZB* strain compared with the *Lan10* strain, indicating that these results significantly expand the coverage of proteins and transcripts of the posterior silk glands in the silkworm. Of the 89 differently changed proteins, 23 were increased, and 66 were decreased. Of the 788 transcripts, 779 were up regulated, and 9 were down regulated. These results confirm that decreased energy utilization / protein translation and enhanced protein degradation are the key factors in lower silk production. Moreover, this study provides novel insight into the molecular changes that may result in lower silk production, namely, a combination of impaired transcription activity, missed protein folding/transport, and lowered yields of the main components of fibroin, along with weakened growth/development of the posterior silk gland.

Published On: Journal of Proteome Research, 2016, 15(1), 15-28.

1 College of Animal Sciences, Zhejiang University, Hangzhou 310058, China.

2 Institute of Apicultural Research/Key Laboratory of Pollinating Insect Biology, Ministry of Agriculture, Chinese Academy of Agricultural Science, Beijing 100093, China.

3 National Institute of Crop Science, NARO, Tsukuba 305-8518, Japan.

*Corresponding author E-mail: bxzhong@zju.edu.cn.

利用基于iTRAQ的蛋白质组学和基于RNA测序的转录组学研究蚕丝产量的分子机制

王少华[1] 尤征英[1] 冯 毛[2] 车家倩[1] 张玉玉[1]
钱秋杰[1] Setsuko Komatsu[3] 钟伯雄[1*]

摘要：家蚕作为模式生物，可以用来分析高、低产丝量两种标准性状。为了解产丝的分子机理，我们用低产丝量*ZB*品系和对照*Lan10*品系开展研究，对五龄第3天的后部丝腺进行了分析。在研究中，我们采用了同位素标记相对与绝对定量的蛋白质组学（iTRAQ）和转录组测序技术进行分析，结果显示，与对照*Lan10*品系相比，在*ZB*品系中新鉴定到了139个蛋白质和630个转录本，研究结果显著增加了家蚕后部丝腺的蛋白质和转录本的覆盖率。在鉴定到的89个差异蛋白中，23个表达上调，66个表达下调，而在788个转录本中，779个表达上调，9个表达下调。这些结果进一步证实了能量的空耗，蛋白质翻译效率降低以及蛋白质降解增强是造成*ZB*品系蛋白质合成减少的关键原因。而且，我们的研究暗示转录功能的受限，蛋白质折叠和转运出错，丝蛋白主要成分表达量的降低以及后部丝腺器官生长发育的减缓都可能导致蚕丝产量的降低。

1 动物科学学院，浙江大学，杭州
2 农业部昆虫生物学研究所/重点实验室，中国农业科学院，北京
3 国立作物科学研究所，国立农业与食品研究组织，茨城，日本

Phosphoproteomic analysis of the posterior silk gland of *Bombyx mori* provides novel insight into phosphorylation regulating the silk production

Song J Che JQ You ZY Ye LP Li JS Zhang YY Qian QJ Zhong BX*

Abstract: To understand phosphorylation event regulating silk synthesis in the posterior silk gland of *Bombyx mori*, phosphoproteome was profiled in a pair of near-isogenic lines, a normally cocooning strain (*IC*) and a nakedly pupated strain (*IN*) that the silk production is much lower than *IC*. In the posterior silk gland of the *IC* and *IN*, 714 and 658 phosphosites resided on 554 and 507 phosphopeptides from 431 and 383 phosphoproteins, were identi fi ed, respectively. Of all the phosphosites, the single phosphosite was the dominate phosphorylation form, comprising>60% of all the phosphosites in two phenotypic of silk production. All these phosphosites were classified as acidophilic and proline-directed kinase classes, and three motifs were uniquely identified in the *IC*. The motif S-P-P might be important for regulating phosphorylation network of silk protein synthesis. The dynamically phosphorylated proteins participated in ribosome, protein transport and energy metabolism suggest that phosphorylation may play key roles in regulating silk protein synthesis and secretion. Furthermore, fibroin heavy chain, an important component of silk protein, was specifi cally phosphorylated in the *IC* strain, suggesting its role to ensure the normal formation of silk structure and silk secretion. The data gain new understanding of the regulatory processes of silk protein synthesis and offer as starting point for further research on the silk production at phosphoproteome level.

Published On: Journal of Proteomics, 2016, 148, 194-201.

College of Animal Sciences, Zhejiang University, Hangzhou 310058, China.
*Corresponding author E-mail: bxzhong@zju.edu.cn.

家蚕后部丝腺磷酸化蛋白质组分析揭示蛋白磷酸化调控蚕丝产量机制

宋 佳 车家倩 尤征英 叶露鹏 李季生
张玉玉 钱秋杰 钟伯雄*

摘要：为了了解磷酸化对后部丝腺丝蛋白合成的调控，我们对一对近等位基因品系，即正常的结茧品系（*IC*）和裸蛹品系（*IN*）的磷酸化蛋白质组进行了分析。分别在*IC*和*IN*的后部丝腺鉴定到431个和383个磷酸化蛋白，包含554个和507个磷酸化多肽，以及714个和658个磷酸化位点。其中，单一磷酸化位点的蛋白质是主要的形式，占这两个家蚕品系磷酸化蛋白的60%。所有这些磷酸化位点被分类成嗜酸性的和脯氨酸定向的激酶，其中3个基序唯一地在*IC*品种中被鉴定到，且基序S-P-P有可能是重要的调控丝蛋白合成的磷酸化网络。大量的磷酸化蛋白参与核糖体、蛋白质的运输和能量代谢，说明磷酸化在调控丝蛋白的合成和分泌中起到重要的作用。此外，丝素重链蛋白，是一个重要的丝蛋白组成成分，在*IC*品种中被特异地磷酸化，表明磷酸化的作用是确保丝蛋白的形成和分泌。我们的研究在磷酸化蛋白质水平上，在丝蛋白的调控过程以及丝蛋白的合成过程获得了新的发现。

动物科学学院，浙江大学，杭州

Analysis of proteome dynamics inside the silk gland lumen of *Bombyx mori*

Dong ZM　Zhao P　Zhang Y　Song QR　Zhang XL
Guo PC　Wang DD　Xia QY*

Abstract: The silk gland is the only organ where silk proteins are synthesized and secreted in the silkworm, *Bombyx mori.* Silk proteins are stored in the lumen of the silk gland for around eight days during the fifth instar. Determining their dynamic changes is helpful for clarifying the secretion mechanism of silk proteins. Here, we identified the proteome in the silk gland lumen using liquid chromatography-tandem mass spectrometry, and demonstrated its changes during two key stages. From day 5 of the fifth instar to day 1 of wandering, the abundances of fbroins, sericins, seroins, and proteins of unknown functions increased significantly in different compartments of the silk gland lumen. As a result, these accumulated proteins constituted the major cocoon components. In contrast, the abundances of enzymes and extracellular matrix proteins decreased in the silk gland lumen, suggesting that they were not the structural constituents of silk. Twenty-five enzymes may be involved in the regulation of hormone metabolism for proper silk gland function. In addition, the metabolism of other non-proteinous components such as chitin and pigment were also discussed in this study.

Published On: Scientific Reports, 2016, 6.

State Key Laboratory of Silkworm Genome Biology, Southwest University, Chongqing 400716, China.
*Corresponding author E-mail: xiaqy@swu.edu.cn.

家蚕丝腺腔内蛋白质组学动力学分析

董照明　赵　萍　张　艳　宋倩茹　张晓璐
郭鹏超　王丹丹　夏庆友*

摘要：丝腺是家蚕(*Bombyx mori*)蚕丝蛋白合成和分泌的唯一器官，丝蛋白在家蚕五龄期的丝腺腔内储存大约8 d。确定丝腺腔内丝蛋白的动态变化有助于阐明其分泌机制。我们利用液相色谱-串联质谱鉴定了丝腺腔中的蛋白质组，并在两个关键阶段证明了其变化。从五龄第5天到上蔟第1天，丝素蛋白、丝胶蛋白、seroins蛋白以及未知功能蛋白在丝腺不同区段都有明显的增加，这些不断积累的蛋白是蚕丝的主要成分。相比之下，丝腺腔内的酶和细胞外基质蛋白的丰度降低，表明它们不是丝的结构成分。25种酶可能参与调节激素代谢以确保丝腺功能的正常。此外，我们还讨论了其他非蛋白质成分如几丁质和色素的代谢。

家蚕基因组生物学国家重点实验室，西南大学，重庆

Comparative proteomic analysis of posterior silk glands of wild and domesticated silkworms reveals functional evolution during domestication

Li JY[1,2] Cai F[2] Ye XG[2] Liang JS[3] Li JK[4] Wu MY[2] Zhao D[2]
Jiang ZD[2] You ZY[2] Zhong BX[2*]

Abstract: The wild silkworm *Bombyx mandarina* was domesticated to produce silk in China approximately 5 000 years ago. Silk production is greatly improved in the domesticated silkworm *B. mori*, but the molecular basis of the functional evolution of silk gland remains elusive. We performed shotgun proteomics with label-free quantification analysis and identified 1 012 and 822 proteins from the posterior silk glands (PSGs) of wild silkworms on the third and fifth days of the fifth instar, respectively, with 128 of these differentially expressed. Bioinformatics analysis revealed that, with the development of the PSG, the up-regulated proteins were mainly involved in the ribosome pathway, similar to what we previously reported for *B. mori*. Additionally, we screened 50 proteins with differential expression between wild and domesticated silkworms that might be involved in domestication at the two stages. Interestingly, the up-regulated proteins in domesticated compared to wild silkworms were enriched in the ribosome pathway, which is closely related to cell size and translation capacity. Together, these results suggest that functional evolution of the PSG during domestication was driven by reinforcing the advantageous pathways to increase the synthesis efficiency of silk proteins in each cell and thereby improve silk yield.

Published On: Journal of Proteome Research, 2017, 16(7), 2495-2507.

1 Institute of Life Sciences, College of Life and Environmental Sciences, Hangzhou Normal University, Hangzhou 310036, China.
2 College of Animal Sciences, Zhejiang University, Hangzhou 310058, China.
3 College of Environmental & Resource Sciences, Zhejiang University, Hangzhou 310058, China.
4 Institute of Apicultural Research, Chinese Academy of Agricultural Sciences, Beijing 100081, China.
*Corresponding author E-mail: bxzhong@zju.edu.cn.

比较蛋白质组学分析野蚕和家蚕后部丝腺揭示其在驯养过程中的功能进化

李建营[1,2]　蔡　芳[2]　叶小刚[2]　梁建设[3]　李建科[4]　吴美玉[2]　赵　丹[2]
蒋振东[2]　尤征英[2]　钟伯雄[2*]

摘要：大约在5 000年前，野蚕在中国就已经被驯养用来产丝。尽管家蚕的丝产量提高了很多，但是丝腺功能进化的分子基础仍然不是很清楚。我们利用鸟枪法蛋白质组学结合非标定量的方法，在野蚕五龄第3天和第5天的后部丝腺中分别鉴定到了1 012和822个蛋白质，其中有128个是差异表达的。生物信息学分析发现，随着后部丝腺的发育，上调的蛋白质主要涉及核糖体通路，这与之前我们报道的情况类似。另外，我们筛选到50个在野蚕和家蚕的上述两个时期都差异表达的蛋白质，这些蛋白质有可能与驯养有关。有意思的是，与野蚕相比，在家蚕中上调的蛋白质富集到了核糖体通路，而该通路与细胞大小和翻译能力密切相关。总之，这些结果表明，后部丝腺在野蚕驯养过程中的功能进化是通过强化优势途径来提高丝蛋白合成效率，从而提高丝的产量。

1　生命科学研究院，生命与环境科学学院，杭州师范大学，杭州
2　动物科学学院，浙江大学，杭州
3　环境与资源学院，浙江大学，杭州
4　蜜蜂研究所，中国农业科学院，北京

GC/MS-based metabolomic studies reveal key roles of glycine in regulating silk synthesis in silkworm, *Bombyx mori*

Chen QM[1#] Liu XY[2#] Zhao P[1] Sun YH[1] Zhao XJ[2]
Xiong Y[1] Xu GW[2*] Xia QY[1*]

Abstract: Metabolic profiling of silkworms, especially the factors that affect silk synthesis at the metabolic level, is little known. Herein, metabolomic method based on gas chromatography-mass spectrometry was applied to identify key metabolic changes in silk synthesis deficient silkworms. Forty-six differential metabolites were identified in *Nd* group with the defect of silk synthesis. Significant changes in the levels of glycine and uric acid (up-regulation), carbohydrates and free fatty acids (down-regulation) were observed. The further metabolomics of silk synthesis deficient silkworms by decreasing silk proteins synthesis using knocking out fibroin heavy chain gene or extirpating silk glands operation showed that the changes of the metabolites were almost consistent with those of the *Nd* group. Furthermore, the increased silk yields by supplying more glycine or its related metabolite confirmed that glycine is a key metabolite to regulate silk synthesis. These findings provide important insights into the regulation between metabolic profiling and silk synthesis.

Published On: Insect Biochemistry and Molecular Biology, 2015, 57, 41-50.

1 State Key Laboratory of Silkworm Genome Biology, Southwest University, Chongqing 400716, China.
2 Key Laboratory of Separation Science for Analytical Chemistry, Dalian Institute of Chemical Physics, Chinese Academy of Sciences, Dalian 116023, China.
[#]These authors contributed equally.
[*]Corresponding author E-mail: xugw@dicp.ac.cn; xiaqy@swu.edu.cn.

家蚕幼虫丝腺发育过程中的DNA复制

张春冬[#]　李方方[#]　陈向云　黄茂华　张　军
崔红娟　潘敏慧[*]　鲁　成[*]

摘要: 丝腺是家蚕合成丝蛋白的重要器官,对吐丝至关重要。丝腺细胞的基因组DNA含量在幼虫期通过核内有丝分裂过程急剧地增加20万—40万倍。体外培养时,我们使用BrdU标记的方法测定了幼虫眠期和眠间期DNA的合成,发现细胞周期的核内有丝分裂在眠间期激活,并在眠期被抑制。前部丝腺、中部丝腺和后部丝腺细胞在幼虫五龄第6天陆续退出核内有丝分裂循环,这与细胞周期相关的*cdt1*,*pcna*,*cyclin E*,*cdk2*和*cdk1*的mRNA在上蔟期的表达量降低相关。此外,饥饿处理对新蜕皮幼虫的丝腺DNA的起始合成没有影响。

家蚕基因组生物学国家重点实验室,蚕学与系统生物学研究所,西南大学,重庆

Autophagy, apoptosis, and ecdysis-related gene expression in the silk gland of the silkworm (*Bombyx mori*) during metamorphosis

Li QR[1,2] Deng XJ[2] Yang WY[2] Huang ZJ[2] Gianluca Tettamanti[3] Cao Y[2*] Feng QL[1*]

Abstract: Degeneration of larval-specific tissues during insect metamorphosis has been suggested to be the result of apoptosis and autophagy and is triggered by ecdysteroids. However, the relationship between autophagy and apoptosis pathways and the mechanism of regulation by ecdysteroids remain to be elucidated. This study examined the events of autophagy, apoptosis, and the expression of ecdysis-related genes in the silk gland of the silkworm (*Bombyx mori* L., 1758) during the larval to pupal transformation. The results indicated that autophagic features appeared in the silk gland at the wandering and spinning stages of the larvae, whereas the apoptotic features such as apoptotic bodies and DNA fragmentation occurred at the prepupal or early-pupal stages. The autophagic granules fused with each other to form large vacuoles where the cytoplasmic material was degraded. Autophagosomes, autolysosomes, and apoptotic bodies were found later in the degenerating silk-gland cells. Expression of the ecdysone receptor gene *BmEcR* and the transcription factor genes *BmE74A* and *BmBR-C* preceded the onset of autophagy and apoptosis, indicating that they may be responsible for triggering these programmed cell death pathways in the silk gland. The results suggest that both autophagy and apoptosis occur in the silk gland cells during degeneration, but autophagy precedes apoptosis.

Published On: Canadian Journal of Zoology, 2010, 88(12), 1169-1178.

1 Guangdong Provincial Key Lab of Biotechnology for Plant Development, School of Life Sciences, South China Normal University, Guangzhou 510631, China.

2 Department of Sericulture Science, College of Animal Science, South China Agricultural University, Guangzhou 510642, China.

3 Department of Biotechnology and Molecular Sciences, University of Insubria, Varese 21100, Italy.

*Corresponding author E-mail: caoyang@scau.edu.cn; qlfeng@scnu.edu.cn.

家蚕变态期丝腺中自噬、凋亡和蜕皮相关基因的表达分析

李庆荣[1,2] 邓小娟[2] 杨婉莹[2] 黄志君[2] Gianluca Tettamanti[3] 曹 阳[2*] 冯启理[1*]

摘要：昆虫在变态期间幼虫特殊组织的退化是由蜕皮激素诱导的细胞凋亡和自噬引起的。然而，自噬与凋亡通路的关系以及蜕皮激素调控机制仍有待阐明。研究调查了从幼虫到蛹期转变过程中家蚕丝腺细胞凋亡、细胞自噬及蜕皮相关基因的表达。结果表明，在幼虫上蔟期和吐丝期的丝腺出现细胞自噬特征，而在预蛹期及早期蛹期则出现凋亡小体及DNA断裂等细胞凋亡特征。自噬颗粒聚集一起形成更大的液泡，在液泡中降解细胞质物质。自噬体、自噬溶酶体和凋亡小体随后出现在正在降解的丝腺细胞中。蜕皮激素受体基因*BmEcR*及转录因子基因*BmE74A*和*BmBR-C*在细胞自噬和细胞凋亡之前表达，表明其可能负责触发丝腺中这些程序性细胞死亡途径。这些结果表明细胞自噬和细胞凋亡都发生在退化过程的丝腺中，而细胞自噬先于细胞凋亡发生。

1 广东省植物发育生物工程重点实验室，生命科学学院，华南师范大学，广州
2 动物科学学院，华南农业大学，广州
3 生物技术与分子科学系，英苏布里亚大学，瓦雷泽，意大利

Proteomic analysis of silk gland programmed cell death during metamorphosis of the silkworm *Bombyx mori*

Jia SH[1] Li MW[1] Zhou B[1] Liu WB[1] Zhang Y[1] Miao XX[1] Zeng R[2] Huang YP[1*]

Abstract: The silk gland of the silkworm *Bombyx mori* undergoes programmed cell death (PCD) during pupal metamorphosis. On the basis of their morphological changes and the occurrence of a DNA ladder, the tissue cells were categorized into three groups: intact, committed, and dying. To identify the proteins involved in this process, we conducted a comparative proteomic analysis. Protein expression changes among the three different cell types were examined by two-dimensional gel electrophoresis. Among 1 000 reproducibly detected protein spots on each gel, 43 were down-regulated and 34 were up-regulated in PCD process. Mass spectrometry identified 17 differentially expressed proteins, including some well-studied proteins as well as some novel PCD related proteins, such as caspases, proteasome subunit, elongation factor, heat shock protein, and hypothetical proteins. Our results suggest that these proteins may participate in the silk gland PCD process of *B. mori* and, thus, provide new insights for this mechanism.

Published On: Journal of Proteome Research, 2007, 6(8), 3003-3010.

1 Institute of Plant Physiology and Ecology, Shanghai Institutes for Biological Sciences, Chinese Academy of Sciences, Shanghai 200032, China.

2 Research Center for Proteome Analysis, Shanghai Institutes for Biological Sciences, Chinese Academy of Sciences, Shanghai 200031, China.

*Corresponding author E-mail: yongping@sippe.ac.cn.

家蚕变态过程中丝腺程序性细胞死亡的蛋白组学分析

贾世海[1] 李木旺[1] 周 博[1] 刘文彬[1] 张 勇[1] 苗雪霞[1] 曾 荣[2] 黄勇平[1*]

摘要:家蚕丝腺在蛹变态过程中发生程序性细胞死亡(PCD)。根据组织细胞形态变化和出现的DNA降解条带,将组织细胞的状态分为三种类型:完整、定型和死亡。为了鉴定参与该过程的蛋白质,我们进行了比较蛋白质组学分析,通过双向电泳检测了这三种不同细胞类型中蛋白质的表达变化。在每块胶上,约有1 000个可重复检测到的蛋白点,在PCD过程中有43个下调,34个上调。通过质谱分析鉴定了17种差异表达的蛋白质,包括一些研究较多的蛋白质以及一些新的PCD相关蛋白,如caspases、蛋白酶亚基、延伸因子、热激蛋白及未知蛋白。我们的研究结果表明,这些蛋白质可能参与家蚕丝腺的PCD过程,从而为此机制提供了新的理论基础。

1 上海生命科学研究院植物生理生态研究所,中国科学院,上海
2 上海生命科学研究院蛋白质研究分析中心,中国科学院,上海

Role of *BmDredd* during apoptosis of silk gland in silkworm, *Bombyx mori*

Chen RT Jiao P Liu Z Lu Y Xin HH Zhang DP Miao YG*

Abstract: Silk glands (SGs) undergo massive apoptosis driven degeneration during the larval-pupal transformation. To better understand this event on molecular level, we investigated the expression of apoptosis-related genes across the developmental transition period that spans day 4 in the fifth instar *Bombyx mori* larvae to day 2 pupae. Increases in the expression of *BmDredd* (an initiator caspase homolog) closely followed the highest *BmEcR* expression and resembled the expression trend of *BmIcE*. Simultaneously, we found that *BmDredd* expression was significantly higher in SG compared to other tissues at 18 h post-spinning, but reduced following injection of the apoptosis inhibitor (Z-DEVD-fmk). Furthermore, *BmDredd* expression correlated with changes of caspase3-like activities in SG and RNAi-mediated knockdown of *BmDredd* delayed SG apoptosis. Moreover, caspase3-like activity was increased in SG by overexpression of *BmDredd*. Taken together, the results suggest that *BmDredd* plays a critical role in SG apoptosis.

Published On: PLoS ONE, 2017, 12(1).

Institute of Sericulture and Apiculture, College of Animal Sciences, Zhejiang University, Hangzhou 310058, China.
*Corresponding author E-mail: miaoyg@zju.edu.cn.

家蚕变态期丝腺凋亡过程中*BmDredd*基因的作用

陈瑞婷　矫　鹏　刘　震　陆　嵒　辛虎虎　张登攀　缪云根*

摘要： 家蚕由幼虫到蛹的变态时期，丝腺（SGs）经历凋亡过程而解离。为了更好地从分子水平上阐明这一现象，我们调查了凋亡相关基因在五龄第4天到化蛹第2天的表达量。*BmDredd*（一个caspase同源起始因子）的表达增加紧随着*BmEcR*的最高表达，并且类似于*BmIcE*的表达趋势。同时我们发现，吐丝18 h后的丝腺组织中*BmDredd*的表达显著高于其他组织，但在注射凋亡抑制制后，BmDredd的表达降低。此外，*BmDredd*表达与丝腺中caspase3样的活性相关。RNAi介导的*BmDredd*基因的敲低，延迟了丝腺凋亡进程。而当过表达*BmDredd*时，丝腺组织中caspase3样的活性增加。研究结果表明*BmDredd*在丝腺细胞凋亡中起关键作用。

蚕蜂研究所，动物科学学院，浙江大学，杭州

$Ras1^{CA}$-upregulated bcpi inhibits cathepsin activity to prevent tissue destruction of the *Bombyx* posterior silk gland

Ma L[1,2] Liu SM[1,2] Shi M[3] Chen XX[3*] Li S[1*]

Abstract: Using the *GAL4/UAS* transgenic system established in the silkworm, *Bombyx mori,* we have previously reported that overexpression of the $Ras1^{CA}$ oncogene specifically in the posterior silk gland (PSG) resulted in improved fibroin synthesis, silk yield, and other phenotypic effects. However, the detailed molecular mechanism remains to be fully elucidated. Using 2D-DIGE-MS/MS analyses, we compared the proteomic profiles of PSGs from the wild type (WT) and $Ras1^{CA}$-overexpressed silkworms. Among the 24 $Ras1^{CA}$-enhanced proteins, the *Bombyx* cysteine protease inhibitor (BCPI) was increased 2.4-fold at the protein level and 3.4-fold at the mRNA level. Consistent with the developmental profiles, injection of recombinant BCPI into the WT silkworms at the early wandering stage inhibited cathepsin activity, prevented tissue destruction of the PSG, and delayed pupation. Moreover, injection of small-molecule inhibitors of cathepsin into the WT silkworms prevented PSG destruction and delayed pupation, confirming the role of BCPI in inhibiting cathepsin activity. Furthermore, injection of chemical inhibitors of the Ras downstream effectors into the $Ras1^{CA}$-overexpressed and WT silkworms revealed that both Raf-MAPK and PI3K-TORC1 pathways were required for Ras1 to induce *bcpi* expression. Taken together, we conclude that via the downstream Raf-MAPK and PI3K-TORC1 pathways, $Ras1^{CA}$ upregulates *bcpi*, which inhibits cathepsin activity thus preventing PSG destruction in *Bombyx*.

Published On: Journal of Proteome Research 2013, 12(4), 1924-1934.

1 Key Laboratory of Insect Developmental and Evolutionary Biology, Institute of Plant Physiology and Ecology, Chinese Academy of Sciences, Shanghai 200032, China

2 State Key Laboratory of Silkworm Genome Biology, Southwest University, Chongqing 400716, China.

3 State Key Laboratory of Rice Biology and Ministry of Agriculture Key Laboratory of Molecular Biology of Crop Pathogens and Insects, Institute of Insect Sciences, Zhejiang University, Hangzhou 310029, China.

*Corresponding author E-mail: xxchen@zju.edu.cn; lisheng01@sibs.ac.cn; shengli@sippe.ac.cn

$Ras1^{CA}$诱导的bcpi蛋白抑制组织蛋白酶活性从而阻止家蚕后部丝腺组织解体

马 俐[1,2] 刘淑敏[1,2] 时 敏[3] 陈学新[3*] 李 胜[1*]

摘要：利用*GAL4/UAS*转基因系统，我们成功地在家蚕后部丝腺（PSG）特异性地过表达了致癌基因$Ras1^{CA}$，改善了丝素蛋白合成，提高了产丝量及其他表型效应。然而，相关分子机制并不清楚。利用2D-DIGE-MS/MS质谱分析技术，我们比较了野生型和$Ras1^{CA}$过表达的家蚕PSG的蛋白质组学图谱。在24个$Ras1^{CA}$过表达的蛋白质中，半胱氨酸蛋白酶抑制剂（BCPI）在蛋白质水平增加了2.4倍，其mRNA水平增加了3.4倍。与发育概况一致，在上蔟早期阶段向野生型蚕注射重组BCPI，可抑制组织蛋白酶活性，防止PSG的组织解离和延迟蛹化。此外，野生型家蚕注射组织蛋白酶的小分子抑制剂会阻碍家蚕后部丝腺的解离，并使化蛹推迟，从而证实了BCPI抑制组织蛋白酶活性的作用。此外，进一步在野生型和过表达了$Ras1^{CA}$的家蚕幼虫体内注射Ras下游效应因子的抑制剂，Ras1诱导bcpi的表达需要Raf-MAPK和PI3K-TORC1两种途径。总的来说，$Ras1^{CA}$通过其下游Raf-MAPK和PI3K-TORC1信号通路上调*bcpi*的表达，从而抑制家蚕后部丝腺的组织蛋白酶活性从而防止其解离。

1 昆虫发育与进化生物学重点实验室，上海生命科学研究院植物生理生态研究所，中国科学院，上海

2 家蚕基因组生物学国家重点实验室，西南大学，重庆

3 农业部作物病虫分子生物学重点开放实验室，昆虫科学研究所，浙江大学，杭州

A juvenile hormone transcription factor *Bm*dimm-fibroin H chain pathway is involved in the synthesis of silk protein in silkworm, *Bombyx mori*

Zhao XM[1,2,3] Liu C[1,2*] Jiang LJ[1,2] Li QY[1,2]
Zhou MT[1,2] Cheng TC[1,2] Kazuei Mita[2] Xia QY[1,2*]

Abstract: The genes being responsible for silk biosynthesis are switched on and off at particular times in the silk glands of *Bombyx mori*. This switch appears to be under the control of endogenous and exogenous hormones. However, the molecular mechanisms by which silk protein synthesis is regulated by the juvenile hormone (JH) are largely unknown. Here, we report a basic helix-loop-helix transcription factor, *Bm*dimm, its silk gland-specific expression, and its direct involvement in the regulation of *fibroin-H chain* (*fib–H*) by binding to an E-box (CAAATG) element of the *fib-H* gene promoter. Far-Western blots, Enzyme-Linked Immunosorbent Assay, and co-immunoprecipitation assays revealed that *Bm*dimm protein interacted with another basic helix-loop-helix transcription factor, *Bm*sage. Immunostaining revealed that *Bm*dimm and *Bm*sage proteins are co-localized in nuclei. *Bmdimm* expression was induced in larval silk glands *in vivo*, in silk glands cultured *in vitro*, and in *B. mori* cell lines after treatment with a JH analog. The JH effect on *Bm*dimm was mediated by the JH-Met-Kr-h1 signaling pathway, and *Bmdimm* expression did not respond to JH by RNA interference with double-stranded *BmKr-h1* RNA. These data suggest that the JH regulatory pathway, the transcription factor *Bm*dimm, and the targeted *fib-H* gene contribute to the synthesis of fibroin-H chain protein in *B. mori*.

Published On: Journal of Biological Chemistry, 2015, 290(2), 972-986.

1 State Key Laboratory of Silkworm Genome Biology, Southwest University, Chongqing 400716, China.
2 Key Sericultural Laboratory of the Ministry of Agriculture, College of Biotechnology, Southwest University, Chongqing 400716.
3 Research Institute of Applied Biology, Shanxi University, Taiyuan 030006, China.
*Corresponding author E-mail: mlliuchun@163.com; xiaqy@swu.edu.cn.

保幼激素转录因子*Bm*dimm-丝素重链通路参与家蚕丝蛋白的合成

赵小明[1,2,3]　刘　春[1,2*]　姜莉君[1,2]　李琼艳[1,2]
周梦婷[1,2]　程廷才[1,2]　Kazuei Mita[2]　夏庆友[1,2*]

摘要：家蚕丝腺中负责家蚕丝蛋白合成的基因在特定的时间被开启和关闭，并且这种转换似乎由内源和外源激素控制。然而，家蚕丝蛋白合成受保幼激素（JH）调控的分子机制在很大程度上是未知的。在我们的研究中，报道了一种碱性的helix-loop-helix转录因子*Bm*dimm，其在丝腺组织特异表达，并且通过结合*fib-H*基因启动子上的E-box（CAAATG）元件直接参与调节丝素蛋白重链（fib-H）的合成。Far-Western blots、ELISA和免疫共沉淀实验显示*Bm*dimm蛋白与另一种碱性helix-loop-helix转录因子*Bm*sage相互作用。免疫染色显示*Bm*dimm和*Bm*sage蛋白共定位在细胞核中。用JH类似物处理体外培养的丝腺和家蚕的细胞系发现*Bm*dimm均被诱导表达。JH对*Bm*dimm的作用由JH-Met-Kr-h1信号通路介导，并且用双链RNA干涉*BmKr-h1*后，*Bmdimm*不会响应JH。这些数据表明，JH调节通路的转录因子*Bm*dimm通过调节其靶基因*fib-H*的表达从而参与家蚕丝素蛋白重链的合成。

1　家蚕基因组生物学国家重点实验室，西南大学，重庆
2　农业部蚕桑学重点实验室，生物技术学院，西南大学，重庆
3　应用生物学研究所，山西大学，太原

Transcription factor *Bm*sage plays a crucial role in silk gland generation in silkworm, *Bombyx mori*

Xin HH　Zhang DP　Chen RT　Cai ZZ　Lu Y　Liang S　Miao YG*

Abstract: Salivary gland secretion is altered in *Drosophila* embryos with loss of function of the *sage* gene. Saliva has a reduced volume and an increased electron density according to transmission electron microscopy, resulting in regions of tube dilation and constriction with intermittent tube closure. However, the precise functions of *Bm*sage in silkworm (*Bombyx mori*) are unknown, although its sequence had been deposited in SilkDB. From this, *Bm*sage is inferred to be a transcription factor that regulates the synthesis of silk fibroin and interacts with another silk gland-specific transcription factor, namely, silk gland factor-1. In this study, we introduced a germline mutation of *Bmsage* using the Cas9/sgRNA system, a genome-editing technology, resulting in deletion of *Bmsage* from the genome of *B. mori*. Of the 15 tested samples, seven displayed alterations at the target site. The mutagenesis efficiency was about 46.7% and there were no obvious off-target effects. In the screened homozygous mutants, silk glands developed poorly and the middle and posterior silk glands (MSG and PSG) were absent, which was significantly different from the wild type. The offspring of G_0 mosaic silkworms had indel mutations causing 2 bp or 9 bp deletions at the target site, but exhibited the same abnormal silk gland structure. Mutant larvae containing different open-reading frames of *Bmsage* had the same silk gland phenotype. This illustrated that the mutant phenotype was due to *Bmsage* knockout. We conclude that *Bm*sage participates in embryonic development of the silk gland.

Published On: Archives of Insect Biochemistry and Physiology, 2015, 90(2), 59-69.

Institute of Sericulture and Apiculture, College of Animal Sciences, Zhejiang University, Hangzhou 310058, China.
*Corresponding author E-mail: miaoyg@zju.edu.cn.

转录因子*Bm*sage在家蚕丝腺发育中发挥关键作用

辛虎虎　张登攀　陈瑞婷　蔡自峥　陆　岳　梁　爽　缪云根*

摘要：果蝇中*sage*基因功能丧失会引起其唾液腺分泌物的变化，表现为唾液分泌量减少，透射电子显微镜下可观察到唾液电子密度的增加，从而导致腺管部分的扩张和收缩，并伴随间断性的关闭。尽管家蚕*Bmsage*基因的序列在SilkDB已明确公布，但其功能仍未知。有报道推测*Bm*sage是一个参与调控家蚕丝素蛋白合成的转录因子，与另一个丝腺特异型转录因子——丝腺因子-1（silk gland factor-1）有互作关系。在我们的研究中，利用Cas9/sgRNA基因组编辑技术删除了家蚕*Bmsage*基因，获得了*Bmsage*基因突变的品系。检测了15个注射样本，其中有7个在靶位点发生改变，其突变率约为46.7%，且未见明显的脱靶效应。筛选出的纯合突变体中，其丝腺发育不完全，与野生型丝腺形态完全不同，表现为中部和后部丝腺（MSG和PSG）缺失。G_0嵌合突变体家蚕的子代在靶位点处发生了2 bp或9 bp的碱基删除，但是两者均表现为相同的丝腺结构异常的表型。*Bmsage*不同开放阅读框的突变体幼虫同样具有相同的丝腺表型。这些结果表明，突变体家蚕的丝腺异常表型是由*Bmsage*基因的敲除导致的。因此，我们推断，*Bmsage*基因参与了家蚕胚胎期丝腺的发育，其在家蚕丝腺的形成过程中起着极其重要的作用。

蚕蜂研究所，动物科学学院，浙江大学，杭州

Multiprotein bridging factor 2 regulates the expression of the *fibroin heavy chain* gene by interacting with *Bm*dimmed in the silkworm *Bombyx mori*

Zhou CY# Zha XF# Shi PP Wei SG Wang H Zheng RW Xia QY*

Abstract: Multiprotein bridging factor 2 (MBF2) was first isolated from the posterior silk gland of *Bombyx mori.* However, its function in *B. mori* is still unknown. Herein, *MBF2* transcripts were detected mainly in the posterior silk gland and Malpighian tubules of *B.mori* larvae via a quantitative PCR analysis. An analysis of temporal expression patterns showed that the expression pattern of *MBF2* was the opposite of that of the *fibroin heavy chain* (*fibH*) gene, as its expression was high during the fourth-instar moulting stage, decreased gradually during the fifth-instar feeding stage and disappeared at the end of the fifth-instar phase. Furthermore, bimolecular fluorescent complementation and Far-Western blot assays showed that MBF2 interacted with the basic helix-loop-helix transcription factor *Bm*dimmed. Dual luciferase reporter assays showed that MBF2 down-regulated the promoter activity of *fibH* and inhibited the effect of *Bmdimmed (Bmdimm)* on *fibH* expression. *MBF2* expression was induced in silk glands after treatment with 20-hydroxyecdysone *in vivo* and *in vitro*. These findings suggest that MBF2 is a transcriptional repressor that is involved in controlling the regulation of the *fibH* gene in the posterior silk gland by interacting with *Bm*dimm.

Published On: Insect Molecular Biology, 2016, 25(4), 509-518.

State Key Laboratory of Silkworm Genome Biology, Southwest University, Chongqing 400716, China.

#These authors contributed equally.

*Corresponding author E-mail: xiaqy@swu.edu.cn.

多蛋白桥连因子2通过与转录因子*Bm*dimm互作调控丝素重链基因的表达

周春燕[#] 查幸福[#] 石盼盼 位曙光 王 鹤 郑人文 夏庆友[*]

摘要:多蛋白桥连因子2(MBF2)最初从家蚕的后部丝腺中分离,然而它在家蚕中的功能还不明确。在研究中,我们通过qPCR检测了*MBF2*在家蚕各组织中的表达情况,发现其主要在家蚕幼虫的后部丝腺和马氏管中表达。时期表达谱分析结果显示*MBF2*的表达模式和家蚕丝素重链基因(*fibH*)的表达模式相反,即*MBF2*在四龄眠起阶段高量表达,在五龄初期表达量逐渐下降并至五龄末期消失。进一步通过双分子荧光互补实验和Far-Western blot实验显示MBF2与转录因子*Bm*dimm存在相互作用。双荧光素酶报告实验结果显示MBF2下调*fibH*启动子的活性并且可以抑制*Bmdimmed*(*Bmdimm*)对*fibH*的调控。无论在体内还是在体外,20-羟基蜕皮酮均可以诱导*MBF2*在后部丝腺的表达。综上表明MBF2是一种转录抑制因子,可以通过与*Bm*dimm互作调控*fibH*在后部丝腺的表达。

家蚕基因组生物学国家重点实验室,西南大学,重庆

Biochemical characterization and functional analysis of the POU transcription factor POU-M2 of *Bombyx mori*

Liu LN[1] Li Y[1] Wang YJ[1,2] Zhao P[1] Wei SG[1] Li ZZ[1] Chang HP[1,2] He HW[1,2*]

Abstract: POU-M2 is a homeodomain transcription factor which plays important roles in the development and silk synthesis of *Bombyx mori*. In this study, we expressed, purified and characterized POU-M2 and studied its transcription regulation on *fibroin heavy chain* gene of *Bombyx mori*. Gel filtration showed POU-M2 existed as a dimer in solution. Far-UV circular dichroism spectra indicated POU-M2 had a well-defined α-helix structure and the α-helix content was about 26.4%. The thermal unfolding transition of POU-M2 was a cooperative process. T_m, $\triangle H$ and $\triangle S$ were (45.15 ± 0.2)℃, (138.4 ± 0.5) kJ/mol and (0.4349 ± 0.04) kJ/(mol · K), respectively. Western blot analysis indicated the expression level of POU-M2 increased slightly from day 3 to day 7 of the fifth instar larvae in the posterior silk gland. POU-M2 was positioned in the nucleus of cells. The luciferase reporter assay demonstrated POU-M2 could stimulate the promoter activity of *fibroin heavy chain* gene, and the activation effect was dependent on the amount of POU-M2. Our study suggested POU-M2 may be involved in the transcriptional regulation of *fibroin heavy chain* gene. These findings expand toward a better understanding of the structure of POU-M2 and its function in silk synthesis of *Bombyx mori*.

Published On: International Journal of Biological Macromolecules, 2016, 86, 701-708.

1 State Key Laboratory of Silkworm Genome Biology, Southwest University, Chongqing 400716, China.
2 Key Sericultural Laboratory of Ministry of Agriculture, College of Biotechnology, Southwest University, Chongqing 400716, China.
*Corresponding author E-mail: hehuawei@swu.edu.cn.

家蚕POU家族转录因子POU-M2的生化表征和功能分析

刘莉娜[1]　李　瑜[1]　王叶菁[1,2]　赵　朋[1]　位曙光[1]
李珍珍[1]　常怀普[1,2]　何华伟[1,2*]

摘要：POU-M2是同源结构域转录因子，在家蚕的发育和丝蛋白合成中起重要作用。在研究中，我们表达、纯化并表征了POU-M2蛋白的理化性质，研究了其对家蚕丝素蛋白重链基因的转录调控作用。凝胶过滤分析显示POU-M2蛋白在溶液中以二聚体形式存在，远紫外圆二色光谱分析表明，POU-M2具有典型的α-螺旋结构，经计算其α-螺旋含量约为26.4%。POU-M2的热变性过程是一个多因素协作的过程，T_m、ΔH和ΔS分别为(45.15±0.2)℃、(138.4±0.5)kJ/mol和(0.4349±0.04)kJ/(mol·K)。免疫印迹分析显示，在家蚕五龄幼虫的后部丝腺中，从五龄第3天至五龄第7天，POU-M2的表达水平逐渐上调。免疫荧光分析显示，POU-M2蛋白定位于细胞核中。荧光素酶报告系统分析显示，POU-M2可上调丝素蛋白重链基因的启动子活性，且该激活作用的程度取决于POU-M2的含量。我们的研究表明POU-M2可能参与丝素蛋白重链基因的转录调控，这些发现使我们对POU-M2的结构及其在蚕丝蛋白合成中的功能有了更深入的理解。

1　家蚕基因组生物学国家重点实验室，西南大学，重庆
2　农学部蚕桑学重点实验室，生物技术学院，西南大学，重庆

Structures of an all-α protein running along the DNA major groove

Yu LY[1#] Cheng W[1#] Zhou K[1#] Li WF[1] Yu HM[1]
Gao XL[1] Shen XD[2] Wu QF[1] Chen YX[1*] Zhou CZ[1*]

Abstract: Despite over 3 300 protein-DNA complex structures have been reported in the past decades, there remain some unknown recognition patterns between protein and target DNA. The silk gland-specific transcription factor FMBP-1 from the silkworm *Bombyx mori* contains a unique DNA-binding domain of four tandem STPRs, namely the score and three amino acid peptide repeats. Here we report three structures of this STPR domain (termed *Bm*STPR) in complex with DNA of various lengths. In the presence of target DNA, *Bm*STPR adopts a zig-zag structure of three or four tandem α-helices that run along the major groove of DNA. Structural analyses combined with binding assays indicate *Bm*STPR prefers the AT-rich sequences, with each α-helix covering a DNA sequence of 4 bp. The successive AT-rich DNAs adopt a wider major groove, which is in complementary in shape and size to the tandem α-helices of *Bm*STPR. Substitutions of DNA sequences and affinity comparison further prove that *Bm*STPR recognizes the major groove mainly via shape readout. Multiple-sequence alignment suggests this unique DNA-binding pattern should be highly conserved for the STPR domain containing proteins which are widespread in animals. Together, our findings provide structural insights into the specific interactions between a novel DNA-binding protein and a unique deformed B-DNA.

Published On: Nucleic Acids Research, 2016, 44(8): 3936-3945.

1 Hefei National Laboratory for Physical Sciences at the Microscale, University of Science and Technology of China, Hefei 230027, China.

2 School of Information Science and Technology, University of Science and Technology of China, Hefei 230027, China.

[#]These authors contributed equally.

[*]Corresponding author E-mail: cyxing@ustc.edu.cn; zcz@ustc.edu.cn.

DNA大沟的全α类蛋白的结构解析

余立艳[1#]　成　望[1#]　周　康[1#]　李卫芳[1]　于红美[1]
郜鑫磊[1]　沈旭东[2]　吴清发[1]　陈宇星[1*]　周丛照[1*]

摘要：尽管目前已有超过3 300个蛋白-DNA复合物结构被解析，但蛋白质与DNA之间还有很多全新的识别模式尚未发现。来自家蚕后部丝腺特有的调节因子FMBP-1包含一种新型的DNA结合结构域，该结构域由4个23肽重复序列串联组成。这里，我们利用X-射线晶体学解析了3个该DNA结合结构域（简称*Bm*STPR）与不同长度靶DNA结合的复合物结构。结构显示，*Bm*STPR的每个重复单元以串联α-螺旋的形式结合在DNA的大沟中。进一步的生化实验和结构分析表明，*Bm*STPR偏好富含AT的序列，每个螺旋对应识别4 bp的DNA双链序列，*Bm*STPR通过形貌特异性而非碱基特异性识别具有较高柔性和较宽大沟的富含AT的DNA序列。这种新型的DNA结合结构域广泛存在于各种动物中，暗示它们可能具有更加保守而重要的生物学功能。

1　合肥微尺度物质科学国家研究中心，中国科学技术大学，合肥
2　信息科学技术学院，中国科学技术大学，合肥

Effect of *BBX-B8* overexpression on development, body weight, silk protein synthesis and egg diapause of *Bombyx mori*

Zheng XJ[1#] Gong YC[1#] Dhiraj Kumar[1] Chen F[1] Kuang SL[1] Liang Z[1]

Hu XL[1,2] Cao GL[1,2] Xue RY[1,2] Gong CL[1,2,3*]

Abstract: Bombyxin (BBX) is an insulin-like peptide exists in the silkworm *Bombyx mori*. Our previous studies on the effects of inhibiting *BBX-B8* expression found that *BBX-B8* is important for the development of organ, reproduction and trehalose metabolism in the silkworms. In this paper, we investigated the expression profile of the *BBX-B8* gene and effect of *BBX-B8* overexpression on the development, body weight, silkprotein synthesis and egg diapause of *B. mori* to further understand *BBX-B8* functions. *BBX-B8* gene expression could be detected in the brains, midguts, anterior silk glands, ovaries, testes, fat bodies, hemolymph, Malpighian tubules and embryos by RT-PCR, however it was mainly expressed in the brain. Western blots showed that the change in *BBX-B8* expression was not obvious in the brain of 1- to 4-day-old larvae of fifth instar silkworms, but expression increased substantially at 5- to 6-day-old larvae of fifth instar silkworms. Transgenic silkworms overexpressing *BBX-B8* were obtained by introducing non-transposon transgenic vector *p*IZT-B8 containing a *BBX-B8* gene driven by *Orgyia pseudotsugata* nucleopolyhedrovirus IE2 promoter into the genome. Development duration of the transgenic silkworms was delayed by 2.5-3.5 days. Cocoon shell weight of transgenic silkworms was reduced by 4.79 % in females and 7.44 % in males, pupal weight of transgenic silkworms was reduced 6.75 % in females and 13.83 % in males compared to non-transgenic silkworms, and 5.56% - 14.29 % of transgenic moths laid nondiapausing eggs. All results indicated that *BBX-B8* plays an important role in the development, silk protein synthesis and egg diapause of silkworm.

Published On: Transgenic Research 2016, 25(4), 507-516.

1 School of Biology and Basic Medical Sciences ,Soochow University, Suzhou 215123, China.

2 National Engineering Laboratory for Modern Silk, Soochow University, Suzhou 215123, China.

3 Pre-clinical Medical and Biological Science College, Soochow University, Suzhou 215123, China.

#These authors contributed equally.

*Corresponding author E-mail: gongcl@suda.edu.cn.

过表达*BBX-B8*对家蚕发育、体重增加、丝蛋白合成和卵滞育的影响

郑小坚[1#]　龚永昌[1#]　Dhiraj Kumar[1]　陈　菲[1]　匡苏兰[1]
梁　子[1]　胡小龙[1,2]　曹广力[1,2]　薛仁宇[1,2]　贡成良[1,2,3*]

摘要：蚕素(BBX)是一种存在于家蚕体内的胰岛素样多肽。以前的研究发现抑制*BBX-B8*基因的表达对家蚕的器官发育、繁殖及海藻糖的代谢有重要影响。我们研究了*BBX-B8*的表达谱及过表达*BBX-B8*对家蚕发育、体重增长、丝蛋白合成和家蚕卵滞育的影响，从而进一步了解了*BBX-B8*基因的功能。RT-PCR结果显示，*BBX-B8*基因在家蚕脑、中肠、前部丝腺、卵巢、精巢、脂肪体、血、马氏管和蚕卵中都有表达，但主要表达于脑中。Western blot结果显示，五龄第1至4天家蚕脑中*BBX-B8*表达不明显，但五龄第5至6天家蚕脑中表达量显著增加。利用舞毒蛾核型多角体病毒(*Orgyia pseudotsugata* nucleopolyhedrovirus)IE2启动子控制*BBX-B8*基因的非转座子转基因载体*p*IZT-B8获得了过表达*BBX-B8*的转基因蚕。我们发现过表达*BBX-B8*的转基因蚕相对于正常蚕来说发育期延迟了2.5—3.5d，转基因蚕的茧层重相对于正常蚕的茧层重，雄性的下降了7.44%，雌性的下降了4.79%，转基因蚕的蛹重相对于正常蚕的蛹重，雄性的下降了13.83%，雌性的下降了6.75%，并且有5.56%—14.29%的转基因蚕蛾产下非滞育卵。所有的结果表明，*BBX-B8*基因在家蚕的发育、丝蛋白合成和卵滞育方面起着重要的作用。

1　生命科学与基础医学系，苏州大学，苏州
2　现代丝绸国家工程实验室，苏州大学，苏州
3　基础医学与生物科学学院，苏州大学，苏州

Comparative proteomics reveal diverse functions and dynamic changes of *Bombyx mori* silk proteins spun from different development stages

Dong ZM[#] Zhao P[#] Wang C Zhang Y

Chen JP Wang X Lin Y Xia QY[*]

Abstract: Silkworms (*Bombyx mori*) produce massive amounts of silk proteins to make cocoons during the final stages of larval development. Although the major components, fibroin and sericin, have been the focus for a long time, few researchers have realized the complexity of the silk proteome. We collected seven kinds of silk fibers spun by silkworm larvae at different developmental stages: the silks spun by new hatched larvae, second instar day 0 larvae, third instar day 0 larvae, fourth instar day 0 larvae, and fourth instar molting larvae, the scaffold silk used to attach the cocoon to the substrate and the cocoon silk. Analysis by liquid chromatography−tandem mass spectrometry identified 500 proteins from the seven silks. In addition to the expected fibroins, sericins, and some known protease inhibitors, we also identified further protease inhibitors, enzymes, proteins of unknown function, and other proteins. Unsurprisingly, our quantitative results showed fibroins and sericins were the most abundant proteins in all seven silks. Except for fibroins and sericins, protease inhibitors, enzymes, and proteins of unknown function were more abundant than other proteins. We found significant change in silk protein compositions through development, being consistent with their different biological functions and complicated formation.

Published On: Journal of Proteome Research, 2013, 12(11), 5213-5222.

State Key Laboratory of Silkworm Genome Biology, Southwest University, Chongqing 400716, China.

[#]These authors contributed equally.

[*]Corresponding author E-mail: xiaqy@swu.edu.cn.

比较蛋白质组学揭示不同发育阶段家蚕纺出的蚕丝蛋白的不同功能和动态变化

董照明[#] 赵 萍[#] 王 晨 张 艳
陈建平 王 鑫 林 英 夏庆友[*]

摘要:家蚕在幼虫的最后一个龄期产生大量的丝蛋白以形成蚕茧。多年来,蚕丝中主要的两种蛋白质——丝素和丝胶一直备受关注,而很少有研究者意识到蚕丝蛋白质组的复杂性。我们收集了家蚕幼虫不同发育时期纺出的7种丝纤维:蚁蚕丝、二龄起蚕丝、三龄起蚕丝、四龄起蚕丝、四龄眠蚕丝、用于固定蚕茧的茧网丝、蚕茧丝。利用液相色谱串联质谱法从以上7种蚕丝中鉴定出500种蛋白质,除了已知的丝素蛋白、丝胶蛋白和几种蛋白酶抑制剂,我们发现了更多的蛋白酶抑制剂、酶类、未知功能的蛋白质以及其他类型蛋白质。不出所料,定量分析发现在上述7种蚕丝中,丝素和丝胶蛋白都是含量最丰富的,除丝素和丝胶蛋白外,蛋白酶抑制剂、酶类和未知功能的蛋白质比其他类型蛋白质的含量更为丰富。我们发现不同发育阶段家蚕的丝蛋白组成有显著的变化,这与它们生物功能的不同和复杂的形成过程相一致。

家蚕基因组生物学国家重点实验室,西南大学,重庆

Comparative proteome analysis of multi-layer cocoon of the silkworm, *Bombyx mori*

Zhang Y[#] Zhao P[#] Dong ZM Wang DD Guo PC
Guo XM Song QR Zhang WW Xia QY[*]

Abstract: *Bombyx mori* cocoon has a multi-layer structure that provides optimal protection for silkworm pupa. Research on the mechanical properties of the multi-layer structure revealed structure-property relationships of the cocoon. Here, we investigated the protein components of the *B. mori* cocoon in terms of its multi-layer structure. Liquid chromatography tandem mass spectrometry identified 286 proteins from the multiple cocoon layers. In addition to fibroins and sericins, we identified abundant protease inhibitors, seroins and proteins of unknown function. By comparing protein abundance across layers, we found that the outermost layer contained more sericin1 and protease inhibitors and the innermost layer had more seroin1. As many as 36 protease inhibitors were identified in cocoons, showing efficient inhibitory activities against a fungal protease. Thus, we propose that more abundant protease inhibitors in the outer cocoon layers may provide better protection for the cocoons. This study increases our understanding of the multi-layer mechanism of cocoons, and helps clarify the biological characteristics of cocoons.

Published On: PLoS ONE, 2015, 10(4).

State Key Laboratory of Silkworm Genome Biology, Southwest University, Chongqing 400716, China.
[#]These authors contributed equally.
[*]Corresponding author E-mail: xiaqy@swu.edu.cn.

家蚕不同茧层的比较蛋白质组学分析

张　艳[#]　赵　萍[#]　董照明　王丹丹　郭鹏超
郭晓朦　宋倩茹　张薇薇　夏庆友[*]

摘要：家蚕蚕茧具有多层结构，为蚕蛹提供了最好的保护。对多层结构力学性能的研究揭示了蚕茧结构和性能的关系。在研究中，我们分析了家蚕蚕茧多层结构中的蛋白质成分。利用LC-MS在茧层中鉴定到286种蛋白质。除了丝素蛋白和丝胶蛋白，我们还鉴定到了大量的蛋白酶抑制剂和seroins及未知功能蛋白。通过比较茧层间的蛋白质丰度，我们发现最外层含有较多的丝胶1和蛋白酶抑制剂，而最内层含有较多的seroin1。在蚕茧中鉴定到了36种蛋白酶抑制剂，对真菌蛋白酶表现出高效的抑制活性。因此，我们猜测外层茧层中更高含量的蛋白酶抑制剂可能为蚕茧提供更好的保护。这项研究增进了对蚕茧多层结构机制的认识，有助于阐明蚕茧的生物学特性。

家蚕基因组生物学国家重点实验室，西南大学，重庆

Protease inhibitors in *Bombyx mori* silk might participate in protecting the pupating larva from microbial infection

Li YS[1,2] Liu HW[1] Zhu R[3] Xia QY[1] Zhao P[1*]

Abstract: Pupae inside cocoons rarely suffer from disease. It is apparent that some factors in the cocoon exert antimicrobial effects whereby the pupae inside can be protected from microbial infection. In the present study, we investigated the expression of cocoon protease inhibitors using immunoblotting and activity staining. Enzymatic hydrolysis of cocoon proteins *in vitro* was performed to characterize their roles in protecting the cocoon from microbial proteases. We found that some protease inhibitors, particularly trypsin inhibitor-like (TIL)-type protease inhibitors, can be secreted into the cocoon layer during the spinning process, thereby providing effective protection to the cocoon and pupa by inhibiting the extracellular proteases that can be secreted by pathogens.

Published On: Insect Science, 2016, 23(6), 835-842.

1 State Key Laboratory of Silkworm Genome Biology, Southwest University, Chongqing 400716, China.

2 Vitamin D Research Institute, Shaanxi University of Technology, Hanzhong 723001, China.

3 School of Management, Shaanxi University of Technology, Hanzhong 723001, China.

*Corresponding author E-mail: zhaop@swu.edu.cn

蚕丝中的蛋白酶抑制剂可能参与保护化蛹的幼虫免于病原物的感染

李游山[1,2] 刘华伟[1] 朱 瑞[3] 夏庆友[1] 赵 萍[1*]

摘要：茧里面的蛹很少被病原微生物感染，很显然是茧里面的一些成分发挥了抗菌作用，因此里面的蛹能够被保护而免受病原菌的感染。在现阶段研究中，我们用免疫印迹和活性染色方法调查了茧蛋白酶抑制剂的表达。通过茧丝蛋白体外的酶促水解反应来表征它保护茧不受微生物蛋白酶的作用。我们发现一些蛋白酶抑制剂，尤其是类胰蛋白酶（TIL）型的抑制剂，在吐丝过程中能够被分泌到茧层中，从而抑制病原菌分泌的胞外蛋白酶，对茧和蛹提供有效的保护。

1 家蚕基因组生物学国家重点实验室，西南大学，重庆
2 维生素D生理与应用研究所，陕西理工大学，汉中
3 管理学院，陕西理工大学，汉中

Proteins in the cocoon of silkworm inhibit the growth of *Beauveria bassiana*

Guo XM[#] Dong ZM[#] Zhang Y Li YS Liu HW Xia QY Zhao P[*]

Abstract: Silk cocoons are composed of fiber proteins (fibroins) and adhesive glue proteins (sericins), which provide a physical barrier to protect the inside pupa. Moreover, other proteins were identified in the cocoon silk, many of which are immune related proteins. In this study, we extracted proteins from the silkworm cocoon by Tris-HCl buffer (pH7.5), and found that they had a strong inhibitory activity against fungal proteases and they had higher abundance in the outer cocoon layers than in the inner cocoon layers. Moreover, we found that extracted cocoon proteins can inhibit the germination of *Beauveria bassiana* spores. Consistent with the distribution of protease inhibitors, we found that proteins from the outer cocoon layers showed better inhibitory effects against *B. bassiana* spores than proteins from the inner layers. Liquid chromatography tandem mass spectrometry was used to reveal the extracted components in the scaffold silk, the outermost cocoon layer. A total of 129 proteins were identified, 30 of which were annotated as protease inhibitors. Protease inhibitors accounted for 89.1% in abundance among extracted proteins. These protease inhibitors have many intramolecular disulfide bonds to maintain their stable structure, and remained active after being boiled. This study added a new understanding to the antimicrobial function of the cocoon.

Published On: PLoS ONE. 2016, 11(3).

State Key Laboratory of Silkworm Genome Biology, Southwest University, Chongqing 400716, China.

[#]These authors contributed equally.

[*]Corresponding author E-mail:zhaop@swu.edu.cn.

蚕茧中的蛋白质抑制白僵菌的生长

郭晓朦[#] 董照明[#] 张 艳 李游山 刘华伟 夏庆友 赵 萍[*]

摘要：蚕丝由纤维蛋白(丝素)和具有黏性的胶蛋白(丝胶)组成,其提供物理屏障以保护内部的蛹。此外,我们在蚕丝中还鉴定到其他蛋白质,包括许多免疫相关蛋白。在研究中,我们利用Tris-HCl缓冲液(pH 7.5)从蚕茧中提取蛋白质,发现它们对真菌蛋白酶具有很强的抑制活性,并且其在外茧层的含量较在内茧层高。此外,我们发现提取的茧丝蛋白可以抑制白僵菌孢子的萌发。与蛋白酶抑制剂的分布一致,我们发现外茧层的蛋白质比内茧层蛋白质对白僵菌孢子的抑制效果更好。利用液相色谱串联质谱法对最外层茧层的蛋白质成分进行鉴定,共鉴定到129个蛋白质,其中30个为蛋白酶抑制剂。蛋白酶抑制剂在提取的蛋白质中的丰度为89.1%。这些蛋白酶抑制剂由于具有许多分子内二硫键,所以结构十分稳定,即使在煮沸后仍可保持活性。研究结果使我们对蚕茧的抗菌功能有了新的认识。

家蚕基因组生物学国家重点实验室,西南大学,重庆

Structure, evolution, and expression of antimicrobial silk proteins, seroins in Lepidoptera

Dong ZM[1,2#] Song QR[1,2#] Zhang Y[1,2] Chen SY[1,2] Zhang XL[1,2] Zhao P[1,2*] Xia QY[1,2]

Abstract: The silks of silkworm and waxworm contain abundant antimicrobial proteins, including protease inhibitors and seroins. Protease inhibitors have antifungal activities, whereas seroins have antiviral and antibacterial activities. In order to obtain insights into the structure, evolution, and expression of seroins, we performed an extensive survey based on the available genome, transcriptome, and expressed sequence tags datasets. Sixty-four seroins were identified in 32 lepidopteran species. The phylogenetic and structural analyses revealed that seroins can be classified into five subfamilies: seroin 1, seroin 2, seroin 3, seroin 2+1, and seroin 3+3. It is interesting that seroin 2+1 contains two tandem seroin domains, seroin 2 and seroin 1, whereas seroin 3+3 has two tandem seroin 3 domains. Each seroin domain contains a proline-rich N-terminal motif and a conserved C-terminal motif. The transcriptome and EST data indicated that *seroin 1* and *seroin 2* genes were expressed in the silk gland but *seroin 3* genes were not. Semi-quantitative RT-PCR and Western blot analyses suggested that seroin 1 and seroin 2 were constantly accumulated in the silk gland of silkworm during the fifth instar, and then secreted into cocoon silk during spinning. Immunofluorescence analyses indicated that seroin 1 was secreted into the fibroin and sericin layers, whereas seroin 2 protein was only secreted into the sericin layer. However, the antimicrobial activity of seroin 2 was more effective than that of seroin 1. The presence of seroin 1 in the fibroin layer suggested that this protein not only acts as an antimicrobial protein, but might also play a role in the assembly and secretion of fibroins. Seroin 3, which was first identified here, might be related to pheromone synthesis or recognition, as it was highly expressed in male antennae and in the pheromone gland.

Published On: Insect Biochemistry and Molecular Biology, 2016, 75, 24-31.

1 State Key Laboratory of Silkworm Genome Biology, Southwest University, Chongqing 400716, China.

2 Chongqing Engineering and Technology Research Center for Novel Silk Materials, Southwest University, Chongqing 400716, China.

[#]These authors contributed equally.

[*]Corresponding author E-mail: zhaop@swu.edu.cn.

鳞翅目昆虫抗菌丝蛋白Seroins的结构、进化与表达分析

董照明[1,2#]　宋倩茹[1,2#]　张　艳[1,2]　陈诗懿[1,2]
张晓璐[1,2]　赵　萍[1,2*]　夏庆友[1,2]

摘要：家蚕和蜡螟的丝中含有丰富的抗菌蛋白，包括蛋白酶抑制剂和Seroins。蛋白酶抑制剂具有抗真菌活性，而Seroins具有抗病毒和抗细菌活性。为了解Seroins的结构、进化和表达情况，我们基于可用的基因组、转录组和表达序列标签数据进行了调查分析，在32个鳞翅目物种中共鉴定到64个Seroins。系统发育和结构分析显示，Seroins可以分为5个亚科：Seroin1，Seroin 2，Seroin 3，Seroin 2+1和Seroin 3+3。有趣的是，Seroin 2+1含有两个串联的Seroin结构域，Seroin 2和Seroin 1，而Seroin 3+3具有两个串联的Seroin 3结构域。每个Seroin结构域含有富含脯氨酸的N末端基序和保守的C末端基序。转录组和EST数据表明*Seroin 1*和*Seroin 2*基因在丝腺中表达，而*Seroin 3*基因在丝腺中不表达。半定量RT-PCR和蛋白质印迹分析表明，Seroin 1和Seroin 2在家蚕五龄期的丝腺中不断积累，然后分泌到茧丝中。免疫荧光分析表明，Seroin1分泌到丝素蛋白和丝胶蛋白层中，而Seroin2蛋白仅分泌到丝胶蛋白层中。Seroin2的抗菌活性比Seroin1的抗菌活性更好。Seroin1存在于丝素蛋白层中，表明该蛋白质不仅起抗菌作用，还可能在丝素蛋白的分泌和组装中起作用。在本研究中，我们首次鉴定到Seroin 3，由于其在雄性触角和附腺中高量表达，推测Seroin 3可能与信息素合成或识别有关。

1　家蚕基因组生物学国家重点实验室，西南大学，重庆
2　重庆市蚕丝纤维新材料工程技术研究中心，西南大学，重庆

Modifying the mechanical properties of silk fiber by genetically disrupting the ionic environment for silk formation

Wang X[1#] Zhao P[1#] Li Y[1] Yi QY[2] Ma SY[1] Xie K[1] Chen HF[1] Xia QY[1*]

Abstract: Silks are widely used in biomaterials, but there are still weaknesses in their mechanical properties. Here we report a method for improving the silk fiber mechanical properties by genetic disruption of the ionic environment for silk fiber formation. An anterior silk gland (ASG) specific promoter was identified and used for overexpressing ion-transporting protein in the ASG of silkworms. After isolation of the transgenic silkworms, we found that the metal ion content, conformation and mechanical properties of transgenic silk fibers changed accordingly. Notably, overexpressing endoplasmic reticulum Ca^{2+}-ATPase in ASG decreased the calcium content of silks. As a consequence, silk fibers had more α-helix and β-sheet conformations, and their tenacity and extension increased significantly. These findings represent the *in vivo* demonstration of a correlation between metal ion content in the spinning duct and the mechanical properties of silk fibers, thus providing a novel method for modifying silk fiber properties.

Published On: Biomacromolecules, 2015, 16(10), 3119-3125.

1 State Key Laboratory of Silkworm Genome Biology, Southwest University, Chongqing 400716, China.

2 Animal Center, Chongqing Medical University, Chongqing 400016, China.

#These authors contributed equally.

*Corresponding author E-mail: xiaqy@swu.edu.cn.

通过遗传改造家蚕丝纤维形成的离子环境以改良蚕丝纤维的力学性能

王　鑫[1#]　赵　萍[1#]　李　懿[1]　衣启营[2]　马三垣[1]
谢　康[1]　陈慧芳[1]　夏庆友[1*]

摘要：蚕丝广泛地应用于生物材料，但其在力学性能方面仍然存在缺点，本文我们报道了通过基因干扰蚕丝纤维形成的离子环境的这种方法，来改良蚕丝纤维的力学性能。鉴定得到的前部丝腺特异的启动子用于前部丝腺离子转运蛋白的过量表达。分离出转基因家蚕后，我们发现转基因家蚕蚕丝纤维的金属离子含量、构象以及力学性能都相应地发生了改变。尤其是过量表达内质网上的Ca^{2+}-ATP酶后，蚕丝中的Ca^{2+}浓度降低了。结果表明，蚕丝纤维中有更多的α-螺旋和β-折叠的构象，并且韧性和延展性都显著地增强了，这些发现证明了在体内吐丝管中的金属离子浓度与蚕丝纤维的力学性能相关，从而提供了一种对蚕丝纤维的性能进行修饰的新方法。

1　家蚕基因组生物学国家重点实验室，西南大学，重庆
2　重庆医科大学动物中心，重庆

Construction of transgenic silkworm spinning antibacterial silk with fluorescence

Li Z[1] Jiang Y[1] Cao GL[1,2]
Li JZ[1] Xue RY[1,2] Gong CL[1,2*]

Abstract: A targeting vector consisting of a fusion gene of the green fluorescent protein (GFP) gene *gfp* and the antimicrobial peptide cecropin gene *cec* flanked by pieces of the 5' and 3' sequences of the fibroin L chain gene *fib-L* of the silkworm (*Bombyx mori*) and a negative selection DsRed marker gene driven by the baculovirus immediate early gene 1 (*i. e. - 1*) promoter, was used to target the silkworm genome in order to explore the possibility of improving the performance of silk. A transgenic silkworm with a green fluorescent cocoon was obtained and PCR analysis of its genome confirmed that the target genes had been integrated into the silkworm genome correctly. Furthermore, in the posterior silk glands of the G_6 generation transformation silkworm, a band representing the fusion protein Fib-L-GFP-Cec with a molecular mass of 68.7 kDa was detected by Western blot with an antibody against GFP. An investigation of the number of bacteria attached to a cocoon showed the transgenic silkworm cocoon possessed antibacterial properties. These results suggested the performance of silk can be improved by modifying the fibroin gene.

Published On: Molecular Biology Reports, 2015, 42(1), 19-25.

1 School of Biology and Basic Medical Sciences, Soochow University, Suzhou 215123, China.
2 National Engineering Laboratory for Modern Silk, Soochow University, Suzhou 215123, China.
*Corresponding author E-mail: gongcl@suda.edu.cn.

绿色荧光抗菌蚕丝的转基因家蚕的制备

李　珍[1]　江　月[1]　曹广力[1,2]
李静芝[1]　薛仁宇[1,2]　贡成良[1,2*]

摘要：构建一个基因靶向载体，该载体含有绿色荧光蛋白基因*gfp*和抗菌肽天蚕素基因*cec*的融合基因，两侧分别是家蚕丝素轻链基因*fib-L*的5′和3′序列，此外，还包含杆状病毒*ie-1*启动子驱动的作为负选择的红色荧光标记基因以探索提高蚕丝性能的可能性。将该载体导入家蚕基因组获得可生产绿色荧光茧的转基因蚕。对其基因组进行了PCR分析，证实目标基因已被正确地整合到家蚕基因组中。此外，在G_6代转基因蚕的后部丝腺中，用GFP抗体通过Western blot检测到分子质量为68.7 kDa的融合蛋白Fib-L-GFP-Cec。对茧附着的细菌数量的调查结果显示，转基因蚕茧具有抗菌性能。这些结果表明，通过改变丝素蛋白基因可以改善蚕丝的性能。

1　基础医学与生物科学学院，苏州大学，苏州
2　现代丝绸国家工程实验室，苏州大学，苏州

Characterization of transgenic silkworm yielded biomaterials with calcium-binding activity

Wang SH[1] Zhang YY[1] Yang MY[1] Ye LP[1] Gong L[2]
Qian QJ[1] Shuai YJ[1] You ZY[1] Chen YY[1] Zhong BX[1*]

Abstract: Silk fibers have many inherent properties that are suitable for their use in biomaterials. In this study, the silk fibroin was genetically modified by including a Ca-binding sequence, $[(AGSGAG)_6ASEYDYDDDSDDDDEWD]_2$ from shell nacreous matrix protein. It can be produced as fibers by transgenic silkworm. The Ca-binding activity and mineralization of the transgenic silk fibroin were examined *in vitro*. The results showed that this transgenic silk fibroin had relatively higher Ca-binding activity than unmodified silk fibroin. The increased Ca-binding activity could promote the usage of silk fibroin as a biomaterial in the pharmaceutical industry. This study shows the possibility of using silk fibroin as a mineralization accelerating medical material by generating genetically modified transgenic silkworm.

Published On: PLoS ONE, 2016, 11(7).

1 College of Animal Sciences, Zhejiang University, Hangzhou 310058, China.
2 College of Life Sciences, Zhejiang University, Hangzhou 310058, China.
*Corresponding author E-mail: bxzhong@zju.edu.cn.

生产钙结合活性生物材料的转基因家蚕的表征

王少华[1] 张玉玉[1] 杨明英[1] 叶露鹏[1] 龚 璐[2]
钱秋杰[1] 帅亚俊[1] 尤征英[1] 陈玉银[1] 钟伯雄[1*]

摘要：家蚕丝蛋白与生俱来的一些优越品质使其非常适合应用于各种生物材料。在研究中，我们利用家蚕转基因技术，以家蚕后部丝腺作为生物反应器生产了包含有珍珠母贝基质蛋白的丝蛋白，该珍珠母贝基质蛋白具有钙结合活性序列$[(AGSGAG)_6ASEYDYDDDSDDDDEWD]_2$。转基因丝蛋白的钙结合活性和矿化作用的体外检测结果表明，转基因家蚕蚕丝的钙结合活性较转基因对照品系增强。这种具有钙离子结合活性的蚕丝的生产，促进了蚕丝作为一种生物材料在医药工业领域里的应用。这些研究表明了通过培育转基因家蚕，将蚕丝作为一种促进矿化的医用材料的可能性。

1 动物科学学院，浙江大学，杭州
2 生命科学学院，浙江大学，杭州

Synthesis of silk fibroin-insulin bioconjugates and their characterization and activities *in vivo*

Zhang YQ[1] Ma Y[1] Xia YY[2] Shen WD[1*] Mao JP[1]
Zha XM[3] Koji Shirai[3] Kenji Kiguchi[3]

Abstract: The regenerated liquid silk fibroin with an average molecular mass of about 60 kDa consists of 18 kinds of amino acids containing ~10% of polar amino acids with hydroxyl and amino groups such as serine and lysine. The liquid silk fibroin is coupled covalently with insulin molecules through these strongly polar side groups by using glutaraldehyde. The physicochemical properties of the silk fibroin-insulin (SF-Ins) bioconjugates were investigated by Enzyme-Linked Immunosorbent Assay (ELISA) for the quantitative measurement of insulin. The biological activities of the insulin bioconjugates were characterized *in vitro* and *in vivo*. The SF-Ins constructs obtained by 5 h of covalent crosslinking showed much higher recovery (about 70%) and *in vitro* stability in human serum than bovine serum albumin-insulin (BSA-Ins) derivatives. The results in human serum indicated that the half-life *in vitro* of the biosynthesized SF-Ins derivatives was 2.1 and 1.7 times more than that of BSA-Ins conjugates and native insulin, respectively. The immunogenicity of the regenerated silk fibroin and the antigenicity of silk fibroin-modified insulin were not observed in both rabbits and rats. The pharmacological activity of the SF-Ins bioconjugates in diabetic rats evidently lengthened and was about 3.5 times as long as that of the native insulin, nearly 21 h. The bioconjugation of insulin with the regenerated silk fibroin greatly improved its physicochemical and biological stability.

Published On: Journal of Biomedical Materials Research Part B—Applied Biomaterials, 2006, 79(2), 275-283.

1 Silk Biotechnol. Lab., School of Life Science, Soochow University, Suzhou 215123, China.
2 No.1 Peoples' Hospital of Soochow University, Suzhou 215006, China.
3 Department of Applied Biology, Faculty of Textile Science and Technology, Shinshu University, Naganoken 386-8567, Japan.
*Corresponding author E-mail: shenwd@suda.edu.cn.

蚕丝丝素-胰岛素生物结合物的合成及其理化性能与体内的生物活性

张雨青[1] 马 燕[1] 夏运岳[2] 沈卫德[1] 毛建萍[1]
查新民[3] Koji Shirai[3] Kenji Kiguchi[3]

摘要:再生的液体丝素平均分子质量约为60 kDa,由18种氨基酸组成,其中包含10%的带有羟基和氨基的极性氨基酸,如丝氨酸和赖氨酸。用戊二醛将液体丝素与胰岛素分子通过强极性侧链进行共价结合。通过酶联免疫吸附实验对这种胰岛素生物结合物的理化特性进行定量分析,并对其生物活性进行体内和体外的调查分析。经5 h共价交联得到的丝素-胰岛素结合物具有较高的回收率(约70%),同时,在体外的人血清中具有比牛血清白蛋白-胰岛素结合物更好的稳定性。在人血清中的结果表明,生物生成的丝素-胰岛素结合物体外半衰期分别是牛血清白蛋白-胰岛素结合物和天然胰岛素的2.1倍和1.7倍;在兔和大鼠中均没有观察到再生丝素蛋白的免疫原性和丝素-胰岛素生物结合物的抗原性;另外,丝素-胰岛素生物结合物在糖尿病大鼠体内的药理活性明显延长,约为天然胰岛素的3.5倍,近21 h。胰岛素与再生丝素蛋白的生物结合大大提高了其物理化学和生物稳定性。

1 蚕丝生物技术实验室,生命科学学院,苏州大学,苏州
2 苏州大学第一人民医院,苏州
3 应用生物实验室,纺织科技学院,信州大学,日本

Formation of silk fibroin nanoparticles in water-miscible organic solvent and their characterization

Zhang YQ[1*] Shen WD[1] Xiang RL[1] Zhuge LJ[2] Gao WJ[2] Wang WB[2]

Abstract: When silk fibre derived from *Bombyx mori*, a native biopolymer, was dissolved in highly concentrated neutral salts such as $CaCl_2$, the regenerated liquid silk, a gradually degraded peptide mixture of silk fibroin, could be obtained. The silk fibroin nanoparticles were prepared rapidly from the liquid silk by using water-miscible protonic and polar aprotonic organic solvents. The nanoparticles are insoluble but well dispersed and stable in aqueous solution and are globular particles with a range of 35-125 nm in diameter by means of TEM, SEM, AFM and laser sizer. Over one half of the ε-amino groups exist around the protein nanoparticles by using a trinitrobenzenesulfonic acid (TNBS) method. Raman spectra shows the tyrosine residues on the surface of the globules are more exposed than those on native silk fibers. The crystalline polymorph and conformation transition of the silk nanoparticles from random-coil and α-helix form (Silk I) into anti-parallel β-sheet form (Silk II) are investigated in detail by using infrared, fluorescence and Raman spectroscopy, DSC, ^{13}C CP-MAS NMR and electron diffraction. X-ray diffraction of the silk nanoparticles shows that the nanoparticles crystallinity is about four fifths of the native fiber. Our results indicate that the degraded peptide chains of the regenerated silk is gathered homogeneously or heterogeneously to form a looser globular structure in aqueous solution. When introduced into excessive organic solvent, the looser globules of the liquid silk are rapidly dispersed and simultaneously dehydrated internally and externally, resulting in the further chain-chain contact, arrangement of those hydrophobic domains inside the globules and final formation of crystalline silk nanoparticles with β-sheet configuration. The morphology and size of the nanoparticles are relative to the kinds, properties and even molecular structures of organic solvents, and more significantly to the looser globular substructure of the degraded silk fibroin in aqueous solution. It is possible that the silk protein nanoparticles are potentially useful in biomaterials such as cosmetics, anti-UV skincare products, industrial materials and surface improving materials, especially in enzyme/drug delivery system as vehicle.

Published On: Journal of Nanoparticle Research, 2007, 9(5), 885-900.

1 Silk Biotechnol. Lab., School of Life Science, Soochow University, Suzhou 215123, China.

2 Analytical Center, Soochow University, Suzhou 215123, China.

*Corresponding author E-mail: yqzhang@public1.sz.js.cn.

水溶性有机溶剂中丝素纳米颗粒的形成及其特性

张雨青[1*] 沈卫德[1] 相入丽[1] 诸葛兰剑[2] 高伟建[2] 王文宝[2]

摘要：来源于家蚕的蚕丝纤维是一种天然的生物高聚物，能溶于高浓度的中性盐溶液，如氯化钙中，可制成再生的液体丝素，这是一种逐步降解的丝素肽混合物。应用水溶性质子型和极性非质子型有机溶剂可将这种液体丝蛋白快速制成丝素纳米颗粒。这种丝素纳米颗粒不溶于水，但能良好地分散并稳定在水溶液中。通过透射电镜、扫描电镜、原子力显微镜和激光粒度仪分析表明，这种丝素颗粒呈球形，粒径分布为35—125 nm。利用TNBS法分析表明超过一半的ε氨基酸基团存在于蛋白质纳米颗粒表面。拉曼光谱分析揭示纳米颗粒表面的酪氨酸残基较天然丝素纤维暴露的更多。应用红外光谱、荧光光谱、拉曼光谱、DSC、^{13}C CP-MAS NMR和电子衍射等方法，对丝素纳米颗粒从无规则卷曲和α-螺旋（Silk I）到反向平行β-片层（Silk II）的构象转变与结晶态作了详细的研究。X-射线衍射分析表明这种丝素颗粒的结晶度约为天然纤维的4/5。研究结果表明，再生丝素的降解多肽链会在水溶液中均匀或不均匀地聚集在一起形成一个较松散球状结构。当加入过量的有机溶剂时，松散的液体丝素纳米会迅速分散，同时内外脱水导致链与链进一步接触，以及疏水区域在颗粒内的排列，最终形成具有β-片层结构的结晶丝素纳米颗粒。这种丝素纳米颗粒的形状和大小与有机溶剂的种类、性质甚至分子结构有关，更与水溶液中降解丝素的松散球状亚结构密切相关。这种丝素纳米颗粒在化妆品、护肤品、防晒膏、表面改性材料、合成工业材料，尤其是酶和多肽药物的缓释载体等方面有潜在的应用价值。

1 蚕丝生物技术实验室，生命科学学院，苏州大学，苏州
2 苏州大学分析中心，苏州

Preparation of silk fibroin nanoparticles and their application to immobilization of L-asparaginase

Zhang YQ* Xiang RL Yan HB Chen XX

Abstract: After the degummed fiber of silk fibroin derived from *Bombyx mori* cocoon was dissolved in $CaCl_2$ ternary solvent system or highly concentrated LiBr solution, three kinds of silk fibroin in liquid could be obtained by means of dialysis. SDS-PAGE analysis results show that three kinds of silk fibroin are of different molecular ranges. The silk fibroin nanoparticles were prepared rapidly from the liquid silk by using water-miscible organic solvents such as acetone. These nanoparticles are insoluble in water but well dispersed and stable in aqueous solution and are globular particles with a size range of 50-120 nm in diameter by means of SEM. L-asparaginase as a model enzyme was bioconjugate with these nanoparticles by cross-linking agent glutaraldehyde. Activity analysis indicated that silk fibroin nanoparticles derived from the fibroin by less breakage of peptide chain are more suitable for the bioconjugation of enzymes. The results showed that the recovery of the immobilized L-asparaginase was about 44%. Its thermal stability increased evidently and the optimal scale of pH was much wider (pH 6.0-8.0) than that of native L-asparaginase. And the optimal reaction temperature of the modified enzyme was increased about 10 ℃. These preliminary results above indicated that the silk protein nanoparticles are also a good support as silk fibroin membrane. Therefore, the silk fibroin nanoparticles as a new drug release system are of potential values for study and development.

Published On: Chemical Journal of Chinese Universities, 2008, 29(3), 628-633.

Silk Biotechnol. Lab., School of Life Science, Soochow University, Suzhou 215123, China.

*Corresponding author E-mail: yqzhang@public1.sz.js.cn.

丝素纳米颗粒的制备及应用于 L-天冬酰胺酶的固定化

张雨青* 相入丽 阎海波 陈晓晓

摘要：丝素蛋白纤维溶于高浓度中性盐溴化锂溶液或氯化钙-乙醇-水三元溶剂中，经过透析和纯化可以制成3种液态丝素。SDS-PAGE分析表明其分子量分布范围明显不同。采用能与水混溶的有机溶剂如丙酮等可将这种丝素制成丝素纳米颗粒，SEM观察表明丝素纳米颗粒粒径分布范围为50—120 nm。以戊二醛为交联剂，将治疗急性淋巴性白血病常用酶制剂——L-天冬酰胺酶共价结合在丝素纳米颗粒上。酶活性分析表明由肽链断裂较少的丝素制备的纳米颗粒更适合于酶的生物结合。酶动力学研究表明，这种固定化酶活性回收率约为44%，较天然L-天冬酰胺酶的热稳定性明显提高，最适pH范围明显加宽，为6.0—8.0，最适反应温度提高10 ℃，抗胰蛋白酶水解能力明显增强。这些结果说明，丝素纳米颗粒也像丝素蛋白膜一样，是一种酶固定化的良好载体，在药物缓释系统方面具有潜在的研究和开发价值。

蚕丝生物技术实验室，生命科学学院，苏州大学，苏州

Biosynthesis of insulin-silk fibroin nanoparticles conjugates and *in vitro* evaluation of a drug delivery system

Yan HB[1,2] Zhang YQ[1,2*] Ma YL[1,2] Zhou LX[1,2]

Abstract: Silk fibroin derived from *Bombyx mori* is a biomacromolecular protein with outstanding biocompatibility. When it was dissolved in highly concentrated $CaCl_2$ solution and then the mixture of the protein and salt was subjected to desalting treatments for long time in flowing water, the resulting liquid silk was water-soluble polypeptides with different molecular masses, ranging from 8 to 70 kDa. When the liquid silk was introduced rapidly into acetone, silk protein nanoparticles with a range of 40-120 nm in diameter could be obtained. The crystalline silk nanoparticles could be conjugated covalently with insulin alone with cross-linking reagent glutaralde-hyde. *In vitro* properties of the insulin-silk fibroin nanoparticles (Ins-SFN) bioconjugates were determined by Enzyme-Linked Immunosorbent Assay (ELISA). The optimal conditions for the biosynthesis of Ins-SFN bioconjugates were investigated. The Ins-SFN constructs obtained by 8 h of covalent cross- linking with 0.7% cross-linking reagent and the proportion of insulin and SFN being 30 IU∶15 mg showed much higher recoveries (90% - 115%). When insulin was coupled covalently with silk nanoparticles, the resistance of the modified insulin to trypsin digestion and *in vitro* stability in human serum were greatly enhanced as compared with insulin alone. The results in human serum indicated that the half-life *in vitro* of the biosynthesized Ins-SFN derivatives was about 2.5 times more than that of native insulin. Therefore, the silk protein nanoparticles have the potential values for being studied and developed as a new bioconjugate for enzyme/polypeptide drug delivery system.

Published On: Journal of Nanoparticle Research, 2009, 11(8), 1937-1946.

1 The State Engineering Laboratory of Modern Silk, Soochow University, Suzhou 215123, China.

2 Silk Biotechnology Key Laboratory of Suzhou City, Medical College of Soochow University, Suzhou 215123, China.

*Corresponding author E-mail: yqzhang@public1.sz.js.cn.

胰岛素-丝素纳米颗粒生物结合物的合成及其药物递送系统的体外评估

阎海波[1,2]　张雨青[1,2*]　马永雷[1,2]　周丽霞[1,2]

摘要: 蚕丝丝素是一种具有生物相容性的高分子量蛋白，经过高浓度的$CaCl_2$溶液溶解和中性盐透析脱盐后获得再生液态丝素，其分子质量为8—70 kDa。利用水溶性有机溶剂丙酮制备结晶性丝素纳米颗粒，经电镜和激光粒度仪分析，这种呈球体的丝素颗粒粒径在40—120 nm，平均粒径为60—80 nm。以戊二醛为交联剂，结晶丝素纳米颗粒可以和胰岛素发生共价交换。应用酶联免疫吸附实验在体外测定了这种胰岛素-丝素纳米颗粒生物结合物中胰岛素的活性，并探讨了这种生物结合物制备的最佳条件。实验表明，在浓度为0.7%的戊二醛交联剂及胰岛素和丝素纳米颗粒的比例为30 IU∶15 mg的情况下交联8 h后获得的生物结合物具有更高的回收率，约90%—115%。当胰岛素与丝素纳米颗粒共价结合时，与单独胰岛素相比，修饰后的胰岛素对胰蛋白酶消化的抵抗力及人血清的体外稳定性大大增强。人血清的体外实验表明，体外生物合成的Ins-SFN衍生物的半衰期约为天然胰岛素的2.5倍。因此，这种丝素纳米颗粒作为多肽药物的载体具有潜在的研究与开发价值。

1　现代丝绸国家工程实验室，苏州大学，苏州
2　苏州市蚕丝生物技术重点实验室，苏州大学药学院，苏州

Silk sericin-insulin bioconjugates: Synthesis, characterization and biological activity

Zhang YQ[1*] Ma Y[1] Xia YY[2] Shen WD[1] Mao JP[1] Xue RY[1]

Abstract: When silk fiber derived from *Bombyx mori* was subjected to degumming treatments twice in water and subsequent degraded processing in slightly alkaline aqueous solution under high-temperature and high-pressure, the water-soluble silk sericin peptides (SS) with different molecular mass from 10 to 70 kDa were obtained. The sericin peptides could be conjugated covalently with insulin alone with cross-linking reagent glutaraldehyde. The physicochemical properties of the silk sericin-insulin (SS-Ins) conjugates were determined by Enzyme-Linked Immunosorbent Assay (ELISA). The biological activities of SS-Ins bioconjugates were investigated *in vitro* and *in vivo*. The results in human serum *in vitro* indicated that the half-life of the synthesized SS-Ins derivatives was 2.3 and 2.7 times more than that of bovine serum albumin-insulin (BSA-Ins) conjugates and intact insulin, respectively. The pharmacological activity of SS-Ins bioconjugates lengthened to 21 h in mice *in vivo*, which was over 4 times longer than that of the native insulin. The immunogenicity of silk sericin and the antigenicity of SS-Ins derivatives were not observed in both rabbits and mice. The bioconjugation of insulin with silk sericin protein evidently improved both physicochemical and biological stability of the polypeptide.

Published On: Journal of Controlled Release, 2006, 115(3), 307-315.

1 Silk Biotechnology Laboratory, School of Life Science, Soochow University, Suzhou 215123, China.
2 No.1 Peoples' Hospital of Soochow University, Suzhou 215006, China.
*Corresponding author E-mail: yqzhang@public1.sz.js.cn.

蚕丝丝胶肽–胰岛素生物结合物：合成、特性及其生物活性

张雨青[1*] 马 燕[1] 夏运岳[2] 沈卫德[1] 毛建萍[1] 薛仁宇[1]

摘要：家蚕蚕丝纤维经过水中的二次脱胶处理，并在弱碱性水溶液中高温高压处理后，获得分子质量为10—70kDa的水溶性丝胶肽。以戊二醛为交联剂将这种丝胶肽与天然胰岛素共价结合制成丝胶肽–胰岛素生物结合物，应用酶联免疫吸附法（ELISA）对这种胰岛素结合物的理化特性做了分析，并对这种丝胶肽修饰的胰岛素的生物学活性做了体内和体外调查。结果表明在体外人血清中这种合成胰岛素的半衰期比牛血清白蛋白（BSA）–胰岛素结合物和天然胰岛素分别高2.3倍和2.7倍。丝胶肽–胰岛素生物结合物在鼠体内的药理活性延长到21 h，是天然胰岛素的4倍以上，在兔和小鼠体内没有观察到这种修饰胰岛免疫的抗原性。所以，天然胰岛素与丝胶蛋白的生物结合，显著地提高了这种多肽药物的理化稳定性和生物学稳定性。

1 蚕丝生物技术实验室，生命科学学院，苏州大学，苏州
2 苏州大学第一人民医院，苏州

Preparation and characterization of silver nanoparticles composited on polyelectrolyte film coated sericin gel for enhanced antibacterial application

Tao G[1] Wang YJ[1,2] Liu LN[1] Chang HP[1] Zhao P[1] He HW[1,2,3*]

Abstract: Sericin has shown a great potential as a biomaterial in biomedical application due to its good hydrophilicity, reactivity and biodegradability. To solve the adherence and growth of microorganisms on sericin gel, here we prepared a polyelectrolyte film with interactive arrangement of opposite charges poly (dimethyldiallylammonium chloride) (PDDA) and poly(acrylic acid) (PAA) coated on the surface of sericin gel via electrostatic interaction. AgNPs was then synthesized with the assistance of UV *in situ* on the polyelectrolyte film to endow the antibacterial activity of sericin gel. SEM revealed the polyelectrolyte film could effectively facilitate the high density growth of AgNPs as a 3-D matrix and the modification of polyelectrolyte and AgNPs did not alter the porous characteristics of sericin gel. XRD studies suggested AgNPs synthesized on the polyelectrolytes film coated sericin gel had good crystal structure. DSC analysis showed AgNPs could tightly bind with the polyelectrolytes film and sericin gel, and the binding may improve the melting temperature of AgNPs. Both bacterial inhibition zone and growth curve assay demonstrated the sericin gel coated with AgNPs-polyelectrolytes film had an excellent and long-lasting antibacterial activity. This novel sericin gel has shown great potentials in biomedical application such as wound healing.

Published On: Science of Advanced Materials, 2016, 8(8), 1547-1552.

1 State Key Laboratory of Silkworm Genome Biology, Southwest University, Chongqing 400716, China.

2 Chongqing Engineering and Technology Research Center for Novel Silk Materials, College of Biotechnology, Southwest University, Chongqing 400716, China.

3 National Laboratory of Biomacromolecules, Institute of Biophysics, Chinese Academy of Sciences, Beijing 100101, China.

*Corresponding author E-mail: hehuawei@swu.edu.cn.

制备和表征纳米银修饰的聚电解质薄膜包被的丝胶凝胶及其强力抗菌应用

陶　刚[1]　王叶菁[1,2]　刘莉娜[1]　常怀普[1]　赵　萍[1]　何华伟[1,2,3*]

摘要：丝胶由于具有良好的亲水性、反应活性和生物降解性，在生物医学应用中显示出巨大的潜力。为了解决微生物在丝胶凝胶上的黏附和生长问题，我们设计了由聚（二甲基二烯丙基氯化铵）与聚（丙烯酸）通过静电相互作用在丝胶凝胶表面形成的聚电解质膜。接着在聚电解质膜表面利用原位UV法合成纳米银颗粒并赋予丝胶凝胶抗菌活性。扫描电镜观察显示，聚电解质膜能作为三维基底有效地促进高密度的纳米银生长及聚电解质膜的包被，且不改变丝胶凝胶多孔结构的特性。XRD研究表明纳米银在聚合电解质薄膜上合成且具有良好的晶体结构。差式热量分析表明纳米银可以与聚合电解质膜紧密结合并提高凝胶的热稳定性。细菌生长曲线和抑菌圈测定表明，纳米银修饰聚电解质薄膜包被的丝胶凝胶对革兰氏阴性菌和革兰氏阳性菌具有长效抗菌活性，这种新型强力抗菌丝胶凝胶在生物医学上显示出巨大的潜力。

1　家蚕基因组生物学国家重点实验室，西南大学，重庆
2　重庆市蚕丝纤维新材料工程技术研究中心，生物技术学院，西南大学，重庆
3　生物大分子国家重点实验室，生物物理研究所，中国科学院，北京

Characterization of silver nanoparticle *in situ* synthesis on porous sericin gel for antibacterial application

Tao G[1] Liu LN[1] Wang YJ[1,2] Chang HP[1]
Zhao P[1] Zuo H[3*] He HW[1,2,4*]

Abstract: Sericin from *Bombyx mori* cocoon has good hydrophilicity, reaction activity, biocompatibility, and biodegradability, which has shown great potentials for biomedical materials. Here, an ultraviolet light-assisted *in situ* synthesis approach is developed to immobilize silver nanoparticles on the surface of sericin gel. The amount of silver nanoparticles immobilized on the surface of sericin gel could be regulated by the irradiation time. The porous structure and property of sericin gel were not affected by the modifcation of AgNPs, as evidenced by the observation of scanning electron microscopy, X-ray diffractometry, and Fourier transform infrared spectroscopy. Differential scanning calorimetry analysis showed that the modifcation of AgNPs increased the thermal stability of sericin gel. The growth curve of bacteria and inhibition zone assays suggested that the sericin gel modifed with AgNPs had good antimicrobial activities against both Gram-negative and Gram-positive bacteria. This novel sericin has shown a great potential for biomedical purpose.

Published On: Journal of Nanomaterials, 2016, 2016.

1 State Key Laboratory of Silkworm Genome Biology, Southwest University, Chongqing 400716, China.

2 Chongqing Engineering and Technology Research Center for Novel Silk Materials, College of Biotechnology, Southwest University, Chongqing 400716, China.

3 College of Pharmaceutical Sciences, Southwest University, Chongqing 400716, China.

4 National Laboratory of Biomacromolecules, Institute of Biophysics, China Acadmy of Sciences, Beijing 100101, China.

*Corresponding author E-mail: zuohua@swu.edu.cn; hehuawei@swu.edu.cn.

原位制备纳米银杂化的丝胶多孔凝胶及其抗菌性能

陶　刚[1]　刘莉娜[1]　王叶菁[1,2]　常怀普[1]
赵　萍[1]　左　华[3*]　何华伟[1,2,4*]

摘要：家蚕丝胶具有良好的亲水性、反应活性、生物相容性和生物可降解性，显示了其作为生物医学材料的巨大潜力。本研究开发了一种紫外光辅助原位合成的方法将银纳米颗粒杂化在丝胶凝胶表面。可以通过辐照时间来调节表面固定纳米银颗粒的数量。扫描电子显微镜、X射线衍射和傅里叶红外分析结果证实了纳米银杂化丝胶蛋白后丝胶凝胶多孔结构和丝胶性能不发生改变。差式热量分析显示纳米银杂化之后增加了丝胶凝胶的热稳定性。细菌生长曲线和抑菌圈测定表明纳米银杂化后的丝胶凝胶对革兰氏阴性菌和革兰氏阳性菌具有良好的抗菌活性。这种新型的丝胶蛋白在生物医学领域有巨大的应用潜力。

1　家蚕基因组生物学国家重点实验室，西南大学，重庆
2　重庆市蚕丝纤维新材料工程技术研究中心，生物技术学院，西南大学，重庆
3　药学院，西南大学，重庆
4　生物大分子国家重点实验室，生物物理研究所，中国科学院，北京

第二章

家蚕变态发育机制研究

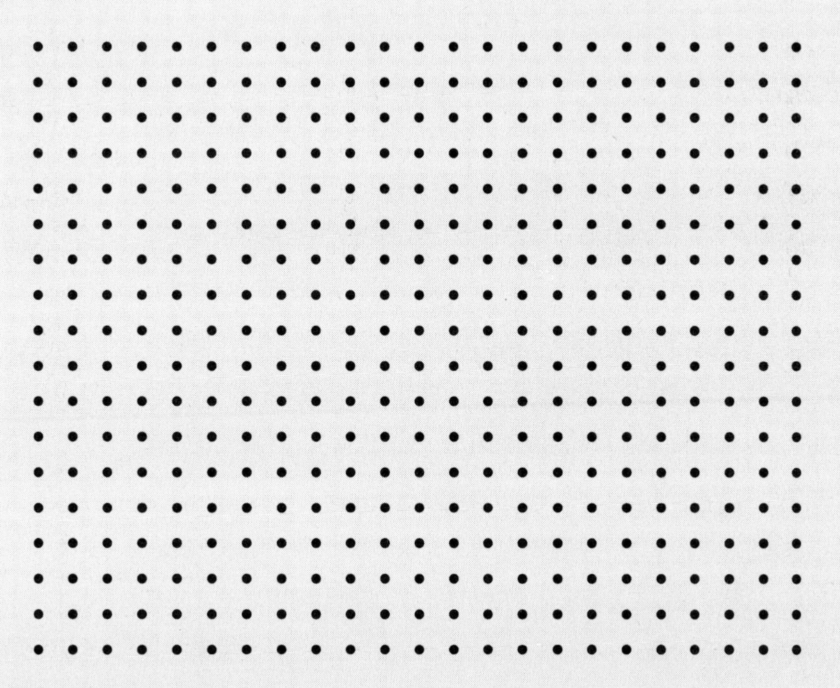

Chapter 2

家蚕是典型的完全变态昆虫，其一生经历卵、幼虫、蛹和成虫四个发育阶段。家蚕幼虫蜕皮次数、化蛹及化蛾时期、个体大小等生长发育特征既决定了蚕丝合成特性及蚕业生产体系，也是研究昆虫变态发育调节机理的重要切入点。近年来，随着家蚕基因组计划的实施，项目组对家蚕变态发育的分子调控机制进行了深入探究，特别是在蜕皮激素（20-hydroxyecdysone，20E）和保幼激素（Juvenile hormone，JH）等内分泌激素对家蚕变态发育的协同调节方面取得了一系列突破性发现。

家蚕变态发育过程中的基因表达。基于转录组及蛋白质组学分析手段，项目组构建了家蚕胚胎发育及中肠、脂肪体、表皮、精巢等组织在变态发育进程中的全基因组表达谱。此外，参与变态发育期诸多生物学过程的一些关键基因或基因家族也相继被鉴定和分析，例如，直接参与蜕皮过程的 *Cathepsin-B* 和 *PPO* 等、翅发育相关的 *Achaete-scute* 基因家族、胚胎发育相关的 *Hox* 基因家族（如 *Ubx*、*abd-A*、*Abd-B* 和 *Antp* 等）、卵巢发育途径 *OVO-1* 基因、卵发育相关的30K蛋白以及卵黄原蛋白及其受体等、表皮和中肠发育相关的几丁质结合蛋白、脂肪体代谢的基因 *BmPLA2*、体色黑化突变决定基因 *AANAT*、幼虫石蚕体形突变基因 *CPR2*、调控神经发育的 *SGF1* 等。

20E 对家蚕变态发育的调控。20E 起始家蚕幼虫生长蜕皮及化蛹化蛾变态蜕皮，20E 合成、信号传导及功能是家蚕变态发育分子调控机制研究的核心内容之一。（1）在20E 合成方面，项目组发现 homeodomain 转录因子即 Antp 和 POU-M2 发生蛋白互作，并通过结合 20E 合成途径限速酶基因 *Phantom* 启动子区的相邻基序来正调控其转录，从而影响20E 合成及发育进程。（2）在20E 信号转导方面，鉴定了参与20E 信号转导的核受体基因家族，证实活化的蛋白激酶 C 受体 1（RACK1）与 20E 信号转导关键分子 BR-C 的蛋白互作控制了 BR-C 的入核、*E75* 基因不同异构体对20E 的应答具有时空特异性且能反馈调节 20E 的合成，以及 JH 潜在受体 Met 参与了 20E 信号的最大化。（3）在 20E 调控变态发育的分子机制方面，重点研究了变态发育进程中的细胞自噬和细胞凋亡，并取得了如下重要发现：鉴定了细胞自噬和细胞凋亡相关的基因、解析了细胞自噬蛋白 ATG8 的结构、证实 20E 在变态期通过上调细胞自噬蛋白基因的表达来诱导细胞自噬发生、发现细胞自噬的发生先于细胞凋亡、确定细胞自噬蛋白 ATG5 和 ATG6 参与调控细胞凋亡。除此之外，还发现蛹期特异的20E应答因子E93促进了幼虫向蛹的转变、Ras1诱导的BCPI蛋白通过抑制组织蛋白酶活性来阻止家蚕后部丝腺组织降解、20E诱导的POU-M2与Hox基因家族成员Abd-A协同调控了表皮蛋白基因的蛹期特异表达、20E应答因子BR-C调控了卵黄原蛋白合成。

JH对家蚕变态发育的调节。项目组系统鉴定获得了参与家蚕JH合成、代谢和信号转导的55个基因，发现参与JH合成的甲羟戊酸（MVA）途径的大部分基因及与JH合成神经肽调节或JH信号传导相关的所有基因都是单拷贝，进化上比较保守；而参与JH生物合成和代谢的类异戊二烯分支途径中的每类基因在鳞翅目昆虫中发生复制，可能与不同昆虫中JH类型的差异有关。以家蚕显性3眠突变（*M*³）为材料，通过定位克隆分析发现homeodomain转录因子Scr与*M*³突变相关，该转录因子对参与JH合成的神经肽及相关酶类基因的转录具有正调控作用。此外，全基因组基因芯片分析显示，JH类似物（JHA）对参与基础代谢和蛋白质加工过程的大部分基因的转录具有上调作用，这为研究JH在代谢稳态维持中的作用机理提供了线索。

程道军、钱文良

The synthesis, transportation and degradation of *Bm*LP3 and *Bm*LP7, two highly homologous *Bombyx mori* 30K proteins

Zhang Y　Zhao P　Liu HL　Dong ZM
Yang Q　Wang DD　Xia QY*

Abstract: The 30K proteins comprise about 35% of the total embryo yolk proteins and function as storage proteins during embryonic development of the domesticated silkworm, *Bombyx mori*. The most abundant components of hemolymph are 30K proteins in the early and middle pupal stages. In the present study, the 30K protein *Bm*LP7 was purified from larval hemolymph by chromatography. We prepared the antibody of this protein and found that it could bind to both *Bm*LP3 and *Bm*LP7. We used western blotting to analyze the dynamic change of *Bm*LP3 and *Bm*LP7 proteins in the hemolymph during development and found their concentration decreased dramatically from day 4 pupae, which appears to be linked to their accumulation in the oocyte for forming yolk granule since then. We found *Bm*LP3 and *Bm*LP7 proteins reduced significantly in day 10 eggs (the day before hatching). The crude extract of the newly hatched larvae showed proteolytic activity against *Bm*LP3 and *Bm*LP7 and immunohistochemistry showed *Bm*LP3 and *Bm*LP7 were degraded in the embryonic gut lumen in day 10 eggs. These systematic studies of *Bm*LP3 and *Bm*LP7 reveal their synthesis, transportation and degradation, which could represent the experience of all 30K proteins.

Published On: Insect Biochemistry and Molecular Biology, 2012, 42(11), 827-834.

State Key Laboratory of Silkworm Genome Biology, Southwest University, Chongqing 400716, China.
*Corresponding author E-mail: xiaqy@swu.edu.cn.

两种高度同源的家蚕30K蛋白：*Bm*LP3和*Bm*LP7的合成、转运和降解

张 艳 赵 萍 刘鸿丽 董照明
杨 强 王丹丹 夏庆友*

摘要：30K蛋白约占总卵黄蛋白的35%，在家蚕胚胎发育过程中起着贮藏蛋白的作用。家蚕早、中期蛹的血淋巴中含量最丰富的成分也是30K蛋白。在此研究中，通过层析法从幼虫血淋巴中纯化获得30K蛋白*Bm*LP7。我们制备了*Bm*LP7的抗体，发现它能够与*Bm*LP7及*Bm*LP3结合。我们利用蛋白质免疫印迹法分析了血淋巴中*Bm*LP7和*Bm*LP3蛋白在发育过程中的动态变化，发现它们的浓度从蛹期第4天开始显著下降，这可能与其在卵母细胞中积累形成卵黄粒有关。我们也发现，*Bm*LP3和*Bm*LP7蛋白在第10天（孵化前一天）的卵中显著减少。新孵化幼虫的粗提取物中对*Bm*LP3和*Bm*LP7蛋白具有水解活性，免疫组化分析表明在第10天的卵中，*Bm*LP3和*Bm*LP7在胚胎肠腔内被降解。这一系列对*Bm*LP3和*Bm*LP7的系统研究揭示了其合成、转运和降解过程，可能代表所有30K蛋白的情况。

家蚕基因组生物学国家重点实验室，西南大学，重庆

Serine protease P-IIc is responsible for the digestion of yolk proteins at the late stage of silkworm embryogenesis

Wang DD Zhang Y Dong ZM Guo PC
Ma SY Guo KY Xia QY Zhao P*

Abstract: In silkworms, yolk proteins comprise vitellin, egg-specific protein and 30K proteins, which are sequentially degraded by endogenous proteases strictly regulated during embryogenesis. Although the process has been extensively investigated, there is still a gap in the knowledge about the degradation of silkworm yolk proteins on the last two days of embryonic development. In the present study, we isolated and purified a gut serine protease P-IIc, which demonstrated optimal activity at 25 ℃ and pH 11. Semiquantitative RT-PCR combined with Western blot showed that P-IIc was actively expressed and significantly accumulated in the gut on the last two days of embryogenesis. When natural yolk proteins were incubated with P-IIc *in vitro*, vitellin and ESP were selectively degraded. P-IIc also demonstrated activity towards 30K proteins as evidenced by rapid and complete digestion of *Bm*LP1 and partial digestion of *Bm*LP2 and *Bm*LP3. Furthermore, RNAi knockdown of P-IIc in silkworm embryos significantly reduced the degradation rate of residual yolk proteins on embryonic day 10. Taken together, our results indicate that P-IIc represents an embryonic gut protease with a relatively broad substrate specificity, which plays an important role in the degradation of yolk proteins at the late stage of silkworm embryogenesis.

Published On: Insect Biochemistry and Molecular Biology, 2016, 74, 42-49.

State Key Laboratory of Silkworm Genome Biology, Southwest University, Chongqing 400716, China.
*Corresponding author E-mail: zhaop@swu.edu.cn.

丝氨酸蛋白酶P-IIc参与家蚕胚胎发育后期卵黄蛋白的降解

王丹丹　张　艳　董照明　郭鹏超
马三垣　郭凯雨　夏庆友　赵　萍*

摘要：在家蚕中，卵黄蛋白主要由卵黄磷蛋白（Vtn）、卵特异性蛋白（ESP）和30K蛋白组成。在胚胎发育中，卵黄蛋白被严格调控的内源性蛋白酶顺序降解。虽然研究者对家蚕卵黄蛋白的降解过程已进行了广泛的研究，但在胚胎发育的后期，卵黄蛋白的降解过程仍然不清楚。在本研究中，我们分离纯化了家蚕肠道蛋白酶P-IIc，测定发现其在温度为25℃且缓冲液pH为11时酶解活性最强。利用半定量逆转录-聚合酶链反应（RT-PCR）和蛋白质免疫印迹技术，我们发现在胚胎发育的最后两天，蛋白酶P-IIc在中肠内活跃表达且大量积累。在体外，我们利用天然卵黄蛋白与蛋白酶P-IIc孵育，发现P-IIc能选择性降解Vtn和ESP。通过体外孵育实验，我们还证明了P-IIc能迅速降解30K蛋白中的*Bm*LP1、部分降解*Bm*LP2和*Bm*LP3。进一步，我们利用RNAi技术降低家蚕胚胎中P-IIc的表达，发现胚胎第10天残余卵黄蛋白的降解效率显著降低。总之，我们的研究结果表明，蛋白酶P-IIc是一个底物广谱性的胚胎肠道蛋白酶，在家蚕胚胎发育后期卵黄蛋白的降解中起着重要作用。

家蚕基因组生物学国家重点实验室，西南大学，重庆

Identification and function of *Abdominal-A* in the silkworm, *Bombyx mori*

Pan MH　Wang XY　Chai CL　Zhang CD　Lu C*　Xiang ZH

Abstract: *Abdominal-A* (*adb-A*) is a key gene in the development of insects. To understand its function in the silkworm, we cloned 1 193 bp of the *abd-A* gene of *Bombyx mori* (*Bmabd-A*), including the complete coding sequence and part of the 3' untranslated region sequence. *Bmabd-A* has at least three mRNA splice variants with coding sequences of lengths 1 032, 1 044 and 1 059 bp, encoding 343, 347 and 352 amino acids, respectively. Each splice variant of *Bmabd-A* has three exons and differs only in second exon size. *Bmabd-A* was expressed at low levels in unfertilized eggs, but increased gradually in fertilized eggs after laying 22 h. *Bmabd-A* expression decreased in ant silkworms (newly hatched silkworms). After RNA interference for *Bmabd-A*, the embryos had two mutant phenotypes, either completely or partially absent abdominal feet from the third to sixth abdominal segments, suggesting that *Bmabd-A* is responsible for normal development of the third to sixth abdominal segments during embryonic development.

Published On: Insect Molecular Biology, 2009, 18(2), 155-160.

Key Sericultural Laboratory of Agricultural Ministry, Southwest University, Chongqing 400716, China.
*Corresponding author E-mail: lucheng@swu.edu.cn.

家蚕*Abdominal-A*的鉴定和功能分析

潘敏慧　王先燕　柴春利　张春冬　鲁　成*　向仲怀

摘要：*Abdominal-A*（*Adb-A*）是昆虫发育的关键基因之一。为了了解其在家蚕中的功能，我们克隆了长度为1 193 bp的家蚕*abd-A*基因（*Bmabd-A*），它包括完整的编码序列和3′非编码区序列的一部分。*Bmabd-A*具有至少3个序列长度为1 032 bp，1 044 bp和1 059 bp的mRNA剪接变体，分别编码343，347和352个氨基酸。*Bmabd-A*的每个剪接变体具有3个外显子，仅在第二外显子上不同。*Bmabd-A*在未受精卵中呈低水平表达，但在受精22 h后卵中表达量逐渐增加。在蚁蚕（新孵出的蚕）中*Bmabd-A*表达又降低。在RNA干扰*Bmabd-A*之后，胚胎具有两个突变表型，即第三至第六腹部段的腹脚完全消失或部分消失，表明*Bmabd-A*在胚胎发育过程中负责第三至第六腹部的正常发育。

农业部蚕桑学重点实验室，西南大学，重庆

Identification and expression of the *achaete-scute* complex in the silkworm, *Bombyx mori*

Tong XL Dai FY Su MK Ma Y Tan D Zhang Z
He NJ Xia QY Lu C* Xiang ZH

Abstract: Recently, the study of *achaete-scute* (*AS-C*) homologues has contributed enormously to understanding of gene duplication and function evolution, particularly in Diptera. We identified four *AS-C* homologue genes in the silkworm, *Bombyx mori*, referred to as *BmASH*, *BmASH2*, *BmASH3*, and *Bmase*. The complex displayed tandem array structure in the genome. Analysis of spatial expression profiles showed that they all were expressed in obviously higher levels in wing disc than in other tissues, suggesting that they might play important roles in the development of the wing. Furthermore, we found that their expression profiles in the wing discs were mostly correlated with the development of the scales, especially the *BmASH* gene. RNA interference results further indicated that *BmASH* was necessary for scale formation in silkworm wing.

Published On: Insect Molecular Biology, 2008, 17(4), 395-404.

The Key Sericultural Laboratory of Agricultural Ministry, Institute of Sericulture and Systems Biology, Southwest University, Chongqing 400716, China.

*Corresponding author E-mail: lucheng@swu.edu.cn.

家蚕*achaete-scute*复合物的鉴定及其表达谱分析

童晓玲　代方银　苏茂科　马　艳　谭　端　张　泽
何宁佳　夏庆友　鲁　成*　向仲怀

摘要：近年来，对*achaete-scute*（*AS-C*）同系物的研究对理解基因复制和功能进化，特别是双翅目昆虫的功能进化有着重要意义。我们鉴定了4个家蚕*AS-C*同源基因，称为*BmASH*、*BmASH2*、*BmASH3*和*Bmase*，它们在基因组中呈串联排列。空间表达谱分析表明这4个基因在翅原基中的表达量明显高于其他组织，说明这些基因可能在翅原基的发育过程中发挥重要作用。此外，我们发现*AS-C*同源基因在翅原基中的表达与鳞毛的生长相关，尤其是*BmASH*基因。RNA干扰结果进一步表明，*BmASH*是家蚕翅膀鳞毛形成的必需基因。

农业部蚕桑学重点实验室，蚕学与系统生物学研究所，西南大学，重庆

Cathepsin B protease is required for metamorphism in silkworm, *Bombyx mori*

Wang GH[#] Liu C[#] Xia QY[*] Zha XF Chen J Jiang L

Abstract: Cathepsin B belongs to lysosomal cysteine protease of the papain family. Temporal and spatial expression analysis of cathepsin B of *Bombyx mori* (*BmCtB*) was carried out based on Expression Sequence Tags (ESTs) data, oligonucleotide microarray, reverse transcription polymerase chain reaction (RT-PCR) and quantitative real-time PCR. Expression of *BmCtB* was observed in all of the tissues and stages. Among the 10 tested tissues, the fat body and posterior silk gland are the two most enriched tissues with *BmCtB*. During *Bombyx* development, there was an expression fastigium of *BmCtB* during metamorphosis. RNA interference was used to suppress the expression of cathepsin B during metamorphosis. Significant developmental defective phenotypes were obtained in the RNAi treated group. The dramatically reduced expression of *BmCtB* was confirmed by Northern blot and quantitative real-time PCR. These evidences strongly suggest cathepsin B proteinase was predominantly involved in the metabolism process of fat body and the posterior silk gland and was critical for metamorphism and development of silkworm, *Bombyx mori*.

Published On: Insect Science, 2008, 15(3), 201-208.

The Key Sericultural Laboratory of the Agricultural Ministry, College of Biotechnology, Southwest University, Chongqing 400716, China.
[#]These authors contributed equally.
[*]Corresponding author E-mail: xiaqy@swu.edu.cn.

组织蛋白酶B在家蚕变态发育过程中的功能研究

王根洪[#] 刘 春[#] 夏庆友[*] 查幸福 陈 杰 蒋 亮

摘要:组织蛋白酶B属于木瓜蛋白酶家族中的溶菌型半胱氨酸蛋白酶的一员。为探索家蚕组织蛋白酶B(*BmCtB*)基因的功能,我们首先通过EST数据、全基因组芯片、RT-PCR和荧光定量PCR等分析,检测了*BmCtB*的时空表达模式。结果显示,*BmCtB*在各组织器官和各发育阶段均有表达,且*BmCtB*在脂肪体和后部丝腺中富集表达。在家蚕发育过程中,*BmCtB*的表达在变态发育过程中明显高于其他时期。*BmCtB*基因的RNAi引起化蛹发育严重缺陷,其基因表达抑制通过Northern杂交及荧光定量PCR分析得到验证。这些证据有力地表明,*BmCtB*参与了家蚕脂肪体和后部丝腺的变态发育过程,对家蚕的变态发育至关重要。

农业部蚕桑学重点实验室,生物技术学院,西南大学,重庆

Hedgehog signaling pathway regulated the target genes for adipogenesis in silkworm *Bombyx mori*

Liang S Chen RT Zhang DP Xin HH
Lu Y Wang MX Miao YG*

Abstract: Hedgehog (Hh) signals regulate invertebrate and vertebrate development, yet the role of the pathway in adipose development remains poorly understood. In this report, we found that Hh pathway components are expressed in the fat body of silkworm larvae. Functional analysis of these components in a *BmN* cell line model revealed that activation of the *Hh* gene stimulated transcription of Hh pathway components, but inhibited the expression of the adipose marker gene *AP2*. Conversely, specific RNA interference mediated knockdown of *Hh* resulted in increased *AP2* expression. This further showed the regulation of Hh signal on the adipose marker gene. In silkworm larval models, enhanced adipocyte differentiation and an increase in adipocyte cell size were observed in silkworms that had been treated with a specific Hh signaling pathway antagonist, cyclopamine. The fat-body-specific Hh blockade tests were consistent with Hh signaling inhibiting silkworm adipogenesis. Our results indicate that the role of Hh signaling in inhibiting fat formation is conserved in vertebrates and invertebrates.

Published On: Insect Science, 2015, 22(5), 587-596.

Key Laboratory of Animal Epidemic Etiology and Immunological Prevention of Ministry of Agriculture, College of Animal Sciences, Zhejiang University, Hangzhou 310058, China.
*Corresponding author E-mail: miaoyg@zju.edu.cn.

Hedgehog信号通路调控家蚕脂肪体形成的靶基因研究

梁　爽　陈瑞婷　张登攀　辛虎虎
陆　喦　王梅仙　缪云根*

摘要：Hedgehog(Hh)信号通路调控无脊椎动物和脊椎动物的发育，但该信号通路在脂肪发育中的作用仍然知之甚少。本文研究发现Hh信号通路基因在家蚕幼虫脂肪体中有表达。在*BmN*细胞系中对这些基因的功能分析，显示*Hh*基因的激活刺激了Hh通路基因的转录，但抑制了脂肪标记基因*AP2*的表达。相反，特异性RNA干扰抑制*Hh*表达导致*AP2*表达增加，这进一步说明了Hh信号对脂肪标记基因的调整作用。在家蚕幼虫中，当用特异性Hh信号通路拮抗剂环巴胺处理家蚕时，观察到脂肪细胞分化的增强和脂肪细胞体积的增大。脂肪体特异性Hh的阻断实验与Hh信号传导抑制家蚕脂肪生成一致。研究结果表明，Hh信号传导在抑制脂肪形成中的作用在脊椎动物和无脊椎动物中是保守的。

农业部动物疫病病原学与免疫控制重点开放实验室，动物科学学院，浙江大学，杭州

*Bm*PLA2 containing conserved domain WD40 affects the metabolic functions of fat body tissue in silkworm, *Bombyx mori*

Orville Singh Chabungbam Xin HH Chen RT Wang MX
Liang S Lu Y Cai ZZ Miao YG*

Abstract: PLA2 enzyme hydrolyzes arachidonic acid, and other polyunsaturated fatty acids, from the sn-2 position to release free arachidonic acid and a lysophospholipid. Previous studies reported that the PLA2 in invertebrate organisms participates in lipid signaling molecules like arachidonic acid release in immune-associated tissues like hemocytes and fat bodies. In the present study, we cloned the *BmPLA2* gene from fat body tissue of silkworm *Bombyx mori*, which has a total sequence of 1.031 kb with a 31.90 kDa protein. *In silico* results of *BmPLA2* indicated that the protein has a putative WD40 conserved domain and its phylogeny tree clustered with *Danaus plexippus* species. We investigated the transcriptional expression in development stages and tissues. The highest expression of *BmPLA2* was screened in fat body among the studied tissues of third day fifth instar larva, with a high expression on third day fifth instar larva followed by a depression of expression in the wandering stage of the fifth instar larva. The expression of *BmPLA2* in female pupa was higher than that of male pupa. Our RNAi-mediated gene silencing results showed highest reduction of *BmPLA2* expression in post-24h followed by post-48 and post-72h. The *BmPLA2*-RNAi larvae and pupa could be characterized by pharate adult lethality and underdevelopment. The phenotypic characters of fat body cells in RNAi-induced larva implied that *BmPLA2* affects the metabolic functions of fat body tissue in silkworm *Bombyx mori*.

Published On: Insect Science, 2016, 23(1), 28-36.

College of Animal Sciences, Zhejiang University, Hangzhou 310058, China.
*Corresponding author E-mail: miaoyg@zju.edu.cn.

含WD40保守结构域的*Bm*PLA2影响家蚕的脂肪体代谢

Orville Singh Chabungbam　辛虎虎　陈瑞婷　王梅仙
梁　爽　陆　嵒　蔡自峥　缪云根*

摘要：PLA2是参与水解花生四烯酸以及其他多种不饱和脂肪酸的超级酶家族中的一员，这些酶从sn-2位置释放出自由的花生四烯酸和一个溶血磷脂。早期研究表明，PLA2在无脊椎动物体内参与脂质信号分子如血细胞和脂肪体等免疫相关组织中的花生四烯酸的释放。在本研究中，我们从家蚕的脂肪体组织中克隆了*BmPLA2*基因，其总序列长1.031 kb，编码31.90 kDa的蛋白质。*BmPLA2*的分析结果表明，蛋白质具有预测的WD40保守结构域，其系统发生树与黑脉金斑蝶聚为一类。我们调查该基因在不同发育阶段和不同组织中的转录情况：在五龄第3天幼虫所有检测组织中，*BmPLA2*在脂肪体中的表达量最高；五龄3天幼虫高表达，随后从五龄幼虫到上蔟阶段表达量下降。*BmPLA2*在雌蛹中的表达量高于在雄蛹。RNAi介导的基因沉默结果显示，注射24h后*BmPLA2*表达量下降最明显，其次是48h后和72h后。幼虫和蛹的*BmPLA2*-RNAi均出现蜕皮致死和发育不完全的表型。RNAi诱导的幼虫脂肪体细胞的表型特征表明*BmPLA2*影响家蚕脂肪体组织的代谢功能。

动物科学学院，浙江大学，杭州

Identification and characterization of novel chitin-binding proteins from the larval cuticle of silkworm, *Bombyx mori*

Dong ZM[1,2#] Zhang WW[1,2#] Zhang Y[1,2,3]
Zhang XL[1,2] Zhao P[1,2*] Xia QY[1,2]

Abstract: Cuticle is mainly made of chitin filaments embedded in a matrix of cuticular proteins (CPs). Cuticular chitins have minor differences, whereas CPs are widely variable with respect to their sequences and structures. To understand the molecular basis underlying the mechanical properties of cuticle, it is necessary to know which CPs interact with chitin and how they are assembled into the cuticle structure. In the present study, a chitin-binding assay was performed followed by liquid chromatography-tandem mass spectrometry to identify the extracted proteins from the larval cuticle of silkworm, *Bombyx mori.* There were 463 proteins identified from the silkworm larval cuticle, 200 of which were recovered in the chitin-binding fraction. A total of 103 proteins were annotated as CPs, which were classified into 11 CP families based on their conserved motifs, including CPR, CPAP, CPT, CPF and CPFL, CPCFC, chitin_bind 3, *Bm*CPH2 homologues, *Bm*CPH9 homologues, *Bm*CPG1 homologues, *Bm*CPG20 homologues, and *Bm*CPG21 homologues. A total of five CP families were newly identified in the chitin-binding fraction, thereby providing new information and insight into the composition, structure, and function of the silkworm larval cuticle.

Published On: Journal of Proteome Research, 2016, 15(5), 1435-1445.

1 State Key Laboratory of Silkworm Genome Biology, Southwest University, Chongqing 400716, China.
2 Chongqing Engineering and Technology Research Center for Novel Silk Materials, Southwest University, Chongqing 400716, China.
3 College of Biotechnology, Southwest University, Chongqing 400716, China.
#These authors contributed equally.
*Corresponding author E-mail: zhaop@swu.edu.cn.

家蚕幼虫表皮中几丁质结合蛋白的鉴定与分析

董照明[1,2#] 张薇薇[1,2#] 张 艳[1,2,3]
张晓璐[1,2] 赵 萍[1,2*] 夏庆友[1,2]

摘要：表皮主要由几丁质嵌入表皮蛋白(CPs)基质中形成。表皮中几丁质层差异较小，而CPs在其序列和结构方面差异较大。为了解表皮机械性能的分子基础，确定CPs与几丁质相互作用以及它们如何组装到表皮结构中是很有必要的。在本研究中，利用液相色谱-串联质谱法鉴定从家蚕幼虫表皮中提取的蛋白质。共鉴定到463个蛋白质，有200个为几丁质结合蛋白。总共103个蛋白被注释为CPs，根据其保守基序，包括CPR，CPAP，CPT，CPF和CPFL，CPCFC，chitin_bind 3，*Bm*CPH2同源物，*Bm*CPH9同源物，*Bm*CPG1同源物，*Bm*CPG20同源物和*Bm*CPG21同源物。本研究鉴定到5个新的表皮蛋白家族，为家蚕幼虫表皮的组成、结构和功能的研究提供了新的信息，并为深入研究奠定了基础。

1 家蚕基因组生物学国家重点实验室，西南大学，重庆
2 重庆市蚕丝纤维新材料工程技术研究中心，西南大学，重庆
3 生物技术学院，西南大学，重庆

Identification and molecular characterization of a chitin deacetylase from *Bombyx mori* peritrophic membrane

Zhong XW[1,2] Wang XH[1] Tan X[1]
Xia QY[1] Xiang ZH[1] Zhao P[1*]

Abstract: The insect midgut epithelium is generally lined with a unique chitin and protein structure, the peritrophic membrane (PM), which facilitates food digestion and protects the gut epithelium. PM proteins are important determinants for PM structure and formation. In this study, the silkworm *Bombyx mori* midgut PM protein *Bm*CDA7 was identified by proteomic tools. The full-length *BmCDA7* cDNA is 1 357 bp; the deduced protein is composed of 379 amino acid residues and includes a 16 amino acid residue signal peptide, a putative polysaccharide deacetylase-like domain and 15 cysteine residues present in three clusters. The heterologously expressed proteins of the *BmCDA7* gene in yeast displayed chitin deacetylase activity. Expression of *B. mori BmCDA7* was detected in the midgut at both the transcriptional and translational levels. The *BmCDA7* gene was expressed by the newly hatched silkworm larvae until day seven of the fifth instar and was expressed at a high level in the newly exuviated larvae of different instars. The functions and regulatory mechanism of *BmCDA7*, however, need further investigation.

Published On: International Journal of Molecular Science, 2014, 15(2), 1946-1961.

1 State Key Laboratory of Silkworm Genome Biology, Southwest University, Chongqing 400716, China.
2 Translational Medicine Research Center, North Sichuan Medical College, Nanchong 637000, China.
*Corresponding author Email: zhaop@swu.edu.cn.

家蚕围食膜几丁质脱乙酰基酶的鉴定和分子表征

钟晓武[1,2]　王晓欢[1]　谭　祥[1]
夏庆友[1]　向仲怀[1]　赵　萍[1*]

摘要：昆虫中肠上皮细胞通常按照由内至外的顺序与几丁质结构蛋白、围食膜排列在一起，围食膜促进食物的消化并保护肠道上皮细胞。围食膜蛋白是其结构形成的重要决定因素。在本研究中，通过蛋白质组学手段鉴定到了家蚕围食膜蛋白*Bm*CDA7。*BmCDA7*基因cDNA全长为1 357 bp，预测蛋白质由379个氨基酸残基组成，包括一个含有16个氨基酸残基的信号肽，此外还含有类多聚糖脱乙酰酶结构域和15个保守的半胱氨酸残基（分为3个簇）。在酵母中异源表达了具有几丁质脱乙酰基酶活性的*Bm*CDA7蛋白。*BmCDA7*基因在转录水平和翻译水平上都在中肠中有表达。*BmCDA7*基因从蚁蚕到五龄第7天都有表达且在各龄期初表达量较高。然而，*BmCDA7*基因的功能和调控机制有待于进一步研究。

1　家蚕基因组生物学国家重点实验室，西南大学，重庆
2　转化医学研究中心，川北医学院，南充

*Bm*Robo2/3 is required for axon guidance in the silkworm *Bombyx mori*

Li XT[#] Yu Q[#] Zhou QS Zhao X
Liu ZY Cui WZ[*] Liu QX[*]

Abstract: Axon guidance is critical for proper wiring of the nervous system. During the neural development, the axon guidance molecules play a key role and direct axons to choose the correct way to reach the target. Robo, as the receptor of axon guidance molecule Slit, is evolutionarily conserved from planarians to humans. However, the function of Robo in the silkworm, *Bombyx mori*, remained unknown. In this study, we cloned *robo2/3* from *B. mori* (*Bmrobo2/3*), a homologue of *robo2/3* in *Tribolium castaneum*. Moreover, *Bm*Robo2/3 was localized in the neuropil, and RNAi-mediated knockdown of *Bmrobo2/3* resulted in the longitudinal connectives forming closer to the midline. These data demonstrate that *Bm*Robo2/3 is required for axon guidance in the silkworm.

Published On: Gene, 2016, 577(2), 174-179.

Laboratory of Developmental Genetics, Shandong Agricultural University, Taian 271018, China.
[#]These authors contributed equally.
[*]Corresponding author E-mail:wzcui88@163.com; liuqingxin@sdau.edu.cn.

家蚕轴突引导需要*Bm*Robo2/3

李晓童[#] 于 奇[#] 周启升 赵 晓
刘朝阳 崔为正[*] 刘庆信[*]

摘要：轴突引导对神经系统的正确连接至关重要。在神经发育过程中，轴突引导分子发挥关键作用，并引导轴突选择正确的方式达到目标。作为轴突引导分子的受体，Robo在进化上从涡虫到人类保守。然而，Robo在家蚕(*Bombyx mori*)中的功能依然未知。在这项研究中，我们克隆了来自*B. mori*的*Robo2/3*，是赤拟谷盗中的同源物。此外，*Bm*Robo2/3位于神经纤维中，RNAi介导的*BmRobo2/3*敲低导致纵向结缔组织更接近中线。这些数据表明，*Bm*Robo2 / 3是家蚕轴突引导所必需的。

发育遗传学实验室，山东农业大学，泰安

*Bm*Robo1a and *Bm*Robo1b control axon repulsion in the silkworm *Bombyx mori*

Li XT# Yu Q# Zhou QS Zhao X

Liu ZY Cui WZ* Liu QX*

Abstract: The development of the nervous system is based on the growth and connection of axons, and axon guidance molecules are the dominant regulators during this course. Robo, as the receptor of axon guidance molecule Slit, plays a key role as a conserved repellent cue for axon guidance during the development of the central nervous system. However, the function of Robo in the silkworm *Bombyx mori* is unknown. In this study, we cloned two novel *robo* genes in *B. mori* (*Bmrobo1a* and *Bmrobo1b*). *Bm*Robo1a and *Bm*Robo1b lack an Ig and a FNIII domain in the extracellular region and the CC0 and CC2 motifs in the intracellular region. *Bm*Robo1a and *Bm*Robo1b were colocalized with *Bm*Slit in the neuropil. Knock-down of *BmRobo1a* and *BmRobo1b* by RNA interference (RNAi) resulted in abnormal development of axons. Our results suggest that *Bm*Robo1a and *Bm*Robo1b have repulsive function in axon guidance, even though their structures are different from Robo1 of other species.

Published On: Gene, 2016, 577(2), 215-220.

Laboratory of Developmental Genetics, Shandong Agricultural University, Taian 271018, China.

#These authors contributed equally.

*Corresponding author E-mail:wzcui88@163.com; liuqingxin@sdau.edu.cn.

家蚕*Bm*Robo1a和*Bm*Robo1b控制轴突排斥

李晓童[#]　于　奇[#]　周启升　赵　晓
刘朝阳　崔为正[*]　刘庆信[*]

摘要：神经系统的发展是基于轴突的生长和轴突之间的连接，轴突导向分子是发展过程中的主要调节因子。Robo作为轴突导向分子的剪切受体，在中枢神经系统发育过程中对轴突引导起关键的保守排斥作用。然而，Robo在家蚕中的功能还是未知的。在本研究中，我们克隆了家蚕两个新的*Robo*基因（*Bmrobo1a*和*Bmrobo1b*）。*Bm*Robo1a和*Bm*Robo1b在细胞外区域缺乏一个Ig和一个FNIII结构域，在细胞内区域缺乏CC0和CC2结构。通过*Bm*Slit将*Bm*Robo1a和*Bm*Robo1b共定位在神经纤维中。RNA干扰（RNAi）*BmRobo1a*和*BmRobo1b*导致轴突异常发育。研究结果表明，尽管*Bm*Robo1a和*Bm*Robo1b的结构与其他物种的Robo1不同，但在轴突导向中具有排斥功能。

发育遗传学实验室，山东农业大学，泰安

Bmaly is an important factor in meiotic progression and spermatid differentiation in *Bombyx mori* (Lepidoptera: Bombycidae)

Zhang PJ[1] Zhong JF[1] Cao GL[1,2] Xue RY[1,2] Gong CL[1,2*]

Abstract: The *Drosophila melanogaster* "*always early*" gene (*Dmaly*), which is required for G_2/M cell-cycle control and spermatid differentiation, is one of the meiotic arrest genes. To study the *Bombyx mori aly* gene (*Bmaly*), the cDNA of *Bmaly* was cloned and sequenced, and the results showed that the open reading frame of *Bmaly* is 1 713 bp in length, encoding 570 amino acid residues, in which a domain in an Rb-related pathway was found. Phylogenetic analysis based on the amino acid sequences of conserved regions showed that Aly from different insects gathered together, except for *Dm*aly and *Culex quinquefasciatus* Aly, which were not clustered to a subgroup according to insect order. The *Bmaly* gene was inserted into expression vector pGS-21a(+) and then the recombinant protein was expressed in *Escherichia coli* and used to immunize mice to prepare the antibody against *Bm*aly. Immunofluorescence examination showed that *Bm*aly was distributed in both the cytoplasm and nucleus of *BmN* cell. The *Bmaly* gene expression could not be detected in the silk gland, Malpighian tubule, fat body, or midgut of the silkworm. Expression level of the *Bmaly* gene were detected in the gonadal tissues, where the level in the testes was 10 times higher than that in the ovaries. Moreover, *Bmaly* expression was detected by quantitative polymerase chain reaction at different stages of *B. mori* testis development, at which fifth instar was relatively grossly expressed. The result suggested *Bmaly* was abundantly expressed in primary spermatocytes and prespermatids. To further explore the function of *Bmaly*, *Bmaly* siRNA was injected into the third and fourth instar silkworm larvae, which markedly inhibited the development of sperm cells. These results suggest that *Bmaly* is a meiotic arrest gene that plays an important role in spermatogenesis.

Published On: Journal of Insect Science, 2014, 14.

1 School of Biology and Basic Medical Sciences ,Soochow University, Suzhou 215123, China.

2 National Engineering Laboratory for Modern Silk, Soochow University, Suzhou 215123, China.

*Corresponding author E-mail: gongcl@suda.edu.cn.

家蚕（鳞翅目：蚕蛾科）*Bmaly* 基因在减数分裂进程和精子分化中起关键作用

张鹏杰[1] 钟金凤[1] 曹广力[1,2] 薛仁宇[1,2] 贡成良[1,2*]

摘要：果蝇 *Dmaly* 基因是减数分裂阻滞基因之一，是减数分裂 G_2/M 细胞周期控制和精子分化所必需的。为了研究家蚕 *Bmaly* 基因，对其进行了克隆和序列测定，结果表明 *Bmaly* 基因的开放阅读框为 1 713 bp，编码 570 个氨基酸残基，其中发现了一个与 Rb 相关通路的蛋白结构域。基于保守区域的氨基酸序列的系统发育分析表明除了 *Dm*Aly 和致倦库蚊 Aly 外，不同昆虫的 Aly 均聚集在一起。这种聚集并未按昆虫所在的目聚类。将 *Bmaly* 基因克隆入 pGS-21a（+）载体进行原核表达，重组蛋白免疫小鼠，获得了鼠源 *Bm*aly 多抗。免疫荧光显示 *Bm*aly 在 *BmN* 细胞的细胞质和细胞核中均有分布。在家蚕丝腺、马氏管、脂肪体和中肠中均不能检测到 *Bmaly* 基因的表达。在性腺组织中检测到 *Bm*aly 基团的表达水平，精算中的表达水平比卵巢高 10 倍。定量聚合酶链式反应检测了蚕精巢发育不同阶段的 *Bmaly* 的表达，其中五龄期相对高表达。结果表明 *Bm*aly 在初级精母细胞和精细胞中大量表达。为了进一步探讨 *Bmaly* 的功能，将 *Bmaly* siRNA 注入家蚕三龄和四龄幼体中，发现明显抑制精子细胞的发育。这些结果显示 *Bmaly* 是一个在精子发生过程中起重要作用的减数分裂阻滞基因。

1 基础医学与生物科学学院，苏州大学，苏州

2 现代丝绸国家工程实验室，苏州大学，苏州

Lysine acetylation stabilizes SP2 protein in the silkworm, *Bombyx mori*

Zhou Y[1] Wu CC[1] Sheng Q[1] Jiang CY[1]
Chen Q[1] Lü ZB[1] Yao JM[2] Nie ZM[1,2*]

Abstract: Lysine acetylation (Kac) is a vital post-translational modification that plays an important role in many cellular processes in organisms. In the present study, the nutrient storage proteins in hemolymph were first found to be highly acetylated-particularly SP2 protein, which contains 20 potential Kac sites. Further results confirmed that lysine acetylation could stabilize and up-regulate the protein level of anti-apoptosis protein SP2, thereby improving the survival of H_2O_2-treated *BmN* cells and suppressing the apoptosis induced by H_2O_2. The potential mechanism involved in the inhibition of ubiquitin-mediated proteasomal degradation by crosstalk between lysine acetylation and ubiquitination. Our results showed that the increase in the acetylation level by TSA could decrease the ubiquitination and improve the protein level of SP2, indicating that lysine acetylation could influence the SP2 protein level through competition between ubiquitination and the suppression of ubiquitin-mediated proteasomal degradation, thereby stabilizing the protein. SP2 is a major nutrient storage protein from hemolymph for amino acid storage and utilization. The crosstalk between lysine acetylation and ubiquitination of SP2 might imply an important role of lysine acetylation for nutrient storage and utilization in silkworm.

Published On: Journal of Insect Physiology, 2016, 91-92, 56-62.

1 College of Life Sciences, Zhejiang Sci-Tech University, Hangzhou 310018, China.
2 College of Materials and Textile, Zhejiang Sci-Tech University, Hangzhou 310018, China.
*Corresponding author E-mail: wuxinzm@126.com.

家蚕赖氨酸乙酰化修饰可以提高SP2蛋白稳定性

周 雍[1] 吴程程[1] 盛 清[1] 蒋彩英[1]
陈 琴[1] 吕正兵[1] 姚菊明[2] 聂作明[1,2*]

摘要：赖氨酸乙酰化修饰（Kac）是一类关键的蛋白翻译后修饰，其在生物体的多个生理过程中起重要的调控作用。本研究首次发现，家蚕血淋巴营养储藏蛋白存在高度乙酰化修饰，尤其是SP2蛋白，其存在20个潜在的Kac位点。进一步研究结果表明，赖氨酸乙酰化修饰能够稳定上调抗凋亡蛋白SP2的蛋白水平，从而提高H_2O_2处理*BmN*细胞的存活率，即抑制H_2O_2诱导的细胞凋亡，其潜在机制涉及到乙酰化和泛素化修饰竞争交互，导致泛素介导的蛋白酶体降解途径被抑制。研究结果显示，提高TSA乙酰化水平后可以降低SP2蛋白的泛素化水平并提高其蛋白水平，表明赖氨酸乙酰化修饰能够通过竞争泛素化修饰抑制泛素介导的蛋白酶体降解途径，从而提高SP2蛋白的稳定性。SP2蛋白是家蚕血淋巴中氨基酸储藏和利用的主要营养储存蛋白，该蛋白的赖氨酸乙酰化和泛素化修饰的交互竞争，暗示了赖氨酸乙酰化修饰可能在家蚕营养物质储藏和利用过程中扮演着重要的角色。

1 生命科学学院，浙江理工大学，杭州
2 材料与纺织学院，浙江理工大学，杭州

Bmovo-1 regulates ovary size in the silkworm, *Bombyx mori*

Xue RY[1,2#] Hu XL[1,2#] Cao GL[1,2] Huang ML[1]
Xue GX[1] Qian Y[1] Song ZW[1] Gong CL[1,2*]

Abstract: The regulation of antagonistic OVO isoforms is critical for germline formation and differentiation in *Drosophila*. However, little is known about genes related to ovary development. In this study, we cloned the *Bombyx mori ovo* gene and investigated its four alternatively spliced isoforms. *Bm*OVO-1, *Bm*OVO-2 and *Bm*OVO-3 all had four C2H2 type zinc fingers, but differed at the N-terminal ends, while *Bm*OVO-4 had a single zinc finger. *Bmovo-1*, *Bmovo-2* and *Bmovo-4* showed the highest levels of mRNA in ovaries, while *Bmovo-3* was primarily expressed in testes. The mRNA expression pattern suggested that *Bmovo* expression was related to ovary development. RNAi and transgenic techniques were used to analyze the biological function of *Bmovo*. The results showed that when the *Bmovo* gene was downregulated, oviposition number decreased. Upregulation of *Bmovo-1* in the gonads of transgenic silkworms increased oviposition number and elevated the trehalose contents of hemolymph and ovaries. We concluded that *Bmovo-1* was involved in protein synthesis, contributing to the development of ovaries and oviposition number in silkworms.

Published On: PLoS ONE, 2014, 9(8).

1 School of Biology and Basic Medical Science, Soochow University, Suzhou 215123, China.
2 National Engineering Laboratory for Modern Silk, Soochow University, Suzhou 215123, China.
#These authors contributed equally.
*Corresponding author E-mail: gongcl@suda.edu.cn.

*Bmovo-1*调节家蚕卵巢的大小

薛仁宇[1,2#]　胡小龙[1,2#]　曹广力[1,2]　黄茉莉[1]
薛高旭[1]　钱　莹[1]　宋作伟[1]　贡成良[1,2*]

摘要：对拮抗性OVO亚型的调节对果蝇的种系形成和分化至关重要。然而，对与卵巢发育相关的基因知之甚少。在这项研究中，我们克隆了家蚕*ovo*基因，并研究了其4个可变剪接的亚型。*Bm*OVO-1，*Bm*OVO-2和*Bm*OVO-3均具有4个C2H2型锌指，但在N末端不同，然而*Bm*OVO-4具有单个锌指。*Bmovo-1*，*Bmovo-2*和*Bmovo-4*在卵巢中mRNA表达水平最高，而*Bmovo-3*主要在精巢中表达。mRNA表达模式暗示*Bmovo*表达与卵巢发育有关。使用RNAi和转基因技术分析*Bmovo*的生物学功能，结果显示：当转基因家蚕性腺中*Bmovo*基因表达下调时，产卵数减少；而*Bmovo-1*的上调则增加了产卵数，并提高了血淋巴和卵巢中海藻糖含量。由此推测*Bmovo-1*参与了蛋白质合成，有助于家蚕卵巢的发育和产卵数的提高。

1　基础医学与生物科学学院，苏州大学，苏州
2　现代丝绸国家工程实验室，苏州大学，苏州

RNA interference-mediated silencing of the *bursicon* gene induces defects in wing expansion of silkworm

Huang JH[1#] Zhang Y[1#] Li MH[1] Wang SB[1]
Liu WB[1] Pierre Couble[2] Zhao GP[1] Huang YP[1*]

Abstract: We studied the role of the *bursicon* gene in wing expansion. First, we investigated its expression at different developmental stages in the silkworm, *Bombyx mori bursicon* gene was expressed at low levels in larvae, high levels in pupae, and low levels again in adults. Then, we injected the double-stranded *bursicon* RNA into *B. mori* pupae to test RNA interference. The level of *bursicon* mRNA was reduced significantly in pupae, and a deficit in wing expansion was observed in adults. In addition, the differential display reverse transcription polymerase chain reaction (DD-RT-PCR) was used to reveal differences in the expression of transcripts in response to the inhibition of *bursicon*. In conclusion, *bursicon* plays a key role in the stereotyped behavioral program involved in wing expansion.

Published On: FEBS Letters, 2007, 581(4), 697-701.

1 Institute of Plant Physiology and Ecology, Shanghai Institutes for Biological Sciences, The Chinese Academy of Sciences, Shanghai 200032, China.

2 CNRS/Université Claude Bernard de Lyon, Vileurbanne 69622, France.

#These authors contributed equally.

*Corresponding author E-mail: yphuang@sibs.ac.cn.

RNA 干扰介导的 *bursicon* 基因沉默诱导家蚕翅伸展缺陷

黄建华[1#]　张　勇[1#]　李明辉[1]　王四宝[1]
刘文彬[1]　Pierre Couble[2]　赵国屏[1]　黄勇平[1*]

摘要：我们研究了 *bursicon* 基因在家蚕翅发育过程中的作用。首先，分析了 *bursicon* 基因在家蚕不同发育阶段的表达。*bursicon* 基因在幼虫期表达量低，蛹期表达量高，成虫期表达量较低。其次，我们将 *bursicon* 基因的双链 RNA 注入家蚕蛹中，来检测 RNA 干扰的效果。结果发现 *bursicon* 的 mRNA 表达水平明显降低，并且成虫的翅出现了卷曲状缺陷。此外，我们利用 DD-RT-PCR 技术检测 *bursicon* mRNA 在干扰组中的表达，发现有明显的下调表达。我们的结果表明，*bursicon* 基因在家蚕翅伸展中起着重要的调控作用。

1　植物生理与生态研究所，中国科学院上海生命科学研究院，上海
2　里昂第一大学，维勒班，法国

Molecular characterization and immunohistochemical localization of a novel troponin C during silkworm development

Chen JQ[1] Chen J[1] Gai QJ[1] Lü ZB[1] Wang D[1] Nie ZM[1]
Wang J[1] Wang XD[1] Wu XF[1,2] Zhang YZ[1*]

Abstract: We have cloned and sequenced a novel *Bombyx mori* gene that encodes a protein having a high degree of homology with other known troponin C (TnC) proteins. The amino acid sequence, DX[DN]X[DSG]X_6E, a highly conserved putative Ca^{2+}-binding motif found in loops within the globular domains of previously identified TnC proteins, is also present in *Bm*TnC. We have expressed and purified to homogeneity a His-tagged *Bm*TnC fusion protein having a molecular weight of approximately 21.6 kDa. We have used this purified fusion protein to produce polyclonal antibodies against *Bm*TnC for Western blot analyses. These analyses have revealed that *Bm*TnC is expressed in the larval head, the Malpighian tubule, the epidermis, the testis, and the gut, as has been confirmed by immunohistochemistry. In addition, real-time reverse transcription/polymerase chain reaction has shown that *Bm*TnC mRNA levels differ substantially among these tissues. Our findings indicate that *Bm*TnC is selectively expressed in the muscular tissues of the silkworm, including portions of the head, the Malpighian tubule, the body wall, and the gut.

Published On: Cell and Tissue Research, 2008, 331(3), 725-738.

1 Institute of Biochemistry, Zhejiang Sci-Tech University, Hangzhou 310018, China.
2 Shanghai Institute of Biochemistry and Cell Biology, Chinese Academy of Sciences, Shanghai 200031, China.
*Corresponding author E-mail: yaozhou@chinagene.com.

一种新的家蚕肌钙蛋白的分子鉴定与免疫组化定位分析

陈剑清[1] 陈 健[1] 盖其静[1] 吕正兵[1] 王 丹[1] 聂作明[1]
王 江[1] 王雪冬[1] 吴祥甫[1,2] 张耀洲[1*]

摘要：我们在家蚕中鉴定了一个与其他物种中肌钙蛋白C（TnC）具有高度同源性的蛋白质，并对该蛋白质的基因进行克隆与测序分析。*Bm*TnC蛋白的氨基酸序列中含有Ca^{2+}结合基序DX[DN]X[DSG]X_6E，它在其他物种TnC蛋白质中存在于球形结构域的环上。我们克隆并原核表达纯化得到了融合His标签的*Bm*TnC蛋白，它的分子质量约为21.6 kDa。用纯化获得的融合蛋白得到*Bm*TnC的多克隆抗体，用于Western blot检测分析。实验结果表明，*Bm*TnC蛋白质在家蚕幼虫的头部、马氏管、表皮、精巢、肠道中都有表达，同时在免疫组化实验中也得到了进一步的证实。此外，RT-PCR结果显示，*Bm*TnC mRNA在这些组织中的水平差异很大。我们发现*Bm*TnC的mRNA在家蚕幼虫的头部、马氏管、体壁以及肠道的肌肉组织中选择性表达。

1 生物化学研究所，浙江理工大学，杭州
2 上海生物化学与细胞生物学研究所，中国科学院，上海

Systematic cloning and analysis of autophagy-related genes from the silkworm *Bombyx mori*

Zhang X[1] Hu ZY[2] Li WF[1] Li QR[2] Deng XJ[2]
Yang WY[2] Cao Y[2] Zhou CZ[1*]

Abstract: Background: Through the whole life of eukaryotes, autophagy plays an important role in various biological events including development, differentiation and determination of lifespan. A full set of genes and their encoded proteins of this evolutionarily conserved pathway have been identified in many eukaryotic organisms from yeast to mammals. However, this pathway in the insect model organism, the silkworm *Bombyx mori*, remains poorly investigated. Results Based on the autophagy pathway in several model organisms and a series of bioinformatic analyses, we have found more than 20 autophagy-related genes from the current database of the silkworm *Bombyx mori*. These genes could be further classified into the signal transduction pathway and two ubiquitin-like pathways. Using the mRNA extracted from the silk gland, we cloned the full length cDNA fragments of some key genes via reverse transcription PCR and 3' rapid amplification of cDNA ends (RACE). In addition, we found that the transcription levels of two indicator genes *BmATG8* and *BmATG12* in the silk gland tend to be increased from 1st to 8th day of the fifth instar larvae. Conclusion Bioinformatics in combination with RT-PCR enable us to remodel a preliminary pathway of autophagy in the silkworm. Amplification and cloning of most autophagy-related genes from the silk gland indicated autophagy is indeed an activated process. Furthermore, the time-course transcriptional profiles of *BmATG8* and *BmATG12* revealed that both genes are up-regulated along the maturation of the silk gland during the fifth instar. These findings suggest that the autophagy should play an important role in *Bombyx mori* silk gland.

Published On: BMC Molecular Biology, 2009, 10.

1 Hefei National Laboratory for Physical Sciences at the Microscale and School of Life Sciences, University of Science and Technology of China, Hefei 230027, China.

2 Department of Sericulture Science, College of Animal Science, South China Agricultural University, Guangzhou 510642, China.

*Corresponding author E-mail: zcz@ustc.edu.cn.

家蚕自噬相关基因的系统克隆和分析

张 璇[1] 胡占英[2] 李卫芳[1] 李庆荣[2] 邓小娟[2]
杨婉莹[2] 曹 阳[2] 周丛照[1*]

摘要:背景——自噬贯穿真核生物的整个生命周期,并在各种生物学事件,如发育、分化和寿命决定中起着重要的作用。自噬通路在进化上是保守的,其相关基因及其编码蛋白已在酵母至哺乳动物中被鉴定到。然而,家蚕作为昆虫模式生物的代表,目前关于其自噬通路的研究较少。结果——基于几种模式生物中的自噬通路和一系列生物信息学分析,我们从家蚕现有的数据库中找到20多个自噬相关基因,这些基因可以归类到信号转导通路和两个泛素样通路。以丝腺中提取的mRNA为模板,通过RT-PCR和3'cDNA末端快速扩增(RACE)克隆到一些关键基因的全长cDNA片段。此外,我们发现丝腺中的两个指示基因*BmATG8*、*BmATG12*的转录水平从幼虫五龄第1天到第8天逐渐增加。总结——结合生物信息学与RT-PCR验证等手段,我们对家蚕自噬初级通路进行了重构。家蚕丝腺中大部分自噬相关基因的扩增、克隆表明自噬的确是一个活化过程。此外,时程转录谱显示家蚕*BmATG8*和*BmATG12*基因在五龄期随着丝腺的成熟而上调表达。这些发现表明自噬在家蚕丝腺发育中扮演着重要的角色。

1 合肥微尺度物质科学国家实验室(筹),生命科学学院,中国科学技术大学,合肥
2 蚕丝科学系,动物科学学院,华南农业大学,广州

Influence of RNAi knockdown for *E-complex* genes on the silkworm proleg development

Xiang H[1] Li MW[1,2] Guo JH[1] Jiang JH[1] Huang YP[1*]

Abstract: Larvae of many holometabolous insects possess abdominal appendages called prolegs. Lepidoptera larvae have prolegs in the segments A3-A6. Functions of Lepidoptera *hox* genes on these abdominal appendages development is still a controversial issue. In this article, we report the use of double strand RNA (dsRNA)-mediated interference (RNAi) to dissect the function of some *hox* genes, specifically *E-complex* genes *Ubx*, *abd-A*, and *Abd-B*, in the ventral appendage development of the Lepidoptera silkworm, *Bombyx mori*. We found that *Ubx* RNAi caused leg identity in A1 segment, *abd-A* RNAi caused severe defect of abdominal prolegs and *Abd-B* RNAi allowed proleg identity in more posterior abdominal segments. These results confirm that Lepidoptera *hox* genes *Ubx* and *Abd-B* have evolved the repressing function to ventral appendage development, which is similar to those of *Drosophila*. However, Lepidoptera *abd-A* might have been modified distinctively during evolution, and has important roles in directing the development of prolegs.

Published On: Archives of Insect Biochemistry and Physiology, 2011, 76(1), 1-11.

1 Shanghai Institute of Plant Physiology and Ecology, Shanghai Institutes for Biological Sciences, Chinese Academy of Sciences, Shanghai 200032, China.

2 Sericultural Research Institute, Chinese Academy of Agricultural Sciences, Zhenjiang 212018, China.

*Corresponding author E-mail: yphuang@sibs.ac.cn.

RNA 干扰敲除 *E* 复合体基因对家蚕腹足发育的影响

相　辉[1]　李木旺[1,2]　郭金华[1]　蒋剑豪[1]　黄勇平[1*]

摘要：在许多完全变态昆虫的幼虫中，腹部由许多称之为腹足的附肢所占据。鳞翅目幼虫腹足分布在 A3—A6 体节上。在这些腹足发育过程中，鳞翅目的 *hox* 基因的功能仍存在争议。在本文中，我们运用双链 RNA（dsRNA）介导干扰（RNAi）的技术来研究一些 *hox* 基因，特别是 *E* 复合基因 *Ubx*，*Abd-A* 和 *Abd-B* 等在鳞翅目昆虫家蚕腹足发育时的功能。我们发现 RNA 干扰 *Ubx* 会导致 A1 体节中的腹足的同一性，RNA 干扰 *Abd-A* 导致许多腹足发生缺陷，RNA 干扰 *Abd-B* 导致更多在腹部后部体节的腹足的同一性。这些结果证实鳞翅目 *hox* 基因 *Ubx* 和 *Abd-B* 参与抑制腹部附肢发育，与其在果蝇中的功能相似。鳞翅目 *Abd-A* 或许在进化过程中发生区别于其他物种的改变，并且在引导腹足发育中起到重要作用。

1　上海植物生理生态研究所，上海生命科学研究院，中国科学院，上海

2　蚕业研究所，中国农业科学院，镇江

Cloning and characterization of *Bmrunt* from the silkworm *Bombyx mori* during embryonic development

Liu WB　Yang F　Jia SH　Miao XX　Huang YP*

Abstract: Pair-rule genes (genes that are expressed only in alternate segments, odd or even) play an important role in translating the broad gradients of upstream genes into dual segment periodicity for body plan patterning in *Drosophila*. However, homologues of pair-rule genes show a remarkable diversity of expression patterns and functions in other insects. We cloned the homologue of *runt* in the silkworm *Bombyx mori*, an intermediate germband-type insect. Whole-mount in situ hybridization revealed three stripes arose one by one before gastrulation at the blastoderm stage. Five additional stripes were then generated sequentially as the growth zone elongated. Eight stripes appeared in a pair-rule manner with two-segment periodicity, each of which was confined to the posterior of an odd-numbered parasegment. The weaker segmental secondary stripes emerged de novo in even-numbered parasegments. The *Bmrunt* transcript vanished before blastokinesis and was then expressed again in the whole embryo. RNA interference for *Bmrunt* caused severely truncated, almost completely asegmental defects. This *cadual*-like phenotype suggests that *Bmrunt* does not function as a pair-rule gene in silkworm segmentation. *Bmrunt* is required for formation of most body segments and axis elongation in *B. mori.*

Published On: Archives of Insect Biochemistry and Physiology, 2008, 69(2), 47-59.

Institute of Plant Physiology and Ecology, Shanghai Institutes for Biological Sciences, Chinese Academy of Sciences, Shanghai 200032, China.

*Corresponding author E-mail: yphuang@sibs.ac.cn.

家蚕胚胎发育时期*Bmrunt*的克隆和鉴定

刘文彬　杨　斐　贾世海　苗雪霞　黄勇平*

摘要：成对规则基因(只在奇数或者偶数交替排列的体节中表达的基因)在果蝇体节形态建成中十分重要，能够将调控体节发育的上游基因信号传递到下游靶基因上。然而，成对规则基因的同源基因在昆虫中表现出表达模式和功能的多样性。我们克隆了家蚕成对规则同源基因，通过全组织包埋原位杂交技术证明其在原肠胚形成前出现了3条连续的条状分布带。随着生长区的延长，逐渐形成了另外5条*Bmrunt*基因的分布条纹。8个条纹以对称的方式出现在两个体节发育周期，每一个都在奇数副体节的后部，较弱的条纹片段出现在副体节的偶数段。*Bmrunt*的转录本在胚动前消失，在胚胎期再次表达。RNAi使*Bmrunt*下调表达后，绝大多数体节发育都出现缺陷。这种*cadual*-like的表型表明*Bmrunt*作为成对规则基因在家蚕的体节发育和胚轴的伸长过程中是至关重要的。

植物生理生态研究所，上海生命科学研究院，中国科学院，上海

Identification and characterization of two chitin-binding proteins from the peritrophic membrane of the silkworm, *Bombyx mori* L.

Yang HJ[1] Zhou F[1] Firdose Ahmad Malik[1] Roy Bhaskar[1]
Li XH[1] Hu JB[1] Sun CG[1] Miao YG[1,2*]

Abstract: The peritrophic membrane (PM) is a semi-permeable lining of the insect midgut, broadly analogous to the mucous lining of vertebrate gut. The PM proteins are important achievements for the function of the PM. In this study, two chitin-binding proteins (*Bm*PM-P43 and *Bm*PM-P41) from the PM of the silkworm, *Bombyx mori*, were identified and cloned. These proteins showed the molecular mass of 43 and 41 kDa, respectively. The deduced amino acid sequences codes for a protein of 381 amino acid residues and 364 amino acid residues, containing 12 and 14 cysteine residues followed by similar domain, both of them have 5 cysteine residues in similar position in the C-terminal. The confirmation of these proteins was performed by western blot analysis of recombinant *Bm*PM-P43 and *Bm*PM-P41. The chitin-binding activity analysis showed that the *Bm*PM-P43 and *Bm*PM-P41 could bind to chitin strongly. It is concluded that *Bm*PM-P43 and *Bm*PM-P41 contains a polysaccharide deacetylase domain instead of peritrophin domain, indicated that these two proteins may belong to a new chitin-binding protein family.

Published On: Archives of Insect Biochemistry and Physiology, 2010, 75(4), 221-230.

1 Key Laboratory of Animal Epidemic Etiology & Immunological Prevention of Ministry of Agriculture, Zhejiang University, Hangzhou 310029, China.

2 College of Animal Sciences, Zhejiang University, Hangzhou 310029, China.

*Corresponding author E-mail: miaoyg@zju.edu.cn.

家蚕围食膜两种几丁质结合蛋白的鉴定及表征

杨华军[1] 周 芳[1] Firdose Ahmad Malik[1] Roy Bhaskar[1]
李兴华[1] 胡嘉彪[1] 孙春光[1] 缪云根[1,2*]

摘要:昆虫围食膜(PM)是昆虫中肠的半渗透膜,类似于脊椎动物的黏液外膜。围食膜蛋白是围食膜发挥功能的重要组成成分。本研究中,我们从家蚕围食膜中克隆和鉴定了两个几丁质结合蛋白(*Bm*PM-P43和*Bm*PM-41)。生物信息学分析表明其分子质量分别为43 kDa和41 kDa。推测其序列分别编码381个氨基酸残基和364个氨基酸残基,含12和14个半胱氨酸残基,两者都在蛋白质的C末端相似位置含5个半胱氨酸残基。Western blotting分析确认重组蛋白*Bm*PM-P43和*Bm*PM-P41的表达。几丁质结合活性分析表明*Bm*PM-P43和*Bm*PM-P41能牢固结合几丁质。研究表明*Bm*PM-P43和*Bm*PM-P41含有一种多糖脱乙酰酶结构域,而不是围食膜因子结构域,揭示这两种蛋白质可能属于新的几丁质结合蛋白家族。

1 农业部动物疫病病原与免疫控制重点实验室,浙江大学,杭州
2 动物科学学院,浙江大学,杭州

Cloning and expression pattern of 3–dehydroecdysone 3β-reductase (3DE 3β-reductase) from the silkworm, *Bombyx mori* L.

Yang HJ[1] Liu XJ[2] Zhou F[1] Hu JB[1] Bhaskar Roy[1]
Li XH[1] Sun CG[1] Firdose Ahmad Malik[1] Niu YS[1] Miao YG[1*]

Abstract: Molting in insects is regulated by molting hormones (ecdysteroids), which are also crucial to insect growth, development, reproduction, etc. Ecdysone was inactivated to 3-dehydroecdysone (3DE) under ecdysone oxidase (EO), and followed by NAD(P)H-dependent irreversible reduction to 3-epiecdysteroid under 3DE 3α-reductase. On the other hand, 3-dehydroecdysone undergoes reversible reduction to ecdysone by 3DE 3β-reductase in the hemolymph. In this article, we cloned and characterized 3-dehydroecdysone 3β-reductase (3DE 3β-reductase) in the different tissues and the developing stage from the silkworm, *Bombyx mori* L. The *B. mori* 3DE 3β-reductase cDNA contains an ORF 972 bp and the deduced protein sequence containing 323 amino acid residues. Analysis showed that the deduced 3DE 3β-reductase belongs to the aldo-keto reductase (AKR) superfamily, which has the NAD(P)-binding domain, indicating that the function of 3DE 3β-reductase depends on the existence of NAD(P)H. Using *Escherichia coli*, a high level expression of a fusion polypeptide band of approx. 40 kDa was observed. The high transcription of 3DE 3β-reductase was mainly observed in the genitalia and fatty bodies in the third day of the fifth-instar larvae, followed next in the head, epidermis, and hemocytes. The expression of 3DE 3β-reductase in the early of every instar was lower than that in the late of instar. When the titer of 3DE is low, higher expression of 3DE 3β-reductase is necessary to maintain the ecdysone titer in body through converting 3DE to ecdysone, while the 3DE titer is high, the expression of 3DE 3β-reductase showed feedback inhibition.

Published On: Archives of Insect Biochemistry and Physiology, 2011, 76(1), 55-66.

1 Key Laboratory of Animal Epidemic Etiology and Immunological Prevention of Ministry of Agriculture, Zhejiang University, Hangzhou 310000, China.

2 Sericultural Research Institute, Zhejiang Academy of Agricultural Sciences, Hangzhou 310000, China.

*Corresponding author E-mail: miaoyg@zju.edu.cn.

家蚕3-脱氢蜕皮激素-3β还原酶基因的克隆及其在家蚕不同发育时期和不同组织中的表达模式

杨华军[1] 柳新菊[2] 周 芳[1] 胡嘉彪[1] Bhaskar Roy[1]
李兴华[1] 孙春光[1] Firdose Ahmad Malik[1] 牛艳山[1] 缪云根[1*]

摘要:昆虫蜕皮由蜕皮激素(蜕皮甾醇)调控,蜕皮激素对昆虫生长、发育和繁殖等也至关重要。蜕皮激素在蜕皮激素氧化酶(EO)作用下失活形成3-脱氢蜕皮激素(3DE),并进一步在NAD(P)H-依赖性的3DE-3α-还原酶的作用下生成3-异构蜕皮激素。另一方面,3-脱氢蜕皮激素可在血淋巴3DE-3β-还原酶下可逆性地还原为蜕皮激素。为此我们克隆并研究了3DE-3β-还原酶在家蚕不同组织不同发育阶段的表达特性。家蚕3DE-3β-还原酶基因cDNA共972 bp,其推导的蛋白质序列包含323个氨基酸。分析表明3DE-3β-还原酶属于醛酮还原酶(AKR)超家族,含有NAD(P)结合的结构域,表明3DE-3β-还原酶的功能依赖于NAD(P)H的存在。我们在大肠杆菌高量表达了3DE-3β-还原酶蛋白,融合蛋白分子质量大约为40 kDa。3DE-3β-还原酶的高转录主要出现在五龄第3天的幼虫生殖器和脂肪体中,其次是在头部、表皮和血细胞中。3DE-3β-还原酶在每龄的初期表达量较低,后期升高。当3DE的滴度较低时,更高的3DE-3β-还原酶的表达将3DE转换为蜕皮激素以维持身体蜕皮激素滴度是必须的,而当3DE滴度高时,3DE-3β-还原酶的表达则显示出反馈抑制。

1 动物疫病病原学与免疫学重点实验室,浙江大学,杭州
2 蚕桑研究所,浙江省农业科学院,杭州

Expression of open reading frames in silkworm pupal cDNA library

Zhang YZ* Chen J Nie ZM Lü ZB Wang D Jiang CY
He PA Liu LL Lou YL Song L Wu XF

Abstract: A cDNA library containing 2 409 singletons was constructed from whole silkworm pupae (*Bombyx mori*). In addition, the types of genes overexpressed in pupa were analyzed. These genes contained 79 types of proteins with the exception of enzyme, mitochondrial DNA, and ribosomal protein. We also analyzed the expression and nonexpression of open reading frame (ORF) sequences in *Escherichia coli*. cDNA sequences were compared to the silkworm (*B. mori*) genome in the GenBank database and the silkworm cDNA database including the SilkBase and KAIKOBLAST databases and 498 novel expressed sequence tags (ESTs) and 217 unknown ESTs were found. After comparison with all available ORF-complete mRNA sequences from the same organism (fruitfly, mosquito, and apis) in the RefSeq collection, 1 659 full-length cDNA were identified. In addition, the structure of silkworm mRNA was analyzed, and it was found that 66.8% of silkworm mRNA tailed with poly(A) contained the highly conserved AAUAAA signal and the signal located 10-17 nucleotides upstream of the putative poly(A). Finally, the composition of nucleotides in promoter region for all ESTs was surveyed. The results imply that the TTTTA box may possess some functions in regulating transcription and expression of some genes.

Published On: Applied Biochemistry and Biotechnology, 2007,136(3), 327-343.

College of Life Sciences, Zhejiang Sci-Tech University, Hangzhou 310018, China.
*Corresponding author E-mail: yaozhou@zist.edu.cn.

蚕蛹cDNA文库中开放阅读框的表达

张耀洲* 陈 健 聂作明 吕正兵 王 丹 蒋彩英
贺平安 刘立丽 楼玉兰 宋 力 吴祥甫

摘要：在整个家蚕（*Bombyx mori*）蛹中构建了含有2 409条序列的cDNA文库。此外，我们还分析了在蛹中过量表达的基因类型。除了酶，线粒体DNA和核糖体蛋白外，这些基因还编码79种蛋白质。我们还分析了开放阅读框（ORF）序列在大肠杆菌中的表达情况。将cDNA序列与GenBank数据库中的家蚕（*B.mori*）基因组、SilkBase和KAIKOBLAST数据库中的cDNA序列进行比对，发现498个新的表达序列标签（EST）和217个未知的EST。在与RefSeq数据库中相似生物（果蝇、蚊子和蚜虫）的所有可用ORF完整的mRNA序列比对后，鉴定出含1 659个核苷酸的全长cDNA。另外，我们还分析了家蚕mRNA的结构，发现66.8%的家蚕mRNA尾部含有poly（A）及其高度保守的AAUAAA信号，该信号位于poly（A）上游10—17个核苷酸的位置。最后，对所有EST的启动子区域中核苷酸的组成进行了分析。结果表明，TTTTA盒可能具有调节某些基因转录和表达的功能。

生命科学学院，浙江理工大学，杭州

Spatiotemporal expression profile of the *Pumilio* gene in the embryonic development of silkworm

Chen L   You ZZ   Xia HC   Tang Q
Zhou Y   Yao Q   Chen KP*

Abstract: We previously identified a *pumilio* gene in silkworm (*Bombyx mori* L.), designated as *BmPUM*, which was specifically expressed in the ovary and testis. To further characterize this gene's involvement in silkworm development, we have determined the spatiotemporal expression pattern of *BmPUM* during all embryonic stages. Real-time polymerase chain reaction (RT-PCR) analysis revealed that *BmPUM* was expressed in all stages of silkworm embryos and that its transcript levels displayed two distinct peaks. The first was observed at the germ-band formation stage (1 d after oviposition) and dropped to a low level at the gonad formation stage (5 d after oviposition). The second was detected at the stage of bristle follicle occurrence (6 d after oviposition), which was confirmed by Western blot analysis and immunohistochemistry. Nanos (Nos), functioning together with Pum in abdomen formation of *Drosophila* embryos, was also highly expressed at the beginning (0 h to 1 d after oviposition) of embryogenesis, but its transcript levels were very low after the stage of germ-band formation. These results suggest that *BmPUM* functions with *Bombyx mori* nanos (*Bm*-nanos) at the early stages of silkworm embryonic development, and may play a role in gonad formation and the occurrence of bristle follicles. Our data thus provide a foundation to uncover the role of *Bm*PUM during silkworm development.

Published On: Zeitschrift fur Naturforschung C—A Journal of Biosciences, 2014, 69(7-8), 317-324.

Institute of Life Sciences, Jiangsu University, Zhenjiang 212013, China.
*Corresponding author E-mail address: kpchen@ujs.edu.cn.

家蚕胚胎发育中*Pumilio*基因的时空表达谱

陈　亮　尤在芝　夏恒传　唐　琦
周　阳　姚　勤　陈克平*

摘要：我们前期，在家蚕中鉴定出了一个名为*BmPUM*的基因，其在卵巢和睾丸中特异性表达。为了进一步阐明该基因参与家蚕的发育，我们研究了胚胎阶段*BmPUM*的时空表达模式。RT-PCR分析显示，*BmPUM*在家蚕胚胎的各个阶段均表达，其转录水平显示出两个不同的峰。观察到第一个峰在胚芽形成阶段(产卵后1 d)，并且在性腺形成阶段(产卵后5 d)下降到低水平。通过蛋白质印迹分析和免疫组织化证实，第二次峰在胚胎毛囊发生阶段(产卵后6 d)。Nanos(Nos)与Pum共同作用于果蝇胚胎的形成，在胚胎发生的开始(产卵后0 d至1 d)高度表达，但在胚芽形成阶段后，其转录水平非常低。这些结果表明，*Bm*PUM在蚕胚胎发育的早期阶段与(*Bm*-nanos)具有相同功能，然后可能在性腺形成和毛囊发生中发挥作用。因此，我们的数据为揭示家蚕发育过程中*BmPUM*的作用提供了基础。

生命科学研究院，江苏大学，镇江

Cloning and expression characteristics of the Notch-associated gene *BmE(spl)mγ* from silkworm, *Bombyx mori*

Liu M[1#] Wang C[1#] Li D[1] Liu Y[2] Sheng Q[1]
Lü ZB[1] Yu W[1] Wang D[1] Zhang YZ[1] Nie ZM[1*]

Abstract: The *E(spl)mγ* gene in *Drosophila* is a regulatory target gene downstream of the Notch pathway. *BmE(spl)mγ* [*Bombyx mori*, *E(spl)mγ*] is an ortholog of the *Drosophila E(spl)mγ* gene, and the gene encodes a protein with 248 amino acid residues. This gene was cloned and overexpressed in *Escherichia coli* BL21 (DE3). The recombinant protein was purified and subsequently used to generate a rabbit polyclonal antibody. Western blot analyses showed that *Bm*E(spl)mγ expression is high in pupa and egg, and low in larva and moth. In the fifth instar larva, the protein levels are high in head, epidermis, sexual gland, trachea, and the fattybody and low in the Malpighian tubule, hemolymph, gut, and silk gland. The further immunohistochemical analyses also showed higher *Bm*E(spl)mγ expression in the head of fifth instar larva and pupa. Of the four moth parts studied, the thorax had the highest expression level. Thus, *BmE(spl)mγ* might be associated with neurogenesis in silkworm. Furthermore, DAPT (a γ-secretase inhibitor and an indirect inhibitor of Notch) blocking experiments showed that higher concentrations of the blocking agent and a longer processing time reduce the transcription levels of the *BmE(spl)mγ* gene, demonstrating that the silkworm *BmE(spl)mγ* gene is associated with the Notch signal pathway. These findings suggest that the function of *BmE(spl)mγ* may be similar to that of its *Drosophila* homolog.

Published On: Applied Biochemistry and Biotechnology, 2014, 173(8), 2065-2075.

1 College of Life Sciences, Zhejiang Sci-Tech University, Hangzhou 310018, China.
2 Zhejiang Economic and Trade Polytechnic, Hangzhou 310018, China.
#These authors contributed equally.
*Corresponding author E-mail: wuxinzm@zstu.edu.cn.

家蚕 Notch 相关基因 *BmE(spl)mγ* 的克隆和表达分析

刘 敏[1#] 王 婵[1#] 李 丹[1] 刘 悦[2] 盛 清[1]
吕正兵[1] 于 威[1] 王 丹[1] 张耀洲[1] 聂作明[1*]

摘要：*E*(*spl*)*mγ* 基因是果蝇中 Notch 信号通路的下游调控靶基因。*BmE*(*spl*)*mγ* 是家蚕中果蝇 *E*(*spl*)*mγ* 的同源基因，编码 248 个氨基酸残基。克隆并通过大肠杆菌 BL21(DE3)表达该蛋白质，纯化后免疫新西兰大白兔获得多克隆抗体，Western blot 结果显示，*Bm*E(spl)mγ 蛋白在家蚕蛹和卵时期表达较高，而在幼虫和成虫时期表达较低。在五龄幼虫时期，*Bm*E(spl)mγ 蛋白在头部、表皮、性腺、气管和脂肪体中的表达较高，而在马氏管、血淋巴、中肠和丝腺中的表达较低。进一步免疫组化分析也表明 *Bm*E(spl)mγ 蛋白在家蚕蛹和五龄幼虫的头部表达较高。成虫 4 个部位的表达检测表明，该蛋白在胸部的表达最高。因此，我们推测 *BmE*(*spl*)*mγ* 可能和家蚕的神经生成和发育相关。进一步采用 Notch 信号途径阻断剂 DAPT 处理家蚕 *BmN* 细胞，结果显示，随着阻断剂浓度的增加和处理时间的延长，*BmE*(*spl*)*mγ* 基因的转录水平呈现下降的趋势，推测 *BmE*(*spl*)*mγ* 基因和家蚕 Notch 信号途径相关。上述结果表明 *BmE*(*spl*)*mγ* 基因的功能可能和果蝇同源基因类似。

1 生命科学学院，浙江理工大学，杭州
2 浙江经贸职业技术学院，杭州

Homeodomain POU and Abd-A proteins regulate the transcription of pupal genes during metamorphosis of the silkworm, *Bombyx mori*

Deng HM[1] Zhang JL[1] Li Y[1] Zheng SC[1] Liu L[1]
Huang LH[1] Xu WH[2] Subba R. Palli[3] Feng QL[1*]

Abstract: A cascade of 20-hydroxyecdysone-mediated gene expression and repression initiates larva-to-pupa metamorphosis. We recently showed that two transcription factors, *Bm*POUM2 and *Bm*βFTZ-F1,bind to the *cis*-regulatory elements in the promoter of the gene coding for cuticle protein, *Bm*WCP4, and regulate its expression during *Bombyx mori* metamorphosis. Here we show that downregulation of *BmPOUM2* expression by RNA interference during the wandering stage resulted in failure to complete metamorphosis. The thorax epidermis of RNA interference-treated larvae became transparent, wing disc growth and differentiation were arrested, and the larvae failed to spin cocoons. Quantitative real-time PCR analysis showed that expression of the genes coding for pupal-specific wing cuticle proteins *Bm*WCP1, *Bm*WCP2, *Bm*WCP3, *Bm*WCP4, *Bm*WCP5, *Bm*WCP6, *Bm*WCP8, and *Bm*WCP9 were down-regulated in *BmPOUM2* dsRNA-treated animals, whereas overexpression of *Bm*POUM2 protein increased the expression of *BmWCP4*, *BmWCP5*, *BmWCP6*, *BmWCP7*, and *BmWCP8*. Pull-down assays, far-Western blot, and electrophoretic mobility shift assay showed that the *Bm*POUM2 protein interacted with another homeodomain transcription factor, *Bm*Abd-A, to induce the expression of *BmWCP4*. Immunohistochemical localization of *Bm*POUM2, *Bm*Abd-A, and *Bm*WCP4 proteins revealed that *Bm*Abd-A and *Bm*POUM2 proteins are colocalized in the wing disc cell nuclei, whereas *Bm*WCP4 protein is localized in the cytoplasm. Together these data suggest that *Bm*POUM2 interacts with the homeodomain transcription factor *Bm*Abd-A and regulates the expression of *BmWCP4* and probably other *BmWCPs* to complete the larva-to-pupa transformation. Although homeodomain proteins are known to regulate embryonic development, this study showed that these proteins also regulate metamorphosis.

Published On: Proceedings of the National Academy of Sciences of the United states of America, 2012, 109(31), 12598-12603.

1 Guangdong Provincial Key Laboratory of Biotechnology for Plant Development, School of Life Sciences, South China Normal University, Guangzhou 510631, China.
2 State Key Laboratory of Biocontrol, School of Life Sciences, Sun-Yat Sen University, Guangzhou 510275, China.
3 Department of Entomology, University of Kentucky, Lexington, KY 40546.
*Corresponding author E-mail: qlfeng@scnu.edu.cn.

同源域类转录因子POU和Abd-A蛋白调控家蚕变态发育过程中蛹期基因的转录

邓惠敏[1] 章佳玲[1] 李 勇[1] 郑思春[1] 刘 琳[1]
黄立华[1] 徐卫华[2] Subba R. Palli[3] 冯启理[1*]

摘要：一系列20–羟基蜕皮酮(20E)介导的基因表达和抑制启动了昆虫幼虫向蛹的变态发育。最近我们发现转录因子*Bm*POUM2和*Bm*βFTZ-F1与家蚕表皮蛋白编码基因*BmWCP4*的启动子中的顺式调控元件结合，并在家蚕变态过程中调控其表达。本文我们利用RNA干扰使*BmPOUM2*在家蚕上蔟期下调表达，导致家蚕不能完成变态发育。RNA干扰处理的幼虫的胸节表皮呈透明状，翅原基的生长分化受到抑制，并且幼虫不能正常结茧。qRT-PCR的结果显示，抑制*BmPOUM2*的表达可下调编码蛹期特异的翅原基表皮蛋白*Bm*WCP1、*Bm*WCP2、*Bm*WCP3、*Bm*WCP4、*Bm*WCP5、*Bm*WCP6、*Bm*WCP8及*Bm*WCP9的基因表达，而过表达*Bm*POUM2蛋白可增强*BmWCP4*、*BmWCP5*、*BmWCP6*、*BmWCP7*及*BmWCP8*的启动子活性。Pull-down实验、far-Western blot和EMSA的结果表明*Bm*POUM2与另一同源域转录因子*Bm*Abd-A相互作用可诱导*BmWCP4*的表达。免疫组化分析显示，*Bm*POUM2和*Bm*Abd-A共定位于翅原基细胞的细胞核中，而*Bm*WCP4定位于细胞质中。总之，这些数据表明，*Bm*POUM2与另一同源域转录因子*Bm*Abd-A互作并调控*BmWCP4*基因的表达，同时也可能参与其他翅原基表皮蛋白基因的表达，从而导致家蚕完成幼虫向蛹的转变。之前的研究表明同源域转录因子参与调控胚胎发育，在本文中我们发现它们同样参与调控昆虫的变态发育。

1 广东省植物发育生物工程重点实验室，生命科学学院，华南师范大学，广州
2 生物防治国家重点实验室，生命科学学院，中山大学，广州
3 昆虫学系，肯塔基大学，列克星敦，美国

Autophagy and its physiological relevance in arthropods: Current knowledge and perspectives

Davide Malagoli[1*] Fabio C. Abdalla[2] Cao Y[3] Feng QL[4] Kozo Fujisaki[5] Ales Gregorc[6] Tomohide Matsuo[7] Ioannis P. Nezis[8] Issidora S. Papassideri[9] Miklós Sass[10] Elaine C.M. Silva-Zacarin[2] Gianluca Tettamanti[11] Rika Umemiya-Shirafuji[5]

Abstract: The autophagic process is one of the best examples of a conserved mechanism of survival in eukaryotes. At the molecular level there are impressive similarities between unicellular and multicellular organisms, but there is increasing evidence that the same process may be used for different ends, i.e., survival or death, at least at the cellular level. Arthropods encompass a wide variety of invertebrates such as insects, crustaceans and spiders, and thus represent the taxon in which most of the investigations on autophagy in nonmammalian models are performed. The present review is focused on the genetic basis and the physiological meaning of the autophagic process in key models of arthropods. The involvement of autophagy in programmed cell death, especially during oogenesis and development, is also discussed.

Published On: Autophagy, 2010, 6(5), 575-588.

1 Department of Biology, University of Modena and Reggio Emilia, Modena 41125, Italy.

2 Laboratory of Structural and Functional Biology (LABEF), Federal University of São Carlos (UFSCar), Campus Sorocaba, São Paulo 18052-780, Brazil.

3 Department of Sericulture Science, College of Animal Science, South China Agricultural University, Guangzhou 510642, China.

4 School of Life Sciences, South China Normal University, Guangzhou 510631, China.

5 Laboratory of Emerging Infectious Diseases, Department of Frontier Veterinary Medicine, Kagoshima University, Kagoshima 890-0065, Japan.

6 Agricultural Institute of Slovenia, Ljubljana 1000, Slovenia.

7 Department of Infectious Diseases, Kyorin University School of Medicine, Tokyo 181-8611, Japan.

8 Centre for Cancer Biomedicine, University of Oslo and Institute for Cancer Research, Department of Biochemistry, The Norwegian Radium Hospital, Montebello N-0310, Oslo Norway.

9 Faculty of Biology, Department of Cell Biology and Biophysics, University of Athens, Panepistimiopolis 15701, Athens Greece.

10 Department of General Zoology, Eötvös Loránd University, Budapest 1117, Hungary.

11 Department of Biotechnology and Molecular Sciences, University of Insubria, Varese 21100, Italy.

*Corresponding author E-mail: davide.malagoli@unimore.it.

关于细胞自噬及其同节肢动物生理关联性的当前认知和观点

Davide Malagoli[1*] Fabio C. Abdalla[2] 曹 阳[3] 冯启理[4] Kozo Fujisaki[5]
Ales Gregorc[6] Tomohide Matsuo[7] Ioannis P.Nezis[8] Issidora S. Papassideri[9]
Miklós Sass[10] Elaine C.M. Silva-Zacarin[2] Gianluca Tettamanti[11] Rika Umemiya-Shirafuji[5]

摘要：自噬过程是真核生物保守机制的最佳实例之一。在分子水平上，单细胞和多细胞生物体之间具有令人印象深刻的相似之处，但是越来越多的证据表明，相同的过程可产生不同的结果，即存活或死亡。至少在细胞水平上是这样的。节肢动物包括各种各样的无脊椎动物，如昆虫、甲壳类动物和蜘蛛，代表了在非哺乳动物模型中进行大部分自噬调查的分类群。本综述主要讨论了节肢动物关键模型上的自噬过程的遗传基础和生理学意义，还讨论了自噬参与细胞程序性死亡，特别是在卵子发生和发育过程中的作用。

1 生物系，摩德纳雷焦艾米利亚大学，摩德纳，意大利
2 结构与功能生物学实验室，圣卡洛斯联邦大学，圣保罗，巴西
3 蚕丝科学系，动物科学学院，华南农业大学，广州
4 生命科学学院，华南师范大学，广州
5 新兴传染病实验室，国立兽医学系，鹿儿岛大学，鹿儿岛，日本
6 斯洛文尼亚农业研究所，卢布尔雅那，斯洛文尼亚
7 传染病学系，杏林大学医学院，东京，日本
8 癌症生物医药中心，奥斯陆大学癌症研究所，生物化学系，挪威镭锭医院，蒙特贝罗，奥斯陆，挪威
9 生物学系，细胞生物学和生物物理学系，雅典大学，雅典，希腊
10 普通动物学系，罗兰大学，布达佩斯，匈牙利
11 生物技术与分子科学系，英苏布里亚大学，瓦雷泽，意大利

20-hydroxyecdysone upregulates *Atg* genes to induce autophagy in the *Bombyx* fat body

Tian L[1,2] Ma L[1] Guo EE[1,3] Deng XJ[3]
Ma SY[2] Xia QY[2] Cao Y[3] Li S[1*]

Abstract: Autophagy is finely regulated at multiple levels and plays crucial roles in development and disease. In the fat body of the silkworm, *Bombyx mori*, autophagy occurs and *Atg* gene expression peaks during the nonfeeding molting and pupation stages when the steroid hormone (20-hydroxyecdysone; 20E) is high. Injection of 20E into the feeding larvae upregulated *Atg* genes and reduced TORC1 activity resulting in autophagy induction in the fat body. Conversely, RNAi knockdown of the 20E receptor partner (*USP*) or targeted overexpression of a dominant negative mutant of the 20E receptor (EcR^{DN}) in the larval fat body reduced autophagy and downregulated the *Atg* genes, confirming the importance of 20E-induction of *Atg* gene expression during pupation. Moreover, *in vitro* treatments of the larval fat body with 20E upregulated the *Atg* genes. Five *Atg* genes were potentially 20E primary-responsive, and a 20E response element was identified in the *Atg1* (ortholog of human *ULK1*) promoter region. Furthermore, RNAi knockdown of 4 key genes (namely *Br-C*, *E74*, *HR3* and *βftz-F1*) in the 20E-triggered transcriptional cascade reduced autophagy and downregulated *Atg* genes to different levels. Taken together, we conclude that in addition to blocking TORC1 activity for autophagosome initiation, 20E upregulates *Atg* genes to induce autophagy in the *Bombyx* fat body.

Published On: Autophagy, 2013, 9(8), 1172-1187.

1 Key Laboratory of Insect Developmental and Evolutionary Biology, Institute of Plant Physiology and Ecology, Shanghai Institutes for Biological Sciences, Chinese Academy of Sciences, Shanghai 210037, China.

2 State Key Laboratory of Silkworm Genome Biology, Southwest University, Chongqing 400716, China.

3 Laboratory of Insect Molecular Biology and Biotechnology, Guangdong Provincial Key Laboratory of Agro-animal Genomics and Molecular Breeding, College of Animal Sciences, South China Agricultural University, Guangzhou 510642, China.

*Corresponding author E-mail: lisheng01@sibs.ac.cn.

20-羟基蜕皮酮通过上调家蚕脂肪体中自噬相关基因*Atg*的表达诱导自噬的发生

田 铃[1,2] 马 俐[1] 郭恩恩[1,3] 邓小娟[3]
马三垣[2] 夏庆友[2] 曹 阳[3] 李 胜[1*]

摘要：细胞自噬在多个水平被精密调控，并在个体发育和疾病发生中具有重要作用。在家蚕脂肪体中，自噬发生和Atg基因的表达高峰处在蜕皮激素（20–羟基蜕皮酮，20E）水平高的非摄食的蜕皮和化蛹时期。在家蚕幼虫摄食时期，注射20E上调*Atg*基因的表达，并抑制TOR信号的活性从而诱导脂肪体细胞自噬的发生。相反地，通过RNA干扰20E受体*USP*或是通过转基因在家蚕脂肪体特异性超表达*EcR*的显性失活形式（EcR^{DN}）可抑制上蔟时期的自噬和*Atg*基因的表达，证实了20E信号诱导的*Atg*基因表达在家蚕蛹期的重要性。此外，用20E体外处理家蚕幼虫脂肪体，证明了其能诱导上调*Atg*基因。5个*Atg*基因具有20E初级应答基因的表达模式，在*Atg1*（人类*ULK1*的同源基因）的启动子区域还鉴定到一个20E的响应元件。此外，RNA干扰20E信号触发的转录级联中的4个关键基因（*Br-C*、*E74*、*HR3*和*βftz-F1*）后，不同程度地抑制细胞自噬和*Atg*基因的表达。总的来说，在家蚕脂肪体中，20E除了能抑制TORC1活性从而启动自噬外，还能上调*Atg*基因的表达诱导自噬。

1 昆虫发育与进化生物学重点实验室，上海植物生理生态研究所，上海生命科学研究院，中国科学院，上海
2 家蚕基因组生物学国家重点实验室，西南大学，重庆
3 昆虫分子生物学和生物技术实验室，广东省农业动物基因组学与分子育种重点实验室，动物科学学院，华南农业大学，广州

*Bm*ATG5 and *Bm*ATG6 mediate apoptosis following autophagy induced by 20-hydroxyecdysone or starvation

Xie K[1,2#] Tian L[1,3#] Guo XY[1] Li K[1,3] Li JP[1] Deng XJ[1] Li QR[4] Xia QY[5]
Zhong YJ[1] Huang ZJ[1] Liu JP[1] Li S[3*] Yang WY[1*] Cao Y[1*]

Abstract: Autophagy and apoptosis, which could be induced by common stimuli, play crucial roles in development and disease. The functional relationship between autophagy and apoptosis is complex due to the dual effects of autophagy. In the *Bombyx Bm-12* cells, 20-hydroxyecdysone (20E) treatment or starvation induced cell death with autophagy preceding apoptosis. In response to 20E or starvation, *Bm*ATG8 was rapidly cleaved and conjugated with PE to form *Bm*ATG8-PE; subsequently, *Bm*ATG5 and *Bm*ATG6 were cleaved into *Bm*ATG5-tN and *Bm*ATG6-C, respectively. Reduced of expression of *BmAtg5* or *BmAtg6* by RNAi decreased the proportion of cells undergoing both autophagy and apoptosis after 20E treatment or starvation. Overexpression of *BmAtg5* or *BmAtg6* induced autophagy but not apoptosis in the absence of the stimuli, but promoted both autophagy and apoptosis induced by 20E or starvation. Notably, overexpression of cleavage site-deleted *BmAtg5* or *BmAtg6* increased autophagy but not apoptosis induced by 20E or starvation, whereas overexpression of *BmAtg5-tN* and *BmAtg6-C* was able to directly trigger apoptosis or promote the induced apoptosis. In conclusion, being cleaved into *Bm*ATG5-tN and *Bm*ATG6-C, *Bm*ATG5 and *Bm*ATG6 mediate apoptosis following autophagy induced by 20E or starvation in *Bombyx Bm-12* cells, reflecting that autophagy precedes apoptosis in the midgut during *Bombyx* metamorphosis.

Published On: Autophagy, 2016, 12(2), 381-396.

1 Guangdong Provincial Key Laboratory of Agro-animal Genomics and Molecular Breeding, Guangdong Provincial Sericulture and Mulberry Engineering Research Center, College of Animal Science, South China Agricultural University, Guangzhou 510642, China.

2 Key Laboratory of Crops with High Quality and Efficient Cultivation and Security Control, Yunnan Higher Education Institutions, College of Life Science and Technology, HongHe University, Mengzi 661199, China.

3 Key Laboratory of Insect Developmental and Evolutionary Biology, Institute of Plant Physiology and Ecology, Shanghai Institutes for Biological Sciences, Chinese Academy of Sciences, Shanghai 200032, China.

4 The Sericultural and Agri-Food Research Institute of the Guangdong Academy of Agricultural Sciences, Guangzhou 510610, China.

5 State Key Laboratory of Silkworm Genome Biology; Southwest University, Chongqing 400716, China.

[#]These authors contributed equally.

[*]Corresponding author E-mail: lisheng01@sibs.ac.cn; yywy@scau.edu.cn; caoyang@scau.edu.cn.

家蚕*Bm*ATG5、*Bm*ATG6蛋白介导蜕皮激素或饥饿诱导的细胞自噬后的细胞凋亡

谢　昆[1,2#]　田　铃[1,3#]　郭新雨[1]　李　康[1,3]　李建平[1]　邓小娟[1]　李庆荣[4]　夏庆友[5]
钟仰进[1]　黄志君[1]　刘吉平[1]　李　胜[3*]　杨婉莹[1*]　曹　阳[1*]

摘要：细胞自噬和细胞凋亡可以响应某些共同的上游信号，并在个体发育和疾病发生中具有重要作用。由于细胞自噬的双重作用，使得其与凋亡之间的功能互作非常复杂。研究发现，在家蚕中蜕皮激素（20E）或是饥饿处理*Bm-12*细胞，都可以诱导细胞自噬先于凋亡启动的细胞死亡。为了响应20E或是饥饿信号，*Bm*ATG8快速裂解后与PE结合形成*Bm*ATG8-PE；随后*Bm*ATG5和*Bm*ATG6蛋白分别被酶切，分别形成*Bm*ATG5-tN和*Bm*ATG6-C。RNA干扰*BmAtg5*或*BmAtg6*降低它们的表达，导致20E或是饥饿信号诱导发生的自噬和凋亡的细胞比例下降；在没有上游信号存在的情况下，过表达*BmAtg5*或*BmAtg6*基因只能诱导自噬，但在上游信号存在的情况下可以同时促进自噬和凋亡。值得注意的是，过表达*BmAtg5*或*BmAtg6*的酶切位点突变形式，只促进20E或饥饿诱导的细胞自噬而非细胞凋亡，而过表达*BmAtg5*-tN和*BmAtg6-C*，则可以直接触发并促进20E或饥饿诱导的凋亡。综上所述，*Bm*ATG5和*Bm*ATG6蛋白被酶切后形成*Bm*ATG5-tN和*Bm*ATG6-C形式，介导了20E或是饥饿诱导的家蚕*Bm*-12细胞中自噬后的凋亡，阐释了在家蚕变态发育时期中肠中自噬先于凋亡发生的分子机制。

1　广东省农业动物基因组学与分子育种重点实验室，广东省蚕桑工程技术研究中心，动物科学学院，华南农业大学，广州
2　云南省高校农作物优质高效栽培与安全控制重点实验室，云南高等院校生命科学与技术学院，红河学院，蒙自
3　昆虫发育与进化生物学重点实验室，植物生理生态研究所，中国科学院，上海
4　广东省农业科学院蚕业与农产品加工研究所，广州
5　家蚕基因组生物学国家重点实验室，西南大学，重庆

Functional analysis of insect molting fluid proteins on the protection and regulation of ecdysis

Zhang J[1] Lu AR[1] Kong LL[2] Zhang QL[1] Ling EJ[1*]

Abstract: Molting fluid accumulates between the old and new cuticles during periodical ecdysis in Ecdysoza. Natural defects in insect ecdysis are frequently associated with melanization (an immunity response) occurring primarily in molting fluids, suggesting that molting fluid may impact immunity as well as affect ecdysis. To address this hypothesis, proteomic analysis of molting fluids from *Bombyx mori* during three different types of ecdysis was performed. Many proteins were newly identified, including immunity-related proteins, in each molting fluid. Molting fluids inhibited the growth of bacteria *in vitro*. The entomopathogenic fungi *Beauveria bassiana*, which can escape immune responses in feeding larvae, is quickly recognized by larvae during ecdysis, followed by melanization in molting fluid and old cuticle. Fungal conidia germination was delayed, and no hyphae were detected in the hemocoels of pharate instar insects. Molting fluids protect the delicate pharate instar insects with extremely thin cuticles against microorganisms. To explore the function of molting fluids in ecdysis regulation, based on protein similarity, 32 genes were selected for analysis in ecdysis regulation through RNAi in *Tribolium castaneum*, a model commonly used to study integument development because RNAi is difficult to achieve in *Bombyx mori*. We identified 24 molting proteins that affected ecdysis after knock down, with different physiological functions, including old cuticle protein recycling, molting fluid pressure balance, detoxification, and signal detection and transfer of molting fluids. We report that insects secrete molting fluid for protection and regulation of ecdysis, which indicates a way to develop new pesticides through interrupting insect ecdysis in the future.

Published On: Journal of Biological Chemistry, 2014, 289(52), 35891-35906.

1 Key Laboratory of Insect Developmental and Evolutionary Biology, Shanghai Institute of Plant Physiology and Ecology, Shanghai Institutes for Biological Sciences, Chinese Academy of Sciences, Shanghai 200032, China.

2 College of Agriculture and Biology, Shanghai Jiaotong University, Shanghai 200240, China.

*Corresponding author E-mail: erjunling@sippe.ac.cn.

昆虫蜕皮液蛋白在保护和调控蜕皮时的功能分析

张 洁[1] 路岸瑞[1] 孔璐璐[2] 张巧利[1] 凌尔军[1*]

摘要：蜕皮液在蜕皮周期中聚集在新旧表皮之间。昆虫蜕皮的天然缺陷通常与黑化作用(一种免疫反应)相关,这种黑化作用主要发生在蜕皮液中,暗示蜕皮液可能会影响免疫功能以及蜕皮。为了回答这一假说,我们对家蚕的3种不同类型的蜕皮液进行了蛋白质组学分析。在每种蜕皮液中,许多新的蛋白质得到了鉴定,其中包括免疫相关蛋白。蜕皮液在体外能够抑制细菌的生长。昆虫病原真菌球孢白僵菌(*Beauveria bassiana*)可逃过食桑期幼虫的免疫反应,但却会被蜕皮时期的幼虫迅速识别,并触发蜕皮液与旧表皮中的黑化反应。真菌分生孢子萌发延迟,在蜕皮时的昆虫幼虫体腔中检测不到菌丝的生成。蜕皮液在极薄的表皮中保护脆弱的正在脱皮的昆虫免于微生物的入侵。为了探讨蜕皮液在蜕皮时的调节功能,基于蛋白质的相似性,32个基因被选定在赤拟谷盗中通过RNAi的方法进行蜕皮调节的分析。赤拟谷盗是一种普遍用于表皮发育研究的模型,而家蚕很难实现RNAi。我们鉴定了24种在敲除后影响蜕皮的蛋白,这些蛋白具有不同的生理功能,包括旧角质蛋白回收、蜕皮液压力平衡、解毒以及蜕皮液的信号检测和转换。我们揭示了昆虫分泌蜕皮液保护和调控蜕皮的机制,提示在未来可以通过阻断昆虫蜕皮的途径开发新型杀虫剂。

1 昆虫发育与进化生物学重点实验室,上海植物生理生态研究所,上海生命科学研究院,中国科学院,上海

2 农业与生物学院,上海交通大学,上海

The homeodomain transcription factors Antennapedia and POU-M2 regulate the transcription of the steroidogenic enzyme gene *Phantom* in the silkworm

Meng M[1,2] Cheng DJ[1,2*] Peng J[1,2] Qian WL[1,2]
Li JR[1,2] Dai DD[1,2] Zhang TL[1,2] Xia QY[1,2*]

Abstract: The steroid hormone ecdysone, which controls insect molting and metamorphosis, is synthesized in the prothoracic gland (PG), and several steroidogenic enzymes that are expressed specifically in the PG are involved in ecdysteroidogenesis. In this study, we identified new regulators that are involved in the transcriptional control of the silkworm steroidogenic enzyme genes. *In silico* analysis predicted several potential *cis*-regulatory elements (CREs) for the homeodomain transcription factors Antennapedia (*Antp*) and *POU-M2* in the proximal promoters of steroidogenic enzyme genes. *Antp* and *POU-M2* are expressed dynamically in the PG during larval development, and their overexpression in silkworm embryo-derived (*BmE*) cells induced the expression of steroidogenic enzyme genes. Importantly, luciferase reporter analyses, electrophoretic mobility shift assays, and chromatin immunoprecipitation assays revealed that Antp and POU-M2 promote the transcription of the silkworm steroidogenic enzyme gene *Phantom* (*Phm*) by binding directly to specific motifs within overlapping CREs in the *Phm* promoter. Mutations of these CREs in the *Phm* promoter suppressed the transcriptional activities of both *Antp* and *POU-M2* in *BmE* cells and decreased the activities of mutated *Phm* promoters in the silkworm PG. In addition, pull-down and co-immunoprecipitation assays demonstrated that Antp can interact with POU-M2. Moreover, RNA interference-mediated down-regulation of either *Antp* or *POU–M2* during silkworm wandering not only decreased the ecdysone titer but also led to the failure of metamorphosis. In summary, our results suggest that *Antp* and *POU-M2* coordinate the transcription of the silkworm *Phm* gene directly, indicating new roles for homeodomain proteins in regulating insect ecdysteroidogenesis.

Published On:Journal of Biological Chemistry, 2015, 290(40), 24438-24452.

1 State Key Laboratory of Silkworm Genome Biology, Southwest University, Chongqing 400716, China.

2 Key Sericultural Laboratory of the Ministry of Agriculture, College of Biotechnology, Southwest University, Chongqing 400716, China.

*Corresponding author E-mail: chengdj@swu.edu.cn; xiaqy@swu.edu.cn.

家蚕同源结构域转录因子Antennapedia(Antp)和POU-M2调控蜕皮激素合成酶基因*Phantom*的转录

孟 勐[1,2] 程道军[1,2*] 彭 健[1,2] 钱文良[1,2]
李加瑞[1,2] 代丹丹[1,2] 张天镭[1,2] 夏庆友[1,2*]

摘要：前胸腺(PG)中合成的蜕皮激素控制着昆虫的蜕皮和变态发育，在前胸腺中特异地表达了一系列的蜕皮激素合成酶参与了蜕皮过程。本研究鉴定了参与家蚕蜕皮激素合成酶基因转录调控的新调控因子。在线预测到在蜕皮激素合成酶基因的近端启动子区域存在许多潜在的同源结构域转录因子Antennapedia(Antp)和POU-M2的顺式调控元件(CRE)。*Antp*和*POU-M2*在幼虫发育过程中的前胸腺动态表达，并且在家蚕胚胎细胞(*BmE*)中过表达能诱导蜕皮激素合成酶基因的表达。重要的是荧光素酶报告实验、凝胶迁移实验和染色质免疫共沉淀实验表明，Antp和POU-M2通过直接与家蚕蜕皮激素合成酶基因*Phantom*(*Phm*)启动子CRE区域的特异基序结合促进其转录。突变*Phm*启动子的CRE区域可以抑制*BmE*细胞中Antp和POU-M2的转录活性，并且能够降低家蚕前胸腺中*Phm*基因的启动子活性。此外，Pull-down和免疫共沉淀实验证明了Antp可以与POU-M2相互作用。在家蚕上蔟期，利用RNAi下调*Antp*或*POU-M2*的表达，不仅能降低蜕皮激素滴度，还会导致家蚕变态发育受阻。总之，该研究结果表明，Antp和POU-M2能够直接协同调控家蚕*Phm*基因的转录，暗示了同源结构域蛋白在调控昆虫蜕皮发育中发挥了新的作用。

1 家蚕基因组生物学国家重点实验室，西南大学，重庆
2 农业部蚕桑学重点实验室，生物技术学院，西南大学，重庆

20-hydroxyecdysone (20E) primary response gene *E93* modulates 20E signaling to promote *Bombyx* larval–pupal metamorphosis

Liu X[1,2] Dai FY[2] Guo EE[1,3] Li K[1,3] Ma L[1] Tian L[1]
Cao Y[3] Zhang GZ[4] Subba R. Palli[5] Li S[1*]

Abstract: As revealed in a previous microarray study to identify genes regulated by 20-hydroxyecdysone (20E) and juvenile hormone (JH) in the silkworm, *Bombyx mori*, *E93* expression in the fat body was markedly low prior to the wandering stage but abundant during larval-pupal metamorphosis. Induced by 20E and suppressed by JH, *E93* expression follows this developmental profile in multiple silkworm alleles. The reduction of *E93* expression by RNAi disrupted 20E signaling and the 20E-induced autophagy, caspase activity and cell dissociation in the fat body. Reducing *E93* expression also decreased the expression of the 20E-induced pupal-specific cuticle protein genes and prevented growth and differentiation of the wing discs. Importantly, the two HTH domains in E93 are critical for inducing the expression of a subset of 20E response genes including *EcR*, *USP*, *E74*, *Br–C*, and *Atg1*. By contrast, the LLQHLL and PLDLSAK motifs in *E93* inhibit its transcriptional activity. E93 binds to the EcR-USP complex via a physical association with USP through its LLQHLL motif; and this association is enhanced by 20E-induced EcR-USP interaction, which attenuates the transcriptional activity of *E93*. E93 acts through the two HTH domains to bind to GAGA-containing motifs present in the *Atg1* promoter region for inducing gene expression. In conclusion, E93 transcriptionally modulates 20E signaling to promote *Bombyx* larval-pupal metamorphosis.

Published On: Journal of Biological Chemistry, 2015, 290(45), 27370-27383.

1 Key Laboratory of Insect Developmental and Evolutionary Biology, Institute of Plant Physiology and Ecology, Chinese Academy of Sciences, Shanghai 200032, China.

2 State Key Laboratory of Silkworm Genome Biology and College of Biotechnology, Southwest University, Chongqing 400716, China.

3 Laboratory of Insect Molecular Biology and Biotechnology, Guangdong Provincial Key Laboratory of Agro-animal Genomics and Molecular Breeding, College of Animal Sciences, South China Agricultural University, Guangzhou 510642, China.

4 Sericultural Research Institute, Chinese Academy of Agricultural Sciences, Zhenjiang 212018, China.

5 Department of Entomology, College of Agriculture, University of Kentucky, Lexington 40546, Kentucky.

*Corresponding author E-mail: lisheng01@sibs.ac.cn.

20E的初级应答基因*E93*通过调节20E信号参与调控家蚕幼虫-蛹变态发育

刘 曦[1,2] 代方银[2] 郭恩恩[1,3] 李 康[1,3] 马 俐[1] 田 铃[1]
曹 阳[3] 张国政[4] Subba R. Palli[5] 李胜[1*]

摘要：先前筛选受蜕皮激素(20E)和保幼激素(JH)调控基因的芯片结果发现，家蚕(*Bombyx mori*)脂肪体中*E93*的表达在上蔟期之前都是处于非常低的水平，而在幼虫-蛹变态时期则具有非常高的水平。在20E的诱导和JH的抑制下，*E93*在不同品系的家蚕中的表达遵循了这种规律。*E93* RNAi阻断了家蚕脂肪体内20E信号、20E诱导的自噬、含半胱氨酸的天冬氨酸蛋白水解酶活性以及细胞解离。*E93*下调后降低20E诱导的蛹期特异的表皮蛋白基因的表达、阻止翅原基的生长与分化，E93的两个HTH结构域对于20E应答基因*EcR*、*USP*、*E74*、*Br-C*、*Atg1*的诱导表达至关重要。相反地，E93的LLQHLL和PLDLSAK两个基序则抑制其自身的转录活性。E93通过LLQHLL基序与EcR-USP复合体中的USP结合，这种结合受到20E诱导的EcR-USP互作的提升，进而削弱*E93*的转录活性。E93通过两个HTH结构域与*Atg1*启动子区域存在的GAGA基序结合诱导其基因表达。总之，E93在转录水平调节20E信号促进家蚕幼虫-蛹变态发育。

1 昆虫发育与进化生物学重点实验室，上海植物生理生态研究所，中国科学院，上海
2 家蚕基因组生物学国家重点实验室与生物技术学院，西南大学，重庆
3 昆虫分子生物学与生物技术实验室，广东省农业动物基因组学与分子育种重点实验室，动物科学学院，华南农业大学，广州
4 蚕业研究所，中国农业科学院，镇江
5 昆虫系，农学院，肯塔基大学，列克星敦市，美国

20-Hydroxyecdysone (20E) primary response gene *E75* isoforms mediate steroidogenesis autoregulation and regulate developmental timing in *Bombyx*

Li K[1,2] Tian L[2,3] Guo ZJ[4] Guo SY[3] Zhang JZ[5]
Gu SH[6] Subba Reddy Palli[7] Cao Y[3] Li S[1,2*]

Abstract: The temporal control mechanisms that precisely control animal development remain largely elusive. The timing of major developmental transitions in insects, including molting and metamorphosis, is coordinated by the steroid hormone 20-hydroxyecdysone (20E). 20E involves feedback loops to maintain pulses of ecdysteroid biosynthesis leading to its upsurge, whereas the underpinning molecular mechanisms are not well understood. Using the silkworm *Bombyx mori* as a model, we demonstrated that *E75*, the 20E primary response gene, mediates a regulatory loop between ecdysteroid biosynthesis and 20E signaling. E75 isoforms A and C directly bind to retinoic acid receptor-related response elements in *Halloween* gene promoter regions to induce gene expression and promote ecdysteroid biosynthesis and developmental transition, whereas isoform B antagonizes the transcriptional activity of isoform A/C through physical interaction. As the expression of E75 isoforms are differentially induced by 20E, the E75-mediated regulatory loop represents a fine autoregulation of steroidogenesis, which contributes to the precise control of developmental timing.

Published On: Journal of Biological Chemistry, 2016, 291(35), 18163-18175.

1 Guangzhou Key Laboratory of Insect Development Regulation and Application Research, Institute of Insect Sciences and School of Life Sciences, South China Normal University, Guangzhou 510631, China.
2 Key Laboratory of Insect Developmental and Evolutionary Biology, Institute of Plant Physiology and Ecology, Chinese Academy of Sciences, Shanghai 200032, China.
3 Laboratory of Insect Molecular Biology and Biotechnology, Guangdong Provincial Key Laboratory of Agro-animal Genomics and Molecular Breeding, College of Animal Sciences, South China Agricultural University, Guangzhou 510642, China.
4 Institute of Life Sciences, Jiangsu University, Zhenjiang 212013, China.
5 Research Institute of Applied Biology, College of Life Science, Shanxi University, Taiyuan, Shanxi, 030006, China.
6 Department of Biology, National Museum of Natural Science, 1 Kuan-Chien Road, Taichung 404, Taiwan.
7 Department of Entomology, College of Agriculture, University of Kentucky, Lexington, Kentucky 40546, America.
*Corresponding author E-mail: lisheng01@sibs.ac.cn.

家蚕中20E的初级应答基因*E75*三个异构体调控20E合成的自我调节和发育时序的机制

李 康[1,2] 田 铃[2,3] 郭忠建[4] 郭三友[3] 张建珍[5]
顾世红[6] Subba Reddy Palli[7] 曹 阳[3] 李 胜[1,2*]

摘要：精确控制动物发育的时间控制机制是非常复杂的。昆虫最主要的发育转变包括蜕皮和变态，而它们发生的过程受到蜕皮激素（20E）的调控。20E反馈循环调控蜕皮酮的生物合成导致其含量的大幅度提高，但其详细机制尚不清楚。利用家蚕作为实验模型，我们证明：作为20E的初级应答基因，*BmE75*调节20E和蜕皮酮生物合成之间的调控环路。*Bm*E75A和*Bm*E75C直接结合在*Halloween*基因启动子区域的视黄酸受体相关反应元件上以此提升蜕皮酮的合成和变态发育，而异构体*Bm*E75B则通过与*Bm*E75A和*Bm*E75C的物理结合拮抗它们的转录激活。与20E差异诱导*BmE75*异构体的表达一样，*Bm*E75调控了蜕皮酮的合成环路途径，从而精密地调控发育时钟。

1 广州市昆虫发育调控与应用研究重点实验室，生命科学学院，华南师范大学，广州
2 中国科学院昆虫发育与进化生物学重点实验室，上海植物生理生态研究所，中国科学院，上海
3 昆虫分子生物学和生物技术实验室，广东省农业动物基因组与分子育种重点实验室，动物科学学院，华南农业大学，广州
4 生命科学研究院，江苏大学，镇江
5 应用生物学研究所，生命科学学院，山西大学，太原
6 生物系，自然科学博物馆，台北
7 昆虫学部，农学院，肯塔基大学，列克星敦市，美国

*Bm*BR-C Z4 is an upstream regulatory factor of *Bm*POUM2 controlling the pupal specific expression of *BmWCP4* in the silkworm, *Bombyx mori*

Deng HM[#] Niu KK[#] Zhang JL Feng QL[*]

Abstract: 20-hydroxyecdysone (20E)-induced expression of the wing disc cuticle protein gene *BmWCP4* was mediated by the transcription factor *Bm*POUM2, which binds to the cis-response elements (CREs) of *BmWCP4* gene in *Bombyx mori*. In this study we report the regulation of *BmPOUM2*. RT-PCR analysis indicated that in response to 20E, *BmPOUM2* was expressed at higher levels in the wing discs during the pre-pupal and mid-pupal stages than other stages and the expression pattern of *BmBR-C Z1*, *BmBR-C Z2* and *BmBR-C Z4* was in tandem with the expression of *BmPOUM2*. *BmBR-C Z4* was induced by 20E in the wing discs, whereas *BmBR-C Z1* and *BmBR-C Z2* were not. Three potential BR-C Z4 *cis*-response elements (CREs) were identified in the promoter region of *BmPOUM2*. The expression of *BmPOUM2* mRNA and protein was increased by the over-expression of *BmBR-C Z4* in *BmN* cells, which acted in the promoter of *BmPOUM2*. Electrophoretic mobility shift assay (EMSA) and luciferase activity analysis under the control of wild-type and mutants of the BR-C Z4 CREs suggested that *Bm*BR-C Z4 protein bound to the predicted BRC-Z4 CRE C (— 684—— 660). Taken together, the data suggest that *BmBR–C Z4* is a direct upstream regulator of *BmPOUM2* and regulates the pupal-specific expression of *BmWCP4* through *BmPOUM2*.

Published On: Insect Biochemistry and Molecular Biology, 2015, 66, 42-50.

Laboratory of Molecular and Developmental Entomology, School of Life Sciences, South China Normal University, Guangzhou 510631, China.

[#]These authors contributed equally.

[*]Corresponding author E-mail: qlfeng@scnu.edu.cn.

家蚕转录因子*Bm*BR-C Z4通过*Bm*POUM2来调控翅原基表皮蛋白基因*BmWCP4*在蛹期的专一性表达

邓惠敏[#] 牛康康[#] 章佳玲 冯启理[*]

摘要：昆虫蜕皮激素（20E）通过转录因子*Bm*POUM2结合到翅原基表皮蛋白基因*BmWCP4*启动子的顺式应答元件上来调控*BmWCP4*的转录。本文揭示了*BmPOUM2*的转录调控机制。RT-PCR分析显示，*BmPOUM2*响应20E，在预蛹和蛹中期的家蚕翅原基中有较高水平的表达，该表达趋势与*BmBR-C Z1*、*BmBR-C Z2*和*BmBR-C Z4*类似。在翅原基中*BmBR-C Z4*的表达受到20E的诱导，而*BmBR-C Z1*和*BmBR-C Z2*不受诱导。在*BmPOUM2*的上游启动子区域鉴定出了3个潜在的BR-C Z4的顺式响应元件（CERs）。在*BmN*细胞中过表达*BmBR-C Z4*可以显著提高*BmPOUM2*启动子的活性，并增加*BmPOUM2*在mRNA和蛋白水平的表达。对对照与突变的BR-C Z4 CREs的EMSA和荧光素酶活分析表明，*Bm*BR-C Z4蛋白可以结合到预测的BRC-Z4 CRE C（-684—-660）区域。上述结果表明，转录因子*Bm*BR-C Z4是*BmPOUM2*的直接上游调控因子，可调控翅原基表皮蛋白基因*BmWCP4*在蛹期的专一性表达。

昆虫分子与发育生物学实验室，生命科学学院，华南师范大学，广州

Transcription factors *Bm*POUM2 and *Bm*βFTZ-F1 are involved in regulation of the expression of the wing cuticle protein gene *BmWCP4* in the silkworm, *Bombyx mori*

Deng HM[#] Zheng SC[#] Yang XH Liu L Feng QL[*]

Abstract: In *Bombyx mori*, the wing cuticle protein gene *BmWCP4* is expressed specifically in the epidermis at the onset and mid-stage of pupation and is responsible for the formation of the pupal cuticle during the larval-pupal metamorphosis. The gene consists of four exons and three introns and is present as a single copy in the genome. Its expression was up-regulated by 20-hydroxyecdysone (20E) and the 20E-induced expression was suppressed by juvenile hormone (JH) Ⅲ. The upstream regulatory sequence region of the *BmWCP4* gene was cloned and the regulatory elements responsible for 20E induction were identified. Two cis-regulatory elements (CREs) bound by the transcription factors *Bm*POUM2 and *Bm*βFTZ-F1 were identified that mediated 20E-regulated expression of this gene. An electrophoretic mobility shift assay detected two nuclear proteins isolated from the epidermis and the *BmN* cell line that specifically bound to the POU and βFTZ-F1 CREs, respectively. *Bm*POUM2 recombinant protein explicitly bound to the POU CRE. Developmental and 20E-induced expression of the *BmWCP4*, *BmPOUM2* and *BmβFTZ-F1* genes showed that *BmPOUM2* and *BmβFTZ-F1* were initially expressed, followed by *BmWCP4*. These data suggest that the 20E-induced expression of *BmWCP4* is mediated by the transcription factors *Bm*POUM2 and *Bm*βFTZ-F1 binding to their CREs in the regulatory sequence region of the *BmWCP4* gene.

Published On: Insect Molecular Biology, 2011, 20(1), 45-60.

Guangdong Provincial Key Lab of Biotechnology for Plant Development, School of Life Sciences, South China Normal University, Guangzhou 510631, China.

[#]These authors contributed equally.

[*]Corresponding author E-mail: qlfeng@scnu.edu.cn.

家蚕转录因子*Bm*POUM2和*Bm*βFTZ-F1参与调控家蚕翅原基表皮蛋白基因*BmWCP4*的表达

邓惠敏[#]　郑思春[#]　杨鑫华　刘　琳　冯启理[*]

摘要：在家蚕中，翅原基表皮蛋白基因*BmWCP4*在蛹前期和中期的表皮中特异表达，主要在幼虫到蛹的变态发育过程中负责蛹表皮的形成。该基因有4个外显子，3个内含子，在基因组中呈现单拷贝。20E可以上调该基因的表达，而且这种上调可以被JHⅢ所抑制。*BmWCP4*基因的上游调控序列已经被克隆，其中应答20E的调控元件也得到了鉴定。在该基因启动子上游鉴定到两个转录因子——*Bm*POUM2和*Bm*βFTZ-F1的顺式作用元件，它们介导20E调控该基因的表达。分别用从表皮和*BmN*细胞中提取的核蛋白进行EMSA实验，结果表明核蛋白可以特异性地结合到POU和βFTZ-F1的顺式作用元件上，而且是*Bm*POUM2重组蛋白明确地结合到POU顺式元件上。发育和20E诱导的*BmWCP4*、*BmPOUM2*和*BmβFTZ-F1*基因的表达图谱显示，*BmPOUM2*和*BmβFTZ-F1*先表达，*BmWCP4*基因随后表达。这些结果表明，20E介导调控*BmWCP4*基因的表达是通过*Bm*POUM2和*Bm*βFTZ-F1蛋白结合到*BmWCP4*基因的调控区域内的顺式元件上来实现的。

广东省植物发育生物工程重点实验室，生命科学学院，华南师范大学，广州

The silkworm homolog of *Methoprene tolerant* (*Met*) gene reveals sequence conservation but function divergence

Li ZQ[1#] Cheng DJ[1#] Wei L[2] Zhao P[1] Shu X[1]
Tang L[1] Xiang ZH[1] Xia QY[1,3*]

Abstract: *Methoprene-tolerant* (*Met*) gene has been found to be involved in juvenile hormone (JH) action in insects. Herein, we isolated a silkworm (*Bombyx mori*) homolog of *Met* gene from *Drosophila melanogaster* using bio-informatics analysis and rapid amplification of cDNA ends-polymerase chain reaction method, and defined it as *BmMet*. The full-length cDNA of *BmMet* gene consists of 1 917 nucleotides and includes a 1 368 bp of open reading frame for a deduced protein of 455 amino acids. All deduced protein sequences from *Met* genes in *B. mori* and other surveyed insects contain four typical domains of bHLH, PAS-A, PAS-B and PAC, highlighting a high sequence conservation of *Met* genes during insect evolution. Also, genomic structure and phylogenic analysis suggested that *Met* in both *B. mori* and *Drosophila* species may originate from an ancestor gene with *gce*, another member of bHLH-PAS family, via gene duplication. In addition, *BmMet* was detected in all surveyed tissues and throughout the whole life of silkworm at transcriptional levels. Furthermore, silkworm individuals with RNAi silencing of *BmMet* gene in the early stage of the fourth instar larvae could molt normally and pupate successfully. This result was different from the observation in *T. castaneum* but similar to that in *D. melanogaster* after *Met* knockdown, revealing that the action mode of Met in *B. mori* and *D. melanogaster* should be divergent with that in other insect species.

Published On: Insect Science, 2010, 17(4), 313-324.

1 The Key Sericultural Laboratory of Agricultural Ministry, Southwest University, Chongqing 400716, China.
2 School of Life Science, Southwest University, Chongqing 400716, China.
3 The Institute of Agronomy and Life Sciences, Chongqing University, Chongqing 400044, China.
#These authors contributed equally.
*Corresponding author E-mail: xiaqy@swu.edu.cn.

家蚕*Met*基因的同源性揭示了序列上的保守性和功能上的差异性

李志清[1#] 程道军[1#] 魏 玲[2] 赵 萍[1] 舒 旭[1]
唐 林[1] 向仲怀[1] 夏庆友[1,3*]

摘要：在昆虫中，已有研究表明*Met*基因参与保幼激素（JH）的信号通路。本研究中，我们通过生物信息学分析和RACE-PCR的方法，从家蚕中分离鉴定到一个与果蝇同源的*BmMet*基因。*BmMet*基因的全长cDNA由1 917个核苷酸组成，含一个1 368 bp的开放阅读框（编码455个氨基酸）。在家蚕和其他昆虫的Met蛋白序列都包含4个典型的结构域，分别是bHLH、PAS-A、PAS-B和PAC，表明*Met*基因在进化过程中具有高度保守性。同时，基因组和系统发生树分析表明，家蚕和果蝇*Met*基因可能来源于同为bHLH-PAS家族成员的祖先基因*gce*，并通过基因复制而来。此外，在转录水平上检测到*BmMet*在所有被检测组织和家蚕整个生命周期中均有表达。当在四龄初期的家蚕个体中用RNAi方法沉默*BmMet*时，家蚕能正常蜕皮和化蛹，该结果与在赤拟谷盗中观察到的表型相反，而与果蝇中*Met*敲除后的表型类似，表明Met在家蚕和果蝇中的作用模式不同于其他昆虫。

1 农业部蚕桑学重点实验室，西南大学，重庆
2 生命科学学院，西南大学，重庆
3 农学及生命科学研究院，重庆大学，重庆

Analysis of expression and chitin-binding activity of the wing disc cuticle protein *Bm*WCP4 in the silkworm, *Bombyx mori*

Deng HM[#] Li Y[#] Zhang JL Liu L Feng QL[*]

Abstract: The insect exoskeleton is mainly composed of chitin filaments linked by cuticle proteins. When insects molt, the cuticle of the exoskeleton is renewed by degrading the old chitin and cuticle proteins and synthesizing new ones. In this study, chitin-binding activity of the wing disc cuticle protein *Bm*WCP4 in *Bombyx mori* was studied. Sequence analysis showed that the protein had a conservative hydrophilic "R&R" chitin-binding domain (CBD). Western blot showed that *Bm*WCP4 was predominately expressed in the wing disc-containing epidermis during the late wandering and early pupal stages. The immunohistochemistry result showed that the *Bm*WCP4 was mainly present in the wing disc tissues containing wing bud and trachea blast during day 2 of wandering stage. Recombinant full-length *Bm*WCP4 protein, "R&R" CBD peptide (CBD), non-CBD peptide (*Bm*WCP4-CBD$^-$), four single site-directed mutated peptides (M_1, M_2, M_3 and M_4) and four-sites-mutated peptide (M_F) were generated and purified, respectively, for *in vitro* chitin-binding assay. The results indicated that both the full-length protein and the "R&R" CBD peptide could bind with chitin, whereas the *Bm*WCP4-CBD$^-$ could not bind with chitin. The single residue mutants M_1, M_2, M_3 and M_4 reduced but did not completely abolish the chitin-binding activity, while four-sites-mutated protein M_F completely lost the chitin-binding activity. These data indicate that *Bm*WCP4 protein plays a critical role by binding to the chitin filaments in the wing during larva-to-pupa transformation. The conserved aromatic amino acids are critical in the interaction between chitin and the cuticle protein.

Published On: Insect Science, 2016, 23(6), 782-790.

Laboratory of Molecular and Developmental Entomology, Guangdong Provincial Key Laboratory of Biotechnology for Plant Development, School of Life Sciences, South China Normal University, Guangzhou 510631, China.

[#]These authors contributed equally.

[*]Corresponding author E-mail: qlfeng@scnu.edu.cn.

家蚕翅原基表皮蛋白*Bm*WCP4的表达与几丁质结合活性的分析

邓惠敏[#] 李 勇[#] 章佳玲 刘 琳 冯启理[*]

摘要：昆虫的外骨骼主要由几丁质细丝和表皮蛋白交联而成。在蜕皮变态过程中，昆虫通过合成新的几丁质和表皮蛋白，降解旧的几丁质和表皮蛋白来实现外骨骼的交替更新。本文研究了家蚕翅原基表皮蛋白*Bm*WCP4与几丁质的结合活性。序列分析结果显示，*Bm*WCP4蛋白具有非常保守的亲水性"R & R"几丁质结合结构域（CBD）。Western blot显示，在上蔟晚期和蛹早期，*Bm*WCP4主要在含表皮的翅原基中表达。免疫组化结果显示*Bm*WCP4主要在上蔟2天的含翅芽的翅原基组织和气管丛中表达。分别生成并纯化重组全长*Bm*WCP4蛋白，"R & R"CBD肽（CBD）、非CBD肽（*Bm*WCP4-CBD$^-$）、4个单位点突变肽（M_1，M_2，M_3和M_4）和四位点突变肽（M_F）并进行体外几丁质结合实验。结果显示，全长蛋白和"R & R"CBD肽能与几丁质结合，但*Bm*WCP4-CBD$^-$不能。单残基突变体M_1，M_2，M_3和M_4与几丁质结合的活性降低了但没有失去，而四位点突变蛋白M_F完全失去了与几丁质结合的活性。这些数据表明，BmWCP4蛋白在幼虫向蛹转变过程中，通过结合几丁质细丝在翅中发挥重要作用。在几丁质和表皮蛋白相互作用中，保守的芳香氨基酸至关重要。

昆虫分子与发育生物学实验室，广东省植物发育生物工程重点实验室，生命科学学院，华南师范大学，广州

Autophagy precedes apoptosis during the remodeling of silkworm larval midgut

Eleonora Franzetti[1#] Huang ZJ[2#] Shi YX[2] Xie K[2] Deng XJ[2] Li JP[2]
Li QR[2,3] Yang WY[2] Zeng WN[2] Morena Casartelli[4] Deng HM[5] Silvia Cappellozza[6]
Annalisa Grimaldi[1] Xia QY[7] Gianluca Tettamanti[1*] Cao Y[2*] Feng QL[5*]

Abstract: Although several features of apoptosis and autophagy have been reported in the larval organs of Lepidoptera during metamorphosis, solid experimental evidence for autophagy is still lacking. Moreover, the role of the two processes and the nature of their relationship are still cryptic. In this study, we perform a cellular, biochemical and molecular analysis of the degeneration process that occurs in the larval midgut of *Bombyx mori* during larval-adult transformation, with the aim to analyze autophagy and apoptosis in cells that die under physiological conditions. We demonstrate that larval midgut degradation is due to the concerted action of the two mechanisms, which occur at different times and have different functions. Autophagy is activated from the wandering stage and reaches a high level of activity during the spinning and prepupal stages, as demonstrated by specific autophagic markers. Our data show that the process of autophagy can recycle molecules from the degenerating cells and supply nutrients to the animal during the non-feeding period. Apoptosis intervenes later. In fact, although genes encoding caspases are transcribed at the end of the larval period, the activity of these proteases is not appreciable until the second day of spinning and apoptotic features are observable from prepupal phase. The abundance of apoptotic features during the pupal phase, when the majority of the cells die, indicates that apoptosis is actually responsible for cell death and for the disappearance of larval midgut cells.

Published On: Apoptosis, 2012, 17(3), 305-324.

1 Department of Biotechnology and Life Sciences, University of Insubria, Varese 21100, Italy.
2 Laboratory of insect Molecular Biology and Biotechnology, Guangdong Provincial Key Laboratory of Agro-animal Genomics and Molecular Breeding, College of Animal Science, South China Agricultural University, Guangzhou 510642, China.
3 The Sericulture & Agri-Food Research Institute, Guangdong Academy of Agricultural Sciences, Guangzhou 510610, China.
4 Department of Biology, Università degli Studi di Milano, Milano 20133, Italy.
5 Guangdong Provincial Key Lab of Biotechnology for Plant Development, School of Life Sciences, South China Normal University, Guangzhou 510631, China.
6 CRA-Honey Bee and Silkworm Research Unit, Padova 35143, Italy.
7 Institute of Sericulture and Systems Biology, Southwest University, Chongqing 400716, China.
[#]These authors contributed equally.
[*]Corresponding author E-mail: gianluca.tettamanti@uninsubria.it; caoyang@scau.edu.cn;qlfeng@scnu.edu.cn.

家蚕幼虫中肠重塑过程中细胞自噬先于细胞凋亡

Eleonora Franzetti[1#] 黄志君[2#] 史艳霞[2] 谢 昆[2] 邓小娟[2] 李建平[2]
李庆荣[2,3] 杨婉莹[2] 曾文年[2] Morena Casartelli[4] 邓惠敏[5] Silvia Cappellozza[6]
Annalisa Grimaldi[1] 夏庆友[7] Gianluca Tettamanti[1*] 曹 阳[2*] 冯启理[5*]

摘要:有关鳞翅目昆虫变态时期幼虫器官发生细胞自噬和凋亡的现象已有一些报道,但仍然缺乏坚实的细胞自噬实验证据。此外,这两个过程和其相互关系仍然不清楚。在本研究中,我们通过细胞生物学、生物化学和分子生物学实验分析了家蚕幼虫中肠在幼虫-蛹转换时期发生的退化过程,目的是分析在生理条件下死亡细胞的自噬或凋亡。我们证明幼虫中肠退化是两种发生在不同时间和具有不同功能的机制协同作用的结果。特定的自噬标记结果证实自噬起始于幼虫上蔟时期,在吐丝期和蛹前期达到高水平。我们的数据显示,自噬过程可以从退化细胞中回收分子,并在非饲喂期间为动物提供营养。细胞凋亡随后进行。事实上,尽管半胱天冬酶(Caspases)的编码基因在幼虫末期已经启动转录,但直到吐丝期第二天才能检测到这些蛋白酶的活性,并且在预蛹阶段可观察到凋亡特征。蛹期大多数死亡的细胞呈现凋亡特征,这表明细胞凋亡实际上是造成细胞死亡和幼虫期中肠细胞消失的原因。

1 生物技术和生命科学系,英苏布里亚大学,瓦雷泽,意大利
2 昆虫分子生物学与生物技术实验室,广东省农业动物基因组学与分子育种重点实验室,动物科学学院,华南农业大学,广州
3 蚕业与农产品加工研究所,广东省农业科学院,广州
4 生物学系,米兰大学,米兰,意大利
5 广东省植物发育生物工程重点实验室,生命科学学院,华南师范大学,广州
6 CRA-蜜蜂和蚕研究所,帕多瓦,意大利
7 蚕学与系统生物学研究所,西南大学,重庆

The Broad Complex isoform 2 (BrC-Z2) transcriptional factor plays a critical role in vitellogenin transcription in the silkworm *Bombyx mori*

Yang CW　Lin Y　Liu HL　Shen GW　Luo J　Zhang HY
Peng ZX　Chen EX　Xing RM　Han CS　Xia QY*

Abstract:

Background: Vitellogenin (Vg) is synthesized in the fat body of the female silkworm *Bombyx mori* and transported to the oocyte as a source of nutrition for embryo development. It is well known that ecdysone regulates physiological, developmental and behavioral events in silkworm. However, it is still not clear how the ecdysone regulates *B. mori Vg* (*BmVg*) transcription.

Methods: Electrophoretic mobility shift assay (EMSA) and cell transfection assay were used to reveal whether *Bm*BrC-Z2 is involved in regulating *BmVg* transcription. RNAi was employed to illustrate the function of *Bm*BrC-Z2 in the silkworm egg formation and development.

Results: (1) The transcription of *BmVg* can be induced by ecdysone in the female fat body. (2) Three putative BrC-Z2 *cis*-response elements were mapped to regions flanking the *BmVg* gene. (3) *Bm*BrC-Z2 required direct binding to the *cis*-response elements on the *BmVg* promoter. (4) Over-expression of three *BmBrC* isoforms in the cell line showed that only *Bm*BrC-Z2 could induce the *BmVg* promoter activity. (5) RNA interference (RNAi) of *BmBrC-Z2* in female remarkably reduced *Bm*Vg synthesis and led to destructive affection on egg formation. The dsRNA of *Bm*BrC-Z2 treated moths laid fewer and whiter eggs compared to the control.

Conclusions: *Bm*BrC-Z2 transported the ecdysone signal then regulated *BmVg* transcription directly to control vitellogenesis and egg formation in the silkworm.

General Significance: The results of this study revealed that *Bm*BrC-Z2 as a key factor to mediate ecdysone regulates reproduction in the silkworm.

Published On: Biochimica et Biophysica Acta-General Subjects, 2014, 1840(9), 2674-2684.

State Key Laboratory of Silkworm Genome Biology, Southwest University, Chongqing 400716, China.
*Corresponding author E-mail: xiaqy@swu.edu.cn.

转录因子BrC-Z2在家蚕卵黄原蛋白(Vg)基因的转录过程中扮演重要角色

杨从文　林　英　刘红玲　沈关望　罗　娟　张海燕
彭芷昕　陈恩祥　邢润苗　韩超珊　夏庆友*

摘要：背景——家蚕卵黄原蛋白(*Bm*Vg)在雌性家蚕脂肪体中合成，随后运输到卵母细胞中，作为胚胎发育的重要营养来源。蜕皮激素能够调节家蚕的生理、发育、行为等一系列过程。然而，蜕皮激素如何调节*BmVg*的转录仍不清楚。方法——EMSA和细胞转染实验用来检测*Bm*BrC-Z2是否参与*BmVg*的转录。利用RNAi实验来说明*Bm*BrC-Z2在家蚕卵的形成和发育中的作用。结果——(1)雌性家蚕脂肪体中*Vg*的转录受蜕皮激素的诱导。(2)*BmVg*基因上游调控区域预测到3个BrC-Z2的顺式作用元件。(3)BrC-Z2能直接结合到*BmVg*启动子上的顺式作用元件。(4)真核细胞中过表达3种形式的BrC，只有*Bm*BrC-Z2能上调*BmVg*启动子的活性。(5)RNA干涉雌蚕体内*BmBrC-Z2*基因后能显著降低*Bm*Vg的合成，导致卵发育受损。*Bm*BrC-Z2的双链RNA处理家蚕后，蛾产下的卵较正常组偏少偏白。结论——*Bm*BrC-Z2响应蜕皮激素信号进而调控*BmVg*的转录，从而调节家蚕卵黄生成和卵的形成。意义——本研究验证了*Bm*BrC-Z2是蜕皮激素调节家蚕繁殖的关键转录因子。

家蚕基因组生物学国家重点实验室，西南大学，重庆

Transcriptomic analysis of developmental features of *Bombyx mori* wing disc during metamorphosis

Ou J Deng HM Zheng SC Huang LH Feng QL* Liu L*

Abstract:

Background: Wing discs of *B. mori* are transformed to pupal wings during the larva-to-pupa metamorphosis with dramatic morphological and structural changes. To understand these changes at a transcriptional level, RNA-seq of the wing discs from 6-day-old fifth instar larvae (L5D6), prepupae (PP) and pupae (P0) was performed.

Results: In total, 12 254 transcripts were obtained from the wing disc, out of which 5 287 were identified to be differentially expressed from L5D6 to PP and from PP to P0. Comprehensive analysis of RNA-seq data showed that during larvae-to-pupae metamorphosis, many genes of 20E signaling pathway were up-regulated and those of JH signaling pathway were down-regulated. Seventeen transcription factors were significantly up-regulated. Cuticle protein genes (especially wing cuticle protein genes)were most abundant and significantly up-regulated at P0 stage. Genes responsible for the degradation and *de novo* synthesis of chitin were significantly up-regulated. There were A and B two types of chitin synthases in *B. mori*, whereas only chitin synthase A was up-regulated. Both trehalose and D-fructose, which are precursors of chitin synthesis, were detected in the hemolymph of L5D6, PP and P0, suggesting *de novo* synthesis of chitin. However, most of the genes, Which are related to early wing disc differentiation, were down-regulated.

Conclusions: Extensive transcriptome and DGE profiling data of wing disc during metamorphosis of silkworm have been generated, which provided comprehensive gene expression information at the transcriptional level. These results implied that during the larva-to-pupa metamorphosis, pupal wing development and transition might be mainly controlled by 20E signaling in *B. mori*. The 17 up-regulated transcription factors might be involved in wing development. Chitin required for pupal wing development might be generated from both degradation of componential chitin and *de novo* synthesis. Chitin synthase A might be responsible for the chitin synthesis in the pupal wing, while both trehalose and D-fructose might contribute to the *de novo* synthesis of chitin during the formation of pupal wing.

Published On: BMC Genomics, 2014, 15.

Laboratory of Molecular and Developmental Entomology, Guangdong Provincial Key Lab of Biotechnology for Plant Development, School of Life Sciences, South China Normal University, Guangzhou 510631, China.

*Corresponding author E-mail: qlfeng@scnu.edu.cn; liul@scnu.edu.cn.

家蚕变态发育时期翅原基的转录组学分析

欧 俊 邓惠敏 郑思春 黄立华 冯启理[*] 刘 琳[*]

摘要:背景——家蚕幼虫翅原基向蛹翅转变的过程中形态和结构都发生了剧烈变化。为从转录水平探究家蚕翅原基向蛹翅转变过程中的发育特点,我们利用RNA-seq的方法,对五龄第6天(L5D6)、预蛹期(PP)和蛹期开始(P0)的家蚕翅原基进行了高通量转录组测序。结果——在不同时期翅原基中总共鉴定到12 254个基因,其中有5 287个是从L5D6到PP和PP到P0时期差异表达的基因。对RNA-seq数据的综合分析结果显示在幼虫翅原基向蛹翅的转变过程中,大量20E信号通路的相关基因表达上调,同时许多JH信号通路的相关基因表达下调,并且有17个转录因子表达显著上调。表皮蛋白基因,尤其是翅原基表皮蛋白基因在P0期大量富集。负责几丁质从头合成和降解的基因显著上调。家蚕中存在几丁质合成酶A和B,但只有几丁质合成酶A表达上调。几丁质合成的前体海藻糖和D-果糖在L5D6,PP和P0时期的血淋巴中大量表达,提示几丁质是从头合成的。然而,许多和早期翅原基分化相关的基因却是表达下调。结论——丰富的蚕变态期间翅盘的转录组和DGE图谱数据,在转录水平上提供了全面的基因表达信息。在幼虫向蛹的变态发育过程中,蛹翅的发育和转变可能主要受到20E信号通路的调控。17个上调表达的转录因子可能参与了翅的发育。蛹期翅膀发育所需的几丁质可能来自成熟几丁质的降解和从头合成。几丁质合成酶A可能主要负责蛹翅几丁质的合成,同时需要利用海藻糖和D-果糖从头开始合成几丁质。

昆虫分子与发育生物学实验室,广东省植物发育生物工程重点实验室,生命科学学院,华南师范大学,广州

Bombyx E75 isoforms display stage-and tissue-specific responses to 20-hydroxyecdysone

Li K[1,2] Guo EE[1,2] Muktadir S. Hossain[2] Li QR[3]
Cao Y[1] Tian L[2*] Deng XJ[1*] Li S[2*]

Abstract: Resulted from alternative splicing of the 5' exons, the nuclear receptor gene *E75* in the silkworm, *Bombyx mori*, processes three mRNA isoforms, *BmE75A*, *BmE75B* and *BmE75C*. From the early 5th larval instar to the prepupal stages, *BmE75A* mRNA and protein levels in the prothoracic glands display developmental profiles similar to ecdysteroid titer. In the fat body, mRNA levels but not protein levels of all three *Bm*E75 isoforms correlate with ecdysteroid titer; moreover, proteins of all three *Bm*E75 isoforms disappear at the prepupal stages, and a modified *Bm*E75 protein with smaller molecular weight and cytoplasm localization occurs. At the early 5th larval instar stage, treatment of the prothoracic glands and fat body with 20-hydroxyecdysone (20E) and/or cycloheximide (CHX) revealed that *Bm*E75A is 20E primary-responsive at both mRNA and protein levels, while *Bm*E75B and *Bm*E75C exhibit various responses to 20E. At the early wandering stage, RNAi-mediated reduction of gene expression of the 20E nuclear receptor complex, EcR-USP, significantly decreased mRNA and protein levels of all three *Bm*E75 isoforms in both tissues. In conclusion, *Bm*E75 isoforms display stage- and tissue-specific responses to 20E at both mRNA and protein levels; moreover, they are regulated by other unknown factors at the protein level.

Published On: Scientific Reports, 2015, 5.

1 Guangdong Provincial Key Laboratory of Agro-animal Genomics and Molecular Breeding, Guangdong Provincial Sericulture and Mulberry Engineering Research Center, College of Animal Sciences, South China Agricultural University, Guangzhou 510642, China.

2 Key Laboratory of Insect Developmental and Evolutionary Biology, Institute of Plant Physiology and Ecology, Chinese Academy of Sciences, Shanghai 200032, China.

3 The Sericultural and Agri-Food Research Institute of the Guangdong Academy of Agricultural Sciences, Guangzhou 510610, China.

*Corresponding author E-mail: tianling@sibs.ac.cn; dengxj@scau.edu.cn; lisheng01@sibs.ac.cn.

家蚕E75异构体展现出对20-蜕皮激素的时期组织特异性响应

李 康[1,2] 郭恩恩[1,2] Muktadir S. Hossain[2] 李庆荣[3]
曹 阳[1] 田 铃[2*] 邓小娟[1*] 李 胜[2*]

摘要：由于5′端外显子的选择性剪切，家蚕的核受体基因E75有3种mRNA异构体，*BmE75A*、*BmE75B*和*BmE75C*。从五龄早期至预蛹期，前胸腺内*Bm*E75A的mRNA与蛋白水平表现出与蜕皮激素的滴度相似的模式，而在脂肪体内3种*Bm*E75异构体的mRNA水平而非蛋白质水平与蜕皮激素滴度趋势一致。此外，*Bm*E75的3种异构体蛋白在预蛹时期均消失，出现一种分子质量较小且定位在细胞质中的*Bm*E75修饰蛋白。环已酰亚胺（CHX）或20E处理五龄早期幼虫的前胸腺或脂肪体显示*Bm*E75A在mRNA和蛋白水平上是20E的初级应答因子，而*Bm*E75B和*Bm*E75C则展现出对20E不同的响应。在上蔟早期，20E受体*EcR-USP*在RNAi后显著地降低前胸腺和脂肪体中*Bm*E75的3种异构体的mRNA和蛋白质水平。总之，*Bm*E75的3种异构体无论是在mRNA水平上还是在蛋白水平上都表现出对20E的时期和组织特异响应，而它们的蛋白水平又受到未知因子的调控。

1 广东省农业动物基因组学与分子育种重点实验室，广东省蚕桑工程技术研究中心，动物科学学院，华南农业大学，广州
2 昆虫发育与进化生物学重点实验室，植物生理生态研究所，中国科学院，上海
3 广东省农业科学院蚕业与农产品加工研究所，广州

Transcription factor SGF1 is critical for the neurodevelopment in the silkworm, *Bombyx mori*

Liu ZY Yu Q Yang CH Meng M
Ren CJ Mu ZM Cui WZ Liu QX*

Abstract: FoxA transcription factors play vital roles in regulating the expression of organ-specific genes. *Bm*SGF1, the sole FoxA family member in *Bombyx mori*, is required for development of the silk gland. However, the function of *Bm*SGF1 in development of the nervous system in the silkworm remains unknown. Here, we show that the amino acids sequence of *Bm*SGF1 is evolutionarily conserved in its middle region from Trichoplax adhaerens to human and diverged from the homologues in most other species in its N-terminal region. *Bm*SGF1 was expressed in the nervous system at the embryonic stage. Knockdown of *Bmsgf1* by RNA interference (RNAi) results in abnormal development of axons. Therefore, our results demonstrate that *Bm*SGF1 is an indispensable regulator for neurodevelopment.

Published On: Gene, 2016, 587(1), 70-75.

Laboratory of Developmental Genetics, Shandong Agricultural University, Tai'an, Shandong 271018, China.
*Corresponding author E-mail: liuqingxin@sdau.edu.cn.

SGF1是家蚕神经发育过程中关键的转录因子

刘朝阳　于　奇　杨春红　孟　苗
任春久　牟志美　崔为正　刘庆信*

摘要：FoxA转录因子在调控器官特异性基因表达中起着至关重要的作用。*Bm*SGF1是家蚕中唯一的FoxA家族成员，对丝腺的发育是必需的。然而，*Bm*SGF1在神经系统发育中的功能仍然是未知的。本文证明了*Bm*SGF1氨基酸序列的中间区从丝盘虫到人类是进化保守的，而在N端区域则与大多数其他物种的同系物发生了分化。*BmSGF1*在胚胎期的神经系统中表达。通过RNA干涉对*Bmsgf1*基因表达进行下调导致轴突发育异常。因此，我们的结果证明，*Bm*SGF1是神经发育必不可少的调节因子。

发育遗传学研究室，山东农业大学，泰安

SAGE tag based cDNA microarray analysis during larval to pupal development and isolation of novel cDNAs in *Bombyx mori*

Zhang Y[1] Huang JH[1] Jia SH[1] Liu WB[1] Li MW[2]
Wang SB[1] Miao XX[1] Xiao HS[3] Huang YP[1*]

Abstract: Many genes act together during the complex process of insect larval and pupal development. 20-hydroxyecdysone interacts with juvenile hormone to control insect growth and development and then activates several transcription factors, i.e., Broad, E74, and E75, and, subsequently, the late target genes. To investigate this phenomenon, we used serial analysis of gene expression (SAGE) tag-based cDNA microarray analysis to monitor the global gene expression profile during larval development and larva-pupa metamorphosis of the silkworm *Bombyx mori*. Of the 330 clones that were dotted to the chip, 267 were obtained by generating longer cDNA fragments from SAGE tags for gene identification, and the others were obtained from SAGE tag-matched genes or expressed sequence tags from public databases. According to the gene expression profile, the genes were classified into 12 clusters using a self-organizing map analysis. The results were partially confirmed using real-time reverse transcription-polymerase chain reaction. We obtained 22 full-length cDNAs using rapid amplification of 5' cDNA ends, of which eight genes were novel in the silkworm. Our results indicated that use of a cDNA microarray based on SAGE tags is effective for identifying and examining some low-expression genes associated with insect development.

Published On: Genomics, 2007, 90(3), 372-329.

1 Institute of Plant Physiology and Ecology, Shanghai Institutes for Biological Sciences, Chinese Academy of Science, Shanghai 200032, China.

2 Sericultural Research Institute, Chinese Academy of Agricultural Sciences, Zhenjiang 212018, China.

3 National Engineering Center for Biochips at Shanghai, Shanghai 201203, China

*Corresponding author E-mail: yphuang@sibs.ac.cn.

利用基于SAGE标签的芯片技术分析并分离家蚕从幼虫到蛹发育过程的新cDNA

张　勇[1]　黄健华[1]　贾世海[1]　刘文彬[1]　李木旺[2]
王四宝[1]　苗雪霞[1]　肖华胜[3]　黄勇平[1*]

摘要:昆虫幼虫和蛹的变态发育过程是许多基因相互作用的结果。20-羟基蜕皮激素与保幼激素的相互作用调控了昆虫的生长和发育,能够激活下游转录因子,包括Broad、E74和E75以及后期20E的响应基因。为了弄清楚这种现象,我们使用基于基因表达(SAGE)标签的cDNA芯片进行分析,检测家蚕幼虫期发育和幼虫向蛹变态过程中的全基因表达谱。采用基因芯片检测到330个克隆,其中267个是通过SAGE标签产生更长的cDNA片段用于基因鉴定获得的,其他的来自与公共数据库的SAGE标签或表达的序列标签相匹配的基因。根据基因表达谱,使用自组织图分析将基因分类为12个簇,其中部分基因用实时荧光定量反应来证实。我们利用5′ cDNA末端的快速扩增技术得到22个全长cDNA,其中8个基因在家蚕中是首次发现的。我们的研究结果表明基于SAGE标签的cDNA芯片对于识别和检测一些与昆虫发育相关的低表达基因是有效的。

1　植物生理生态研究所,上海生命科学研究院,中国科学院,上海
2　蚕业研究所,中国农业科学院,镇江
3　生物芯片上海国家工程研究中心,上海

Ecdysone receptor (EcR) is involved in the transcription of cell cycle genes in the silkworm

Qian WL Kang LX Zhang TL Meng M
Wang YH Li ZQ Xia QY* Cheng DJ*

Abstract: EcR (ecdysone receptor) - mediated ecdysone signaling pathway contributes to regulate the transcription of genes involved in various processes during insect development. In this work, we detected the expression of *EcR* gene in silkworm ovary-derived *BmN4* cells and found that *EcR* RNAi result in an alteration of cell shape, indicating that EcR may orchestrate cell cycle progression. *EcR* RNAi and *EcR* overexpression analysis revealed that in the cultured *BmN4* cells, EcR respectively promoted and suppressed the transcription of *E2F-1* and *CycE*, two genes controlling cell cycle progression. Further examination demonstrated that ecdysone application in *BmN4* cells not only changed the transcription of these two cell cycle genes like that under *EcR* overexpression, but also induced cell cycle arrest at G_2/M phase. *In vivo* analysis confirmed that *E2F-1* expression was elevated in silk gland of silkworm larvae after ecdysone application, which is same as its response to ecdysone in *BmN4* cells. However, ecdysone also promotes *CycE* transcription in silk gland, and this is converse with the observation in *BmN4* cells. These results provide new insights into understanding the roles of EcR-mediated ecdysone signaling in the regulation of cell cycle.

Published On: International Journal of Molecular Science, 2015, 16, 3335-3349.

State Key Laboratory of Silkworm Genome Biology, Southwest University, Chongqing 400716, China.
*Corresponding author E-mail: xiaqy@swu.edu.cn; chengdj@swu.edu.cn.

家蚕蜕皮激素受体(EcR)参与细胞周期基因转录

钱文良　康丽霞　张天镭　孟　勐
王永虎　李志清　夏庆友*　程道军*

摘要：EcR(蜕皮激素受体)介导的蜕皮激素信号通路有助于调节昆虫发育过程中的基因转录调控。在这项工作中，我们检测到*EcR*基因在家蚕卵巢衍生的*BmN4*细胞中表达，发现*EcR* RNAi导致细胞形态的改变，表明EcR可能协调细胞周期进程。*EcR* RNAi和*EcR*过表达分析显示，在培养的*BmN4*细胞中，EcR分别促进和抑制细胞周期进程的两个基因*E2F-1*和*CycE*的转录。进一步的研究表明，蜕皮激素在*BmN4*细胞中的应用不仅改变了这两种细胞周期基因的转录，如*EcR*过表达，类似还可诱导细胞周期阻滞在G_2/M期。体内分析证实，添加蜕皮激素后，家蚕丝腺中*E2F-1*表达升高，与*BmN4*细胞中蜕皮激素的反应相同。然而，蜕皮激素还促进丝腺中的*CycE*转录，这与在*BmN4*细胞中观察的情况相反。这些结果为理解EcR介导的蜕皮激素在细胞周期调控中的作用提供了新的见解。

家蚕基因组生物学国家重点实验室，西南大学，重庆

Antennapedia is involved in the development of thoracic legs and segmentation in the silkworm, *Bombyx mori*

Chen P[1#] Tong XL[1#] Li DD[1] Fu MY[1] He SZ[1]
Hu H[1] Xiang ZH[1] Lu C[1*] Dai FY[1,2*]

Abstract: Homeotic genes, which are associated closely with body patterning of various species, specify segment identity. The *Wedge eye-spot* (*Wes*) is a new homeotic mutant located on the sixth linkage group. Homozygous *Wes/Wes* embryos are lethal and display a pair of antenna-like appendages under the mouthparts as well as fused thoracic segments. These mutants also exhibit a narrower eye-spot at the larval stage compared with the wild type. By positional cloning, we identified the candidate gene of the *Wes* locus, *Bombyx mori Antennapedia* (*BmAntp*). Two *BmAntp* transcripts were identified in the homozygote of the *Wes* mutant, including a normal form and an abnormal form with a 1 570 bp insertion. Our data showed that the insertion element was a long interspersed nuclear element (LINE)-like transposon that destroyed the original open reading frame of *BmAntp*. Quantitative RT-PCR analysis showed that the expression levels of normal *BmAntp* transcripts were increased markedly in the *Wes* heterozygous larvae compared with the wild type. Furthermore, we performed RNAi of *BmAntp* and observed fused thoracic segments and defective thoracic legs in the developing embryos. Our results indicated that *BmAntp* is responsible for the *Wes* mutant and has an important role in determining the proper development of the thoracic segments. Our identification of a homeotic mutation in the silkworm is an important contribution to our understanding of the regulation of *Hox* genes at different levels of expression.

Published On: Heredity, 2013, 111(3), 182-188.

1 State Key Laboratory of Silkworm Genome Biology, Southwest University, Chongqing 400716, China.
2 College of Biotechnology, Southwest University, Chongqing 400716, China.
#These authors contributed equally.
*Corresponding author E-mail:lucheng@swu.edu.cn; fydai@swu.edu.cn.

Antennapedia 参与家蚕胸足的发育及体节的分化

陈　鹏[1#]　童晓玲[1#]　李丹丹[1]　付明月[1]　何松真[1]
胡　海[1]　向仲怀[1]　鲁　成[1*]　代方银[1,2*]

摘要：同源异型基因与各物种的躯体发育模式密切相关。*Wes* 是位于第六连锁群上的新的同源异型突变体。纯合的 *Wes*/*Wes* 胚胎致死，并在口器和融合的胸段下方出现一对触须状附属物。与野生型相比，突变体在幼虫期表现出较窄的眼状纹。通过定位克隆，我们鉴定了 *Wes* 基因座候选基因 *BmAntp*。在 *Wes* 突变体的纯合子中鉴定了两个 *BmAntp* 转录物，包括正常转录本和具有 1 570 bp 插入的异常转录本。我们的数据表明，插入部分是一个类似长散在重复序列转座子（LINE），破坏了 *BmAntp* 的开放阅读框。定量 RT-PCR 分析显示，与野生型相比，*Wes* 杂合幼虫中正常 *BmAntp* 转录本的表达水平显著提高。此外，我们进行了 *BmAntp* 的 RNAi，并观察到发育中的胚胎出现的融合胸段和缺陷胸足。我们的研究结果表明，*BmAntp* 是 *Wes* 突变体的突变基因，其在确保胸部正常发育中起重要作用。我们鉴定到家蚕中的同源异型突变对我们理解不同表达水平的 *Hox* 基因调控很有帮助。

1　家蚕基因组生物学国家重点实验室，西南大学，重庆
2　生物技术学院，西南大学，重庆

MET is required for the maximal action of 20-hydroxyecdysone during *Bombyx* metamorphosis

Guo EE[1,2] He QY[1] Liu SM[1] Tian L[1] Sheng ZT[1] Peng Q[1] Guan JM[1]
Shi MA[1] Li K[2] Lawrence I. Gilbert[3] Wang J[4] Cao Y[2*] Li S[1*]

Abstract: Little is known about how the putative juvenile hormone (JH) receptor, the bHLH-PAS transcription factor MET, is involved in 20-hydroxyecdysone (20E; the molting hormone) action. Here we report that two MET proteins found in the silkworm, *Bombyx mori,* participate in 20E signal transduction. *Met* is 20E responsive and its expression peaks during molting and pupation, when the 20E titer is high. As found with results from RNAi knockdown of *EcR-USP* (the ecdysone receptor genes), RNAi knockdown of *Met* at the early wandering stage disrupts the 20E-triggered transcriptional cascade, preventing tissue remodeling (including autophagy, apoptosis and destruction of larval tissues and generation of adult structures) and causing lethality during the larval-pupal transition. MET physically interacts with EcR-USP. Moreover, MET, EcR-USP and the 20E-response element (EcRE) form a protein-DNA complex, implying that MET might modulate 20E-induced gene transcription by interacting with EcR-USP. In conclusion, the 20E induction of MET is required for the maximal action of 20E during *Bombyx* metamorphosis.

Published On: PLoS ONE, 2012, 7(12).

1 Key Laboratory of Insect Developmental and Evolutionary Biology, Shanghai Institute of Plant Physiology and Ecology, Shanghai Institutes for Biological Sciences, Chinese Academy of Sciences, Shanghai 200032, China.

2 Laboratory of Insect Molecular Biology and Biotechnology, Guangdong Provincial Key Laboratory of Agro-animal Genomics and Molecular Breeding, College of Animal Sciences, South China Agricultural University, Guangzhou 510642, China.

3 Department of Biology, University of North Carolina, North Carolina 27599, United States of America.

4 Department of Entomology, University of Maryland, College Park, Maryland 20742, United States of America.

*Corresponding author E-mail: caoyang@scau.edu.cn; shengli@sippe.ac.cn.

MET蛋白是家蚕变态发育过程中20E最大程度发挥功能的必需蛋白

郭恩恩[1,2] 何倩毓[1] 刘淑敏[1] 田 铃[1] 胜振涛[1] 彭 琴[1] 管静敏[1]
施明安[1] 李 康[2] Lawrence I. Gilbert[3] 王 健[4] 曹 阳[2*] 李 胜[1*]

摘要：对于假定的保幼激素(JH)受体(bHLH-PAS转录因子MET)如何参与20-羟基蜕皮酮(20E；蜕皮激素)信号通路知之甚少。我们在家蚕中发现了两种MET蛋白参与20E信号转导。*Met*基因响应20E信号，并在蜕皮和化蛹这两个20E处于高水平的时期具有表达峰。与蜕皮激素受体基因(*EcR-USP*)的RNAi敲低的结果一样，在家蚕上蔟早期用RNAi敲低*Met*基因，扰乱了20E启动的转录级联反应，阻止幼虫组织的重建(包括自噬、细胞凋亡和幼虫组织的解离以及成虫器官的形成)，进而导致幼虫在幼虫-蛹转变时期死亡。生理上，MET与EcR-USP能够相互作用。此外，MET蛋白、EcR-USP与20E应答元件(EcRE)形成一个蛋白-DNA复合体，显示MET可能通过与EcR-USP的相互作用来调控20E诱导的基因表达。综上所述，20E诱导的MET是家蚕变态发育过程中20E最大程度发挥功能所必需的蛋白。

1 昆虫发育与进化生物学重点实验室，上海植物生理生态研究所，上海生命科学研究院，中国科学院，上海
2 昆虫分子生物学与生物技术实验室，广东省农业动物基因组学与分子育种重点实验室，动物科学学院，华南农业大学，广州
3 生物系，北卡罗来纳大学，北卡罗来纳州，美国
4 昆虫学系，马里兰大学帕克分校，马里兰州，美国

Coevolution of axon guidance molecule Slit and its receptor Robo

Yu Q[1] Li XT[1] Zhao X[1] Liu XL[1]
Kazuho Ikeo[2] Takashi Gojobori[2] Liu QX[1*]

Abstract: Coevolution is important for the maintenance of the interaction between a ligand and its receptor during evolution. The interaction between axon guidance molecule Slit and its receptor Robo is critical for the axon repulsion in neural tissues, which is evolutionarily conserved from planarians to humans. However, the mechanism of coevolution between Slit and Robo remains unclear. In this study, we found that coordinated amino acid changes took place at interacting sites of Slit and Robo by comparing the amino acids at these sites among different organisms. In addition, the high level correlation between evolutionary rate of Slit and Robo was identified in vertebrates. Furthermore, the sites under positive selection of *slit* and *robo* were detected in the same lineage such as mosquito and teleost. Overall, our results provide evidence for the coevolution between Slit and Robo.

Published On: PLoS ONE. 2014, 9(10).

1 Laboratory of Developmental Genetics, Shandong Agricultural University, Tai'an, Shandong 271018, China.
2 Center for Information Biology and DNA Data Bank of Japan, National Institute of Genetics, Mishima 411-8540, Japan.
*Corresponding author E-mail: liuqingxin@sdau.edu.cn

轴突导向因子Slit及其受体Robo的共同进化

于 奇[1] 李晓童[1] 赵 晓[1] 刘训理[1]
Kazuho Ikeo[2] Takashi Gojobori[2] 刘庆信[1*]

摘要:在进化过程中,协同进化对于维持配体与其受体之间的相互作用至关重要。轴突导向因子Slit与其受体Robo之间的相互作用对神经组织中的轴突排斥起关键作用,从涡虫到人类,轴突排斥在进化上是保守的。然而,Slit和Robo之间的协同进化机制尚不清楚。本研究我们通过比较不同生物间的相互作用位点上的氨基酸,发现Slit和Robo的相互作用位点上的氨基酸发生了协同进化。另外,在脊椎动物中还发现了*slit*和*robo*进化率之间的高度相关性。此外,在同一谱系的蚊子和硬骨鱼中也发现了Slit和Robo正向选择的位点。总体而言,本研究结果证明了Slit和Robo之间的协同进化。

1 发育遗传学研究室,山东农业大学,泰安
2 日本信息生物学与DNA数据库中心,国立遗传研究所,三岛市,日本

Nuclear import of transcription factor BR-C is mediated by its interaction with RACK1

Cheng DJ[1#] Qian WL[1#] Wang YH[1] Meng M[1]
Wei L[2] Li ZQ[3] Kang LX[1] Peng J[1] Xia QY[1*]

Abstract: The transcription factor Broad Complex (BR-C) is an early ecdysone response gene in insects and contains two types of domains: two zinc finger domains for the activation of gene transcription and a Bric-a-brac/Tramtrack/Broad complex (BTB) domain for protein-protein interaction. Although the mechanism of zinc finger-mediated gene transcription is well studied, the partners interacting with the BTB domain of BR-C has not been elucidated until now. Here, we performed a yeast two hybrid screen using the BTB domain of silkworm BR-C as bait and identified the receptor for activated C-kinase 1 (RACK1), a scaffolding/anchoring protein, as the novel partner capable of interacting with BR-C. The interaction between BR-C and RACK1 was further confirmed by far-western blotting and pull-down assays. Importantly, the disruption of this interaction, via RNAi against the endogenous *RACK1* gene or deletion of the BTB domain, abolished the nuclear import of BR-C in *BmN4* cells. In addition, RNAi against the endogenous *PKC* gene as well as phosphorylation-deficient mutation of the predicted PKC phosphorylation sites at either Ser373 or Thr406 in BR-C phenocopied *RACK1* RNAi and altered the nuclear localization of BR-C. However, when BTB domain was deleted, phosphorylation mimics of either Ser373 or Thr406 had no effect on the nuclear import of BR-C. Moreover, mutating the PKC phosphorylation sites at Ser373 and Thr406 or deleting the BTB domain significantly decreased the transcriptional activation of a BR-C target gene. Given that RACK1 is necessary for recruiting PKC to close and phosphorylate target proteins, we suggest that the PKC-mediated phosphorylation and nuclear import of BR-C is determined by its interaction with RACK1. This novel finding will be helpful for further deciphering the mechanism underlying the role of BR-C proteins during insect development.

Published On: PLoS ONE, 2014, 9(10).

1 State Key Laboratory of Silkworm Genome Biology, Southwest University, Chongqing 400716, China.
2 School of Life Science, Southwest University, Chongqing 400716, China.
3 Laboratory of Silkworm Science, Kyushu University Graduate School of Bioresource and Bioenvironmental Sciences, Fukuoka 812-8581, Japan.
[#]These authors contributed equally.
[*]Corresponding author E-mail: xiaqy@swu.edu.cn.

家蚕转录因子BR-C通过与RACK1的相互作用来完成其核转位

程道军[1#] 钱文良[1#] 王永虎[1] 孟 勐[1]
魏 玲[2] 李志清[3] 康丽霞[1] 彭 健[1] 夏庆友[1*]

摘要：昆虫转录因子Broad Complex（BR-C）是早期蜕皮激素应答基因，包含两种类型的结构域：激活基因转录的两个锌指结构域和蛋白质-蛋白质相互作用的Bric-a-brac / Tramtrack / Broad复合物（BTB）结构域。尽管锌指介导的基因转录的机制得到了充分的研究，但是与BR-C的BTB结构域相互作用的配体尚未被阐明。本研究，我们利用家蚕BR-C的BTB结构域作为诱饵进行了酵母双杂交筛选，并鉴定到receptor for activated C-kinase 1（RACK1），一种骨架/锚定蛋白，是能够与BR-C相互作用的新型分子伴侣。BR-C和RACK1之间的相互作用进一步通过far-Western blot和Pull-down实验证实。重要的是，通过RNAi干涉内源性*RACK1*基因或删除BR-C的BTB结构域破坏了两者的相互作用后阻止了*BmN4*细胞中BR-C的入核。此外，同*RACK1*基因的RNAi干涉一样，*PKC*的双链干涉和预测的PKC磷酸化位点Ser373或Thr406磷酸化缺陷突变同样改变了BR-C的细胞核定位。然而，当BTB结构域被删除时，Ser373或Thr406的磷酸化对BR-C的入核没有影响。在Ser373和Thr406处突变PKC磷酸化位点或删除BTB结构域显著地降低了BR-C靶基因的转录活性。鉴于RACK1是招募PKC关闭和开启靶蛋白磷酸化所必需的，我们认为PKC介导的BR-C磷酸化和入核是通过其与RACK1的相互作用来完成的。这一新发现将有助于进一步解析BR-C在昆虫发育过程中的作用机制。

1 家蚕基因组生物学国家重点实验室，西南大学，重庆
2 生命科学学院，西南大学，重庆
3 生物资源环境科学研究院家蚕科研室，九州大学，福冈，日本

Preliminary characterization of a death-related gene in silkworm *Bombyx mori*

Chi G[1] Gao L[1] Chen KP[1*] Yao Q[1]
Yang Z[1] Chen HQ[1] Wang L[2]

Abstract: In order to study the mechanism of programmed cell death (PCD) in silkworm *Bombyx mori*, we employed suppressive subtractive hybridization to screen associated genes. In this study, a novel gene was isolated from *B. mori*, which may be involved in PCD as analyzed by bioinformatics and it was named as DRP (death-related protein) gene. RT-PCR analysis showed prokaryotic expression. Through RT-PCR analysis of death-related protein gene in different tissues and different developmental stage of *B. mori*, it showed the distributed condition of the gene. It was widely expressed in various tissues and mainly expressed in testis, malphigian vessels, posterior intestine, silk gland. Meanwhile, it was widely expressed in virous developmental stages and mainly expressed in egg, indicating its possible roles in development of programmed cell death. We also tested the expression of this gene in prokaryotic system, which may be useful for subsequent studies. This gene is registered in GenBank under the accession number FJ447481.

Published On: African Journal of Biotechnology, 2009, 8(10), 2118-2124.

1 Institute of Life Sciences, Jiangsu University, Zhenjiang 212013, China.
2 Beijing Entry-Exit Inspection and Quarantine Bureau, Beijing 100026, China.
*Corresponding author E-mail: kpchen@ujs.edu.cn.

家蚕中死亡相关基因特性的初步研究

池 刚[1] 高 路[1] 陈克平[1*] 姚 勤[1]
杨 哲[1] 陈慧卿[1] 汪 琳[2]

摘要：为了研究家蚕中细胞程序性死亡(PCD)的机制，我们采用抑制差减杂交法筛选相关基因。在本研究中，我们通过生物信息学分析，在家蚕中分离出一种可能参与PCD过程的新型基因，并命名为DRP(死亡相关蛋白)基因。对家蚕不同组织和不同发育阶段的死亡相关蛋白基因的RT-PCR分析，显示了该基因的表达分布状态。该基因在各种组织中广泛地表达，主要在精巢、马氏管、后肠、丝腺中表达。同时，该基因在不同的发育阶段也广泛表达，主要在卵细胞中表达，表明其可能在程序性细胞死亡发展中充当一个重要的角色。我们还在原核系统中检测了该基因的表达情况，为后续的研究提供了基础。该基因在GenBank的登录号为FJ447481。

1 生命科学研究院，江苏大学，镇江
2 北京出入境检验检疫局，北京

Transcriptional regulation of the gene for prothoracicotropic hormone in the silkworm, *Bombyx mori*

Wei ZJ[1*] Yu M[1] Tang SM[2] Yi YZ[2] Hong GY[1] Jiang ST[1]

Abstract: Prothoracicotropic hormone (PTTH) is one of key players in regulation of insect growth, molting, metamorphosis, diapause, and is expressed specifically in the two pairs of lateral PTTH-producing neurosecretory cells in the brain. Analysis of *cis*-regulatory elements of the *PTTH* promoter might elucidate the regulatory mechanism controlling *PTTH* expression. In this study, the *PTTH* gene promoter of *Bombyx mori* (*Bom-PTTH*) was cloned and sequenced. The *cis*-regulatory elements in *Bom-PTTH* gene promoter were predicted using Matinspector software, including myocyte-specific enhancer factor 2, pre-B-cell leukemia homeobox 1, TATA box, etc. Transient transfection assays using a series of fragments linked to the luciferase reporter gene indicated that the fragment spanning –110 to +33 bp of the *Bom-PTTH* promoter showed high ability to support reporter gene expression, but the region of +34 to +192 bp and –512 to –111 bp repressed the promoter activity in the *BmN* and *Bm5* cell lines. Electrophoretic mobility shift assays demonstrated that the nuclear protein could specifically bind to the region spanning –124 to –6bp of the *Bom-PTTH* promoter. Furthermore, we observed that the nuclear protein could specifically bind to the –59 to –30 bp region of the *Bom-PTTH* promoter. A classical TATA box, TATATAA, localized at positions –47 to –41 bp, which is a potential site for interaction with TATA box binding protein (TBP). Mutation of this TATA box resulted in no distinct binding band. Taken together, TATA box was involved in regulation of *PTTH* gene expression in *B. mori*.

Published On: Molecular Biology Reports, 2011, 38(2), 1121-1127.

1 Department of Biotechnology, Hefei University of Technology, Hefei 230009, China.

2 Key Laboratory of Silkworm Biotechnology, Ministry of Agriculture, Sericultural Research Institute, Chinese Academy of Agricultural Sciences, Zhenjiang 212018, China.

*Corresponding author E-mail: zjwei@hfut.edu.cn.

家蚕促前胸腺激素的转录调控

魏兆军[1*]　于　森[1]　唐顺明[2]　易咏竹[2]　洪桂云[1]　姜绍通[1]

摘要：促前胸腺激素(PTTH)是调节昆虫生长、蜕皮、变态、滞育的关键因素之一，并且在脑中两对产生PTTH的外侧神经分泌细胞中特异表达。分析*PTTH*启动子顺式调节元件可以阐明*PTTH*表达的调控机理。本研究克隆了家蚕的*PTTH*(*Bom-PTTH*)基因启动子并进行测序。使用Matinspector软件预测*Bom-PTTH*基因启动子中的顺式作用元件，包括肌细胞特异性增强因子2，前B细胞白血病同源异型盒1，TATA盒等。将一系列的启动子片段连接到荧光素酶报告基因进行转染实验，结果表明，*Bom-PTTH*启动子-110 bp至+33 bp之间的片段有高的活性，明显地促进了报告基因的表达，而+34 bp至+192 bp和-512 bp至-111 bp的区域抑制了启动子在*BmN*和*Bm5*细胞系中的活性。电泳迁移实验表明，核蛋白可以特异性地结合*Bom-PTTH*启动子的-124 bp到-6 bp的区域。此外，我们还观察到核蛋白可以特异性结合*Bom-PTTH*启动子的-59 bp至-30 bp区域。位于-47 bp至-41 bp的TATA盒TATATAA，是与TATA盒结合蛋白(TBP)相互作用的潜在位点，突变该TATA盒没有产生明显的结合带。综上所述，TATA盒参与家蚕*PTTH*基因的表达调控。

1　生物技术系，合肥工业大学，合肥
2　家蚕生物技术重点开放实验室，农业部，中国农业科学院，镇江

Phenol oxidase is a necessary enzyme for the silkworm molting which is regulated by molting hormone

Wang MX Lu Y Cai ZZ Liang S Niu YS Miao YG*

Abstract: Insect molting is an important developmental process of metamorphosis, which is initiated by molting hormone. The molting process includes the activation of dermal cells, epidermal cells separation, molting fluid secretion, the formation of new epidermis and old epidermis excoriation etc. Polyphenol oxidases (PPOs), dopa decarboxylase and acetyltransferase are necessary enzymes for this process. Traditionally, the phenol oxidase was considered as an enzyme for epidermal layer' s tanning and melanization. This work suggested that polyphenol oxidases are one set of the key enzymes in molting, which closely related with the role of ecdysone in regulation of molting processes. The data showed that the expression peak of phenol oxidase in silkworm is higher during molting stage, and decreases after molting. The significant increase in the ecdysone levels of haemolymph was observed in the artificially fed silkworm larvae with ecdysone hormone. Consistently, the phenol oxidase expression was significantly elevated compared to the control. PPO1 RNAi induced phenol oxidase expression obviously declined in the silkworm larvae, and caused the pupae incomplete pupation. Overall, the results described that the phenol oxidase expression is regulated by the molting hormone, and is a necessary enzyme for the silkworm molting.

Published On: Molecular Biology Reports, 2013, 40(5), 3549-3555.

Key Laboratory of Animal Virology of Ministry of Agriculture, College of Animal Sciences, Zhejiang University, Hangzhou 310058, China.

*Corresponding author E-mail: miaoyg@zju.edu.cn.

酚氧化酶是受蜕皮激素调节的家蚕蜕皮必需酶

王梅仙　陆　岳　蔡自峥
梁　爽　牛艳山　缪云根*

摘要：蜕皮是受蜕皮激素调控的重要变态发育过程。蜕皮过程包括真皮细胞活化、表皮细胞分离、蜕皮液分泌、新表皮形成和旧表皮剥落等。多酚氧化酶(PPOs)、多巴脱羧酶和乙酰转移酶是该过程的必需酶。传统上认为，酚氧化酶是促进表皮层硬化和黑化的酶。本研究表明，多酚氧化酶是蜕皮过程中的关键酶之一，其与蜕皮激素在调节蜕皮过程中的作用密切相关。结果表明，家蚕中酚氧化酶在蜕皮期表达峰值较高，蜕皮后降低。人工添加蜕皮激素的家蚕幼虫血淋巴中，蜕皮激素水平显著提高。同时，与对照组相比，酚氧化酶表达显著升高。PPO1 RNAi诱导的家蚕幼虫中酚氧化酶表达明显下降，导致不完全化蛹。结果表明，酚氧化酶表达受蜕皮激素调控，是家蚕蜕皮的必需酶。

农业部动物病毒学重点实验室，动物科学学院，浙江大学，杭州

Expression and functions of dopa decarboxylase in the silkworm, *Bombyx mori* was regulated by molting hormone

Wang MX Cai ZZ Lu Y Xin HH Chen RT Liang S
Chabungbam Orville Singh Jong-Nam Kim Niu YS Miao YG*

Abstract: Insect molting is an important developmental process of metamorphosis, which is initiated by molting hormone. Molting includes the activation of dermal cells, epidermal cells separation, molting fluid secretion, the formation of new epidermis and old epidermis shed and other series of continuous processes. Polyphenol oxidases, dopa decarboxylase and acetyltransferase are necessary enzymes for this process. Traditionally, the dopa decarboxylase (*Bm*Ddc) was considered as an enzyme for epidermal layer's tanning and melanization. This work suggested that dopa decarboxylase is one set of the key enzymes in molting, which closely related with the regulation of ecdysone at the time of biological molting processes. The data showed that the expression peak of dopa decarboxylase in silkworm is higher during molting stage, and decreases after molting. The significant increase in the ecdysone levels of haemolymph was also observed in the artificially fed silkworm larvae with ecdysone hormone. Consistently, the dopa decarboxylase expression was significantly elevated compared to the control. *BmDdc* RNAi induced dopa decarboxylase expression obviously declined in the silkworm larvae, and caused the pupae appeared no pupation or incomplete pupation. *Bm*Ddc was mainly expressed and stored in the peripheral plasma area near the nucleus in *BmN* cells. In larval, *Bm*Ddc was mainly located in the brain and epidermis, which is consisted with its function in sclerotization and melanization. Overall, the results described that the dopa decarboxylase expression is regulated by the molting hormone, and is a necessary enzyme for the silkworm molting.

Published On: Molecular Biology Reports, 2013, 40(6), 4115-4122.

Key Laboratory of Animal Virology of Ministry of Agriculture, College of Animal Sciences, Zhejiang University, Hangzhou 310058, China.
*Corresponding author E-mail: miaoyg@zju.edu.cn.

由蜕皮激素调控的家蚕多巴脱羧酶的表达和功能研究

王梅仙　蔡自峥　陆　嵒　辛虎虎　陈瑞婷　梁　爽
Chabungbam Orville Singh　Jong-Nam Kim　牛艳山　缪云根*

摘要：昆虫蜕皮是蜕皮激素调控的重要变态发育过程。蜕皮包括真皮细胞激活、表皮细胞分离、蜕皮液分泌、新表皮形成和旧表皮脱落等一系列连续过程。多酚氧化酶、多巴脱羧酶和乙酰转移酶是该过程的必需酶。传统上，多巴脱羧酶（*Bm*Ddc）被认为是参与表皮硬化和黑化的酶。本研究表明，多巴脱羧酶是蜕皮过程中的关键酶之一，与蜕皮过程中蜕皮激素的调节密切相关。研究显示多巴脱羧酶在家蚕蜕皮时期的表达达到峰值，蜕皮后降低。在添加蜕皮激素喂养的家蚕幼虫中观察到血淋巴中蜕皮激素水平显著增加。与对照相比，多巴脱羧酶表达显著升高。RNAi 干扰 *BmDdc* 明显降低了家蚕幼虫中多巴脱羧酶的表达，导致出现不化蛹或化蛹不完全现象。*Bm*Ddc 主要表达并存储在 *BmN* 细胞核附近的外周胞质区。在幼虫中，*Bm*Ddc 主要位于大脑和表皮中，这与其在硬化和黑化中的功能相关。总体而言，结果表明多巴脱羧酶表达受蜕皮激素调控，是家蚕蜕皮的必需酶。

农业部动物病毒学重点实验室，动物科学学院，浙江大学，杭州

Molecular cloning and expression characterization of translationally controlled tumor protein in silkworm pupae

Nie ZM Lü ZB Qian JW Chen J Li SS
Sheng Q Wang D Shen HD Yu W Wu XF Zhang YZ*

Abstract: A *Bombyx mori* (*B. mori*) cDNA was isolated from silkworm pupae cDNA library encoding a homologue of translationally controlled tumor protein (*Bm*TCTPk). *Bm*TCTPk was expressed in *E. coli*; SDS-PAGE and Western blot showed the molecular weight of recombinant and native *Bm*TCTPk is approximately 28 and 25 kDa, respectively; they are larger than the theoretical molecular weight. Immunohistochemical studies showed that *Bm*TCTPk is uniformly distributed throughout the cytoplasm of *BmN* cells. In silkworm pupae, *Bm*TCTPk is expressed in the midgut wall, the midgut cavity, and some fat body tissues lying between the midgut wall and body wall. Western blot and ELISAs performed on total protein extracts isolated from silkworm pupae at different development stages showed that, although *Bm*TCTPk is expressed during all pupae stages, its expression level increases dramatically during late pupae stages, suggesting that *Bm*TCTPk may play an important role during the developmental transition from pupa to imago.

Published On: Molecular Biology Reports, 2010, 37(6), 2621-2628.

Institute of Biochemistry, Zhejiang Sci-Tech University, Hangzhou 310018, China.
*Corresponding author E-mail: yaozhou@chinagene.com.

蚕蛹翻译调控肿瘤蛋白的分子克隆及表达特征研究

聂作明　吕正兵　钱家伟　陈　健　李珊珊
盛　清　王　丹　沈红丹　于　威　吴祥甫　张耀洲*

摘要：从家蚕蛹期cDNA文库中分离出一条编码翻译调节肿瘤蛋白同源物（*Bm*TCTPk）的cDNA。在大肠杆菌中表达*Bm*TCTPk。SDS-PAGE和Western blot结果显示，重组和天然的*Bm*TCTPk的分子质量分别约28 kDa和25 kDa，都比理论分子质量大。免疫组化学研究显示，*Bm*TCTPk均匀地分布在*BmN*细胞的整个细胞质中。在蚕蛹中，*Bm*TCTPk在中肠壁、中肠腔和中肠壁与体壁之间的一些脂肪体组织表达。对家蚕蛹不同发育阶段分离的总蛋白提取物进行Western blot和ELISA检测，结果显示尽管*Bm*TCTPk在所有蛹期均表达，但在蛹后期表达水平显著增加，表明*Bm*TCTPk在家蚕从蛹到成虫的发育过程中可能具有重要作用。

生物化学研究所，浙江理工大学，杭州

Molecular cloning and transcription expression of 3-dehydroecdysone 3α-reductase (3DE 3α-reductase) in the different tissues and the developing stage from the silkworm, *Bombyx mori* L.

Yang HJ Xin HH Lu Y Cai ZZ Wang MX Chen RT
Liang S Chabungbam Orville Singh Jong-nam Kim Miao YG*

Abstract: Molting in insects is regulated by molting hormones (ecdysteroids), which are also crucial to insect growth, development, and reproduction etc. The decreased ecdysteroid in titre results from enhanced ecdysteroid inactivation reactions including the formation of 3-epiecdyson under ecdysone oxidase and 3-dehydroecdysone 3α - reductase (3DE 3α - reductase). In this paper, we cloned and characterized 3-dehydroecdysone 3α-reductase (3DE 3α-reductase) in different tissues and developing stage of the silkworm, *Bombyx mori* L. The *B. mori* 3DE 3α-reductase cDNA contains an ORF 783 bp and the deduced protein sequence containing 260 amino acid residues. Analysis showed the deduced 3DE 3α-reductase belongs to SDR family, which has the NAD(P)-binding domain. Using the *Escherichia coli,* a high level expression of a fusion polypeptide band of approx. 33 kDa was observed. High transcription of 3DE 3α-reductase was mainly presented in the midgut and hemolymph in the third day of fifth instar larvae in silkworm. The expression of 3DE 3α-reductase at different stages of larval showed that the activity in the early instar was high, and then reduced in late instar. This is parallel to the changes of molting hormone titer in larval. 3DE 3α-reductase is key enzyme in inactivation path of ecdysteroid. The data elucidate the regulation of 3DE 3α-reductase in ecdyteroid titer of its targeting organs and the relationship between the enzyme and metamorphosis.

Published On: Archives of Insect Biochemistry and Physiology, 2013, 84(2), 78-89.

Key Laboratory of Animal Epidemic Etiology and Immunological Prevention of Ministry of Agriculture, College of Animal Sciences, Zhejiang University, Hangzhou 310058, China.
*Corresponding author E-mail: miaoyg@zju.edu.cn.

家蚕3-DE 3α-还原酶的克隆及其在家蚕不同发育时期和不同组织中的转录表达分析

杨华军　辛虎虎　陆　岳　蔡自峥　王梅仙　陈瑞婷
梁　爽　Chabungbam Orville Singh　Jong-nam Kim　缪云根*

摘要：昆虫蜕皮是通过蜕皮激素（蜕皮甾醇）调节的，蜕皮激素在昆虫生长、发育和繁殖等过程中都起着非常关键的作用。蜕皮激素含量的降低是由蜕皮激素灭活反应增强引起的，包括在蜕皮激素氧化酶和3-脱氢蜕皮激素-3α-还原酶的作用下3α-羟基蜕皮激素的形成。本文中，我们克隆3-DE 3α-还原酶，并分析其在家蚕不同组织和发育阶段的表达情况。家蚕3-DE 3α-还原酶基因的cDNA包含783 bp的ORF，推测其蛋白序列包含260个氨基酸残基。分析显示3-DE 3α-还原酶属于SDR家族，具有NAD(P)结合结构域。利用大肠杆菌高效地表达并获得分子质量为33 kDa的融合蛋白。3-DE 3α-还原酶的高转录主要出现在家蚕五龄第3天幼虫的中肠和血淋巴中。幼虫不同阶段的3-DE 3α-还原酶的表达分析显示，在早期幼虫中的活性高，在后期降低。这与幼虫蜕皮激素含量的变化是一致的，3-DE 3α-还原酶是蜕皮激素失活途径的关键酶。该结果阐明了3-DE 3α-还原酶对其靶向器官中的蜕皮激素的调节作用以及酶与变态发育之间的关系。

农业部动物病毒学重点实验室，动物科学学院，浙江大学，杭州

Characterization of adaptor protein complex-1 in the silkworm, *Bombyx mori*

Niu YS Cai ZZ Lu Y Wang MX Liang S Zhou F Miao YG*

Abstract: To investigate the function of adaptor protein complex-1 (AP-1) in the silkworm, we characterized AP-1 in the silkworm by RNAi technique and co-localization methods. As a result, AP-1 was found to exist as cytosolic form and membrane-bound form distinguished by phosphate status, showing molecular mass difference. There was relatively more cytosolic form of AP-1 than its membrane-bound counterpart in the silkworm. However, AP-1 distributed predominantly as cytosolic form in *BmN* cells. Interruption of AP-1 expression via DsRNA was more efficient in *BmN* cells than in the insect larval, which led to a tendency to dissociation between subcellular organelles like the Golgi apparatus and the mitochondria. Environmental condition changes like relatively higher temperature and treatment with dimethyl sulfoxide can lead to expression variance of AP-1 both in mRNA and protein level. In *BmN* cells, both the heavy chain γ and light chain σ could clearly co-localize with AP-1 β, mostly forming pits in cytoplasm. Two isoforms of AP-1 σ corresponded to distinct subcellular distribution pattern, possibly due to C-terminal amino acids difference.

Published On: Archives of Insect Biochemistry and Physiology, 2013, 82(2), 84-95.

Key Laboratory of Animal Virology of Ministry of Agriculture, College of Animal Sciences, Zhejiang University, Hangzhou 310058, China.

*Corresponding author E-mail: miaoyg@zju.edu.cn.

家蚕衔接蛋白复合物1(AP-1)的表征

牛艳山 蔡自峥 陆 嵒 王梅仙 梁 爽 周 芳 缪云根*

摘要:为了研究家蚕衔接蛋白复合物1(AP-1)的功能,我们用RNAi技术和共定位方法对家蚕中的AP-1进行了表征分析。结果发现AP-1以细胞溶质形式和膜结合形式存在,这两种形式以不同的磷酸盐状态予以区分,并表现出相对分子质量差异。与家蚕的膜结合形式相比,AP-1的细胞溶质形式相对较多。然而,AP-1在*BmN*细胞中主要为细胞溶质形式。通过DsRNA干扰AP-1的表达在*BmN*细胞中比在昆虫幼虫中更有效,导致高尔基体和线粒体等亚细胞器之间出现解离的倾向。环境条件改变,如温度相对较高或用二甲基亚砜处理,也可能导致AP-1在mRNA和蛋白质水平上的表达变化。在*BmN*细胞中,重链γ和轻链σ都可以明显地与AP-1β共定位,多数在细胞质中形成小窝。AP-1σ的两个异构体对应于不同的亚细胞分布模式,可能是由C-末端氨基酸差异引起的。

农业部动物病毒学重点实验室,动物科学学院,浙江大学,杭州

Cloning and characterization of the *Bombyx mori* ecdysone oxidase

Yang HJ Wang MX Zhang P Awquib Sabhat Firdose Ahmad Malik Roy Bhaskar Zhou F
Li XH Hu JB Sun CG Niu YS Miao YG*

Abstract: The physiological titer of molting hormones in insects depends on relative activities of synthesis and degradation pathways. Ecdysone oxidase (EO) is a key enzyme in the inactivation of ecdysteroid. However, there are only a few reports on ecdysteroid inactivation and its enzymes in silkworm. In this study, we cloned and characterized the *Bombyx mori EO* (*BmEO*). The *BmEO* cDNA contains an ORF of 1 695 bp and the deduced protein sequence contains 564 amino acid residues. The deduced protein sequence contains two functional domains of glucose-methanol-choline oxidoreductase in N-terminal and C-terminal. Compared with the expression levels of *BmEO* in different tissues, high transcription was mainly present in hemocytes. Reduced expression of this enzyme is expected to lead to pathological accumulation of ecdysone in the hemolymph of silkworm larvae or pupae. Our data show that RNA inference of *BmEO* transcripts resulted in the accumulation of ecdysteroid and death of larvae or pupae. We infer that EO is a crucial element in the physiology of insect development.

Published On: Archives of Insect Biochemistry and Physiology, 2011, 78(1), 17-29.

Key Laboratory of Animal Epidemic Etiology and Immunological Prevention of Ministry of Agriculture, College of Animal Sciences, Zhejiang University, Hangzhou 310058, China.
*Corresponding author E-mail: miaoyg@zju.edu.cn.

家蚕蜕皮激素氧化酶的克隆与分析

杨华军 王梅仙 张 芃 Awquib Sabhat Firdose Ahmad Malik Roy Bhaskar 周 芳
李兴华 胡嘉彪 孙春光 牛艳山 缪云根*

摘要:昆虫蜕皮激素的生理效价取决于合成和降解途径的相对活性。蜕皮激素氧化酶(EO)是蜕皮激素失活的关键酶。目前对于家蚕蜕皮激素失活及其相关酶的研究较少。本研究中,我们克隆和鉴定了家蚕*EO*(*BmEO*)。*BmEO* cDNA包含1 695个碱基,推导的蛋白序列含有564个氨基酸残基。其蛋白序列N端和C端含有葡萄糖–甲醇–胆碱氧化还原酶的两个功能域。比较不同组织中*BmEO*的表达水平*BmEO*主要在家蚕幼虫的血细胞中表达。*BmEO*的表达降低,可能会导致家蚕幼虫或蛹的血淋巴蜕皮激素滴度的病理积累。研究表明*BmEO*的转录表达在RNA干扰后导致蜕皮激素的滴度升高,引起幼虫或蛹死亡。我们推断EO是昆虫发育生理机能中的一个关键因素。

农业部动物病毒学重点实验室,动物科学学院,浙江大学,杭州

RNAi knockdown of *BmRab3* led to larva and pupa lethality in silkworm *Bombyx mori* L.

Chabungbam Orville Singh　Xin HH　Chen RT　Wang MX

Liang S　Lu Y　Cai ZZ　Zhang DP　Miao YG*

Abstract: Rab3 GTPases are known to play key roles in vesicular trafficking, and are expressed highest in brain and endocrine tissues. In mammals, Rab3 GTPases are paralogs unlike in insect. In this study, we cloned *Rab3* from the silk gland tissue of silkworm *Bombyx mori*, and identified it as *BmRab3*. Our *in silico* analysis indicated that *Bm*Rab3 is an isoform with a theoretical isoelectric point and molecular weight of 5.52 and 24.3 kDa, respectively. Further, *Bm*Rab3 showed the C-terminal hypervariability for GGT2 site but having two other putative guanine nucleotide exchange factor/GDP dissociation inhibitor interaction sites. Multiple alignment sequence indicated high similarities of *BmRab3* with *Rab3* isoforms of other species. The phylogenetic tree showed *BmRab3* clustered between the species of *Tribolium castaneum* and *Aedes aegypti*. Meanwhile, the expression analysis of *BmRab3* showed the highest expression in middle silk glands (MSGs) than all other tissues in the third day of fifth-instar larva. Simultaneously, we showed the differential expression of *BmRab3* in the early instar larva development, followed by higher expression in male than in female pupae. *In vivo* dsRNA interference of *BmRab3* reduced the expression of *BmRab3* by 75% compared to the control in the MSGs in the first day. But as the worm grew to the third day, the difference of *BmRab3* between knockdown and control was only about 10%. The knockdown later witnessed underdevelopment of the larvae and pharate pupae lethality in the overall development of silkworm *B. mori* L.

Published On: Archives of Insect Biochemistry and Physiology, 2015, 89(2), 98-110.

Institute of Sericulture and Apiculture, College of Animal Sciences, Zhejiang University, Hangzhou 310058, China.

*Corresponding author E-mail: miaoyg@zju.edu.cn.

RNAi干涉家蚕*BmRab3*导致家蚕幼虫和蛹致死

Chabungbam Orville Singh　辛虎虎　陈瑞婷　王梅仙
梁　爽　陆　岳　蔡自峥　张登攀　缪云根*

摘要：众所周知，Rab3 GTP酶在膜泡运输中起关键作用，并且在脑和内分泌组织中表达量最高。与昆虫不同，Rab3 GTP酶在哺乳动物中是旁系同源基因。在本研究中，我们从家蚕丝腺组织克隆了*Rab3*，并命名为*BmRab3*。信息分析表明，*Bm*Rab3的理论等电点和分子质量分别为5.52和24.3 kDa。此外，*Bm*Rab3的C端含有高突变性的GGT2位点，但具有另外两个假定的鸟嘌呤核苷酸交换因子/GDP解离抑制剂相互作用位点。多重序列比对表明*BmRab3*与其他物种的*Rab3*具有高度相似性。系统发生树显示*BmRab3*聚在赤拟谷盗和埃及伊蚊之间。同时，表达分析显示在五龄第3天幼虫中，*BmRab3*在中部丝腺（MSGs）的表达高于在其他组织中，而且在早期幼虫发育中的表达存在差异，雄蛹中的表达量高于雌蛹。干涉体内*BmRab3*第1天后与对照组相比，*BmRab3*在MSGs的表达量降低了75%，但是随着幼虫长大到第3天，*BmRab3*的表达量干涉组同对照组相比只降低约10%，且在家蚕的整个发育过程中，干涉组幼虫发育不完全并且因蛹蜕皮障碍而死。

蚕蜂研究所，动物科学学院，浙江大学，杭州

Structure of autophagy-related protein Atg8 from the silkworm *Bombyx mori*

Hu C Zhang X Teng YB Hu HX Li WF*

Abstract: Autophagy-related protein Atg8 is ubiquitous in all eukaryotes. It is involved in the Atg8-PE ubiquitin-like conjugation system, which is essential for autophagosome formation. The structures of Atg8 from different species are very similar and share a ubiquitin-fold domain at the C-terminus. In the 2.40 Å crystal structure of Atg8 from the silkworm *Bombyx mori* reported here, the ubiquitin fold at the C-terminus is preceded by two additional helices at the N-terminus.

Published On: Acta Crystallographica Section F-Structual Biology and Crystallization Communications, 2010, 66, 787-790.

Hefei National Laboratory for Physical Sciences at the Microscale and School of Life Sciences, University of Science and Technology of China, Hefei 230026, China.

*Corresponding author E-mail: liwf@ustc.edu.cn.

家蚕自噬相关蛋白Atg8的结构分析

胡 晨 张 璇 滕衍斌 胡海汐 李卫芳*

摘要：自噬相关蛋白Atg8在所有真核生物中都是普遍存在的。它参与了Atg8-PE泛素样缀合系统，该系统是自噬体形成所必需的。来自不同物种的Atg8的结构非常相似，并且在C末端都存在泛素折叠结构域。在这里报道了来自家蚕的Atg8的晶体结构，分辨率为2.40 Å，其N末端的两个附加螺旋是在C末端的泛素折叠前面。

合肥微尺度物质科学国家实验室（筹），生命科学学院，中国科学技术大学，合肥

Microarray analysis of the juvenile hormone response in larval integument of the silkworm, *Bombyx mori*

Cheng DJ[1#] Peng J[1#] Meng M[1] Wei L[2]
Kang LX[1] Qian WL[1] Xia QY[1*]

Abstract: Juvenile hormone (JH) coordinates with 20-hydroxyecdysone (20E) to regulate larval growth and molting in insects. However, little is known about how this cooperative control is achieved during larval stages. Here, we induced silkworm superlarvae by applying the JH analogue (JHA) methoprene and used a microarray approach to survey the mRNA expression changes in response to JHA in the silkworm integument. We found that JHA application significantly increased the expression levels of most genes involved in basic metabolic processes and protein processing and decreased the expression of genes associated with oxidative phosphorylation in the integument. Several key genes involved in the pathways of insulin/insulin-like growth factor signaling (IIS) and 20E signaling were also upregulated after JHA application. Taken together, we suggest that JH may mediate the nutrient-dependent IIS pathway by regulating various metabolic pathways and further modulate 20E signaling.

Published On: International Journal of Genomics, 2014.

1 State Key Laboratory of Silkworm Genome Biology, Southwest University, Chongqing 400716, China.
2 School of Life Science, Southwest University, Chongqing 400716, China.
[#]These authors contributed equally.
[*]Corresponding author E-mail: xiaqy@swu.edu.cn.

家蚕幼虫表皮保幼激素应答的芯片分析

程道军[1#] 彭 健[1#] 孟 勐[1] 魏 玲[2]
康丽霞[1] 钱文良[1] 夏庆友[1*]

摘要：在昆虫中，保幼激素（JH）与20-羟基蜕皮酮（20E）协同调节幼虫生长和蜕皮。然而，对如何在幼虫阶段实现这种协同调控的机制知之甚少。在这里，我们通过应用JH类似物（JHA）烯虫酯诱导产生超龄幼虫，并使用基因芯片技术检测家蚕表皮中响应JHA的mRNA变化情况。我们发现JHA处理显著提高了基因代谢过程和蛋白质加工中涉及的大多数基因的表达水平，并降低了表皮中氧化磷酸化相关基因的表达。参与胰岛素/胰岛素样生长因子信号通路（IIS）和20E信号通路的几个关键基因的表达也在JHA处理后发生上调。总之，我们认为JH可以通过调节不同的代谢途径并进一步调节20E信号来介导营养依赖的IIS途径。

1 家蚕基因组生物学国家重点实验室，西南大学，重庆
2 生命科学学院，西南大学，重庆

Expression pattern of enzymes related to juvenile hormone metabolism in the silkworm, *Bombyx mori* L.

Yang HJ Zhou F Sabhat Awquib Firdose Ahmad Malik Bhaskar Roy
Li XH Hu JB Sun CG Niu YS Miao YG*

Abstract: The physiological balance of juvenile hormone (JH) in insects depends on its biosynthesis and degradation pathway. Three key enzymes namely, juvenile hormone esterase (JHE), juvenile hormone epoxide hydrolase (JHEH) and juvenile hormone diol kinase (JHDK) are required for degradation in insects. Our present results showed that JHE and JHEH exhibited expression in almost all the tissues. This indicated that JHE and JHEH might degrade JH simultaneously. In addition, the highest levels of JHDK were observed in the midgut, with trace level being found in the malpighian tubule and haemocytes. Since the midgut is a digestive organ and not a JH target, it was hypothesized that both JHE and JHEH hydrolyzed JH to JH diol (JHd) which was then transported to midgut and hydrolyzed further by JHDK, to be finally excreted out of the body. Also the expression studies on JH degradation enzymes in different tissues and stages indicated that the activities of the three enzymes are specific and coincident with the JH functions in silkworm, *Bombyx mori* L.

Published On: Molecular Biology Reports, 2011, 38(7), 4337-4342.

Key Laboratory of Animal Epidemic Etiology and Immunological Prevention of Ministry of Agriculture, College of Animal Sciences, Zhejiang University, Hangzhou 310029, China.
*Corresponding author E-mail: miaoyg@zju.edu.cn.

家蚕保幼激素代谢相关酶的表达模式

杨华军　周　芳　Sabhat Awquib　Firdose Ahmad Malik　Bhaskar Roy
李兴华　胡嘉彪　孙春光　牛艳山　缪云根*

摘要：昆虫体内保幼激素(JH)的生理平衡取决于其生物合成和降解途径。保幼激素酯酶(JHE)、保幼激素环氧化物水解酶(JHEH)和保幼激素二醇激酶(JHDK)是昆虫保幼激素降解所需的三种关键酶。研究表明JHE和JHEH在几乎所有的组织中表达。这表明JHE和JHEH可能同时参与保幼激素的降解。而JHDK在中肠组织中大量表达，在马氏管和血细胞中仅少量表达。由于中肠是消化器官，而不是JH的靶器官，猜测JHE和JHEH首先水解JH成JH二醇(JHd)，然后JHd被转运到中肠组织，在JHDK作用下进一步水解，最终被排出体外。家蚕不同组织和不同发育阶段JH降解酶的表达表明这3种酶是特异性的，并且其与JH的功能是相一致的。

农业部动物疫病病原学与免疫预防重点开放实验室，动物科学学院，浙江大学，杭州

Molecular cloning and characterization of hatching enzyme-like gene in the silkworm, *Bombyx mori*

Lu FH[1] Tang SM[2*] Shen XJ[2*] Wang N[1]

Zhao QL[2] Zhang GZ[2] Guo XJ[2]

Abstract: Hatching is the important process for the life of the metazoan, in which hatching enzyme (HE) plays a key role. In this paper, we cloned the full-length sequence of hatching enzyme-like cDNA from bluish-silkworm-eggs of *Bombyx mori* (*BmHEL*) by the method of in silico cloning, SMART cDNA synthesis and RACE-PCR technique. The *Bm*HEL is 974 bp in length, and contains an ORF of 885 bp, encoding 294 amino acids residues. The deduced amino acid sequence of *Bm*HEL has 30.3%—47.1% identities to that of HE identified in the other species. Two similar signature sequences of *HE* gene family harbor in the *Bm*HEL. The *BmHEL* gene structure is 6-exon—5-intron, and a promoter region with high scores has been predicted, which harbors some basal elements and some embryo-development related transcription factor binding sites. In the silkworm eggs at different developmental stages during incubation, the *BmHEL* transcripts can be detected and keep at a low level during the early stages, increase dramatically since 7th day of incubation, and reach to the maximum on 9th day. Change of *BmHEL* transcripts is in accordance with the process of embryo development and hatching, indicated that it plays an important role in these processes. Moreover, *BmHEL* transcript can be detected in the midgut and testis at larval stage, suggested that *BmHEL* may have other biological functions. To the best of our knowledge, this is the first report on *HE* gene in the Lepidoptera insects and will be helpful to provide a molecular basis for understanding the complicated mechanism underlying silkworm hatching.

Publish On: Molecular Biology Reports, 2009, 37(3): 1175-1182.

1 College of Biotechnology and Environmental Engineering, Jiangsu University of Science and Technology, Zhenjiang 212018, China

2 The Key Laboratory of Silkworm and Mulberry Genetic Improvement, Ministry of Agriculture, Sericultural Research Institute of Chinese Academy of Agricultural Sciences, Zhenjiang 212018, China

3 *Corresponding author E-mail: tangshunming2002@163.com; shenxj63@yahoo.com.cn

家蚕类孵化酶基因的分子克隆和鉴定

卢福浩[1]　唐顺明[2*]　沈兴家[2*]　王　娜[1]
赵巧玲[2]　张国政[2]　郭锡杰[2]

摘要：孵化是后生动物生命中的重要过程，其中孵化酶（HE）起着关键作用。在研究中，通过电子克隆、SMART cDNA 合成以及 RACE-PCR 技术，我们从家蚕的转青卵中克隆到了类孵化酶 *BmHEL* 的全长 cDNA。*BmHEL* 长度为 974bp，包含一个 885bp 的 ORF，编码 294 个氨基酸。*Bm*HEL 的氨基酸序列与其他物种的 HE 序列同源性为 30.3%—47.1%。两个 *HE* 基因家族相似的信号序列位于 *BmHEL* 中。*BmHEL* 基因由 6 个外显子，5 个内含子和已经预测出的一个得分较高的启动子区域构成，该启动子区域含有一些与胚胎发育相关的转录因子结合位点。家蚕卵在孵化过程的不同发育阶段中，*BmHEL* 的转录可在早期被检测到并维持在较低水平，从孵化第 7 天开始急剧上升，在第 9 天达到最高水平。*BmHEL* 转录水平的变化与胚胎发育和孵化过程一致，表明它在这些过程中起着重要作用。此外，在幼虫阶段的中肠和精巢中也可以检测到 *BmHEL* 的转录，这暗示 *Bm*HEL 可能还有其他的生物学功能。我们的研究将为了解蚕卵孵化的复杂机制提供分子基础。

1　生物与环境工程学院，江苏科技大学，镇江
2　农业部蚕桑遗传改良重点实验室，中国农业科学院蚕业研究所，镇江

Molecular cloning and characterization of hatching enzyme-like gene II (*BmHELII*) in the silkworm, *Bombyx mori*

Tang SM[1,2*] Wu J[1] Zhao XH[1] Wang HY[1]
Qiu ZY[1,2] Shen XJ[1,2] Guo XJ[1,2]

Abstract: Hatching enzyme (HE) is an enzyme that digests an egg envelop at the time of embryo hatching. Previously, we have reported a kind of *Bombyx mori* hatching enzyme-like gene (*BmHEL*). In this paper, the full length of another *BmHEL* cDNA sequence (*BmHELII*, GenBank ID: JN627443) was cloned from bluish-silkworm-eggs. The cDNA was 977 bp in length with an open reading frame of 885 bp which encodes a polypeptide of 294 amino acids including a putative signal peptide of 16 amino acid residues and a mature protein of 278 amino acids. The deduced *BmHELII* had a predicted molecular mass of 33.62 kDa, isoelectric point of 5.44 and two conserved signature sequences of astacin family. Bioinformatic analysis results showed that the deduced protease domain amino acid sequence of *Bm*HELII had 29.5% – 87.0% identities to that of HE identified in the other species. The *BmHELII* gene structure was 6-exon-5-intron, and the promoter region harbored some basal promoter elements and some embryo development related transcription factor binding sites. Semi-quantitative RT-PCR analysis revealed that the relative level of *BmHELII* transcripts at different stages during egg incubation increased with the development of embryos and reached to a maximum just before hatching, hence declined gradually after hatching. The spatio-temporal expression pattern of *BmHELII* basically resembled that of hatching enzyme gene. Moreover, the *BmHELII* transcript was detected in testis of the silkworm, and semi-quantitative RT-PCR analysis showed that it kept at the high level in testis of silkworm from larvae to moth, which suggested that *Bm*HELII might take part in the development of sperm. These results will be helpful to provide a molecular basis for understanding the mechanism underlying silkworm hatching as well as spermatogenesis.

Published On: Biochemical and Biophysical Research Communications, 2012, 419(2), 194-199.

1 Jiangsu University of Science and Technology, Zhenjiang 212018, China.

2 Key Laboratory of Silkworm and Mulberry Genetic Improvement, Ministry of Agriculture, Sericultural Research Institute, Chinese Academy of Agricultural Sciences, Zhenjiang 212018, China.

*Corresponding author E-mail: tangshunming2002@163.com.

家蚕类孵化酶基因II(*BmHELII*)的分子克隆及特征分析

唐顺明[1,2*]　吴　俊[1]　赵新慧[1]　王焕英[1]
裘智勇[1,2]　沈兴家[1,2]　郭锡杰[1,2]

摘要：孵化酶(HE)是一种在胚胎孵化时消化卵壳的酶。先前的报道中，我们发现了一种家蚕孵化酶基因(*BmHEL*)。在本文中，从转青时期的蚕卵中克隆了另一个*BmHEL*的全长cDNA序列(*BmHELII*, GenBank ID: JN627443)。其cDNA的长度为977 bp，开放阅读框为885 bp，编码了一条含294个氨基酸的多肽，其中包括16个氨基酸残基的预测信号肽和278个氨基酸的成熟蛋白。*BmHELII*的理论分子质量为33.62 kDa，等电点为5.44，具有虾红素家族的两个保守特征序列。生物信息学分析结果表明，预测的家蚕*Bm*HELII与其他物种HE的蛋白酶结构域氨基酸序列具有29.5%—87.0%的相似性。*BmHELII*基因结构是6-外显子—5-内含子，其启动子区域包含一些基本的启动子元件和胚胎发育相关的转录因子结合位点。半定量RT-PCR分析显示，在卵孵化期间，不同阶段的*BmHELII*转录本相对水平随胚胎发育而增加，孵化前达到最大值，孵化后逐渐下降。并且*BmHELII*的时空表达模式基本类似于孵化酶基因。此外，检测家蚕精巢中的*BmHELII*转录本，半定量RT-PCR分析显示该转录本从家蚕幼虫到蛾的精巢中保持高水平，这表明*Bm*HELII可能参与了精子的发育。这些结果为更好地理解家蚕孵化机制以及精子发生提供了分子基础。

1　江苏科技大学，镇江
2　农业部蚕桑遗传改良重点实验室，蚕业研究所，中国农业科学院，镇江

第三章

家蚕免疫机制研究

Chapter 3

家蚕依赖其先天免疫系统抵御病原微生物的侵袭，其免疫应答反应包括以生成抗菌肽(anti-microbial peptides, AMP)为主要效应分子的体液免疫、以血淋巴细胞吞噬或包被微生物为主的细胞免疫以及酚氧化酶原(prophenoloxidase, PPO)激活后的黑化反应。近年来，项目组鉴定了参与家蚕免疫应答的重要分子并探明了其功能，对决定病毒、微孢子虫等主要病原微生物侵染和增殖能力的分子进行了深入研究，在解析家蚕免疫系统和病原微生物相互作用的基础上确定了分子育种的靶标，通过转基因技术在家蚕个体水平上实现了抗病性的提高。

家蚕免疫相关分子的鉴定和功能研究。凭借快速发展的基因组学、转录组学和蛋白质组学，项目组鉴定了参与家蚕免疫应答的主要分子，包括病原相关分子模式识别受体、Toll基因家族、多种抗菌肽、转录因子、类免疫球蛋白家族、丝氨酸蛋白酶基因家族、丝氨酸蛋白酶抑制剂家族等，梳理了家蚕免疫应答的信号通路；并针对其中具有代表性的分子，构建了家蚕组织表达谱和被不同微生物感染后的诱导表达谱。通过天然纯化蛋白或构建重组蛋白，研究了蛋白的生化功能和在免疫应答中的作用，例如参与抵御球孢白僵菌的TIL类蛋白酶抑制剂*Bm*SPI38、具有抑制多种病原微生物活性的抗菌肽Glover-ins、作为酚氧化酶途径起始的模式识别受体PGRP-S5、参与家蚕细胞免疫应答的粒细胞特异*Bm*integrinaPS3、免疫稳态调控分子*Bm*FAF等。

病原微生物重要分子的鉴定和功能研究。项目组通过比较基因组学研究了家蚕微孢子虫基因组的进化特征和基因分布，鉴定了家蚕微孢子虫孢壁蛋白、蛋白激酶、几丁质酶、蛋白酶等，解析了家蚕微孢子虫孢壁蛋白质组，精细定位了15种孢壁蛋白和3种极管蛋白，绘制了以SWP1为核心的孢壁蛋白及其他侵染相关蛋白间的相互作用图谱，确定了其中6种孢壁蛋白与侵染有关；明确了类枯草杆菌蛋白酶NbSLP1在极管弹出过程中的重要作用。同时，项目组还绘制了黑胸败血芽孢杆菌基因组图谱，鉴定了其毒素蛋白-toxin和伴孢晶体PC蛋白，揭示了这些蛋白导致家蚕病理的机制。针对引起家蚕脓病的核型多角体病毒*Bm*NPV，项目组分析了其多种早期、晚期表达基因，明确了*Bm*NPV *BmP95*、*odv-e25*等基因对病毒粒子的产生发挥重要作用，而*Bm111*、*Bm91*等基因非病毒复制所必需，但可能影响其毒力。

家蚕与病原微生物的互作分析。项目组系统分析了家蚕被黑胸败血芽孢杆菌、微孢子虫、核型多角体病毒 *Bm*NPV、质型多角体病毒 *Bm*CPV 感染后，家蚕肠道、消化液、脂肪体、大脑等全基因组表达水平或蛋白质组分的变化，从代谢途径、免疫信号通路、细胞凋亡、肠道菌群变化等方面阐述了家蚕对感染的应答。并通过比较易感品系和抗性品系在基因表达和蛋白质水平的差异，鉴定了 *Bm*PKCI、丝氨酸蛋白酶、脂肪酶-1 等多个和家蚕抗病毒能力相关的分子。项目组还发现 *Bm*NPV 通过抑制 *Bm*Spry 的表达来激活家蚕的 MAPK/ERK 信号通路从而促进复制，而病毒也能利用 *Bm*EGFR 和 *Bm*PTP-h 等分子增强其感染和增殖。

利用分子靶标转基因提高家蚕抗病性。在对家蚕免疫应答及其病原微生物致病机理的研究基础上，项目组通过 *piggy*Bac 转基因技术利用组织专一性或诱导型启动子在家蚕体内实现了过表达或干涉多个靶标基因，并检测了这些转基因家蚕的抗病性，例如在家蚕中肠过表达抗病毒基因 *Bmlipase-1*、干涉 *Bm*NPV 复制必需基因 *ie-1* 和 *lef-1* 等，深入探索了抗病分子育种的实现途径。

王菲

Structures, regulatory regions, and inductive expression patterns of antimicrobial peptide genes in the silkworm *Bombyx mori*

Cheng TC Zhao P Liu C Xu PZ Gao ZH Xia QY* Xiang ZH

Abstract: Antimicrobial peptides (AMPs) are a group of immune proteins that protect the host from infection. In *Drosophila*, seven groups of inducible AMPs have been identified, with activities against fungi and gram-positive and gram-negative bacteria. On the basis of the silkworm genome sequence and expressed sequence tags, we identified 35 AMP genes, mostly belonging to the *cecropin*, *moricin*, and *gloverin* gene families. We predicted the core promoters required for gene transcription and the *cis*-regulatory elements for NF-κB/Rel and GATA transcription factors. The expression profiles of these genes after an immune challenge with lipopolysaccharide were examined by reverse transcription PCR. Members of the cecropin B and gloverin A subfamilies were intensely expressed in the fat body after induction. In contrast, those of the moricin B subfamily were not expressed under the same conditions. Such results suggest that these regulatory elements and their positions in the upstream regions play an important role in regulating the transcription of these defense genes.

Published On: Genomics, 2006, 87(3), 356-365.

Key Sericultural Laboratory of Agricultural Ministry, Southwest University, Chongqing 400716, China.
*Corresponding author E-mail: xiaqy@swu.edu.cn.

家蚕抗菌肽基因的结构、调控区域以及诱导表达模式

程廷才　赵　萍　刘　春　许平震　高志宏　夏庆友*　向仲怀

摘要：抗菌肽（AMPs）是一类保护宿主免受感染的免疫蛋白。在果蝇中，7类可诱导的AMPs已经被鉴定，它们具有抗真菌，抗革兰氏阴性细菌和抗革兰氏阳性细菌的能力。基于家蚕的全基因组序列和表达序列标签（ESTs），我们鉴定了35个抗菌肽基因，大部分属于*cecropin*、*moricin*和*gloverin*基因家族。我们对基因转录所需的核心启动子，NF-κB/Rel与GATA转录因子结合的顺式调控元件进行预测。通过反转录PCR对这些基因在脂多糖诱导下的表达谱进行检测分析，结果显示cecropin B和gloverin A亚家族的成员在诱导后的脂肪体中高量表达。相比之下，moricin B亚家族在相同的条件下却没有表达。上述结果暗示了这些调控元件及它们的上游区域在这些防御基因的转录调控上起重要作用。

农业部蚕桑学重点实验室，西南大学，重庆

Proteolytic activation of pro-spätzle is required for the induced transcription of antimicrobial peptide genes in lepidopteran insects

Wang Y[1] Cheng TC[2] Subrahmanyam Rayaprolu[1] Zou Z[1]
Xia QY[2] Xiang ZH[2] Jiang HB[1*]

Abstract: Microbial infection leads to proteolytic activation of *Drosophila* spätzle, which binds to the toll receptor and induces the synthesis of immune proteins. To test whether or not this mechanism exists in lepidopteran insects, we cloned the cDNA of *Bombyx mori spätzle-1* and overexpressed the full-length and truncated *BmSpz1* cDNA in *Escherichia coli*. The insoluble fusion proteins were affinity-purified under denaturing condition. After the silkworm larvae were injected with renatured *Bm*Spz1, mRNA levels of antimicrobial peptide genes were greatly increased. Similar transcriptional up-regulation was also found in *Manduca sexta*. Injection of pro-*Bm*Spz1 had no such effect. When pro-*Bm*Spz1 and *Micrococcus luteus* were incubated with the plasma from *M. sexta* larvae, we detected proteolytic processing of pro-*Bm*Spz1. These results suggest that active spätzle is required for the induced production of antimicrobial peptides in *B. mori* and *M. sexta*.

Published On: Developmental & Comparative Immunology, 2007, 31(10), 1002-1012.

1 Department of Entomology and Plant Pathology, Oklahoma State University, Stillwater, OK 74078, USA.
2 Key Sericultural Laboratory of Agricultural Ministry, Southwest University, Chongqing 400716, China.
*Corresponding author E-mail: haobo.jiang@okstate.edu.

裂解活化的pro-spätzle蛋白是鳞翅目昆虫抗菌肽基因诱导上调表达的必需条件

王 杨[1] 程廷才[2] Subrahmanyam Rayaprolu[1] 邹 振[1]
夏庆友[2] 向仲怀[2] 蒋浩波[1*]

摘要：微生物感染能导致果蝇的spätzle蛋白裂解活化，进而结合Toll受体并诱导合成免疫蛋白。为了验证鳞翅目昆虫是否也存在这种机制，我们克隆了家蚕的*spätzle-1*，并在大肠杆菌中过表达它的全长和截短型的*Bm*Spz1蛋白。通过亲和层析纯化不可溶融合蛋白。用复性的*Bm*Spz1蛋白注射家蚕幼虫，导致抗菌肽基因的转录水平显著升高。在鳞翅目昆虫烟青虫中也得到相似的结果。而注射前体蛋白pro-*Bm*Spz1却不能诱导抗菌肽基因表达。当把pro-*Bm*Spz1蛋白和藤黄微球菌共同在烟青虫的血淋巴中孵育后，我们检测到了pro-*Bm*Spz1蛋白有发生水解的现象。这些结果表明活化的spätzle是家蚕和烟青虫抗菌肽基因诱导上调表达所必需的。

1 昆虫学和植物病理学系，俄克拉荷马州立大学，斯蒂尔沃特，美国
2 农业部蚕桑学重点实验室，西南大学，重庆

Smt3 is required for the immune response of silkworm, *Bombyx mori*

Xu HP　Hao W　He D　Xu YS*

Abstract: SUMO works in a similar way as ubiquitin to alter the biological properties of a target protein by conjugation. The homologous gene of SUMO named *BmSmt3* was identified for the first time in silkworm. The expression of *BmSmt3* was enhanced in the fat body of silkworm after immune challenge. However, the expression of *BmSmt3* after immune challenge was almost invariant in silk gland, which is the nonimmune organ in silkworm. In addition, the expression of *BmRelA* and *CecropinB1* was significantly decreased in pupae after the *BmSmt3* was knocked down *in vivo*. According to our results, *BmSmt3* might participate in the immune response through regulating the expression of *BmRelA* gene, which can further regulate the expression of antibacterial peptide subsequently in silkworm.

Published On: Biochimie, 2010, 92(10), 1306-1314.

Institute of Sericulture & Apiculture, College of Animal Sciences, Zhejiang University, Hangzhou 310029, China.
*Corresponding author E-mail: xuyusong@zju.edu.cn.

*Smt3*基因为家蚕免疫反应所必需的基因

徐和平　郝　威　何　达　徐豫松*

摘要：SUMO与泛素的作用方式相似，都是通过共轭结合到靶蛋白上，从而改变靶蛋白的生物学特征。在家蚕中首次鉴定了名为*BmSmt3*的SUMO同源基因。免疫反应激发以后，家蚕脂肪体中*BmSmt3*的表达增强。并且，免疫反应激发后，*BmSmt3*在家蚕非免疫器官丝腺中的表达几乎不变。另外，在*BmSmt3*被敲低后的蛹中，*BmRelA*和*CecropinB1*的表达明显下降。根据本研究结果，*BmSmt3*可能通过调节*BmRelA*基因的表达参与免疫反应，从而进一步调节家蚕抗菌肽的表达。

蚕蜂研究所，动物科学学院，浙江大学，杭州

Gloverins of the silkworm *Bombyx mori*: structural and binding properties and activities

Yi HY[1,2] Deng XJ [1] Yang WY [1]

Zhou CZ [3] Cao Y [1*] Yu XQ [2*]

Abstract: Gloverins are basic, glycine-rich and heat-stable antibacterial proteins (~14 kDa) in lepidopteran insects with activity against *Escherichia coli*, Gram-positive bacteria, fungi and a virus. *Hyalophora gloveri* gloverin adopts a random coil structure in aqueous solution but has a-helical structure in membrane -like environment, and it may interact with the lipid A moiety of lipopolysaccharide (LPS). *Manduca sexta* gloverin binds to the O-specific antigen and outer core carbohydrate of LPS. In the silkworm *Bombyx mori,* there are four gloverins with slightly acidic to neutral isoelectric points. In this study, we investigate structural and binding properties and activities of *B. mori* gloverins (*Bm*Glvs), as well as correlations between structure, binding property and activity. Recombinant *Bm*Glv1-4 was expressed in bacteria and purified. Circular dichroism (CD) spectra showed that all four *Bm*Glvs mainly adopted random coli structure (>50%) in aqueous solution in regardless of pH, but contained α-helical structure in the presence of 1,1,1,3,3,3-hexafluoro-2-propanol (HFIP), smooth and rough mutants (*Ra*, *Rc* and *Re*) of LPS and lipid A. Plate ELISA assay showed that *Bm*Glvs at pH 5.0 bound to rough mutants of LPS and lipid A but not to smooth LPS. Antibacterial activity assay showed that positively charged *Bm*Glvs (at pH 5.0) were active against *E. coli* mutant strains containing rough LPS but inactive against *E. coli* with smooth LPS. Our results suggest that binding to rough LPS is the prerequisite for the activity of *Bm*Glvs against *E. coli*.

Published On: Insect Biochemistry and Molecular Biology, 2013, 43(7), 612-625.

1 Laboratory of Insect Molecular Biology and Biotechnology, Guangdong Provincial Key Laboratory of Agro-animal Genomics and Molecular Breeding, College of Animal Science, South China Agricultural University, Guangzhou 510642, China.

2 Division of Cell Biology and Biophysics, School of Biological Sciences, University of Missouri-Kansas City, Kansas City, MO 64110, USA.

3 School of Life Sciences, University of Science and Technology of China, Hefei, Anhui 230026, China.

*Corresponding author E-mail: caoyang@scau.edu.cn; Yux@umkc.edu.

家蚕葛佬素的结构、结合特性和活性研究

易辉玉[1,2]　邓小娟[1]　杨婉莹[1]
周丛照[3]　曹　阳[1*]　余小强[2*]

摘要：葛佬素(Gloverins)是鳞翅目昆虫中碱性、富含甘氨酸的热稳定抗菌蛋白(分子质量约14 kDa)，对大肠杆菌、革兰氏阳性菌、真菌和病毒均有抵抗活性。黑曲霉葛佬素在水溶液中形成无规则卷曲结构，但是在类细胞膜环境中有一个α-螺旋，并且它可能与脂多糖(LPS)的类脂A单元相互作用。烟草天蛾的葛佬素结合到O型特异的抗原和LPS的核外碳水化合物上。在家蚕中有4个葛佬素，它们的等电点处于弱酸性到中性范围。在本研究中，我们研究了家蚕葛佬素(*Bm*Glvs)的结构、结合性质和活性以及它们之间的相互关系。我们用细菌表达并纯化了重组的*Bm*Glv1-4蛋白。圆二色(CD)光谱结果表明，在任何pH的水溶液中，这4种葛佬素蛋白主要是以无规则卷曲结构(>50%)存在，但是在存在六氟异丙醇(HFIP)、LPS光滑型和粗糙型的突变体(*Ra*，*Rc*和*Re*)及类脂A的时候含有α-螺旋结构。酶联免疫吸附实验表明，*Bm*Glvs在pH为5时与LPS的粗糙型突变体和类脂A结合而不与光滑型的LPS结合。抗菌活性实验表明，带正电的*Bm*Glvs(pH为5时)对含有粗糙型LPS的大肠杆菌突变株有抑制活性，但对含有光滑型LPS的大肠杆菌没有抑制活性。我们的结果表明，结合粗糙型的LPS是*Bm*Glvs抑制大肠杆菌的前提条件。

1　昆虫分子生物学与生物技术实验室，广东省农业动物基因组学与分子育种重点实验室，动物科学学院，华南农业大学，广州
2　细胞生物学与生物物理学系，生物科学学院，堪萨斯城密苏里大学，堪萨斯，美国
3　生命科学学院，中国科学技术大学，合肥

Translationally controlled tumor protein, a dual functional protein involved in the immune response of the silkworm, *Bombyx mori*

Wang F[#] Hu CM[#] Hua XT Song L Xia QY[*]

Abstract: Insect gut immunity is the first line of defense against oral infection. Although a few immune-related molecules in insect intestine has been identified by genomics or proteomics approach with comparison to well-studied tissues, such as hemolymph or fat body, our knowledge about the molecular mechanism underlying the gut immunity which would involve a variety of unidentified molecules is still limited. To uncover additional molecules that might take part in pathogen recognition, signal transduction or immune regulation in insect intestine, a T7 phage display cDNA library of the silkworm midgut is constructed. By use of different ligands for biopanning, Translationally Controlled Tumor Protein (TCTP) has been selected. *Bm*TCTP is produced in intestinal epithelial cells and released into the gut lumen. The protein level of *Bm*TCTP increases at the early time points during oral microbial infection and declines afterwards. *In vitro* binding assay confirms its activity as a multi-ligand binding molecule and it can further function as an opsonin that promotes the phagocytosis of microorganisms. Moreover, it can induce the production of anti-microbial peptide via a signaling pathway in which ERK is required and a dynamic tyrosine phosphorylation of certain cytoplasmic membrane protein. Taken together, our results characterize *Bm*TCTP as a dual-functional protein involved in both the cellular and the humoral immune response of the silkworm, *Bombyx mori*.

Published On: PLoS ONE, 2013, 8(7).

State Key Laboratory of Silkworm Genome Biology, Southwest University, Chongqing 400716, China.

[#]These authors contributed equally.

[*]Corresponding author E-mail: xiaqy@swu.edu.cn.

翻译调控肿瘤蛋白作为双功能蛋白参与家蚕免疫反应

王　菲[#]　胡翠美[#]　化晓婷　宋　亮　夏庆友[*]

摘要：昆虫肠道免疫是经口感染的第一道防线。虽然通过基因组学或蛋白质组学方法已经鉴定到了一些昆虫肠道免疫相关分子，但是与研究较清楚的组织如血淋巴或脂肪体相比较，对肠道免疫未知分子以及其分子机制的了解仍然有限。为了发现可能参与昆虫肠道病原体识别、信号转导或免疫调节的其他分子，我们构建了家蚕中肠T7噬菌体cDNA展示文库。通过使用不同的配体进行生物筛选，获得了翻译调控肿瘤蛋白（TCTP）。*Bm*TCTP在肠上皮细胞中产生并释放到肠腔中。*Bm*TCTP蛋白质水平在经口感染微生物的早期上升，随后下降。体外结合实验确认其具有多配体结合分子的活性，并且具有调理素作用可以促进微生物吞噬。此外，它可以通过某些细胞质膜蛋白的动态酪氨酸磷酸化激活ERK信号通路来诱导抗菌肽的产生。总之，我们的研究结果表明*Bm*TCTP作为一种双功能蛋白，参与家蚕的细胞和体液免疫反应。

家蚕基因组生物学国家重点实验室，西南大学，重庆

Cloning and characterization of *Bombyx mori* PP-BP, a gene induced by viral infection

Hu ZG Chen KP* Yao Q
Gao GT Xu JP Chen HQ

Abstract: The ENF peptide family, so termed after the consensus sequence in their amino termini (Glu-Asn-Phe -), is assumed to play multiple important roles in defense reactions, growth regulation, and homeostasis of Lepidopteran insects. The paralytic peptide of *Bombyx mori* (*Bm*PP) is one such peptide that is involved in the paralytic and plasmatocyte-spreading activities in the hemocyte immune reaction. The growth-blocking peptide of *Pseudaletia separata* (*Ps*GBP), which is also a member of the ENF peptide family, has similar functions that can reportedly be attenuated by the growth-blocking peptide-binding protein (GBP-BP). Using the fluorescent differential display (FDD) technique, the differential expression pattern of genes in highly susceptible silkworm strain *306* were analyzed, following infection with *B. mori* nuclear polyhedrosis virus (*Bm*NPV), and a differential band ($G12_{782}$) was obtained from the hemolymph RNA pools. Using 5′-RACE with a specially designed primer based on the FDD study, a 1 401 bp cDNA clone was obtained containing a 1 311 bp open reading frame (ORF, GenBank accession number DQ306881). The deduced protein was highly homologous in primary structure to GBP-BP and was termed *B. mori* paralytic peptide-binding protein (PP-BP). The *B. mori PP-BP* gene is organized into two exons and only one intron, using bioinformatics searches. Using RT-PCR analysis, it was found that the *B. mori PP-BP* gene was expressed almost exclusively in the hemolymph. Real-time quantitative PCR analysis indicated that the *B. mori PP-BP* mRNA level in *B. mori* strain *306* exposed to *Bm*NPV was much higher than that in *B. mori* strain without the virus infection. This result implies that the *B. mori* PP-BP is related to the cellular immune response after *Bm*NPV invades the hemolymph.

Published On: Acta Genetica Sinica, 2006, 33(11): 975-983.

Institute of Life Sciences, Jiangsu University, Zhenjiang 212013, China
*Corresponding author E-mail: kpchen@ujs.edu.cn.

家蚕瘫痪肽结合蛋白基因克隆与分析

胡志刚　陈克平*　姚　勤
高贵田　徐家萍　陈慧卿

摘要：ENF肽家族具有保守的N末端结构（Glu-Asn-Phe-）。该家族成员肽大多具有重叠功能活性，在鳞翅目昆虫的免疫反应、生长调控和自体调节等方面都发挥着重要的作用。在昆虫的免疫反应中，血细胞尤其是淋巴液的黏附性是针对外来侵入物的免疫应答过程中的重要因素。家蚕瘫痪肽（paralytic peptide）是ENF肽家族的一种，其具有多种的生物学活性，包括致瘫痪性及在家蚕血细胞免疫反应中的促吞噬细胞扩散活性。ENF肽家族的另一成员，黏虫（*Pseudaletia separata*）的生长阻抑肽（Growth-blocking peptide），同家蚕瘫痪肽一样能够在黏虫的血细胞免疫反应中起到调节吞噬细胞的功能。在本研究中，利用荧光差异显示技术（FDD）分析了家蚕*306*品系感染*Bm*NPV病毒后基因表达差异情况，在血淋巴中获得了一条差异条带$G12_{782}$。在FDD的研究基础上，利用5′-RACE技术和特异性引物，首次在家蚕中克隆得到了该基因的全长1 401 bp的cDNA序列，包含1 311 bp的开放阅读框（ORF），GenBank收入号为DQ306881。通过同源性分析得知，该基因所编码的蛋白质与黏虫的生长阻抑肽结合蛋白具有很高的同源性，并被命名为家蚕瘫痪肽结合蛋白（*B. mori* paralytic peptide binding protein，PP-BP）。利用生物信息学方法对该基因的结构进行了分析，发现该基因具有两个外显子和一个内含子。通过RT-PCR研究发现，该蛋白基因在血淋巴中大量表达。同时，利用实时荧光定量PCR（Real-time quantitative PCR）技术分析了该基因在正常饲养家蚕与添食*Bm*NPV病毒的家蚕中的表达差异，结果显示该基因在家蚕添食*Bm*NPV病毒后的表达量大大增强，这就暗示该基因可能与*Bm*NPV病毒刺激后所引起的家蚕血液细胞免疫反应相关。

生命科学研究院，江苏大学，镇江

Molecular cloning and functional characterization of the dual oxidase (*BmDuox*) gene from the Silkworm *Bombyx mori*

Hu XL[1] Yang R[1] Zhang X[2] Chen L[1]
Xiang XW[1] Gong CL[3] Wu XF[1*]

Abstract: Reactive oxygen species (ROS) from nicotinamide adenine dinucleotide phosphate (NADPH) oxidases and their related dual oxidases are known to have significant roles in innate immunity and cell proliferation. In this study, the 5 545 bp cDNA of the silkworm *Bombyx mori dual oxidase* (*BmDuox*) gene containing a full-length open reading frame was cloned. It includes an N-terminal signal peptide consisting of 28 amino acid residues, a 240 bp 5′-terminal untranslated region (5′-UTR), an 802 bp 3′-terminal region (3′-UTR), which contains nine ATTTA motifs, and a 4 503 bp open reading frame encoding a polypeptide of 1 500 amino acid residues. Structural analysis indicated that *Bm*Duox contains a typical peroxidase domain at the N-terminus followed by a calcium-binding domain, a ferric-reducing domain, six transmembrane regions and binding domains for flavin adenine dinucleotide (FAD) and nicotinamide adenine dinucleotide (NAD). Transcriptional analysis revealed that *BmDuox* mRNA was expressed more highly in the head, testis and trachea compared to the midgut, hemocyte, Malpighian tube, ovary, fat bodies and silk glands. *BmDuox* mRNA was expressed during all the developmental stages of the silkworm. Subcellular localization revealed that *BmDoux* was mainly present in the periphery of the cells. Some cytoplasmic staining was detected, with rare signals in the nucleus. Expression of *BmDuox* was significantly induced in the larval midgut upon challenge by *Escherichia coli* and *Bombyx mori* nucleopolyhedrovirus (*Bm*NPV). *BmDuox*-deleted larvae showed a marked increase in microbial proliferation in the midgut after ingestion of fluorescence-labeled bacteria compared to the control. We conclude that reduced *BmDuox* expression greatly increased the bacterial load, suggesting *Bm*Duox plays an important role in inhibiting microbial proliferation and the maintenance of homeostasis in the silkworm midgut.

Published On: PLoS ONE, 2013, 8(8).

1 College of Animal Sciences, Zhejiang University, Hangzhou 310058, China.
2 Institute of Cell Biology, Zhejiang University, Hangzhou 310058, China.
3 National Engineering Laboratory for Modern Silk, Soochow University, Suzhou 215006, China.
*Corresponding author E-mail: wuxiaofeng@zju.edu.cn.

家蚕双重氧化酶*BmDuox*基因的分子克隆和功能特性

胡小龙[1] 杨 锐[1] 张 星[2] 陈 琳[1]
相兴伟[1] 贡成良[3] 吴小锋[1*]

摘要:已知由烟酰胺腺嘌呤二核苷酸磷酸(NADPH)氧化酶及其相关的双重氧化酶所产生的活性氧(ROS)对先天性免疫及细胞增殖具有重要作用。本文成功克隆了家蚕双重氧化酶(*BmDuox*)基因,其cDNA大小为5 545 bp,开放阅读框(ORF)长4 503 bp,编码1 500个氨基酸残基,N端含有28个氨基酸残基的信号肽序列,5′端非编码区(5′-UTR)长240 bp,3′端非编码区(3′-UTR)长802 bp,含有9个典型的ATTTA基序。结构域分析显示*Bm*Duox蛋白的N端含有典型的过氧化物酶结构域、钙离子结合域、铁还原结构域、6个跨膜结构域、黄素腺嘌呤二核苷酸(FAD)和烟酰胺腺嘌呤二核苷酸(NAD)结合结构域。转录分析结果显示,*BmDuox*基因mRNA在家蚕的头部、精巢和气管中高于中肠、血细胞、马氏管、卵巢、脂肪体和丝腺,且其在家蚕的各个发育时期均有表达。亚细胞定位结果表明,*Bm*Doux蛋白主要分布在细胞外周,细胞质中有少量分布,细胞核中则几乎没有信号。*BmDoux*基因在家蚕中肠中的表达水平能被大肠杆菌和家蚕核型多角体病毒(*Bm*NPV)显著地诱导激活。*BmDuox*缺失家蚕突变体实验显示,经口摄入的荧光标记细菌在肠道中存留数量比对照组显著增加,表明抑制*BmDuox*的表达会导致细菌负荷大大增加。由此说明,*Bm*Duox在抑制肠道微生物的过度增殖、维持家蚕肠道微生物稳态等方面发挥着重要作用。

1 动物科学学院,浙江大学,杭州
2 细胞生物学研究所,浙江大学,杭州
3 现代丝绸国家工程实验室,苏州大学,苏州

Activity of fusion prophenoloxidase-GFP and its potential applications for innate immunity study

Yang B[1] Lu AR[1] Peng Q[1] Ling QZ[2] Ling EJ[1*]

Abstract: Insect prophenoloxidase (PPO) is essential for physiological functions, such as melanization of invading pathogens, wound healing and cuticle sclerotization. The insect PPO activation pathway is well understood. However, it is not very clear how PPO is released from hemocytes and how PPO takes part in cellular immunity. To begin to assess this, three *Drosophila melanogaster PPO* genes were separately fused with GFP at the C-terminus (rPPO-GFP) and were over-expressed in S2 cells. The results of staining and morphological observation show that rPPO-GFP expressed in S2 cells has green fluorescence and enzyme activity if Cu^{2+} was added during transfection. Each rPPO-GFP has similar properties as the corresponding rPPO. However, cells with rPPO-GFP over-expressed are easier to trace without PO activation and staining. Further experiments show that rPPO1-GFP is cleaved and activated by *Drosophila* serine protease, and rPPO1-GFP binds to *Micrococcus luteus* and *Beauveria bassiana* spores as silkworm plasma PPO. The above research indicates that the GFP-tag has no influence on the fusion enzyme activation and PPO-involved innate immunity action *in vitro*. Thus, rPPO-GFP may be a convenient tool for innate immunity study in the future if it can be expressed *in vivo*.

Published On: PLoS ONE, 2013, 8(5).

1 Key Laboratory of Insect Developmental and Evolutionary Biology, Institute of Plant Physiology and Ecology, Shanghai Institutes for Biological Sciences, Chinese Academy of Sciences, Shanghai 200032, China.

2 Department of Applied Biology, Zhejiang Pharmaceutical College, Ningbo 315100, China.

*Corresponding author E-mail: erjunling@sippe.ac.cn.

PPO-GFP融合蛋白的活性及其在昆虫先天免疫研究中的潜在应用

杨 兵[1] 路岸瑞[1] 彭 琴[1] 凌庆枝[2] 凌尔军[1*]

摘要：昆虫酚氧化酶(PPO)对于入侵病原物的黑化、创伤愈合以及表皮硬化等生理功能至关重要。昆虫PPO的激活通路已经研究得较为透彻。然而，关于PPO如何从血细胞中被释放以及如何参与细胞免疫知之甚少。基于此，我们在果蝇的3种*PPO*基因的C端分别加上GFP蛋白(成为rPPO-GFP融合蛋白并在S2细胞中过表达)。染色及形态学观察结果显示在S2细胞表达的rPPO-GFP在转染过程中，加入Cu^{2+}后发出绿荧光并表现出酶活性。每种rPPO-GFP与对应的rPPO具有相似的性质。然而，在细胞中过表达rPPO-GFP在没有PO活性以及染色时更易于追踪监测。进一步的研究结果显示rPPO1-GFP可以被果蝇的丝氨酸蛋白酶剪切激活，而rPPO1-GFP像家蚕细胞质PPO一样也可以与藤黄微球菌*Micrococcus luteus*和球孢白僵菌*Beauveria bassiana*孢子结合。上述研究结果表明GFP标签对融合的酶激活以及PPO参与的体外先天免疫没有影响。因此，在以后的研究中如果rPPO-GFP能够在体内表达将会是研究昆虫先天免疫非常便利的工具。

1 昆虫发育与进化生物学重点实验室，植物生理生态研究所，上海生命科学研究院，中国科学院，上海

2 应用生物学系，浙江医药高等专科学校，宁波

Paralytic peptide activates insect humoral immune response via epidermal growth factor receptor

Song L Wang F Dong SF
Hu CM Hua XT Xia QY*

Abstract: Paralytic peptide (PP) activates innate immunity of silkworm *Bombyx mori*, inducing production of anti-microbial peptides (AMPs) and phagocytosis-related proteins; however, the signal pathways of PP-dependent immune responses are not clear. In the present study, we characterized *BmE* cells as a PP - responsive cell line by examining the expression of *AMP* genes and activation of p38 mitogen-activated protein kinase (p38 MAPK) under PP stimulation, and we also found PP directly binds to *BmE* cell membrane. Then we found that PP-dependent expression of *AMP* genes is suppressed by tyrosine kinase inhibitor (genistein) both in *BmE* cells and in fat body of silkworm larvae. Moreover, the specific tyrosine kinase epidermal growth factor receptor (EGFR) inhibitor (AG1478) attenuates PP-induced expression of AMP genes in *BmE* cells and fat body of silkworm and RNA interference (RNAi) to *Bm*EGFR also suppresses PP-induced expression of *AMP* genes. Furthermore, the PP-induced p38 MAPK phosphorylation is inhibited by AG1478. Our results suggest that *BmE* cells can be used as a cell model to investigate the signal pathway of PP-dependent humoral immune response and receptor tyrosine kinase EGFR / p38 MAPK pathway is involved in the production of AMPs induced by PP.

Published On: Peptides, 2015, 71, 20-27.

State Key Laboratory of Silkworm Genome Biology, Southwest University, Chongqing 400716, China.
*Corresponding author E-mail: xiaqy@swu.edu.cn.

麻痹肽通过表皮生长因子受体激活昆虫体液免疫反应

宋 亮 王 菲 董世峰
胡翠美 化晓婷 夏庆友*

摘要：麻痹肽(PP)激活家蚕先天性免疫，诱导产生抗菌肽(AMPs)和吞噬相关蛋白，然而PP关联的免疫反应信号通路还不清楚。在本研究中，我们以*BmE*细胞作为PP应答细胞来检测PP刺激下*AMP*基因的表达情况和p38丝裂原活化蛋白激酶(p38 MAPK)活化情况，并发现PP直接结合到*BmE*细胞膜上。接下来的实验我们发现酪氨酸激酶抑制剂(genistein)在*BmE*细胞和家蚕幼虫脂肪体中都能抑制*AMP*基因的PP依赖性表达。此外，特异性的酪氨酸激酶表皮生长因子受体(EGFR)抑制剂(AG1478)和RNA干涉减弱可抑制*BmE*细胞及家蚕脂肪体中*AMP*基因的PP诱导性表达。另外，AG1478也能抑制PP诱导的p38丝裂原活化蛋白激酶磷酸化。我们的结果表明*BmE*细胞可用于PP依赖的体液免疫反应信号通路研究，同时也证明EGFR/p38 MAPK通路通过PP的诱导参与AMPs的生成。

家蚕基因组生物学国家重点实验室，西南大学，重庆

A Fas associated factor negatively regulates anti-bacterial immunity by promoting Relish degradation in *Bombyx mori*

Ma XJ Li XY Dong SF Xia QY Wang F*

Abstract: Negative regulation is required to keep NF-κB-dependent immune response under tight control. Previously, we have identified a Fas associated factor (FAF) family member in *Bombyx mori*, *BmFAF*, and proposed it may act as a negative regulator in immune response. In this study, we found knock-down of *BmFAF* by RNAi led to a remarkable increase in transcriptional level of several antimicrobial peptide genes, including *BmCecropinA1* and *BmMoricin*, and higher survival rate to Gram-negative bacterial infection. We also confirmed the regulatory role of *Bm*FAF in suppressing NF-κB-dependent transcription by employing an inducible promoter in *BmE* cells. Consistent with these physiological phenotypes, *Bm*FAF suppressed the activity of the essential transcription factor, Relish, in IMD signaling pathway by promoting its proteasomal degradation through direct interaction. In addition, by constructing various truncation mutants, we further demonstrated that UBA domain in *Bm*FAF is required for the inhibitory role, and potential ubiquitination also occurs in this domain. Taken together, our results suggest that *Bm*FAF is a negative regulator of IMD pathway by mediating degradation of Relish.

Published On: Insect Biochemistry and Molecular Biology, 2015, 63, 144-151.

State Key Laboratory of Silkworm Genome Biology, Southwest University, Chongqing 400716, China.
*Corresponding author E-mail: fwangswu@gmail.com.

Fas相关因子通过降解Relish负调控家蚕抗菌免疫反应

马晓娟　李显扬　董世峰　夏庆友　王　菲*

摘要：负调控是严格控制NF-κB依赖性免疫应答所必需的。在之前的研究中，我们已经鉴定到一个家蚕Fas相关因子（FAF）家族成员*Bm*FAF，推测其可能作为免疫应答的负调控因子。在本研究中，我们发现通过RNAi干涉*Bm*FAF导致几种抗微生物肽基因*BmCecropinA1*和*BmMoricin*的转录水平显著提高，革兰氏阴性细菌感染的存活率提高了。我们还通过在*BmE*细胞中使用诱导型启动子证实了*Bm*FAF抑制NF-κB依赖性转录的调控作用。与这些生理表型一致，*Bm*FAF通过直接与IMD信号通路中必需转录因子Relish相互作用，促进蛋白酶体降解Relish抑制它的活性。此外，通过构建各种截短突变体，我们进一步证明了*Bm*FAF中的UBA结构域是抑制作用所必需的，并且在该结构域中也存在潜在的泛素化。总之，我们的研究结果表明，*Bm*FAF是通过调控Relish的降解调控IMD途径的负调控因子。

家蚕基因组生物学国家重点实验室，西南大学，重庆

Overexpression of *Bm65* correlates with reduced susceptibility to inactivation by UV light

Tang Q Hu ZY Yang YH Wu HL
Qiu LP Chen KP* Li GH*

Abstract: Ultraviolet (UV) light is one of the factors that causes baculovirus inactivation. However, little is known about the response of baculoviruses to UV light. In the present study, *Bombyx mori* nucleopolyhedrovirus (*Bm*NPV) orf 65 (*Bm*65), the homolog of *Autographa californica* nucleopolyhedrovirus orf 79 (*Ac*79), a predicted endonuclease, was analyzed. Preliminary results indicated that *Bm*65 was mainly accumulated within the nucleus and could improve the survival rate of *Escherichia coli* (*E. coli*) and *Bm*NPV BVs after UV radiation, suggesting that *Bm*65 was involved in the repairment of UV-induced DNA damage.

Published On: Journal of Invertebrate Pathology, 2015, 127, 87-92.

Institute of Life Sciences, Jiangsu University, Zhenjiang 212013, China.
*Corresponding author E-mail: kpchen@ujs.edu.cn; ghli@ujs.edu.cn.

Bm65 的过量表达与紫外线灭活后的敏感性降低有关

唐　琦　胡朝阳　杨艳华　乌慧玲
邱立鹏　陈克平*　李国辉*

摘要：紫外线（UV）光是导致杆状病毒被灭活的因素之一。然而，关于杆状病毒对紫外光的反应知之甚少。在本研究中，分析了家蚕核型多角体病毒（*Bm*NPV）orf 65（*Bm*65），即丫纹夜蛾属核型多角体病毒 orf79（*Ac*79）的同源物。*Bm*65 被预测为核酸内切酶。初步结果表明，*Bm*65 主要积累在核内，可以提高紫外线辐射后大肠杆菌和 *Bm*NPV BVs 的存活率，暗示 *Bm*65 参与了紫外线诱导的 DNA 损伤修复。

生命科学研究院，江苏大学，镇江

Characterization of the *Bombyx mori Cecropin A1* promoter regulated by IMD pathway

Hua XT Ma XJ Xue RJ Cheng TC Wang F* Xia QY

Abstract: *Cecropin A1* (*CecA1*) promoter from *Bombyx mori* was cloned and characterized to provide insight into the transcriptional control of this antimicrobial peptide gene upon immune challenges. Reporter gene assays demonstrated that both *Escherichia coli* and lipopolysaccharide could induce expression in *BmE* cells but *B. bombyseptieus* or peptidogly can failed, and the induction pattern of the reporter gene was coincident with the endogenous *CecA1*. Analysis of deletion and mutation constructs revealed that the regulatory region was the *κB* motif located between -176 and -166, and no other predicted elements on *CecA1* promoter affected its inducibility. Insertion of additional *κB* motifs increased the activity of *CecA1* promoter. Furthermore, binding of Relish to *κB* motif was confirmed by electrophoretic mobility shift assay. These findings indicate the regulatory mechanism of *CecA1* expression in IMD pathway and suggest an approach of engineering antimicrobial peptide promoter with enhanced activities that may lead to broad applications.

Published On: Insect Science, 2016, 23(2), 297-304.

State Key Laboratory of Silkworm Genome Biology, Southwest University, Chongqing 400716, China.

*Corresponding author E-mail: fwangswu@gmail.com.

受IMD通路调节的家蚕*Cecropin A1*启动子分析

化晓婷　马晓娟　薛仁举　程廷才　王　菲*　夏庆友

摘要：在本研究中，我们克隆并分析了家蚕的*Cecropin A1*（*CecA1*）启动子，进而深入了解该抗菌肽基因在免疫过程中的转录变化。报告基因检测结果表明，大肠杆菌和脂多糖均可诱导其在*BmE*细胞中表达，而黑胸败血芽孢杆菌或肽聚糖不能诱导，而且报告基因的诱导模式与内源性*CecA1*一致。缺失和突变分析表明调控区是*κB*基序，位于-176和-166之间，且*CecA1*启动子上没有预测到其他能够影响其活性的转录元件。插入额外的*κB*基序能增加*CecA1*启动子的活性。通过凝胶阻滞电泳证实了Relish可以与*κB*基序结合。这些研究结果阐明了*CecA1*在IMD途径中的表达调控机制，并提出了一种可以广泛应用的增强抗菌肽启动子活性的方法。

家蚕基因组生物学国家重点实验室，西南大学，重庆

Functional divergence of six isoforms of antifungal peptide Drosomycin in *Drosophila melanogaster*

Yang WY[1] Wen SY[2] Huang YD[3] Ye MQ[1]
Deng XJ[1] Han D[1] Xia QY[4] Cao Y[1*]

Abstract: *Drosomycin (Drs)* gene encodes a 44-residue inducible antifungal peptide, Drosomycin, in *Drosophila melanogaster.* Six genes, *Drs-lC*, *Drs-lD*, *Drs-lE*, *Drs-lF*, *Drs-lG* and *Drs-lI*, show homology to the *Drs* form in a multigene family on the 3rd chromosome of *D. melanogaster.* It is the first experimental demonstration that the six members in the Drs family act as functional genes. To further delineate the functional divergence of these six members, their cDNA sequences were cloned into the pET-3C vector and expressed in the *E. coli*. The antifungal activity of the expression products was assayed using the Cerletti's method. The results showed a difference among the six isoforms in antifungal activity against the tested fungal strains: in which Drs was most effective and showed antifungal activity to all seven fungal strains, whereas isoform Drs-lC was effective to six strains, Drs-lD was effective to five strains, Drs-lG was effective to four strains, and Drs-lE and Drs-lF were effective to only three strains. Drs-lI had no activity against any tested fungal strains. By comparing the variable residue sites of these six isoforms to that of Drosomycin in the three-dimensional structure, we suggested that the reduction in the antifungal activity was due to the variable residues that were not in the alpha-helix. In addition, two inserted residues (RV) in Drs-lI may affect the dimensional structure and resulted in a functional change. These results may explain the evolution of the *Drosomycin* multigene family and its functional divergence.

Published On: Gene, 2006, 379, 26-32.

1 Department of Sericulture Science, College of Animal Science, South China Agricultural University, Guangzhou 510642, China.

2 Department of Entomology, College of Natural Resource and Environment, South China Agricultural University, Guangzhou 510642, China.

3 Biopharmaceutical Research and Development Center, Jinan University, Guangzhou 510632, China.

4 Key Sericultural Laboratory of Agriculture Ministry, College of Sericulture and Biotechnology, Southwest University, Chongqing 400716, China.

*Corresponding author E-mail: caoyang@scau.edu.cn.

黑腹果蝇中抗真菌肽Drosomycin的六种异构体的功能分化

杨婉莹[1] 温硕洋[2] 黄亚东[3] 叶明强[1]
邓小娟[1] 韩 冬[1] 夏庆友[4] 曹 阳[1*]

摘要：黑腹果蝇中 *Drosomycin*（*Drs*）基因编码44个氨基酸诱导型抗真菌肽Drosomycin。在黑腹果蝇第三号染色体上的多基因家族中，6个基因 *Drs-lC*，*Drs-lD*，*Drs-lE*，*Drs-lR*，*Drs-lgG* 和 *Drs-lI* 显示了与 *Drs* 形式的同源性。这是Drs家族中6位成员作为功能基因的第一个实验证明。为了进一步描述这6个成员的功能差异，将它们的cDNA序列分别克隆到pET-3C载体中，并在大肠杆菌中表达。使用Cerletti方法测定表达产物的抗真菌活性。结果表明，6种异构体在抗真菌活性中与测试真菌菌株之间存在差异：其中Drs最有效，且对于所有7种真菌菌株都显示出抗真菌活性，而同种型Drs-lC对6种菌株有效，Drs-lD对5种菌株有效，Drs-lG对4种菌株有效，Drs-lE和Drs-lF仅对3种菌株有效。Drs-lI对任何测试的真菌菌株都没有抗菌活性。通过比较这6种亚型与Drosomycin在三维结构中的可变残基位点，我们认为抗真菌活性的降低是由于α-螺旋中不存在可变残基。此外，Drs-1I中的两个插入残基（RV）可能影响空间结构并导致功能改变。这些结果可能解释了 *Drosomycin* 多基因家族的进化和功能分化。

1 蚕丝科学系，动物科学学院，华南农业大学，广州
2 昆虫学系，自然资源与环境学院，华南农业大学，广州
3 生物制药研究与开发中心，暨南大学，广州
4 农业部蚕桑学重点实验室，生物技术学院，西南大学，重庆

Inducing and isolation of antibacterial peptides from oriental fruit fly, *Bactrocera dorsalis* Hendel

Dang XL[1] Tian JH[2] Yi HY[1] Wang WX[1]
Zheng M[1] Li YF[1] Cao Y[3] Wen SY[1*]

Abstract: One antibacterial activity fraction from an immunized dipteran insect, *Bactrocera dorsalis*, was isolated and purified by prepurification, ion-exchange chromatography, gel filtration chromatography and reverse-phase high performance liquid chromatography (HPLC). The final purified fraction was checked on the Smart system HPLC and was judged as a pure fraction. The results of physical and biological analysis revealed that this fraction is heat stable and showed strong activities against Gram-positive bacterial growth. It possesses antibicrobial peptide properties and is worth further investigation.

Published On: Insect Science, 2006,13(4), 257-262.

1 Department of Entomology, South China Agricultural University, Guangzhou 510642, China.
2 Department of Material Science and Engineering, Jinan University, Jinan 510632, China.
3 Department of Sericulture Science, College of Animal Science, South China Agricultural University, Guangzhou 510642, China.
*Corresponding author E-mail: shywen@scau.edu.cn.

桔实蝇中抗菌肽的诱导和分离

党向利[1]　田金环[2]　易辉玉[1]　王文献[1]
郑　敏[1]　李怡峰[1]　曹　阳[3]　温硕洋[1*]

摘要：通过预纯化、离子交换色谱、凝胶过滤色谱和反相高效液相色谱（HPLC）分离纯化了免疫双翅目昆虫桔实蝇体内的一种抗菌活性成分。在Smart系统HPLC上检查最后纯化的组分，并将其判断为纯组分。物理和生物学分析的结果表明，该组分是热稳定的，并且显示出对革兰氏阳性菌生长的强烈抑制活性。它具有抗菌肽的性质，值得进一步研究。

1　昆虫学系，华南农业大学，广州
2　材料科学与工程系，暨南大学，广州
3　蚕丝科学系，动物科学学院，华南农业大学，广州

A silkworm hemolymph protein is a prophenoloxidase activation blocker

Liu BL Qi XW Han Q
Jia L Xiang ZH He NJ*

Abstract: Melanization in insect hemolymph is triggered by the recognition of pathogen-associated molecular patterns via pattern recognition receptors. The signal transduction leads to the activation of the prophenoloxidase and hence the generation of melanin. The proPO activation process must be tightly controlled to minimize the host damage caused by reactive intermediates during melanin synthesis. The full-length cDNA sequence of a 20 kDa hemolymph protein from *Bombyx* (*Bm*hp20) was determined. *Bmhp20* gene was expressed in larval fat body, integument, trachea, and ovary and was induced by the challenge of *B. bombyseptieus*. Binding of recombinant *Bm*hp20 to microbial cell wall components as well as gram-positive bacteria and fungi was confirmed. Phenoloxidase activity assay indicated that recombinant *Bm*hp20 blocked the proPO activation in hemolymph that was triggered by peptidoglycan or beta-1, 3-glucan. Our data suggest that *Bm*hp20 plays bifunctional roles in silkworm humoral responses: to participate in pattern recognition and to block the activation of proPO.

Published On: American Journal of Molecular Biology, 2012, 2, 332-340.

State Key Laboratory of Silkworm Genome Biology, Southwest University, Chongqing 400716, China.
*Corresponding author E-mail: hejia@swu.edu.cn.

家蚕血淋巴蛋白是一种酚氧化酶活化阻断剂

刘碧朗　亓希武　韩　琦
贾　凌　向仲怀　何宁佳*

摘要：通过模式识别受体识别与病原体相关的分子模式，引发了昆虫血淋巴中的黑化。信号转导引起酚氧化酶活化，进而导致黑色素的产生。proPO活化过程必须被严格控制，以最大限度地减少黑色素合成期间由反应性中间体引起的宿主损伤。测定家蚕20 kDa血淋巴蛋白（*Bm*hp20）的全长cDNA序列。*Bmhp20*基因在幼虫脂肪体、皮肤、气管和卵巢中表达，并被黑胸败血芽孢杆菌诱导。重组*Bm*hp20与微生物细胞壁组分以及革兰氏阳性细菌和真菌的结合也得到了证实。酚氧化酶活性测定表明重组*Bm*hp20阻断由肽聚糖或β-1，3-葡聚糖触发的血淋巴中的proPO活化。我们的数据表明，*Bm*hp20在家蚕体液应答中起双功能作用：参与模式识别并阻止proPO的激活。

家蚕基因组生物学国家重点实验室，西南大学，重庆

Molecular cloning, characterization and expression analysis of *Cathepsin O* in silkworm *Bombyx mori* related to bacterial response

Zhang K# Su JJ# Chen SY Yu S Tan J Xu M
Liang HH Zhao YZ Chao HJ Yang LQ Cui HJ*

Abstract: Cathepsins are the main members of the cysteine family and play important roles in immune response in vertebrates. The *Cathepsin O* of *Bombyx mori* (*BmCathepsin O*) was cloned from the hemocytes by the rapid amplification of cDNA ends (RACE). The genomic DNA was 6 131 bp long with a total of six exons and five introns. Its pre-mRNA was spliced to generate two spliceosomes. By comparisons with other reported *Cathepsins O*, it was concluded that the identity between them ranged from 29% to 39%. Expression analysis indicated that *BmCathepsin O* was specific-expressed in hemocytes, and highly expressed at the 4th molting and metamorphosis stages. Immunofluorescence assay and qRT-PCR showed that *BmCathepsin O* was expressed in granulocytes and plasmatocytes. Interestingly, *BmCathepsin O* was significantly up-regulated after stimulated by 20-hydroxyecdysone (20-E) *in vivo*, which suggested that *BmCathepsin O* may be regulated by 20E. Moreover, activation of *BmCathepsin O* was also observed in hemocytes challenged by *Escherichia coli*, indicating its potential involvement in the innate immune system of silkworm, *B. mori*. In summary, our studies provide a new insight into the functional features of Cathepsin O.

Published On: Molecular Immunology, 2015, 66(2), 409-417.

State Key Laboratory of Silkworm Genome Biology, Southwest University, Chongqing 400716, China.
#These authors contributed equally.
*Corresponding author E-mail: hongjuan.cui@gmail.com; hcui@swu.edu.cn.

家蚕细菌应答相关基因*Cathepsin O*的分子克隆、鉴定和表达分析

张 奎[#] 苏晶晶[#] 陈思源 余 霜 谈 娟
徐 曼 梁航华 赵羽卒 晁会娟 杨丽群 崔红娟[*]

摘要：组织蛋白酶是半胱氨酸家族的主要成员，其在脊椎动物的免疫应答中起重要作用。利用cDNA末端快速扩增（RACE）技术，从血细胞中克隆了家蚕组织蛋白酶O（*BmCathepsin O*）。*BmCathepsin O*基因全长为6 131 bp，共包含6个外显子和5个内含子。它的前体mRNA被剪切产生了两个剪切体。它们与其他已报道的组织蛋白酶O基因的相似度为29%—39%。表达分析表明*BmCathepsin O*在血细胞中特异表达，并在四眠期和变态期高量表达。免疫荧光实验和qRT-PCR结果显示，*BmCathepsin O*在粒细胞和浆细胞中表达。有趣的是，当家蚕受到20-羟基蜕皮激素（20-E）刺激后，*BmCathepsin O*表达量显著上调，这表明*BmCathepsin O*可能受20E调节。此外，经大肠杆菌处理后，血细胞中的*BmCathepsin O*也被激活，表明其在家蚕的先天免疫系统中发挥潜在的功能。总之，本研究为组织蛋白酶O的功能提供了新的认识。

家蚕基因组生物学国家重点实验室，西南大学，重庆

Molecular cloning of *Bombyx mori* cytochrome *P-450* gene and its involvement in fluoride resistance

Zhou HL　Chen KP*　Yao Q　Gao L　Wang Y

Abstract: To investigate the effects of fluorosis on development and gene expression profiles of silkworm, highly resistant silkworm strain 441, and highly susceptible silkworm strain 440 were treated with 200ppm fluoride (designated as 440F and 441F) and water (designated as 440DZ and 441DZ). Fluorotic silkworm showed body color and behavior changes. Statistical analysis indicated that growth index of 440F was lower than 440DZ, 441DZ, and 441F. The mortality of 440F was higher than others. Fluorescent differential display enabled us to obtain a differentially expressed cDNA. Bioinformatics analyses indicated that it belonged to cytochrome *P-450* family, denoted *Bmcyp306a1*, which contained seven exons and six introns. Phylogenetic tree showed *Bm*CYP306A1 had high homology with *Manduca sexta*' P-450 protein. Expression analysis indicated that *Bmcyp306a1* was exclusively expressed in 441DZ and 441F and was down-regulated under fluoride treatment. The tissue-specific expression indicated *Bmcyp306a1* had high-expression level in midgut and ovary in 441F. The data revealed that there was obvious dose-effect and times relationship with the pathological changes and gene expression. Expression profiles of *Bmcyp306a1* suggested that *P-450* gene was crucial to physiological modification and might be involved in fluoride resistance.

Published On: Journal of Hazardous Materials, 2008, 160(2-3), 330-336.

Institute of Life Sciences, Jiangsu University, Zhenjiang 212013, China.
*Corresponding author E-mail: kpchen@ujs.edu.cn.

家蚕细胞色素*P-450*基因的分子克隆及其在抵抗氟化物中毒中的作用

周红亮 陈克平* 姚 勤 高 路 王 勇

摘要：为了研究氟中毒对家蚕发育及基因表达特征的影响，采用200 ppm（百万分之一）氟化物分别对高耐受蚕品系441和高敏感蚕品系440进行处理，以水处理作为对照，处理后的群体分别命名为440F和441F、440DZ和441DZ。氟中毒的家蚕在身体颜色和行为上均有所变化。统计分析表明，440F的生长指数低于440DZ、441DZ和441F。440F的死亡率高于其他组。通过荧光差异技术，我们在抗性品系中获得了一条差异表达的cDNA。生物信息学分析表明，差异表达基因属于细胞色素*P-450*家族，命名为*Bmcyp306a1*，其含有7个外显子和6个内含子。系统进化分析显示*Bm*CYP306A1与烟草天蛾P-450蛋白具有高度的同源性。表达分析表明，*Bmcyp306a1*仅在441DZ和441F中表达，且在氟化物处理后表达下调。*Bmcyp306a1*在441F的中肠和卵巢中高量表达。结果表明，家蚕的病理变化和基因表达水平均具有明显的剂量效应和时间关系。*Bmcyp306a1*的表达特征显示*P-450*基因对生理修饰至关重要，可能参与家蚕对氟化物的抗性。

生命科学研究院，江苏大学，镇江

A short-type peptidoglycan recognition protein from the silkworm: expression, characterization and involvement in the prophenoloxidase activation pathway

Chen KK[1] Liu C[1] He Y[2] Jiang HB[2] Lü ZQ[1,3*]

Abstract: Recognition of invading microbes as non-self is the first step of immune responses. In insects, peptidoglycan recognition proteins (PGRPs) detect peptidoglycans (PGs) of bacterial cell wall, leading to the activation of defense responses. Twelve PGRPs have been identified in the silkworm, *Bombyx mori*, through bioinformatics analysis. However, their biochemical functions are mostly uncharacterized. In this study, we found *PGRP-S5* transcript levels were up-regulated in fat body and midgut after bacterial infection. Using recombinant protein isolated from *Escherichia coli*, we showed that PGRP-S5 binds to PGs from certain bacterial strains and induces bacteria agglutination. Enzyme activity assay confirmed PGRP-S5 is an amidase; we also showed it is an antibacterial protein effective against both Gram-positive and - negative bacteria. Additionally, we demonstrated that speci fi c recognition of PGs by PGRP-S5 is involved in the prophenoloxidase activation pathway. Together, these data suggest the silkworm PGRP-S5 functions as a pattern recognition receptor for the prophenoloxidase pathway initiation and as an effecter to inhibit bacterial growth as well. We finally discussed possible roles of PGRP-S5 as a receptor for antimicrobial peptide gene induction and as an immune modulator in the midgut.

Published On: Developmental and Comparative Immunology, 2014, 45(1), 1-9.

1 Department of Entomology, College of Plant Protection, Northwest A&F University, Yangling 712100, China.

2 Department of Entomology and Plant Pathology, Oklahoma State University, OK 74078, USA.

3 Key Laboratory of Plant Protection Resources and Pest Management, Ministry of Education, Northwest A&F University, Yangling 712100, China.

*Corresponding author E-mail: zhiqiang.lu@nwsuaf.edu.cn.

家蚕肽聚糖识别蛋白PGRP-S5的表达纯化及其在酚氧化酶级联反应中的作用研究

陈康康[1] 刘 晨[1] 何 燕[2] 蒋浩波[2] 吕志强[1,3*]

摘要：识别非己入侵病原物是免疫反应的第一步。在昆虫中，肽聚糖识别蛋白（peptidoglycan recognition proteins，PGRPs）可以识别细菌细胞壁的肽聚糖（peptidoglycans，PGs），并激活昆虫的免疫反应以抵御入侵病原物。通过生物信息学分析发现在家蚕（*Bombyx mori*）中有12种肽聚糖识别蛋白，但这些肽聚糖识别蛋白的生理功能大多数仍未知。在本研究中，我们发现细菌感染后的家蚕中*PGRP-S5*在脂肪体和中肠中都上调表达。通过体外重组表达的家蚕PGRP-S5与不同类型肽聚糖的结合实验表明，PGRP-S5能够结合某些细菌的肽聚糖并能促使细菌凝集。酶活分析证实家蚕PGRP-S5具有酰胺酶活性，对革兰氏阳性菌和革兰氏阴性菌均有高效的抗菌作用。另外，我们发现家蚕PGRP-S5特异性识别肽聚糖后可以参与酚氧化酶激活途径。这些结果证明家蚕PGRP-S5可以作为模式识别受体引发酚氧化酶途径的激活，同时家蚕PGRP-S5还可以作为效应因子抑制细菌的生长。最后，我们讨论了PGRP-S5作为抑菌肽基因诱导受体和中肠免疫调节剂的可能作用。

1 植物保护学院，西北农林科技大学，杨凌
2 昆虫学与植物病理学系，俄克拉何马州立大学，俄克拉何马州，美国
3 植保资源与病虫害治理重点实验室，西北农林科技大学，杨凌

Expression and localization of silkworm adaptor protein complex-1 subunits, which were down-regulated post baculovirus infection

Niu YS Wang MX Liang S Zhou F Miao YG*

Abstract: Adaptor protein complexes (APs) function as vesicle coat components in different membrane traffic pathways. In this study the subunits of adaptor protein complex-1 (AP-1) of silkworm *Bombyx mori* were molecularly characterized. All coding genes for the four subunits were cloned and sequenced. Phylogenic tree for each adaptin was constructed and all subunits were found to be conserved in respective group among organisms. The mRNA expression pattern for each adaptin was similar among tissues. Alternative splicing event was observed in genes encoding both the heavy chain gamma and beta adaptin and the light chain subunit, which could generate other possible adaptin forms. GFP-tagged fusion proteins indicated that AP-1 located in the peripheral plasma area. Furthermore, the *Bm*NPV infection in *B. mori* cells had differentiated effect on the expression level of AP-1 subunits.

Published On: Molecular Biology Reports, 2012, 39(12), 10775-10783.

Key Laboratory of Animal Virology of Ministry of Agriculture, College of Animal Sciences, Zhejiang University, Hangzhou 310058, China.

*Corresponding author E-mail: miaoyg@zju.edu.cn.

杆状病毒感染后下调表达的家蚕衔接蛋白复合物-1的亚基分子的表达和定位

牛艳山　王梅仙　梁　爽　周　芳　缪云根*

摘要：衔接蛋白复合体（APs）是不同膜交换途径中的囊泡外壳的组成成分。本研究对家蚕衔接蛋白复合物-1（AP-1）的亚基分子进行表征。对编码APs 4个亚基的基因进行克隆和序列分析。构建了不同物种衔接蛋白的进化发育树，发现所有亚基在物种中保守，衔接蛋白在不同的组织中表达模式相似。结果表明：在编码重链γ和β衔接蛋白以及轻链亚基的基因中存在选择性剪接，其可以产生其他可能的衔接蛋白形式。GFP标记的融合蛋白表明AP-1位于细胞质中。此外，*Bm*NPV感染家蚕细胞，导致AP-1亚基的表达水平发生改变。

农业部动物病毒学重点实验室，动物科学学院，浙江大学，杭州

Characterization of *CIb1* gene promoter from silkworm, *Bombyx mori*

Zhao QL[1,2,3] Shen XJ[2,3] Zhu LJ[1*] Yi YZ[2,3]
Tang SM[2,3] Zhang GZ[2,3] Guo XJ[2,3]

Abstract: The hemolymph chymotrypsin inhibitor b1 (CIb1) of silkworm, *Bombyx mori*, plays an important role in innate immunity. In order to study its encoding gene *CIb1*, five heterogeneous promoter fragments of 844 bp, 682 bp, 516 bp, 312 bp and 82 bp in length were cloned from genomic DNA of the *p50* silkworm strain. Characterization of the *CIb1* promoter was performed *in vitro* using the firefly *luciferase* gene as reporter. The results showed that *CIb1* promoter fragments have transcription activities in the *B. mori* ovary-derived *BmN* cell line. The 82 bp fragment (-72 to +10 nt) containing the eukaryotic core promoter elements revealed a basic transcription activity. The *Bm1* element, upstream the transcription initiation site, showed a positive regulation function to the *CIb1* promoter. *CIb1* promoter-like fragments from the genomic DNA of the tetra hybrid silkworm *Suju* × *Minghu* provided a natural deletion model for the study of the *CIb1* promoter. *In vitro* analysis indicated that the 132 bp fragment from -517 nt to -386 nt upstream of the transcription initiation site strongly suppressed the transcription activity of the *CIb1* promoter, suggesting that the 132 bp fragment harbours strong negative *cis*-acting elements. Infection of *Bombyx mori* nucleopolyhedrovirus (*Bm*NPV) increased the activity of the *CIb1* promoter, having provided another evidence to the function of *CIb1* in the innate immunity of silkworm.

Published On: Zeit Schrift Fur Naturforschung Section C—A Journal of Biosciences, 2007, 62 (11-12), 875-880.

1 College of Animal Science, Zhejiang University, Hangzhou 310029, China.
2 The Sericultural Research Institute, Jiangsu University of Science and Technology, Zhenjiang 212018, China.
3 The Key Laboratory of Silkworm Biotechnology, Ministry of Agriculture, Zhenjiang 212018, China.
*Corresponding author E-mail: ljzhu@zjs.edu.cn.

家蚕*CIb1*基因启动子的特征分析

赵巧玲[1,2,3] 沈兴家[2,3] 朱良均[1*] 易咏竹[2,3]
唐顺明[2,3] 张国政[2,3] 郭锡杰[2,3]

摘要:家蚕血淋巴胰凝乳蛋白酶抑制剂b1(CIb1)在先天免疫中起着重要作用。为了研究其编码基因*CIb1*,从*p50*家蚕基因组DNA中克隆了5个长度为844 bp,682 bp,516 bp,312 bp和82 bp的不同的启动子片段。在体外使用萤火虫荧光素酶基因作为报告基因进行*CIb1*启动子的表征。结果表明,*CIb1*启动子片段在家蚕卵巢细胞来源的*BmN*细胞系中具有转录活性。含有真核核心启动子元件的82 bp片段(-72 nt至+10 nt)揭示了其基本的转录活性。转录起始位点上游的*Bm1*元件对*CIb1*启动子显示出正调节功能。来自四元杂交种*Suju* × *Minghu*的基因组DNA的*CIb1*启动子片段为*CIb1*启动子的研究提供了一种天然的缺失模型。体外分析表明,转录起始位点上游的-517 nt至-386 nt的132 bp片段强烈抑制了*CIb1*启动子的转录活性,表明132 bp片段含有强的负顺式作用元件。家蚕核型多角体病毒(*Bm*NPV)的感染增加了*CIb1*启动子的活性,为家蚕先天免疫中*CIb1*的功能提供了另一个证据。

1 动物科学学院,浙江大学,杭州
2 蚕业研究所,江苏科技大学,镇江
3 农业部家蚕生物技术重点实验室,镇江

Molecular cloning, expression and characterization of *Bmserpin-2* gene from *Bombyx mori*

Pan Y[1] Xia HC[1] Lü P[1] Chen KP[1*] Yao Q[1]
Chen HQ[1] Gao L[1,2] He YQ[1] Wang L[3]

Abstract: Serpins are a broadly distributed family of protease inhibitors. In this study, the gene encoding *Bombyx mori* serpin-2 (*Bmserpin-2*) was cloned and expressed in *E. coli*. The *Bmserpin-2* cDNA contains a 1 125 bp open reading frame (ORF). The deduced protein has 374 amino-acid residues, contains a conserved SERPIN domain and shares extensive homology with other invertebrate serpins. RT-PCR analysis showed that *Bmserpin-2* was expressed in all developmental stages of *B. mori* larvae and various larval tissues. Subcellular localization analysis indicated that Bmserpin-2 protein was located in the cytoplasm. Interestingly, real-time quantitative PCR revealed that the expression of *Bmserpin-2* in the midgut of susceptible *B. mori* strain *306* significantly increased at 72 hours post inoculation (h.p.i.) when infected with *Bm*NPV. However, there was no significant increase of the *Bmserpin-2* expression in resistant strain *NB* infected with *Bm*NPV. Thus, our data indicates that *Bm*serpin-2 may be involved in *B. mori* antiviral response.

Published On: Acta Biochimica Polonica, 2009, 56(4), 671-677.

1 Institute of Life Sciences, Jiangsu University, Zhenjiang 212013, China.
2 School of Medical Science and Laboratory Medicine, Jiangsu University, Zhenjiang 212013, China.
3 Beijing Entry-Exit Inspection and Quarantine Bureau, Beijing 100026, China.
*Corresponding author E-mail: kpchen@ujs.edu.cn.

家蚕*Bmserpin-2*基因的分子克隆、表达及特征分析

潘　晔[1]　夏恒传[1]　吕　鹏[1]　陈克平[1*]　姚　勤[1]
陈慧卿[1]　高　路[1,2]　何远清[1]　汪　琳[3]

摘要:Serpins是一种广泛分布的蛋白酶抑制剂家族。在本研究中,我们克隆了*Bmserpin-2*基因并在大肠杆菌中进行表达。*Bmserpin-2*的cDNA包含1个1 125 bp的开放阅读框,预测蛋白质含有374个氨基酸残基,包含一个保守的SERPIN结构域并且和其他无脊椎动物中的serpins广泛同源。RT-PCR分析结果显示*Bmserpin-2*在各个发育阶段的家蚕幼虫各组织中均有表达。亚细胞定位分析表明*Bm*serpin-2蛋白位于细胞质中。有趣的是,qRT-PCR结果显示,易感家蚕品种*306*在感染*Bm*NPV 72 h后,其中肠中的*Bmserpin-2*表达量显著增加;而在抗性品种*NB*中,感染*Bm*NPV之后,*Bmserpin-2*的表达量没有明显变化。以上数据暗示*Bm*serpin-2蛋白可能和家蚕的抗病毒应答有关。

1　生命科学研究院,江苏大学,镇江
2　基础医学与医学技术学院,江苏大学,镇江
3　北京出入境检验检疫局,北京

Expression, purification and characterization of a three-domain Kazal-type inhibitor from silkworm pupae (*Bombyx mori*)

Zheng QL　Chen J　Nie ZM
Lü ZB　Wang D　Zhang YZ*

Abstract: Serine protease inhibitors are essential for host physiological and immunological activities in insects. Analyzing the amino-acid sequence of a cDNA coding for a serine protease inhibitor in *Bombyx mori* (*Bm*SPI), we found that *Bm*SPI contained three homologous domains with a conserved sequence of C-X3-C-X9-C-X6-Y-X7-C-X3-C-X11-C similar to that of Kazal-type serine protease inhibitors, suggesting *Bm*SPI as a new member of the Kazal-type serine protease inhibitor family. To characterize the three-domain Kazal-type inhibitor from silkworm pupae, the recombinant protein was expressed in *Escherichia coli* BL21 (DE3) Star. After purification with affinity and reversed-phase chromatographies, the recombinant *Bm*SPI with a molecular mass of 33.642 kDa was shown to be a specific subtilisin A inhibitor. Further studies indicated that the *Ki* value of the recombinant *Bm*SPI was 3.35 nmol/L and the inhibitor seemed to form a 1∶1 complex with subtilisin A. This is a first description of the structure and characterization of Kazal-type inhibitor with three domains cloned from silkworm pupae, *B.mori.*

Published On: Comparative Biochemistry and Physiology B—Biochemistry & Molecular Biology, 2007,146(2), 234-240.

College of Life Science, Zhejiang Sci-Tech University, Hangzhou 310018, China.
*Corresponding author E-mail: yaozhou@chinagene.com.

家蚕蛹中含有三个结构域的Kazal型蛋白酶抑制剂的表达、纯化和表征

郑青亮　陈　健　聂作明
吕正兵　王　丹　张耀洲*

摘要：丝氨酸蛋白酶抑制剂对昆虫的宿主生理和免疫反应至关重要。通过分析家蚕丝氨酸蛋白酶抑制剂（*Bm*SPI）cDNA编码的氨基酸序列，发现*Bm*SPI含有3个同源结构域，其保守序列为C-X3-C-X9-C-X6-Y-X7-C-X3-C-X11-C，类似于Kazal型丝氨酸蛋白酶抑制剂，这表明*Bm*SPI是Kazal型丝氨酸蛋白抑制剂家族的新成员。为了表征来自家蚕蛹中含有3个结构域的Kazal型抑制剂，我们利用大肠杆菌BL21（DE3）Star进行了重组蛋白的表达。在通过亲和层析和反相色谱法纯化后，得到分子质量为33.642 kDa的重组蛋白，*Bm*SPI被证明是特异性的枯草杆菌蛋白酶A抑制剂。进一步研究表明，重组蛋白*Bm*SPI的*Ki*值为3.35 nmol/L，并且该抑制剂与枯草杆菌蛋白酶A似乎是以1∶1的方式形成复合物。这是首次对从家蚕蛹中克隆出的含有3个结构域的Kazal型抑制剂的结构和特征进行描述。

生命科学学院，浙江理工大学，浙江，杭州

Glutathione S-transferases from the larval gut of the silkworm *Bombyx mori*: cDNA cloning, gene structure, expression and distribution

Gui ZZ[1,2] Bo Yeon Kim[1] Kwang Sik Lee[1] Wei YD[1,2]
Guo XJ[2] Hung Dae Sohn[1] Byung Rae Jin[1*]

Abstract: Two glutathione S-transferase (GST) cDNAs, GSTD2 and GSTS2, were cloned from the silkworm *Bombyx mori*. The *B. mori* GSTD2 (*BmGSTD2*) gene spans 4 371 bp and consists of four introns and five exons that encode 222 amino acid residues. The deduced amino acid sequence of *Bm*GSTD2 showed 58% protein sequence identity to the Delta-class GST of *Maduca sexta*. The *B. mori* GSTS2 (*BmGSTS2*) gene spans 3 470 bp and consists of three introns and four exons that encode 206 amino acid residues. The deduced amino acid sequence of BmGSTS2 revealed 67%, 63%, and 61% protein sequence identities to the Sigma-class GSTs from *B. mori*, *Platynota idaeusalis*, and *M. sexta*, respectively. The *BmGSTD2* and *BmGSTS2* cDNAs were expressed as 25 kDa and 23 kDa polypeptides, respectively, in baculovirus-infected insect *Sf9* cells. Northern blot and Western blot analyses showed that *Bm*GSTD2 and *Bm*GSTS2 were specifically expressed in three gut regions, indicating that the gut is the prime site for *Bm*GSTD2 and *Bm*GSTS2 synthesis in *B. mori* larvae.

Published On: European Journal of Entomology, 2008, 105(4), 567-574.

1 College of Natural Resources and Life Science, Dong-A University, Busan 604-714, Korea.
2 Sericultural Research Institute, Chinese Academy of Agricultural Sciences, Zhenjiang 212018, China.
*Corresponding author E-mail: brjin@dau.ac.kr.

家蚕幼虫肠道谷胱甘肽S转移酶的克隆、基因结构、表达与分布

桂仲争[1,2] Bo Yeon Kim[1] Kwang Sik Lee[1] 韦亚东[1,2]
郭锡杰[2] Hung Dae Sohn[1] Byung Rae Jin[1*]

摘要：从家蚕中克隆得到两个谷胱甘肽S转移酶(GST)，分别命名为GSTD2和GSTS2。家蚕GSTD2(*BmGSTD2*)基因全长4 371 bp，由4个内含子和5个外显子组成，编码222个氨基酸。*Bm*GSTD2的氨基酸序列与烟草天蛾的δ家族GST具有58%的同源性。家蚕*BmGSTS2*基因全长3 470 bp，包含3个内含子和4个外显子，编码206个氨基酸。*Bm*GSTS2的氨基酸序列与家蚕、苹白小卷蛾和烟草天蛾的σ家族GST分别具有67%、63%和61%的同源性。利用杆状病毒侵染*sf9*细胞对这两种蛋白质进行了表达，所得到的蛋白质分别为25 kDa(*Bm*GSTD2)和23 kDa(*Bm*GSTS2)。Northern blot和Western blot显示这两种蛋白都在家蚕肠道3个区段特异表达，暗示了家蚕肠道是GST的主要合成部位。

1 自然资源与生命科学学院，东亚大学，釜山，韩国
2 蚕业研究所，中国农业科学院，镇江

Purification and biochemical characterization of a novel glutathione S-transferase of the silkworm, *Bombyx mori*

Hou CX[1,2] Fu ZQ[3] Byung Rae Jin[4] Gui ZZ[1,2*]

Abstract: A novel glutathione S-transferase has been purified from *Bombyx mori* larvae using affinity chromatography on a glutathione agarose column. The purified enzyme appeared as a single band on SDS-PAGE and had a M_r of 28 kDa. Steady state kinetic assays of the enzyme were conducted with 1-chloro-2,4-dinitrobenzene as a substrate. The K_m, V_{max}, K_{cat} and K_{cat}/K_m for the purified *Bm*GST were 0.494 mmol/L, 72.07 μmol/(min · mg), 65.43 s^{-1} and 132.45 $(mmol/L)^{-1} \cdot s^{-1}$, respectively. The enzyme had a maximum activity at approximately pH 7.1 and 25 °C. *Bm*GST indicated lower inhibitory rate by some inhibitors (albendazol, praziquantel, bile acid and NaCl), suggesting that this novel *Bm*GST could differ structurally or functionally from other animal GSTs.

Published On: African Journal of Biotechnology, 2008, 7(3), 311-316.

1 Jiangsu University of Science and Technology, Zhenjiang 212018, China.
2 Sericultural Research Institute, Chinese Academy of Agricultural Sciences, Zhenjiang 212018, China.
3 Shanghai Veterinary Research Institute, Chinese Academy of Sciences, Shanghai 200232, China.
4 College of Natural Resources and Life Science, Dong-A University, Pusan 604714, Korea.
*Corresponding author E-mail: srizzgui@hotmail.com.

家蚕新型谷胱甘肽S转移酶的纯化和生化特征分析

侯成香[1,2]　付志强[3]　Byung Rae Jin[4]　桂仲争[1,2*]

摘要：本文利用亲和层析和谷胱甘肽琼脂糖柱从家蚕幼虫中纯化出一种新型的谷胱甘肽S转移酶。聚丙烯酰胺凝胶电泳显示，该酶条带单一，分子质量约为28 kDa。利用1-氯-2，4-二硝基苯作为底物对该酶的稳态动力学进行了测定。该酶的K_m，V_{max}，K_{cat}和K_{cat}/K_m分别为0.494 mmol/L，72.07 μmol/(min·mg)，65.43 s^{-1}和132.45 $(mmol/L)^{-1}\cdot s^{-1}$。在25 ℃，pH为7.1时，该酶的活性最高。阿苯达唑、吡喹酮、胆汁酸和NaCl这4种抑制剂对该酶的抑制效应较低，也暗示了这种新型的谷胱甘肽S转移酶在结构和功能上与其他动物的谷胱甘肽S转移酶有所不同。

1　江苏科技大学，镇江
2　蚕业研究所，中国农业科学院，镇江
3　上海兽医研究所，中国科学院，上海
4　自然资源与生命科学学院，东亚大学，釜山，韩国

Identification, genomic organization and expression pattern of glutathione S-transferase in the silkworm, *Bombyx mori*

Yu QY[1,3] Lu C[1*] Li B[1] Fang SM[2]

Zuo WD[1] Dai FY[1] Zhang Z[1,3] Xiang ZH[1]

Abstract: Glutathione *S*-transferases (GSTs) are a multifunctional supergene family and some play an important role in insecticide resistance. We have identified 23 putative cytosolic GSTs by searching the new assembly of the *Bombyx mori* genome sequence. Phylogenetic analyses on the amino acid sequences reveal that 21 of the *B. mori* GSTs fall into six classes represented in other insects, the other two being unclassified. The majority of the silkworm GSTs belong to the Delta, Epsilon, and Omega classes. Most members of each class are tandemly arranged in the genome, except for the Epsilon GSTs. Expressed sequence tags (ESTs) corresponding to 19 of the 23 GSTs were found in available databases. Furthermore, RT-PCR experiments detected expression of all the GSTs in multiple tissues on day 3 of fifth instar larvae. Surprisingly, we found little or no expression of most Delta and Epsilon GSTs in the fat body, which is thought to be the main detoxification organ. This may explain the sensitivity of the silkworm to certain insecticides. Our data provide some insights into the evolution of the *B. mori* GST family and the functions of individual GST enzymes.

Published On: Insect Biochemistry and Molecular Biology, 2008, 38(12), 1158-1164.

1 The Key Sericultural Laboratory of Agricultural Ministry College of Biotechology, Southwest University, Chongqing 400716, China.

2 College of Life Science, China West Normal University, Nanchong 637002, China.

3 The Institute of Agricultural and Life Sciences, Chongqing University, Chongqing 400030, China.

*Corresponding author E-mail: lucheng@swu.edu.cn.

家蚕谷胱甘肽S转移酶基因的克隆、序列分析及其表达模式

余泉友[1,3]　鲁　成[1*]　李　斌[1]　房守敏[2]
左伟东[1]　代方银[1]　张　泽[1,3]　向仲怀[1]

摘要：谷胱甘肽S转移酶(GSTs)是一个多功能超基因家族，其家族中一些蛋白质在杀虫剂抗性中起重要作用。我们通过搜索新组装的家蚕基因组序列，鉴定出23种假定的胞质GSTs。氨基酸序列的系统发育分析显示，21种家蚕GSTs被分为6大类，另外2种未被分类。大多数家蚕GSTs属于Delta，Epsilon和Omega类。除了Epsilon GSTs之外，每个类别的大多数成员都串联排列在基因组中。在可用的数据库中找到了与23个GST中的19个对应的表达序列标签(ESTs)。此外，RT-PCR实验检测了五龄幼虫第3天多个组织中所有GSTs的表达。令人惊讶的是，我们发现大多数Delta和Epsilon GSTs在脂肪体中几乎没有或没有表达，而脂肪体被认为是主要的排毒器官，这可以解释家蚕对某些杀虫剂敏感的原因。我们的数据为研究家蚕GSTs家族的进化和GSTs酶的功能提供了一些见解。

1　农业部蚕桑学重点实验室，生物技术学院，西南大学，重庆

2　生命科学学院，西华师范大学，南充

3　农学与生命科学研究院，重庆大学，重庆

Identification of a novel spore wall protein (SWP26) from microsporidia *Nosema bombycis*

Li YH[1] Wu ZL[1] Pan GQ[1] He WW[1]
Zhang RZ[1] Hu JH[1] Zhou ZY[1,2*]

Abstract: Microsporidia are obligate intracellular parasites related to fungi with resistant spores against various environmental stresses. The rigid spore walls of these organisms are composed of two major layers, which are the exospore and the endospore. Two spore wall proteins (the endosporal protein-SWP30 and the exosporal protein-SWP32) have been previously identified in *Nosema bombycis*. In this study, using the MALDI-TOF-MS technique, we have characterised a new 25.7-kDa spore wall protein (SWP26) recognised by monoclonal antibody 2G10. SWP26 is predicted to have a signal peptide, four potential N-glycosylation sites, and a C-terminal heparin-binding motif (HBM) which is known to interact with extracellular glycosaminoglycans. By using a host cell binding assay, recombinant SWP26 protein (rSWP26) can inhibit spore adherence by 10%, resulting in decreased host cell infection. In contrast, the mutant rSWP26 (rΔSWP26, without HBM) was not effective in inhibiting spore adherence. Immuno-electron microscopy revealed that this protein was expressed largely in endospore and plasma membrane during endospore development, but sparsely distributed in the exospore of mature spores. The present results suggest that SWP26 is a microsporidia cell wall protein that is involved in endospore formation, host cell adherence and infection *in vitro*. Moreover, SWP26 could be used as a good prospective target for diagnostic research and drug design in controlling the silkworm, *Bombyx mori*, pebrine disease in sericulture.

Published On: International Journal for Parasitology, 2009, 39(4), 391-398.

1 Key Sericultural Laboratory of Agricultural Ministry, Southwest University, Chongqing 400716, China.
2 Laboratory of Animal Biology, Chongqing Normal University, Chongqing 400047, China.
*Correspongding author E-mail: zyzhou@swu.edu.cn.

从家蚕微孢子虫中鉴定一种新的孢壁蛋白(SWP26)

李艳红[1] 吴正理[1] 潘国庆[1] 何为韦[1]
张瑞芝[1] 胡军华[1] 周泽扬[1,2*]

摘要:微孢子是一类与真菌有关的细胞内寄生物,可以形成孢子结构来抵抗多种环境压力。微孢子坚硬的孢壁是由两层构成的,称为孢子内壁和孢子外壁。先前已有两种家蚕微孢子孢壁蛋白被鉴定出来,分别是内壁蛋白SWP30和外壁蛋白SWP32。在本项研究中,利用MALDI-TOF-MS质谱技术,我们鉴定到一个新的分子质量为27.5 kDa的孢壁蛋白(SWP26),其可以被单克隆抗体2G10识别。SWP26被预测包含1个信号肽结构域,4个可能的N糖基化位点,以及1个羧基端肝素结合结构域(HBM),该结构域可以与细胞外黏多糖相互作用。利用宿主细胞结合实验,发现重组SWP26蛋白(rSWP26)可以让孢子吸附宿主细胞水平降低约10%,从而降低宿主细胞的感染。相比之下,突变的rSWP26(rΔSWP26,不含有HBM)却不能抑制孢子吸附宿主细胞。免疫电镜结果显示,在孢子内壁发育过程中,该蛋白在孢子内壁和细胞质膜中高表达,但是在成熟孢子的外壁中分布较少。这些结果显示SWP26是一个与孢子内壁形成、宿主细胞吸附和感染相关的微孢子孢壁蛋白。另外,SWP26可以作为控制蚕业重要疫病(家蚕微粒子病)的诊断研究和药物设计的理想靶点。

1 农业部蚕桑学重点实验室,西南大学,重庆
2 生命科学学院,重庆师范大学,重庆

SWP25, a novel protein associated with the *Nosema bombycis* endospore

Wu ZL[1] Li YH[1] Pan GQ[1] Zhou ZY[1,2*] Xiang ZH[1]

Abstract: Microsporidia are eukaryotic, obligate intracellular, spore-forming parasites. The resistant spores, which harbor a rigid cell wall, are critical for their host-to-host transmission and persistence in the environment. The spore wall comprises two major layers: the exospore and the endospore. In *Nosema bombycis*, two spore wall proteins have been characterized--an endosporal protein, SWP30, and an exosporal protein, SWP32. Here, we report the identification of the third spore wall protein of *N. bombycis*, SWP25, the gene of which has no known homologue. SWP25 is predicted to posses a signal peptide and a heparin-binding motif. Immunoelectron microscopy analysis showed that this protein is localized to the endospore. This characterization of a new spore wall protein of *N. bombycis* may facilitate our investigation of the relationship between *N. bombycis* and its host, *Bombyx mori*.

Published On: Journal of Eukaryotic Microbiology, 2009, 56(2), 113-118.

1 Key Sericultural Laboratory of Agricultural Ministry, Southwest University, Chongqing 400716, China.

2 Laboratory of Animal Biology, Chongqing Normal University, Chongqing 400047, China.

*Corresponding author E-mail: zyzhou@swu.edu.cn.

SWP25，一个新的家蚕微孢子虫内壁相关蛋白质

吴正理[1] 李艳红[1] 潘国庆[1] 周泽扬[1,2*] 向仲怀[1]

摘要:微孢子虫是一类细胞内专性寄生的可形成孢子的单细胞真核生物。具有刚性细胞壁结构的孢子对于其在环境中宿主间的传播和存活起到至关重要的作用。孢子壁主要包括两层:孢子外壁和孢子内壁。在家蚕微孢子虫中,已经验证了两种孢壁蛋白质:孢子内壁蛋白质SWP30和外壁蛋白SWP32。本研究中,我们鉴定获得家蚕微孢子虫中的第三种孢壁蛋白SWP25,其基因没有已知的同源物,预测SWP25具有信号肽和肝素结合基序。免疫电子显微观察分析表明,该蛋白质定位于孢子内壁。本研究获得的新孢壁蛋白将有利于我们研究家蚕微孢子虫与其寄主家蚕之间的互作关系机制。

1 农业部蚕桑学重点实验室,西南大学,重庆

2 生命科学学院,重庆师范大学,重庆

Protein profile of *Nomuraea rileyi* spore isolated from infected silkworm

Qin LG[1] Liu XY[1] Li J[2] Chen HQ[1] Yao Q[1]
Yang Z[1] Wang L[3] Chen KP[1*]

Abstract: *Nomuraea rileyi (N. rileyi)* is the causative agent of the silkworm, *Bombyx mori*, green muscardine which can cause severe worldwide economical loss in sericulture. Little is known about *N. rileyi* at the protein level for this entomopathogenic parasite which belongs to the Ascomycota. Here, we employed proteomic-based approach to identify proteins of *N. rileyi* spores collected from the dead silkworm. In all, 252 proteins were separated by two-dimensional gel electrophoresis (2-DE), and were subjected to mass spectrometry (MS) analysis, 121 proteins have good MS signal, and 24 of them were identified due to unavailability of genomic information from *N. rileyi*. This data will be helpful in understanding the biochemistry of *N. rileyi*.

Published On: Current Microbiology, 2009, 58(6), 578-585.

1 Institute of Life Sciences, Jiangsu University, Zhenjiang 212013, China.
2 Center for Physics and Chemistry, Jiangsu University, Zhenjiang, 212013, China.
3 Beijing Entry-Exit Inspection and Quarantine Bureau, Beijing 100026, China.
*Corresponding author E-mail:kpchen@ujs.edu.cn.

侵染家蚕的*Nomuraea rileyi*孢子的蛋白质图谱

覃吕高[1] 刘晓勇[1] 李 军[2] 陈慧卿[1] 姚 勤[1]
杨 哲[1] 汪 琳[3] 陈克平[1*]

摘要:*Nomuraea rileyi*(*N. rileyi*)是家蚕绿僵病的致病病原菌,可对全球蚕业经济造成严重的损害。但在蛋白质水平上对这个属于子囊菌门的昆虫病原知之甚少。在本研究中,我们采用蛋白质组学的方法鉴定从死蚕收集的*N. rileyi*孢子的蛋白质成分。通过二维凝胶电泳(2-DE)总共分离获得252个蛋白质。质谱(MS)分析结果显示,其中121个蛋白质具有良好的MS信号。由于*N. rileyi*基因组信息的缺失,目前我们仅鉴定获得其中的24个蛋白质的序列信息。这些数据将有助于对*N. rileyi*的生理生化性质的进一步分析和理解。

1 生命科学研究院,江苏大学,镇江
2 分析测试中心,江苏大学,镇江
3 北京出入境检验检疫局,北京

Characterization of a transcriptionally active *Tc1*-like transposon in the microsporidian *Nosema bombycis*

Xu JS[1] Luo J[2] Bettina Debrunner-Vossbrinck[3]

Zhang XY[1] Liu HD[2] Zhou ZY[1,2*]

Abstract: The *Tc1* transposable element has been found in a wide variety of organisms including vertebrates, insects and fungi but has not been previously reported in Microsporidia. In this study, we characterize an intact DNA transposon (*NbTc1*) from the microsporidian *Nosema bombycis*. This transposable element encodes a 337 amino acid transposase sequence, which contains the D, D34E functional motif required for transposition. Southern blot of *N. bombycis* DNA separated by pulsedesis shows that copies of the *NbTc1* transposon are present on 10 of the 14 chromosomes of *N. bombycis*. Amino acid sequence variation among copies of the *NbTc1* is low, suggesting a conserved function for this transposon within *N. bombycis*. Phylogenetic analysis indicates that *NbTc1* is a new member of the *Tc1* family lineage, quite distinct from all previously described *Tc1* elements, including those from fungi, indicating that *Nb*Tc1 forms a unique clade of the Tc1 superfamily. However, the *Tc1* transposon is too divergent to resolve the major phylogenetic relationships among these superfamilies. Reverse transcriptase PCR and Solexa sequencing suggest that *NbTc1* possesses transcriptional activity. Considering the interest in Microsporidia as biological control agents, the *NbTc1* transposon may be a useful vector for the efficient transfection of these important parasites into host species.

Published On: Acta Parasitologica, 2010, 55(1), 8-15.

1 Laboratory of Animal Biology, Chongqing Normal University, Chongqing 400047, China.

2 Institute of Sericulture and Systems Biology, Southwest University, Chongqing 400716, China.

3 Department of Health Sciences, Quinnipiac University, Hamden CT 06518-1908, USA.

*Corresponding author E-mail:zyzhou@cqnu.edu.cn.

微孢子虫中的转录活性*Tc1*样转座子的特征分析

许金山[1]　罗　洁[2]　Bettina Debrunner-Vossbrinck[3]
张小燕[1]　刘含登[2]　周泽扬[1,2*]

摘要：*Tc1*转座元件在许多生物体内已被发现，包括脊椎动物、昆虫和真菌，但以前没有报道过微孢子虫中有此种转座子。在本研究中，我们鉴定了来自微孢子虫的完整DNA转座子(*NbTc1*)。该转座元件编码337个氨基酸转座酶序列，其含有转座所需的D，D34E功能基序。通过脉冲法分离*N. bombycis* DNA后进行Southern印迹显示*NbTc1*转座子的拷贝存在于*N.bombycis*的14条染色体中的10条上。*NbTc1*拷贝数之间的氨基酸序列差异较小，表明该转座子在*N.bombycis*中具有保守功能。系统发育分析表明，*Nb*Tc1是Tc1家族谱系的新成员，与以前描述的所有*Tc1*元件(包括真菌)都有很大区别，表明*NbTc1*构成了Tc1超家族的独特进化枝。然而，由于*Tc1*转座子差异太大，故无法确定这些超家族之间的主要系统发育关系。逆转录酶PCR和Solexa测序表明*NbTc1*具有转录活性。鉴于微孢子虫可作为生物控制剂，*NbTc1*转座子可能是这些重要寄生虫有效侵染到宿主物种的有用载体。

1　生命科学学院，重庆师范大学，重庆
2　蚕学与系统生物学研究所，西南大学，重庆
3　健康科学系，昆尼皮亚克大学，哈姆登，美国

Identification of *NbME* MITE families: potential molecular markers in the microsporidia *Nosema bombycis*

Xu JS[1,2] Wang M[1,2] Zhang XY[1,2]
Tang FH[1] Pan GQ[3] Zhou ZY[1,2,3*]

Abstract: Six novel families of miniature inverted-repeat transposable elements (MITEs) were characterized in the microsporidia *Nosema bombycis* and were named *NbMEs*. The structural characteristics and the distribution of *NbME* copies in the *N. bombycis* genome were investigated, and it was found that portions of *NbMEs* are associated with gene sections. Potential molecular markers for various *N. bombycis* strains were identified in this study through utilization of the MITE-AFLP technique. Three distinct pathogenic isolates collected from different areas were distinguished, and polymorphisms were detected using the *NbME5* marker, thereby establishing this *NbME* as a potential marker for studying isolate variation in *N. bombycis*.

Published On: Journal of Invertebrate Pathology, 2010, 103(1), 48-52.

1 Laboratory of Animal Biology, Chongqing Normal University, Chongqing 400047, China.
2 Engineering Research Center of Bioactive Substances, Chongqing Normal University, Chongqing 400047, China.
3 Institute of Sericulture and Systems Biology, Southwest University, Chongqing 400716, China.
*Corresponding author E-mail: zyzhou@cqnu.edu.cn.

鉴定*NbME* MITE家族：家蚕微孢子虫潜在的分子标记物

许金山[1,2]　王　敏[1,2]　张小燕[1,2]
唐发辉[1]　潘国庆[3]　周泽扬[1,2,3*]

摘要：6种新的微型反向重复转座元件(MITEs)在微孢子虫中被鉴定，并命名为*NbMEs*。研究了*NbME*拷贝在*N. bombycis*基因组中的结构特征和分布，发现部分*NbMEs*与基因片段相关。在本研究中，利用MITE-AFLP技术鉴定了各种*N. bombycis*菌株的潜在分子标记。研究区分了来自不同区域的3种不同致病菌，并且用*NbME5*标记检测其多态性，从而确立了以这种*NbME*作为在家蚕微孢子虫中研究分离株变异的潜在标记。

1　生命科学学院，重庆师范大学，重庆
2　生物活性物质工程研究中心，重庆师范大学，重庆
3　蚕学与系统生物学研究所，西南大学，重庆

Characterization of a subtilisin-like protease with apical localization from microsporidian *Nosema bombycis*

Dang XQ[1] Pan GQ[1] Li T[1] Lin LP[1] Ma Q[1]
Geng LN[1] He YL[1] Zhou ZY[1,2*]

Abstract: The microsporidian *Nosema bombycis* is the pathogen causing pébrine leading to heavy economic loss in sericulture. Little is known of the proteases of microsporidia that are important for both parasite development and pathogenesis. Here, we identified a subtilisin-like serine protease *Nb*SLP1 which contains an inhibitor_I9 and a peptidase_S8 domain. Three dimensional modeling of the catalytic domain of the *Nb*SLP1 exhibited a typical 3-layer sandwich structure with S1 pocket substituted by Y^{359}. Phylogenetic analysis confirms that subtilisin-like serine proteases of microsporidia fall into two clades: SLP1 and SLP2, suggesting the initial subtilisin gene duplication events preceded microsporidia speciation. In addition, transcripts of *Nbslp1* were detected in the midgut of *Bombyx mori* infection by *N. bombycis* by RT-PCR. Antibodies against *Nb*SLP1 recognized both the precursor and mature enzyme by 2D Western blot. Besides, indirect immunofluorescence assay revealed that the *Nb*SLP1 is mainly localized at the two poles of spore which make the spore look like "safety pins". Remarkably, the mature protease is only detected in the apical region of the spore after germination. These studies demonstrate that *Nb*SLP1 is a conserved subtilisin protease in microsporidia and suggest that *Nb*SLP1 play a significant role in polar tube extrusion process.

Published On: Journal of Invertebrate Pathology, 2013, 112 (2), 166-174.

1 State Key Laboratory of Silkworm Genome Biology, Southwest University, Chongqing 400716, China.
2 Laboratory of Animal Biology, Chongqing Normal University, Chongqing 400047, China.
*Corresponding author E-mail: zyzhou@swu.edu.cn.

家蚕微孢子虫的一种具有顶端定位特征的类枯草杆菌蛋白酶*Nb*SLP1的鉴定

党晓群[1] 潘国庆[1] 李 田[1] 林立鹏[1] 马 强[1]
耿莉娜[1] 贺元莉[1] 周泽扬[1,2*]

摘要：家蚕微孢子虫(*Nosema bombycis*)是家蚕微粒子病的病原，给蚕业生产造成了巨大的经济损失。但关于家蚕微孢子虫发育和致病有关的蛋白酶的研究还鲜有报道。本研究在微孢子虫中鉴定了一类具有 Inhibitor_I9 结构域和蛋白酶 Peptidase_S8 结构域的类枯草杆菌蛋白酶的丝氨酸蛋白酶，命名为*Nb*SLP1。*Nb*SLP1 的三维结构模拟显示，其结构类似“三明治”，其底物 S1 口袋被 Y^{359} 代替。系统发育分析表明，微孢子虫的类枯草杆菌蛋白酶聚为两支(SLP1 和 SLP2)，这说明最初的枯草杆菌蛋白酶基因复制事件早于微孢子虫的形成。RT-PCR 检测发现，在家蚕微孢子虫感染家蚕后家蚕中肠内有 *Nbslp1* 基因的转录。利用已制备的抗体进行双向 Western blot，检测到了 NbSLP1 的酶原和成熟酶形式。另外，间接免疫荧光实验显示，*Nb*SLP1 主要分布于成熟孢子的两端，使孢子看上去像别针。值得注意的是，当孢子发芽后，*Nb*SLP1 的成熟酶位于孢子顶端。这些研究结果表明，*Nb*SLP1 是一类在微孢子虫中保守存在的类枯草杆菌蛋白酶，推测 *Nb*SLP1 在极管的弹出过程中有重要作用。

1 家蚕基因组生物学国家重点实验室，西南大学，重庆
2 动物生物学实验室，重庆师范大学，重庆

Identification of a novel chitin-binding spore wall protein (*Nb*SWP12) with a BAR-2 domain from *Nosema bombycis* (microsporidia)

Chen J[1] Geng LN[1] Long MX[1] Li T[1] Li Z[1,2]
Yang DL[1] Ma C[1] Wu HJ[1] Ma ZG[1]
Li CF[1] Pan GQ[1] Zhou ZY[1,2*]

Abstract: The spore wall of *Nosema bombycis* plays an important role in microsporidian pathogenesis. Protein fractions from germinated spore coats were analyzed by two-dimensional polyacrylamide gel electrophoresis and MALDI-TOF / TOF mass spectrometry. Three protein spots were identified as the hypothetical spore wall protein *Nb*HSWP12. A BAR-2 domain (e-value: 1.35e-03) was identified in the protein, and an N-terminal protein-heparin interaction motif, a potential N-glycosylation site, and 16 phosphorylation sites primarily activated by protein kinase C were also predicted. The sequence analysis suggested that *Nbhswp12* and its homologous genes are widely distributed among microsporidia. Additionally, *Nbhswp12* gene homologues share similar sequence features. An indirect immunofluorescence analysis showed that *Nb*HSWP12 localized to the spore wall, and thus we renamed it spore wall protein 12 (*Nb*SWP12). Moreover, *Nb*SWP12 could adhere to deproteinized *N. bombycis* chitin coats that were obtained by hot alkaline treatment. This novel *N. bombycis* spore wall protein may function in a structural capacity to facilitate microsporidial spore maintenance.

Published On: Parasitology, 2013, 140(11), 1394-1402.

1 State Key Laboratory of Silkworm Genome Biology, Southwest University, Chongqing 400716, China.
2 College of Life Sciences, Chongqing Normal University, Chongqing 400047, China.
*Corresponding author E-mail: zyzhou@swu.edu.cn.

家蚕微孢子虫的一种具有BAR-2功能域的几丁质结合孢壁蛋白(*Nb*SWP12)的鉴定

陈　洁[1]　耿丽娜[1]　龙梦娴[1]　李　田[1]　李　治[1,2]
杨东林[1]　马　成[1]　吴海晶[1]　马振刚[1]
李春风[1]　潘国庆[1]　周泽扬[1,2*]

摘要：家蚕微孢子虫的孢壁在微孢子虫发病机制方面起着非常重要的作用，通过对发芽后孢壁上蛋白样品的双向电泳以及质谱分析结果发现，从3个蛋白点中都鉴定得到了假定极管蛋白*Nb*HSWP12。在*Nb*HSWP12蛋白中鉴定得到了一个BAR-2功能域、一个N端肝素结合基序、一个潜在的N糖基化位点以及16个主要由蛋白激酶C激活的磷酸化位点。序列分析表明*Nbhswp12*的同源基因在微孢子虫中分布非常广泛，并且具有较高的序列相似性。间接免疫荧光实验表明*Nb*HSWP12定位于孢壁上，所以将该蛋白命名为家蚕微孢子虫孢壁蛋白12(*Nb*SWP12)。另外，在热碱处理中*Nb*SWP12可以结合到家蚕微孢子虫脱蛋白几丁质壳上。该家蚕微孢子虫孢壁蛋白可能在保持微孢子虫孢子的结构完整性方面起作用。

1　家蚕基因组生物学国家重点实验室，西南大学，重庆
2　生命科学学院，重庆师范大学，重庆

Identification of a protein interacting with the spore wall protein SWP26 of *Nosema bombycis* in a cultured *BmN* cell line of silkworm

Zhu F[1#] Shen ZY[1,2#] Hou JG[1] Zhang J[1]
Geng T[1] Tang XD[1,2] Xu L[1,2] Guo XJ [1,2*]

Abstract: *Nosema bombycis* is a silkworm parasite that causes severe economic damage to sericulture worldwide. It is the first microsporidia to be described in the literature, and to date, very little molecular information is available regarding microsporidian physiology and their relationships with their hosts. Therefore, the interaction between the microsporidia *N. bombycis* and its host silkworm, *Bombyx mori*, was analyzed in this study. The microsporidian spore wall proteins (SWPs) play a specific role in spore adherence to host cells and recognition by the host during invasion. In this study, SWP26 fused with enhanced green fluorescence protein (EGFP) was expressed in *BmN* cells by using a Bac-to-Bac expression system. Subsequently, the turtle-like protein of *B. mori* (*Bm*TLP) was determined to interact with SWP26 via the use of anti-EGFP microbeads. This interaction was then confirmed by yeast two-hybrid analysis. The *BmTLP* cDNA encodes a polypeptide of 447 amino acids that includes a putative signal peptide of 27 amino acid residues. In addition, the *Bm*TLP protein contains 2 immunoglobulin (IG) domains and 2 IGc2-type domains, which is the typical domain structure of IG proteins. The results of this study indicated that SWP26 interacts with the IG-like protein *Bm*TLP, which contributes to the infectivity of *N. bombycis* to its host silkworm.

Published On: Infection, Genetics and Evolution, 2013, 17, 38-45.

1 College of Biotechnology and Chemical Engineering, Jiangsu University of Science and Technology, Zhenjiang 212018, China.
2 Sericultural Research Institute, Chinese Academy of Agricultural Sciences, Zhenjiang 212018, China.
[#]These authors contributed equally.
[*]Corresponding author E-mail: guoxijie@126.com.

家蚕微孢子虫孢壁蛋白——SWP26与家蚕*BmN*细胞系互作蛋白的筛选与鉴定

朱　峰[1#]　沈中元[1,2#]　侯建革[1]　张　姣[1]
耿　涛[1]　唐旭东[1,2]　徐　莉[1,2]　郭锡杰[1,2*]

摘要：家蚕微孢子虫是人类最早发现与记载的微孢子虫，其引起的家蚕微粒子病曾给养蚕业带来毁灭性的危害。然而由于该物种的特殊性，迄今有关微孢子虫与其宿主之间相互作用的分子机理依然知之甚少。因此，我们希望通过实验研究探索家蚕微孢子虫与其宿主家蚕之间的相互关系。孢壁蛋白是微孢子虫孢壁的主要组成成分，在微孢子虫孢子与宿主细胞间的黏附和识别中扮演着重要角色。本实验利用家蚕核型多角体病毒Bac-to-Bac表达系统，在家蚕培养细胞*BmN*中成功表达了SWP26-EGFP融合蛋白。利用EGFP作为抗原决定簇通过免疫共沉淀技术筛选与之相互作用的宿主细胞蛋白。通过免疫共沉淀筛选获得一个与SWP26有相互作用的家蚕类海龟蛋白（*Bombyx mori* turtle-like protein，*Bm*TLP），并经酵母双杂交实验验证了二者之间的相互作用。该蛋白基因开放阅读框编码447个氨基酸，在其N-端有一段长为27个氨基酸残基的信号肽。此外，*Bm*TLP含有两个IG结构域和两个IGc2结构域，属于典型的免疫球蛋白家族成员。本研究表明，SWP26与类IG蛋白*Bm*TLP相互作用，有助于微孢子虫对宿主家蚕的感染。

1　生物与化学工程学院，江苏科技大学，镇江
2　蚕业研究所，中国农业科学院，镇江

Molecular characteristics of the alpha- and beta-tubulin genes of *Nosema philosamiae*

Zhu F[1] Shen ZY[1,2] Xu L[1,2] Guo XJ[1,2*]

Abstract: Microsporidia are intracellular parasites of insects and other higher eukaryotes. The microsporidian *Nosema philosamiae* Talukdar, 1961 was isolated from the eri silkworm, *Philosamia cynthia ricini* Grote. In the present study, alpha- and beta-tubulin genes from *N. philosamiae* were characterized. The identity analysis of nucleotide and amino acid sequences indicated high similarity with species of *Nosema Nägeli*, 1857 *sensu lato* (nucleotide sequences, > or = 96.0%; amino acid sequences, > or = 99.0%). However, the tubulin genes of *N. philosamiae* share low sequence similarity with that of *N. ceranae* Fries, Feng, da Silva, Slemenda et Pieniazek, 1996 (strain *BRL01*) and a *Nosema/Vairimorpha* species. Phylogenies based on alpha-, beta- and combined alpha- plus beta-tubulin gene sequences showed that *N. philosamiae*, along with the true *Nosema* species, forms a separate clade with a high bootstrap value, with *N. ceranae BRL01* forming a clade of its own. The results indicated that the alpha- and beta-tubulin sequences may be useful as a diagnostic tool to discriminate the true *Nosema* group from the *Nosema/Vairimorpha* group.

Published On: Folia Parasitologica, 2013, 60(5), 411-415.

1 College of Biotechnology and Chemical Engineering, Jiangsu University of Science and Technology, Zhenjiang 212018, China.

2 Sericultural Research Institute, Chinese Academy of Agricultural Sciences, Zhenjiang 212018, China.

*Corresponding author E-mail: guoxijie@126.com.

蓖麻蚕微孢子虫α-与β-微管蛋白的分子特征

朱　峰[1]　沈中元[1,2]　徐　莉[1,2]　郭锡杰[1,2*]

摘要：作为细胞内寄生的病原生物，微孢子虫可寄生于昆虫和其他高等真核生物体内，1961年Talukdar从蓖麻蚕中分离出了一种昆虫微孢子虫——蓖麻蚕微孢子虫。本研究对蓖麻蚕微孢子虫α-与β-微管蛋白的基因序列进行了克隆与特征分析。序列相似性分析结果显示，蓖麻蚕微孢子虫微管蛋白与*Nosema Nägeli*微孢子虫核苷酸序列相似性在96.0%以上，氨基酸序列相似性在99.0%以上。而与*Nosema/Vairimorpha*属中的东方蜜蜂微孢子虫(*Nosema ceranae BRL01*)的序列相似性很低。基于α-、β-以及α-+β-微管蛋白氨基酸序列的系统发育树分析结果显示，蓖麻蚕微孢子虫与真正*Nosema*属其他微孢子虫聚类在同一分支，而东方蜜蜂微孢子虫单独聚类在一支。本研究结果表明，α-与β-微管蛋白可作为微孢子虫*Nosema*属与*Nosema/Vairimorpha*属之间区分的一个依据。

1　生物与化学工程学院，江苏科技大学，镇江

2　蚕业研究所，中国农业科学院，镇江

*Nb*HSWP11, a microsporidia *Nosema bombycis* protein, localizing in the spore wall and membranes, reduces spore adherence to host cell *BmE*

Yang DL Dang XQ Peng P Long MX Ma C Jia JJ
Qin GW Wu HJ Liu T Zhou XW Pan GQ Zhou ZY*

Abstract: Microsporidia are obligate intracellular parasites, and a derivative of fungi, which harbor a rigid spore wall to resist adverse environmental pressures. The spore wall protein, which is thought to be the first and direct protein interacting with the host cell, may play a key role in the process of microsporidia infection. In this study, we report a protein, *Nb*HSWP11, with a dnaJ domain. The protein has 6 heparin-binding motifs which are known to interact with extracellular glycosaminoglycans. Syntenic analysis indicated that gene loci of *Nbhswp11* are conserved and syntenic between *Nosema bombycis* and *Nosema ceranae*. Phylogenetic tree analysis showed that *Nbhswp11* clusters with fungal dnaJ proteins and has 98% identity with an *N. bombycis* dnaJ protein. *Nbhswp11* was transcribed throughout the entire life stages, and gradually increased during 1 - 7 days, in a silkworm that was infected by *N. bombycis*, as determined by reverse-transcription PCR (RT-PCR). The recombinant protein *Nb*HSWP11 (rSWP11-HIS) was obtained and purified using gene cloning and prokaryotic expression. Western blot analysis displayed *Nb*HSWP11 expressed in the total mature spore proteins and spore coat proteins. Indirect immunofluorescence assay revealed *Nb*HSWP11 located at the spore wall of mature spores and the spore coats. Furthermore, immune electron microscopy showed that *Nb*HSWP11 localized in the cytoplasm of the sporont. Within the developmental process of *N. bombycis*, a portion of *Nb*HSWP11 is targeted to the spore wall of sporoblasts and mature spores. However, most of *Nb*HSWP11 distributes on the membraneous structures of the sporoblast and mature spore. In addition, using a host cell binding assay, native protein *Nb*HSWP11 in the supernatant of total soluble mature spore proteins is shown to bind to the host cell *BmE* surface. Finally, an antibody blocking assay showed that purified rabbit antibody of *Nb*HSWP11 inhibits spore adherence and decreases the adherence rate of spores by 20% compared to untreated spores. Collectively, the present results suggest that *Nb*HSWP11 is involved in host cell adherence *in vitro*. Therefore *Nb*HSWP11, which has a dnaJ domain, may modulate protein assembly, disassembly, and translocation in *N. bombycis*.

Published On: Journal of Parasitology, 2014, 100(5), 623-632.

State Key Laboratory of Silkworm Genome Biology, Southwest University, Chongqing 400716, China.
*Corresponding author E-mail: zyzhou@swu.edu.cn.

定位于孢壁和孢膜上的家蚕微孢子虫蛋白*Nb*HSWP11减弱孢子对宿主细胞*BmE*的黏附

杨东林 党晓群 彭 湃 龙梦娴 马 成 贾俊杰
秦国伟 吴海晶 刘 铁 周小伟 潘国庆 周泽扬*

摘要:微孢子虫是一类专性细胞内寄生的真菌病原微生物,其坚硬的孢壁可抵抗不利的环境压力。孢壁蛋白被认为是第一个直接和宿主细胞相互作用的蛋白,它在微孢子虫感染的过程中发挥着关键的作用。在本研究中,我们报道了一个含dnaJ结构域的*Nb*HSWP11蛋白。该蛋白具有6个可以和细胞外黏多糖相互作用的肝素结合基序。共线性分析表明家蚕微孢子虫和蜜蜂微孢子虫中的*Nbhswp11*基因是保守的、共线性的。系统进化树分析显示*Nbhswp11*和真菌dnaJ蛋白成簇排列且与家蚕微孢子虫dnaJ蛋白具有98%的相似性。反转录PCR(RT-PCR)检测发现,*Nbhswp11*在家蚕的整个生命周期都会发生转录,且在家蚕微孢子虫感染后的1—7d内逐渐增加。通过基因克隆和原核生物表达纯化得到了重组蛋白*Nb*HSWP11(rSWP11-HIS)。蛋白免疫印迹分析显示*Nb*HSWP11在所有的成熟孢蛋白和孢壳蛋白中都有表达。间接免疫荧光实验显示*Nb*HSWP11定位于成熟孢子的孢壁和孢壳上。进一步的免疫电镜观察显示*Nb*HSWP11位于产孢体胞质中。在家蚕微孢子虫的发育过程中,一部分*Nb*HSWP11蛋白处于成孢子细胞和成熟孢子的孢壁,而大部分*Nb*HSWP11分布在成孢子细胞和成熟孢子的膜上。此外,宿主细胞结合实验显示在全溶性成熟孢子蛋白中的天然蛋白*Nb*HSWP11结合到宿主细胞*BmE*表面。最后的抗体封闭实验显示纯化得到的*Nb*HSWP11兔抗抑制孢子黏附。与未处理的孢子相比,*Nb*HSWP11兔抗处理组的孢子黏附率降低20%。总之,这些体外结果表明*Nb*HSWP11参与宿主细胞黏附。因此,具有dnaJ结构域的*Nb*HSWP11可以调节家蚕微孢子虫内蛋白的组装、分离和易位。

家蚕基因组生物学国家重点实验室,西南大学,重庆

Morphological and molecular studies of *Vairimorpha necatrix* BM, a new strain of the microsporidium *V. necatrix* (microsporidia, burenellidae) recorded in the silkworm, *Bombyx mori*

Luo B[1] Liu HD[2] Pan GQ[1*] Li T[1] Li Z[1]
Dang XQ[3] Liu T[1] Zhou ZY[1,3]

Abstract: *Vairimorpha* sp. BM (2012) is a recent isolate of the microsporidia from the silkworm in Shandong, China. The ultrastructure, tissue pathology and molecular characterization of this isolate is described in this study. This pathogenic fungus causes pebrine disease in silkworms which manifests as a systemic infection. Meanwhile, the silkworm eggs produced by the infected moths were examined using a microscope and PCR amplification. Neither spores nor the expected PCR band were observed, suggesting that no vertical transmission occurred in *Bombyx mori*. In addition, the ultrastructure of the isolate was studied by light microscopy and transmission electron microscopy. Two types of spores were observed: diplokaroytic spores with 13-17 coils of polar tubes and monokaryotic spores with less coils of polar tubes which could form octospores; however, no sporophorous vesicles were observed. Finally, phylogenetic analysis of the small subunit rRNA genes of *Vairimorpha* species showed that this isolate has a closer relationship to *Vairimorpha necatrix* than the other species studied. This result also is supported by phylogenetic analysis based on their actin genes, heat shock protein 70 (HSP70) and RNA polymerase II (RPB1). Based on the information gained during this study, we propose that this microsporidian species infecting *B. mori* should be given the name *V. necatrix* BM.

Published On: Experimental Parasitology, 2014, 143, 74-82.

1 State Key Laboratory of Silkworm Genome Biology, Southwest University, Chongqing 400716, China.
2 Experimental Teaching Center, Chongqing Medical University, Chongqing 400016, China.
3 Laboratory of Animal Biology, Chongqing Normal University, Chongqing 400047, China.
*Corresponding author E-mail: gqpan@swu.edu.cn.

一株家蚕中新分离的纳卡变形微孢子虫的形态学以及分子研究

罗　波[1]　刘含登[2]　潘国庆[1*]　李　田[1]　李　曾[1]
党晓群[3]　刘　铁[1]　周泽扬[1,3]

摘要：家蚕变形微孢子虫是2012年在山东地区新分离的一株微孢子虫，本文主要对该分离株的超微结构、组织病理以及分子特征进行了描述。这种致病真菌在家蚕中能够引起微粒子病，表现为全身感染。同时，使用显微镜和PCR扩增法检测感染母蛾产的蚕卵，没有观察到成熟孢子和预期的PCR带，表明在蚕蛾中没有发生垂直传播。此外，通过光学显微镜和透射电子显微镜研究了该分离株的超微结构。观察到两种类型的孢子：双核孢子具有13—17圈极管，单核孢子具有较少的极管圈数，并且可形成八孢子，然而，在八孢子中没有观察到寄生囊泡。最后，变形孢虫属物种rRNA小亚基的系统发育分析显示，与其他物种相比，该分离株与纳卡变形微孢子虫有更密切的关系。该结果还通过基于它们的肌动蛋白基因、热休克蛋白70（HSP70）和RNA聚合酶II（RPB1）的系统发育分析得到了进一步证实。根据本研究获得的信息，我们建议将这种感染家蚕的微孢子虫种命名为纳卡变形微孢子虫家蚕分离株（*V.necatrix* BM）。

1　家蚕基因组生物学国家重点实验室，西南大学，重庆
2　实验教学中心，重庆医科大学，重庆
3　动物生物学实验室，重庆师范大学，重庆

Nosema sp. PM-1, a new isolate of microsporidian from infected *Papilio machaon* Linnaeus, based on ultrastructure and molecular identification

Liu T[1,2] Xu JS[3] Luo B[1] Chen J[1] Li Z[1]
Li T[1] Pan GQ[1] Li XY[4] Zhou ZY[1,3*]

Abstract: A new microsporidium, *Nosema* sp. PM-1, was first isolated from *Papilio machaon* Linnaeus. The spore shape of the PM-1 isolate was a longoval with an average size of 3.22 μm × 1.96 μm. Ultrastructure observation showed that PM-1 had a typical *Nosema* common diplokaryotic nuclei structure with 10 – 13 polar filament coils, spore wall, plasma membrane, and anchoring disk. The complete rRNA gene sequences were obtained by polymerase chain reaction amplification and each rRNA unit was arrayed as follows: 5′-LSU (2 497 bp)-ITS (179 bp)-SSU (1 232 bp)-IGS (278 bp)-5S (115 bp)-3′, which was the same as typical *Nosema*. The phylogenetic trees of rRNA, DNA-directed RNA polymerase II subunit, and tubulin genes all show that PM-1 was a sister to the clade comprising *Nosema bombycis*, *Nosema spodopterae*, and *Nosema* sp. PX-1. The spore morphology, ultrastructure, and complete rRNA structure indicate that this isolate assigned to the "true" *Nosema* group, can parasitized in *Papilio machaon* Linnaeus, which provides a wider host range for *Nosema*.

Published On: Acta Parasitologica, 2015, 60(2), 330-336.

1 State Key Laboratory of Silkworm Genome Biology, Southwest University, Chongqing 400716, China.
2 School of Mathematics and Statistics, Southwest University, Chongqing 400716, China.
3 College of Life Sciences, Chongqing Normal University, Chongqing 400047, China.
4 Kunming Institute of Zoology, Yunnan, 650223, China.
*Corresponding author E-mail:zyzhou@swu.edu.cn.

通过超结构和分子鉴定从被感染的金凤蝶中分离到新型微孢子虫 *Nosema* sp. PM-1

刘 铁[1,2] 许金山[3] 罗 波[1] 陈 洁[1] 李 曾[1]
李 田[1] 潘国庆[1] 李雪艳[4] 周泽扬[1,3*]

摘要：首次从金凤蝶体内分离到一株新型微孢子虫 *Nosema* sp. PM-1。该微孢子虫呈长椭圆形，平均大小为3.22 μm×1.96 μm。超结构观察显示PM-1具有典型的微粒子属双核结构，包含10—13圈极管、孢壁、孢膜和锚定盘。通过聚合酶链式反应扩增得到rRNA基因的全序列，每个rRNA单元按照5′-LSU（2 497 bp）-ITS（179 bp）-SSU（1 232 bp）-IGS（278 bp）-5S（115 bp）-3′排列，这和典型的微粒子属一样。rRNA、依赖DNA的RNA聚合酶II亚单位和微管蛋白基因的系统进化树显示PM-1与家蚕微孢子虫、斜纹夜蛾微孢子虫以及小菜蛾微孢子虫 *Nosema* sp. PX-1构成的进化枝互为姐妹枝。孢子形态、超结构观察和rRNA全序列分析表明归类到“正宗”微粒子属的PM-1能寄生于金凤蝶体内，这也拓宽了微粒子的宿主范围。

1 家蚕基因组生物学国家重点实验室，西南大学，重庆
2 数学与统计学院，西南大学，重庆
3 生命科学学院，重庆师范大学，重庆
4 昆明动物研究所，云南

Interaction and assembly of two novel proteins in the spore wall of the microsporidian species *Nosema bombycis* and their roles in adherence to and infection of host cells

Yang DL[1,2,4] Pan GQ[1,4] Dang XQ[1,3,4] Shi YW[2] Li CF[1,4] Peng P[1,4]
Luo B[1,4] Bian MF[1,4,6] Song Y[1,4] Ma C[1,4] Chen J[1,4] Ma ZG[1,3,4] Geng LN[1,4,5]
Li Z[1,3,4] Tian R[1,4] Wei CF[1,4] Zhou ZY[1,3,4*]

Abstract: Microsporidia are obligate intracellular parasites with rigid spore walls that protect against various environmental pressures. Despite an extensive description of the spore wall, little is known regarding the mechanism by which it is deposited or the role it plays in cell adhesion and infection. In this study, we report the identification and characterization of two novel spore wall proteins, SWP7 and SWP9, in the microsporidian species *Nosema bombycis*. SWP7 and SWP9 are mainly localized to the exospore and endospore of mature spores and the cytoplasm of sporonts, respectively. In addition, a portion of SWP9 is targeted to the spore wall of sporoblasts earlier than SWP7 is. Both SWP7 and SWP9 are specifically colocalized to the spore wall in mature spores. Furthermore, immunoprecipitation, far-Western blot, unreduced SDS-PAGE, and yeast two-hybrid data demonstrated that SWP7 interacted with SWP9. The chitin binding assay showed that, within the total spore protein, SWP9 and SWP7 can bind to the deproteinated chitin spore coats (DCSCs) of *N. bombycis*. However, binding of the recombinant protein rSWP7-His to the DCSCs is dependent on the combination of rSWP9 - glutathione *S*-transferase (GST) with the DCSCs. Finally, rSWP9-GST, anti-SWP9, and anti-SWP7 antibodies decreased spore adhesion and infection of the host cell. In conclusion, SWP7 and SWP9 may have important structural capacities and play significant roles in modulating host cell adherence and infection *in vitro*. A possible major function of SWP9 is as a scaffolding protein that supports other proteins (such as SWP7) that form the integrated spore wall of *N. bombycis*.

Published On: Infection and Immunity, 2015, 83(4), 1715-1731.

1 State Key Laboratory of Silkworm Genome Biology, Southwest University, Chongqing 400716, China.

2 Institute of Biotechnology, Key Laboratory of Chemical Biology and Molecular Engineering of Ministry of Education, Shanxi University, Taiyuan 030006, China.

3 College of Life Sciences, Chongqing Normal University, Chongqing 400047, China.

4 Key Laboratory for Sericulture Functional Genomics and Biotechnology of Agricultural Ministry, Southwest University, Chongqing 400716, China.

5 Chongqing Tobacco Science Research Institute, Southwest University, Chongqing 400716, China.

6 Chongqing Three Gorges Medical College, Chongqing 404000, China.

*Corresponding author E-mail: zyzhou@swu.edu.cn.

两个新型家蚕微孢子虫孢壁蛋白的互作、组装及其在宿主细胞感染和黏附中的作用

杨东林[1,2,4] 潘国庆[1,4] 党晓群[1,3,4] 石亚伟[2] 李春峰[1,4] 彭 湃[1,4]
罗 波[1,4] 边茂飞[1,4,6] 宋 跃[1,4] 马 成[1,4] 陈 洁[1,4] 马振刚[1,3,4] 耿莉娜[1,4,5]
李 治[1,3,4] 田 锐[1,4] 魏翠芳[1,4] 周泽扬[1,3,4*]

摘要:微孢子虫是专性细胞内寄生生物,具有坚硬的可预防各种环境压力的孢壁。目前,关于孢壁的研究虽然很多,但人们对它的形成机制和在细胞黏附、感染中的作用知之甚少。本文对家蚕微孢子虫中的两个新孢壁蛋白SW7和SWP9进行了鉴定和形态特征描述。SWP7、SWP9分别定位于成熟孢子内外壁和产孢体的细胞质中。SWP9的一部分比SWP7更早靶向到成孢子细胞的孢壁上,而它们都能够特异地共定位于成熟孢子的孢壁。此外,免疫共沉淀、far-Western blot、非还原性SDS-PAGE和酵母双杂交实验证实SWP7可以和SWP9发生相互作用。几丁质结合实验显示,全孢子蛋白中的SWP9和SWP7能结合到家蚕微孢子虫的脱蛋白几丁质孢壳(DCSCs)上。而重组蛋白rSWP7-His结合DCSCs取决于rSWP9-谷胱甘肽S转移酶和DCSCs的结合。最后证实rSWP9-GST、anti-SWP9和anti-SWP7抗体都能减弱孢子对宿主细胞的黏附和感染。总之,SWP7和SWP9是重要的结构蛋白,它们在体外调节宿主细胞黏附和感染的过程中发挥着重要的作用。据推测,SWP9的主要功能可能是充当支架蛋白以维持其他蛋白(如SWP7)形成完整的家蚕微孢子虫的孢壁。

1 家蚕基因组生物学国家重点实验,西南大学,重庆
2 生物技术研究所,化学生物学与分子工程教育部重点实验室,山西大学,太原
3 生命科学学院,重庆师范大学,重庆
4 农业部蚕桑功能基因组与生物技术重点实验室,西南大学,重庆
5 重庆烟草科学研究所,西南大学,重庆
6 重庆三峡医药高等专科学校,重庆

Characterization of a novel spore wall protein *Nb*SWP16 with proline-rich tandem repeats from *Nosema bombycis* (microsporidia)

Wang Y[1] Dang XQ[1,2] Ma Q[1] Liu FY[1]
Pan GQ[1] LI T[1] Zhou ZY[1,2*]

Abstract: *Nosema bombycis*, a pathogen of silkworm pebrine, is an obligate unicellular eukaryotic parasite. It is reported that the spore wall proteins have essential functions in the adherence and infection process of microsporidia. To date, the information related to spore wall proteins from microsporidia is still limited. Here, a 44 kDa spore wall protein *Nb*SWP16 was characterized in *N. bombycis*. In *Nb*SWP16, a 25 amino acids signal peptide and three heparin binding motifs were predicted. Interestingly, a region that contains three proline-rich tandem repeats lacking homology to any known protein was also present in this protein. The immunofluorescence analysis demonstrated that distinct fluorescent signals were detected both on the surface of mature spores and the germinated spore coats. Immunolocation by electron microscopy revealed that *Nb*SWP16 localized on the exospore regions. Finally, spore adherence analysis indicated that spore adherence to host cell was decreased more than 20% by anti-*Nb*SWP16 blocking compared with the negative control *in vitro*. In contrast with anti-*Nb*SWP16, no remarkable decrement inhibition was detected when antibodies of *Nb*SWP16 and *Nb*SWP5 were used simultaneously. Collectively, these results suggest that *Nb*SWP16 is a new exospore protein and probably be involved in spore adherence of *N. bombycis*.

Published On: Parasitology, 2015, 142(4), 534-542.

1 State Key Laboratory of Silkworm genome Biology, Southwest University, Chongqing 400716, China.
2 Laboratory of Animal Biology, Chongqing Normal University, Chongqing 400047, China.
*Corresponding author E-mail: zyzhou@swu.edu.cn.

一种新的富含脯氨酸串联重复序列的家蚕微孢子虫孢壁蛋白*Nb*SWP16的鉴定

王 营[1] 党晓群[1,2] 马 强[1] 刘方燕[1]
潘国庆[1] 李 田[1] 周泽扬[1,2*]

摘要：家蚕微孢子虫(*Nosema bombycis*)是一种细胞内专性寄生的单细胞真核生物，它可以诱发家蚕微粒子病。已有研究表明微孢子虫的孢壁蛋白在微孢子虫对宿主的黏附和感染过程中发挥重要作用。但是，目前关于微孢子虫孢壁蛋白的报道还比较少。在本文中，我们鉴定了一个分子质量为44 kDa的家蚕微孢子虫孢壁蛋白*Nb*SWP16。序列分析结果表明，该蛋白质含有一个由25个氨基酸组成的信号肽和3个肝素结合基序。有意思的是，该蛋白质还含有3个富含脯氨酸的串联重复序列，并且这个串联重复序列是家蚕微孢子虫所特有的。免疫荧光结果表明在成熟孢子表面和发芽孢子表面都能检测到*Nb*SWP16的信号。免疫电镜的结果也显示*Nb*SWP16分布在外孢子区。抗体封闭实验表明*Nb*SWP16的抗体可以使家蚕微孢子虫对宿主细胞的黏附率降低20%以上。但是，当我们同时用*Nb*SWP16和*Nb*SWP5的抗体进行封闭实验，发现其对宿主细胞的黏附率并没有显著地降低。以上结果表明*Nb*SWP16是一个新的孢子外壁蛋白，并且这个蛋白质可能参与了家蚕微孢子虫对宿主细胞的黏附过程。

1 家蚕基因组生物学国家重点实验室，西南大学，重庆
2 动物生物学实验室，重庆师范大学，重庆

Characterization of a novel otubain-like protease with deubiquitination activity from *Nosema bombycis* (Microsporidia)

Wang Y[1] Dang XQ[2] Luo B[1] Li CF[1] Long MX[1]
Li T[1] Li Z[2] Pan GQ[1] Zhou ZY[1,2*]

Abstract: Otubains are a recently identified family of deubiquitination enzymes (DUBs). They are involved in diverse biological processes including protein degradation, signal transduction and cell immune response. Several microsporidian genomes have been published in the last decade; however, little is known about the otubain-like protease in these widely-spread obligate intracellular parasites. Here, we characterized a 25 kDa otubain-like protease (*Nb*OTU1) from the microsporidian *Nosema bombycis*, the pathogen causing pebrine disease in the economically important insect *Bombyx mori*. Sequence analysis showed this protein contained a conserved catalytic triad of otubains composed of aspartate, cysteine and histidine residues. The expression of *Nbotu1* began on day 3 post-infection as determined by RT-PCR. Immunofluorescence analysis indicated that *Nb*OTU1 is localized on the spore wall of *N. bombycis*. The subcellular localization of the *Nb*OTU1 was further detected with immunoelectron microscopy, which showed that *Nb*OTU1 is localized at the regions around endospore wall and plasma membrane. Deubiquitination analysis confirmed that the recombinant *Nb*OTU1 possessed deubiquitination activity *in vitro*. Taken together, a novel microsporidian otubain - like protease *Nb*OTU1 was partially characterized in *N. bombycis*, demonstrating its subcellular location and deubiquitination activity. This study provided a basic reference for further dissecting the function of otubains in microsporidia.

Published On: Parasitology Research, 2015, 114(10), 3759-3766.

1 State Key Laboratory of Silkworm genome Biology, Southwest University, Chongqing 400716, China.
2 Laboratory of Animal Biology, Chongqing Normal University, Chongqing 400047, China.
*Corresponding author E-mail: zyzhou@swu.edu.cn.

一种新的家蚕微孢子虫类otubain蛋白酶的鉴定

王　营[1]　党晓群[2]　罗　波[1]　李春峰[1]　龙梦娴[1]
李　田[1]　李　治[2]　潘国庆[1]　周泽扬[1,2*]

摘要：Otubain是去泛素化蛋白酶（DUBs）中的一个家族。Otubain参与多种生物学过程，包括蛋白降解、信号传导和细胞免疫。在过去的十年里，虽然多种微孢子虫的基因组相继被测序，但是在这些广泛传播的细胞内寄生虫中，关于类Otubain蛋白酶我们还知之甚少。在本文中，我们在家蚕微粒子的病原——家蚕微孢子虫中鉴定到了一个分子质量为25 kDa的类Otubain蛋白酶*Nb*OTU1。序列分析结果表明，*Nb*OTU1含有一个由天冬氨酸、半胱氨酸和组氨酸组成的催化三联体结构域。RT-PCR结果表明*Nbotu1*在微孢子虫感染家蚕3天后即开始表达。免疫荧光和免疫电镜结果表明*Nb*OTU1主要定位于家蚕微孢子虫的孢子内壁和细胞膜。体外的去泛素化作用分析表明重组的*Nb*OTU1蛋白具有去泛素化能力。总之，我们鉴定了一个新的微孢子虫类Otubain蛋白酶*Nb*OTU1，并进行了亚细胞定位研究和去泛素化活性分析，这将为微孢子虫Otubain的功能研究提供有力的参考。

1　家蚕基因组生物学国家重点实验室，西南大学，重庆
2　动物生物学实验室，重庆师范大学，重庆

Characterization of the first fungal glycosyl hydrolase family 19 chitinase (*Nb*chiA) from *Nosema bombycis* (*Nb*)

Han B [1#] Zhou K [2#] Li ZH [1] Sun B [1] Ni Q [1] Meng XZ [1]
Pan GQ [1] Li CF [1] Long MX [1] Li T [1] Zhou CZ [2] Li WF [2*] Zhou ZY [1,3*]

Abstract: Chitinases (*EC 3.2.1.14*), as one kind of glycosyl hydrolase, hydrolyze the β-(1, 4) linkages of chitin. According to the sequence similarity, chitinases can be divided into glycoside hydrolase family 18 and family 19. Here, a chitinase from *Nosema bombycis* (*Nb*chiA) was cloned and purified by metal affinity chromatography and molecular exclusion chromatography. Sequence analysis indicated that *Nb*chiA belongs to glycoside hydrolase family 19 class IV chitinase. The optimal pH and temperature of *Nb*chiA are 7.0 and 40 °C, respectively. This purified chitinase showed high activity toward soluble substrates, such as ethylene glycol chitin and soluble chitosan. The degradation of chitin oligosaccharides $(GlcNAc)_{2-5}$ detected by high-performance liquid chromatography showed that *Nb*chiA hydrolyzed mainly the second glycosidic linkage from the reducing end of $(GlcNAc)_{3-5}$. On the basis of structure-based multiple-sequence alignment, Glu51 and Glu60 are believed to be the key catalytic residues. The site-directed mutation analysis revealed that the enzymatic activity was decreased upon mutation of Glu60, whereas mutation of Glu51 totally abolished the enzymatic activity. This is the first report of a GH19 chitinase in fungi and in Microsporidia.

Published On: Journal of Eukaryotic Microbiology, 2016, 63(1), 37-45.

1 State Key Laboratory of Silkworm Genome Biology, Southwest University, Chongqing 400716, China.

2 Hefei National Laboratory for Physical Sciences at the Microscale and School of Life Sciences, University of Science and Technology of China, Hefei 230027, China.

3 College of Life Sciences, Chongqing Normal University, Chongqing 400047, China.

[#]These authors contributed equally.

[*]Corresponding author E-mail: zyzhou@swu.edu.cn; liwf@ustc.edu.cn.

家蚕微孢子虫糖基水解酶第19家族的几丁质酶(*Nb*chiA)的鉴定

韩　冰[1#]　周　康[2#]　李致宏[1]　孙　滨[1]　倪　琪[1]
孟宪志[1]　潘国庆[1]　李春峰[1]　龙梦娴[1]
李　田[1]　周丛照[2]　李卫芳[2*]　周泽扬[1,3*]

摘要：作为一种糖基水解酶，几丁质酶可以水解β-1，4糖苷键连接的几丁质。根据序列相似性，几丁质酶属于糖基水解酶第18和第19家族。在本文中，我们克隆了家蚕微孢子虫的一个几丁质酶(*Nb*chiA)，并利用金属亲和层析以及分子筛层析的方法对该蛋白质进行了纯化。通过序列分析发现，*Nb*chiA属于糖基水解酶第19家族中的Ⅳ簇几丁质酶。*Nb*chiA最适pH为7.0，最适温度为40 ℃。纯化的几丁质酶对可溶的底物，比如乙二醇几丁质和可溶性脱乙酰几丁质具有较高的活性。利用HPLC检测*Nb*chiA对几丁质寡糖$(GlcNAc)_{2-5}$的降解效果，结果表明*Nb*chiA可以从几丁质寡糖$(GlcNAc)_{3-5}$的第二个糖苷键的还原端开始降解。对蛋白质进行多序列比对分析，发现Glu51和Glu60为*Nb*chiA潜在的活性关键位点，对这两个位点进行点突变，发现突变Glu60后，*Nb*chiA酶活降低，而突变Glu51后NbchiA完全失去活性。本研究首次报道了关于微孢子虫以及真菌的GH19几丁质酶。

1　家蚕基因组生物学国家重点实验室，西南大学，重庆
2　合肥微尺度物质科学国家研究中心，中国科技大学，合肥
3　生命科学学院，重庆师范大学，重庆

Exogenous gene can be integrated into *Nosema bombycis* genome by mediating with a non-transposon vector

Guo R[1,2#] Cao GL[1,3#] Lu YH[1#] Xue RY[1,3]
Dhiraj Kumar[1] Hu XL[1,3] Gong CL[1,3*]

Abstract: *Nosema bombycis*, a microsporidium, is a pathogen of pebrine disease of silkworms, and its genomic DNA sequences had been determined. Thus far, the research of gene functions of microsporidium including *N. bombycis* cannot be performed with gain/loss of function. In the present study, we targeted to construct transgenic *N. bombycis*. Therefore, hemocytes of the infected silkworm were transfected with a non-transposon vector pIZT/V5-His vector *in vivo*, and the blood, in which the hemocyte with green fluorescence could be observed, was added to the cultured *BmN* cells. Furthermore, normal *BmN* cells were infected with germinated *N. bombycis*, and the infected cells were transfected with *p*IZT/V5-His. Continuous fluorescence observations exposed that there were *N. bombycis* with green fluorescence in some *N. bombycis*-infected cells, and the extracted genome from the purified *N. bombycis* spore was used as templates. PCR amplification was carried out with a pair of primers for specifically amplifying the green fluorescence protein (GFP) gene; a specific product representing the *gfp* gene could be amplified. Expression of the GFP protein through Western blotting also demonstrated that the *gfp* gene was perfectly inserted into the genome of *N. bombycis*. These results illustrated that exogenous gene can be integrated into *N. bombycis* genome by mediating with a non-transposon vector. Our research not only offers a strategy for research on gene function of *N. bombycis* but also provides an important reference for constructing genetically modified microsporidium utilized for biocontrol of pests.

Published On: Parasitology Research, 2016, 115(8), 3093-3098.

1 School of Biology and Basic Medical Sciences ,Soochow University, Suzhou 215123, China.
2 College of Bee Science, Fujian Agriculture and Forestry University, Fuzhou 350000, China.
3 National Engineering Laboratory for Modern Silk, Soochow University, Suzhou 215123, China.
#These authors contributed equally.
*Corresponding author E-mail: gongcl@suda.edu.cn.

非转座子载体可介导外源基因整合进入家蚕微孢子虫基因组

郭 睿[1,2#] 曹广力[1,3#] 陆亚红[1#] 薛仁宇[1,3]
Dhiraj Kumar[1] 胡小龙[1,3] 贡成良[1,3*]

摘要：*Nosema bombycis* 是一种微孢子虫，是家蚕的病原体，其基因组DNA序列已经确定。到目前为止，包括*N. bombycis*在内的微孢子虫的基因功能研究不能用功能的获得/丧失进行。在本研究中，我们将非转座子载体pIZT/V5-His注射进感染*N. bombycis*的家蚕体内，将血球细胞呈绿色荧光的血液添加到培养的*BmN*细胞中。此外，用发芽的*N. bombycis*感染正常的*BmN*细胞，再用pIZT / V5-His载体转染感染的细胞。连续的荧光观察显示，在一些感染*N. bombycis*的细胞中存在具有绿色荧光的*N. bombycis*，从纯化的*N. bombycis*孢子中提取基因组用作模板，用一对能特异性扩增绿色荧光蛋白（GFP）基因的引物可扩增出*gfp*基因。GFP蛋白的免疫印迹法也证明了*gfp*基因被完全插入到*N. bombycis*的基因组中。这些结果表明，非转座子载体可介导外源基因整合入*N. bombycis*基因组中。我们的研究不仅提供了一种研究*N. bombycis*基因功能的策略，也为构建用于生物防治害虫的转基因微孢子虫提供了重要参考。

1 基础医学与生物科学学院，苏州大学，苏州
2 蜂学学院，福建农林大学，福州
3 现代丝绸国家工程实验室，苏州大学，苏州

Multiple rDNA units distributed on all chromosomes of *Nosema bombycis*

Liu HD[1] Pan GQ[1] Song SH[1] Xu JS[2]
Li T[1] Deng YB[3] Zhou ZY[1,2*]

Abstract: Among Microsporidia, *Nosema bombycis* has a novel arrangement of LSUrRNA, SSUrRNA, ITS, IGS and 5S rRNA. To determine the distribution of rDNA among the chromosomes, we performed genome-wide screening and Southern blotting with three probes (SSU, ITS and IGS). Southern blotting revealed that ribosomal RNA genes are distributed on all chromosomes of *N. bombycis*, which is contrary to the previous result, which concluded that the *N. bombycis* rRNA genes were limited to a single chromosome. This wide distribution is similar to that of the rDNA unit of *Encephalitozoon cuniculi*. Screening of the *N. bombycis* genome detected 53 LSUrRNA elements, 43 SSUrRNA elements and 36 5S rRNA elements. However, it is still difficult to determine their loci on the chromosomes as the genomic map is unfinished.

Published On: Journal of Invertebrate Pathology, 2008, 99(2), 235-238.

1 Key Sericultural Laboratory of Agricultural Ministry, Southwest University, Chongqing 400716, China.
2 Laboratory of Animal Biology, Chongqing Normal University, Chongqing 400047, China.
3 School of Life Science & Technology, Beijing Institute of Technology, Beijing 100081, China.
*Corresponding author E-mail: zyzhou@cqnu.edu.cn.

家蚕微孢子虫全基因组中分布的若干核糖体DNA(rDNA)单元

刘含登[1] 潘国庆[1] 宋世洪[1] 许金山[2]
李 田[1] 邓严波[3] 周泽扬[1,2*]

摘要:在微孢子目中,家蚕微孢子虫(*Nosema bombycis*)含有一系列新的LSUrRNA、SSUrRNA、ITS、IGS和5S rRNA。为了鉴定核糖体DNA(rDNA)在微孢子虫染色体中的分布,我们进行了全基因组范围的筛查以及用SSU、ITS和IGS三个探针进行了Southern blotting检测,结果显示家蚕微孢子虫核糖体RNA基因分布在家蚕微孢子虫的所有染色体上,这与先前的报道称家蚕微孢子虫核糖体RNA基因只分布在单个的染色体上的结论相反。这种广泛的分布与*Encephalitozoon cuniculi* 的rDNA基因的分布相似。筛查家蚕微孢子虫全基因组,我们发现53个LSUrRNA元件、43个SSUrRNA元件以及36个5S rRNA元件。但是,由于家蚕微孢子虫基因组图谱还未绘制完成,目前还很难对这些元件在染色体上进行精确定位。

1 农业部蚕桑学重点实验室,西南大学,重庆
2 动物生物学实验室,重庆师范大学,重庆
3 生命科学与技术学院,北京科技大学,北京

Intraspecific polymorphism of rDNA among five *Nosema bombycis* isolates from different geographic regions in China

Liu HD[1,2] Pan GQ[1*] Luo B[1] Li T[1] Yang Q[4]
Charles R. Vossbrinck[5] Bettina A. Debrunner-Vossbrinck[6] Zhou ZY[1,3]

Abstract: The microsporidian *Nosema bombycis* is the causative agent of pébrine, a highly infectious disease of the silkworm *Bombyx mori*. Three regions of the multicopy rDNA gene were examined in order to investigate the relationships among five *Nosema* isolates from various regions of China. Ribosomal DNA alleles are present on each of the 18 chromosomes of *N. bombycis* and show a high degree of variation. In this study the small subunit (SSU) rDNA, internal transcribed spacer (ITS) and intergenic spacer (IGS) regions for up to 10 different rDNA copies from each *N. bombycis* isolate were cloned and sequenced. As expected we see greater polymorphism in the ITS region (88 variable sites in 179 nucleotides) and IGS (200 variable sites in 279 nucleotides) than in the SSU rDNA (24 variable sites in 1 232 nucleotides). Phylogenetic analysis shows greater differences between alleles within an isolate than between the same alleles from different isolates. The data reveal two very different groups, one from the Sichuan province and the other with a broad distribution including four provinces in southeast China and Japan. The Sichuan isolate does not have any rDNA alleles with sequences identical to those in the other isolates, implying that it is a separate, non-intermixing, population or perhaps a separate species from the other isolates. In light of the polymorphic nature of the *rDNA* alleles in *N. bombycis* and their presence on every chromosome, the rDNA gene may be useful for understanding the movement and ultimately the source of pébrine infections.

Published On: Journal of Invertebrate Pathology, 2013, 113(1), 63-69.

1 State Key Laboratory of Silkworm Genome Biology, Southwest University, Chongqing 400716, China.
2 Experimental Teaching Center, Chongqing Medical University, Chongqing 400016, China.
3 Laboratory of Animal Biology, Chongqing Normal University, Chongqing 400047, China.
4 The Sericulture & Farm Produce Processing Research Institute, Guangdong Academy of Agricultural Sciences, Guangzhou 510610, China.
5 The Connecticut Agricultural Experiment Station, New Haven, CT 06504, USA.
6 Department of Biology, Gateway Community College, New Haven, CT 06504, USA.
*Corresponding author E-mail: gqpan@swu.edu.cn.

来自中国五个不同地域的家蚕微孢子虫rDNA 序列种内多态性研究

刘含登[1,2] 潘国庆[1*] 罗 波[1] 李 田[1] 杨 琼[4]
Charles R. Vossbrinck[5] Bettina A. Debrunner-Vossbrinck[6] 周泽扬[1,3]

摘要:家蚕微粒子病是家蚕的一种重要传染病,家蚕微孢子虫是该病的重要病原体。为了调查来自于中国5个不同地区的微孢子虫株系间的关系,我们检测了多拷贝rDNA基因的三个区段。rDNA序列位于家蚕微孢子虫的所有18条染色体上,序列呈高度多态性。本研究中,我们克隆测序得到小亚基rDNA(SSU rDNA)、内转录间隔区(ITS)和基因间隔区(IGS)分别不少于10条序列。经序列分析发现:ITS区(179个核苷酸中存在88个变异位点)和IGS区(279个核苷酸中存在200个变异位点)的序列多态性远高于SSU rDNA区(1 232个核苷酸中存在24个变异位点)。系统发育分析表明:同一株系内等位基因差异程度高于不同株系间同一等位基因序列差异。分析结果表明其形成了两大不同群组,来自四川省的株系独立形成一群,另一群分布广泛,包括中国东南和日本的四个省。四川株系与其他株系rDNA等位基因序列上几乎没有一致性,这暗示其可能是一个独立非混合或者独立于其他株系外的株系。rDNA等位基因存在于家蚕微孢子虫所有染色体上,且具有明显的多态性,这将有助于我们理解家蚕微粒子病感染的活动与感染来源。

1 家蚕基因组生物学国家重点实验室,西南大学,重庆
2 实验教学中心,重庆医科大学,重庆
3 动物生物学实验室,重庆师范大学,重庆
4 蚕业与农产品加工研究所,广东省农业科学院,广州
5 康涅狄格州农业试验站,纽黑文,美国
6 生物系,盖特威社区学院,纽黑文,美国

Development of an approach to analyze the interaction between *Nosema bombycis* (microsporidia) deproteinated chitin spore coats and spore wall proteins

Yang DL[1] Dang XQ[2] Tian R[1] Long MX[1] Li CF[1]
Li T[1] Chen J[1] Li Z[1,2] Pan GQ[1] Zhou ZY[1,2*]

Abstract: *Nosema bombycis* is an obligate intracellular parasite of the *Bombyx mori* insect. The spore wall of *N. bombycis* is composed of an electron-dense proteinaceous outer layer, an electron-transparent chitinous inner layer. and the spore wall is connected to the plasma membrane. In this study, the deproteinated chitin spore coats (DCSCs) were acquired by boiling *N. bombycis* in 1 mol/L NaOH. Under a transmission electron microscope, the chitin spore coat resembles a loosely curled ring with strong refractivity; organelles and nuclei were not observed inside the spore. The anti-SWP25, 26, 30 and 32 antibodies were used to detect whether spore wall proteins within the total soluble and mature spore proteins could bind to the DCSCs. Furthermore, a chitin binding assay showed that within the total soluble and mature spore proteins, the SWP26, SWP30 and SWP32 spore wall proteins, bound to the deproteinated chitin spore coats, although SWP25 was incapable of this interaction. Moreover, after the DCSCs were incubated with the alkali-soluble proteins, the latter were obtained by treating *N. bombycis* with 0.1 mol/L NaOH. Following this treatment, SWP32 was still capable of binding the DCSCs, while SWP26 and SWP30 were unable to bind. Collectively, the DCSCs are useful for investigating the arrangement of spore wall proteins, and they shed light on how the microsporidia spore wall is self-assembled.

Published On: Journal of Invertebrate Pathology, 2014, 115, 1-7.

1 State Key Laboratory of Silkworm Genome Biology, Southwest University, Chongqing 400716, China.
2 Laboratory of Animal Biology, Chongqing Normal University, Chongqing 400047, China.
*Corresponding author E-mail: zyzhou@swu.edu.cn.

家蚕微孢子虫脱蛋白几丁质孢壳和孢壁蛋白相互作用的分析方法的建立

杨东林[1] 党晓群[2] 田 锐[1] 龙梦娴[1] 李春峰[1]
李 田[1] 陈 洁[1] 李 治[1,2] 潘国庆[1] 周泽扬[1,2*]

摘要：家蚕微孢子虫是家蚕中一类细胞内专性寄生虫。微孢子虫的孢壁由高电子密度的富含蛋白质的外壁和低电子透明的几丁质内壁组成，孢壁与质膜相连。本研究中，脱蛋白WJ质孢壳(DCSCs)需要在1 mol/L NaOH中煮沸家蚕微孢子虫来获得。透射电镜观察显示几丁质孢壳像一个具有强折射率的疏松卷曲环，而没有观察到孢子内的细胞器和细胞核。利用anti-SPW25、26、30、32抗体检测所有可溶性成熟孢子蛋白中孢壁蛋白能否结合到DCSCs上。几丁质结合实验显示，在所有可溶性成熟孢子蛋白中，SWP26、SWP30和SWP32孢壁蛋白可结合到DCSCs上，然而SWP25不能与其发生相互作用。用0.1 mol/L NaOH处理家蚕微孢子虫得到碱溶性蛋白，将该蛋白与DCSCs孵育后发现，SWP32仍旧能结合DCSCs，而SWP26和SWP30不能结合DCSCs。总之，DCSCs有助于研究孢壁蛋白的排列和阐明微孢子虫孢壁是如何自组装的。

1 家蚕基因组生物学国家重点实验室，西南大学，重庆
2 动物生物学实验室，重庆师范大学，重庆

Bm59 is an early gene, but is unessential for the propagation and assembly of *Bombyx mori* nucleopolyhedrovirus

Hu XL[1,2] Shen YW[1] Zheng Q[1] Wang GB[1] Wu XF[1*] Gong CL[2*]

Abstract: *Bombyx mori* nucleopolyhedrovirus (*Bm*NPV)is a major pathogen that specifically infects the domestic silkworm and causes serious economic loss to sericulture around the world. The function of *Bm*NPV *Bm59* gene in the viral life cycle is inconclusive. To investigate the role of *Bm59* during viral infection, the transcription initiation site and temporal expression of *Bm59* were analyzed, and *Bm59*-knockout virus was generated through homologous recombination in *Escherichia coli.* The results showed that *Bm59* is an early transcription gene with an atypia early transcriptional start motif. Budded virion (BV) production and DNA replication in the *BmN* cells transfected with the *Bm59*-knockout virus bacmid were similar to those in the cells transfected with the wild-type virus. Electron microscopy revealed that the occlusion-derived virus can be produced in cells infected with the *Bm59*-knockout virus. These results indicated that *Bm59* is an early gene and isnot essential for viral replication or assembly of *Bm*NPV. These findings suggested that non-essential gene (*Bm59*)remained in the viral genome, which may interact with other viral/host genes in a certain situation.

Published On: Molecular Genetics and Genomes, 2016, 291(1), 145-154.

1 College of Animal Sciences, Zhejiang University, Hangzhou 310058, China.

2 School of Biology and Basic Medical Science, Soochow University, Suzhou 215123, China.

*Corresponding author E-mail: wuxiaofeng@zju.edu.cn; gongcl@suda.edu.cn.

早期基因*Bm59*不是家蚕核型多角体病毒*Bm*NPV传播和组装的必需基因

胡小龙[1,2] 沈运旺[1] 郑 秦[1] 王国宝[1] 吴小锋[1*] 贡成良[2*]

摘要:家蚕核型多角体病毒(*Bm*NPV)是一种特异性感染家蚕的病原体,其能造成世界各地养蚕业严重的经济损失。*Bm*NPV *Bm59*基因在病毒生命周期中的功能尚未确定。为了研究*Bm59*在病毒感染过程中的作用,我们分析了*Bm59*的转录起始位点和时期表达特征,并通过大肠杆菌中的同源重组获得了*Bm59*敲除型病毒。结果表明,*Bm59*是一个具有非典型早期转录起始基序的早期转录基因。用*Bm59*敲除型质粒转染的*BmN*细胞,芽型病毒(BV)和DNA复制情况与用野生型病毒转染的细胞相似。电子显微镜显示,用*Bm59*敲除型病毒感染细胞可以产生ODV病毒。这些结果表明,*Bm59*是一个早期基因,但不是*Bm*NPV病毒复制或组装所必需的。这些发现暗示,非必需基因(*Bm59*)存在于病毒基因组中,可能在某种情况下与其他病毒/宿主基因发生相互作用。

1 动物科学学院,浙江大学,杭州
2 基础医学与生物科学学院,苏州大学,苏州

Characterization of *Bombyx mori* nucleopolyhedrovirus with a knockout of *Bm17*

Shen HX[1] Zhou Y[2] Zhang W[1] Ni B[1] Wang H[1] Wang XC[1]
Shao SH[1] Chen HQ[2] Guo ZJ[2] Liu XY[2] Yao Q[2] Chen KP[2*]

Abstract: Open reading frame 17 (*Bm17*) gene of *Bombyx mori* nucleopolyhedrovirus is a highly conserved gene in lepidopteran nucleopolyhedroviruses, but its function remains unknown. In this report, transient-expression and superinfection assays indicated that Bm17 localized in the nucleus and cytoplasm of infected *BmN* cells. To determine the role of *Bm17* in baculovirus life cycle, we constructed a *Bm17* knockout virus and characterized its properties in cells. Analysis of the production and infection of budded virions, the level of viral DNA replication revealed showed that there was no significant difference among the mutant, the control, and the *Bm17* repaired virus strains. These results suggest that *Bm17* is not essential for virus replication in cultured cells.

Published On: Cytotechnology, 2012, 64(6), 711-718.

1 School of Medical Science and Laboratory Medicine, Jiangsu University, Zhenjiang 212013, China.
2 Institute of Life Sciences, Jiangsu University, Zhenjiang 212013, China.
*Corresponding author E-mail: kpchen@ujs.edu.cn.

敲除家蚕核型多角体病毒基因*Bm17*的特性分析

申红星[1] 周 阳[2] 张 文[1] 倪 斌[1] 王 华[1] 王晓春[1]
邵世和[1] 陈慧卿[2] 郭忠建[2] 刘晓勇[2] 姚 勤[2] 陈克平[2*]

摘要:家蚕核型多角体病毒基因17(*Bm17*)在鳞翅目核型多角体病毒中高度保守,但其功能尚不清楚。在本报告中,瞬时表达和重复感染实验表明Bm17位于感染*BmN*细胞的细胞核和细胞质中。为了确定*Bm17*在杆状病毒生命周期中的作用,我们构建了一种*Bm17*敲除病毒,并在细胞中表征其特性。分析出芽病毒粒子的产生和感染,以及病毒DNA复制水平,发现突变体、对照和*Bm17*修复的病毒株之间没有显著差异。这些结果表明*Bm17*对于培养细胞中的病毒复制不是必需的。

1 基础医学与医学技术学院 ,江苏大学,镇江
2 生命科学研究院,江苏大学,镇江

BM61 of *Bombyx mori* nucleopolyhedrovirus: its involvement in the egress of nucleocapsids from the nucleus

Shen HX[1] Chen KP[2*]

Abstract: All lepidopteran baculovirus genomes sequenced encode a homolog of the *Bombyx mori* nucleopolyhedrovirus *orf61* gene (*Bm61*). To determine the role of *Bm61* in the baculoviral life cycle, we constructed a *Bm61* knockout virus and characterized it in cells. We observed that the *Bm61* deletion bacmid led to a defect in production of infectious budded virus (BV). Quantitative PCR analysis of BV in the media culturing the transfected cell indicated that BV was not produced due to *Bm61* deletion. Electron microscope analysis showed that in the knockout of *Bm61*, nucleocapsids were not transported from the nucleus to the cytoplasm. From these results, we concluded that BM61 is required in the BV pathway for the egress of nucleocapsids from the nucleus to the cytoplasm.

Published On: FEBS Letters, 2012, 586(7), 990-995.

1 School of Medical Science and Laboratory Medicine, Jiangsu University, Zhenjiang 212013, China.

2 Institute of Life Sciences, Jiangsu University, Zhenjiang 212013, China.

*Corresponding author E-mail address: kpchen@ujs.edu.cn.

家蚕核型多角体病毒*BM61*：参与病毒粒子的出核

申红星[1]　陈克平[2*]

摘要：所有鳞翅目杆状病毒基因组序列编码一种家蚕核型多角体病毒*orf61*基因（*Bm61*）的同源物。为了确定*Bm61*在杆状病毒生命周期中的作用，我们构建了一种*Bm61*敲除病毒，并在细胞中表征。我们观察到*Bm61*缺失导致感染性出芽病毒（BV）的生产缺陷。转染细胞的培养基中BV的定量PCR分析表明，*Bm61*缺失不产生BV。电子显微镜分析显示，*Bm61*敲除导致病毒粒子没有从核转运到细胞质。从这些结果我们得出结论，BM61是BV从细胞核到细胞质所必需的。

1　医学院与药学院，江苏大学，镇江

2　生命科学研究院，江苏大学，镇江

Expression of *Bombyx mori* nucleopolyhedrovirus ORF76 in permissive and non-permissive cell lines by a novel Bac-to-Bac/*Bm*NPV baculovirus expression system

Su WJ Wu Y Wu HL

Zhu SY Wang WB*

Abstract: Open reading frame 76 of *Bombyx mori* nucleopolyhedrovirus (*Bm*NPV), designated as *Bm76*, is a gene whose function is completely unknown. With *EGFP* fused to the 3' terminal of *Bm76* as the reporter gene and *Bm*NPV bacmid as the expression vector, a recombinant bacmid was successfully constructed expressing *Bm*76-EGFP fusion protein under the control of *polyhedrin* promoter in *Bombyx mori* cells (*Bm* cells), *Bm*NPV's permissive cell line, laying the foundation for rescue experiment of *Bm76* deletion mutant. Moreover, the supernatant from *Bm* cells transfected with the recombinant bacmid was used to infect *Trichoplusia Ni* cells (*Tn* cells), *Bm*NPV's non-permissive cell line. Unexpectedly, the expression of *Bm*76-EGFP fusion protein in some Tn cells was detected, implying that viral DNA was replicated in these cells. The causes are being studied for the inability of *Bm*NPV to produce enough viable budded viruses in *Tn* cells despite of viral DNA replication.

Published On: Polish Journal of Microbiology, 2008, 57(4), 271-274.

Institute of Life Sciences, Jiangsu University, Zhenjiang 212013, China.

*Corresponding author E-mail: wenbingwang@ujs.edu.cn.

利用新的Bac-to-Bac杆状病毒表达系统在敏感及非敏感细胞中表达家蚕杆状病毒的基因ORF76

苏武杰　吴　岩　乌慧玲　朱姗颖　王文兵*

摘要：家蚕核型多角体病毒的开放阅读框76基因（*Bm76*）功能未知。将*EGFP*与*Bm76*的3′末端融合作为报告基因，以*Bm*NPV bacmid为表达载体，成功构建了表达*Bm*76-EGFP融合蛋白的重组bacmid，该bacmid可以在家蚕*Bm*NPV的敏感细胞系（*Bm*细胞）中在多角体启动子控制下启动融合蛋白表达，为*Bm*76缺失突变体的拯救实验建立了基础。此外，利用重组bacmid转染的*Bm*细胞的上清液感染*Trichoplusia Ni*细胞（*Tn*细胞），即*Bm*NPV的非敏感细胞系。出乎意料的是，检测到*Bm*76-EGFP融合蛋白在一些*Tn*细胞中表达，这意味着病毒DNA在这些细胞中复制。尽管病毒DNA可以复制，为什么*Bm*NPV无法在*Tn*细胞中产生足够的存活出芽病毒，该原因正在研究。

生命科学研究院，江苏大学，镇江

Characterization of a late gene, *ORF67* from *Bombyx mori* nucleopolyhedrovirus

Chen HQ[1] Chen KP[1*] Yao Q[1] Guo ZJ[1] Wang LL[2]

Abstract: Open reading frame 67 of *Bombyx mori* nucleopolyhedrovirus (*BmORF67*) is a homologue of *Autographa californica* multiple NPV *ORF81*. The gene is conserved among all baculoviruses and is thus considered a baculovirus core gene. The transcript of *BmORF67* was detected at 18-72 h post-infection (p.i.). Polyclonal antiserum raised to a His-*Bm*ORF67 fusion protein recognized *Bm*ORF67 in infected cell lysates from 24 to 72 h p.i., suggesting that *BmORF67* is a late gene. *Bm*ORF67 was not detected either in budded viruses or occlusion - derived virus. Immunofluoresence analysis showed that the protein located in the cytoplasm and interacted with host protein actin A3. In conclusion, *Bm*ORF67 is a late protein localized in the cytoplasm of infected cells that interacts with host protein.

Published On: FEBS Letters, 2007, 581(30), 5836-5842.

1 Institute of Life Sciences, Jiangsu University, Zhenjiang 212013, China.

2 College of Life Science, Chongqing Normal University, Chongqing 212013, China.

*Corresponding author E-mail: kpchen@ujs.edu.cn.

家蚕核型多角体病毒晚期基因 *ORF67*的特征分析

陈慧卿[1] 陈克平[1*] 姚 勤[1] 郭忠建[1] 王林玲[2]

摘要:家蚕核型多角体病毒的*BmORF67*与苜蓿银纹夜蛾核型多角体病毒*ORF81*同源。该病毒基因在所有杆状病毒中均是保守的,因此被认为是杆状病毒核心基因。在感染后18—72 h能检测到*BmORF67*的转录本。His-*Bm*ORF67融合蛋白制备的多克隆抗体,可以识别*Bm*ORF67,分析显示在感染24—72 h后的细胞裂解液中均能检测到*Bm*ORF67蛋白表达,这进一步证明*BmORF67*是晚期基因。在出芽型病毒粒子或包涵体型病毒粒子中均不能检测到*Bm*ORF67。免疫荧光分析显示*Bm*ORF67蛋白位于细胞质中并与宿主肌动蛋白A3相互作用。总之,*Bm*ORF67是定位在与宿主蛋白相互作用的感染细胞的细胞质中的晚期表达蛋白质。

1 生命科学研究院,江苏大学,镇江

2 生命科学学院,重庆师范大学,重庆

Characterization of *Bombyx mori* nucleopolyhedrovirus *orf74*, a novel gene involved in virulence of virus

Shen HX　Chen KP*　Yao Q　Yu W

Pan Y　Huo J　Xia HC　Huang GP

Abstract: *orf 74* of *Bombyx mori* nucleopolyhedrovirus(*Bm*NPV) is a homologue of uncharacterized *orf 91* of *Autographa californica* multicapsid NPV, predicted to encode a protein of 17.3 kDa. RT-PCR showed that the transcription of *orf 74* was initiated at 12 h post-infection stage. Expression assay indicated that ORF74 localized in the nucleus of infected *BmN* cells. To study the function of ORF74, an *orf 74*-knockout virus was constructed using bacmid technology. Analysis of the production of budded virions and occlusion bodies, the formation of nucleocapsids, and the level of viral DNA revealed no significant difference among the mutant, the control, and the *orf 74*-repaired virus. Interestingly, bioassay showed that the median lethal time of the *orf 74*-knockout virus was 14.7 h longer than the control virus, but the virus yield was similar to that of control virus. Thus, our data indicated that ORF74 is most likely implicated in the virulence of *Bm*NPV, and is not essential for virus replication.

Published On: Virus Genes, 2009, 38(3), 487-494.

Institute of Life Sciences, Jiangsu University, Zhenjiang 212013, China.

*Corresponding author E-mail:kpchen@ujs.edu.cn.

一个家蚕核型多角体病毒毒力相关新基因*orf74*的鉴定

申红星　陈克平*　姚　勤　余　蔚
潘　晔　霍　娟　夏恒传　黄国平

摘要：家蚕核型多角体病毒（*Bm*NPV）的*orf74*是苜蓿银纹夜蛾核型多角体病毒（*Autographa californica* multicapsid NPV）功能未知基因*orf91*的同源物，预计其编码分子质量为17.3 kDa的蛋白质。RT-PCR显示*orf74*在感染后12 h开始转录。蛋白表达分析表明ORF74位于感染*BmN*细胞的细胞核中。使用bacmid技术构建*orf74*敲除病毒以研究ORF74功能。对发芽病毒粒子和包涵体的产生情况，核衣壳的形成以及病毒DNA水平调查发现，在突变体、对照和*orf74*回复突变病毒株之间，这些性状均没有显著差异。有趣的是，生物测定显示，*orf74*敲除病毒的中位数死亡时间比对照病毒的长14.7 h，但病毒产量与对照病毒的相似。因此，我们的数据表明，ORF74最有可能与*Bm*NPV的毒力相关，并不是病毒复制所必需的。

生命科学研究院，江苏大学，镇江

Characterization of the *Bm61* of the *Bombyx mori* nucleopolyhedrovirus

Shen HX Chen KP* Yao Q Zhou Y

Abstract: *Orf 61 (bm61)* of *Bombyx mori* nucleopolyhedrovirus(*Bm*NPV) is a highly conserved baculovirus gene, suggesting that it performs an important role in the virus life cycle whose function is unknown. In this study, we describe the characterization of *bm61*. Quantitative polymerase chain reaction (qPCR) and Western blot analysis demonstrated that *bm61* was expressed as a late gene. Immunofluorescence analysis by confocal microscopy showed that BM61 protein was localized on nuclear membrane and in intranuclear ring zone of infected cells. Structure localization of the BM61 in BV and ODV by western analysis demonstrated that BM61 was the protein of both BV and ODV. In addition, our data indicated that BM61 was a late structure protein localized in nucleus.

Published On: Current Microbiology, 2009, 59(1), 65-70.

Institute of Life Sciences, Jiangsu University, Zhenjiang 212013, China.
*Corresponding author E-mail:kpchen@ujs.edu.cn.

家蚕核型多角体病毒*Bm61*的特征分析

申红星　陈克平*　姚　勤　周　阳

摘要：家蚕核型多角体病毒（*Bm*NPV）的*orf61*（*bm61*）是一种高度保守的杆状病毒基因，其在病毒生命进程中发挥着重要作用，但具体的功能至今未知。在本研究中，我们初步分析了*bm61*的特征。定量聚合酶链式反应（qPCR）和蛋白质印迹分析表明，*bm61*是一个晚期表达基因。通过共聚焦显微镜的免疫荧光分析显示，BM61蛋白定位于核膜和感染细胞的核内环带区。蛋白质印迹分析显示，BM61在BV和ODV中均表达。此外，我们的数据也表明BM61是定位于细胞核的晚期结构蛋白。

生命科学研究院，江苏大学，镇江

Characterization of a late expression gene of *Bombyx mori* nucleopolyhedrovirus

Zhou Y[1] Chen KP[1*] Yao Q[1] Shen HX[2]
Liang GT[1] Li XG[1] Wang N[1] Li YJ[3]

Abstract: *Bombyx mori* nucleopolyhedrovirus (*Bm*NPV) ORF5 (*Bm5*) is a gene present in many lepidopteran nucleopolyhedroviruses (NPVs), but its function is unknown. In this study, *Bm5* was characterized. The transcript of *Bm5* was detected 12 -72 h post infection (p.i.). Polyclonal antiserum raised to a His-BM5 fusion protein recognized BM5 in infected cell lysates from 24 to 72 h p.i., suggesting that *Bm5* is a late gene. Immunofluorescence analysis by confocal microscopy showed that the BM5 protein is localized primarily in the cytoplasm. Localization of BM5 in budded virion (BV) and occlusion-derived virion (ODV) by Western analyses demonstrated that BM5 is not a structural protein associated with BV or ODV.

Published On:Zeit Schrift Fur Naturforschung Section C—A Journal of Biosciences, 2010, 65(7-8), 508-518.

1 Institute of Life Sciences, Jiangsu University, Zhenjiang 212013,China.
2 School of Medical Sciences and Laboratory Medicine, Jiangsu University, Zhenjiang 212013, China.
3 Zhejiang Province Yuyao No. 2 Middle School, Yuyao 315400, China.
*Corresponding author E-mail:kpchen@ujs.edu.cn.

家蚕核型多角体病毒晚期表达基因的表征

周　阳[1]　陈克平[1*]　姚　勤[1]　申红星[2]
梁桂廷[1]　李晓刚[1]　王　楠[1]　李怡佳[3]

摘要：家蚕核型多角体病毒（*Bm*NPV）ORF5（*Bm5*）是一种存在于许多鳞翅目核型多角体病毒（NPVs）中的基因，但其功能尚不清楚。本研究对*Bm5*进行表征。宿主感染12—72 h（p.i.）后检测到*Bm5*的转录物。使用His-BM5融合蛋白的多克隆抗体在24—72 h（p.i.）的感染细胞裂解物中检测出BM5，表明*Bm5*是晚期基因。通过共焦显微镜的免疫荧光分析显示，BM5蛋白主要定位在细胞质中。通过Western blot分析出芽型病毒粒子（BV）或包涵体型病毒粒子（ODV）中BM5的定位表明，BM5不是与BV或ODV相关的结构蛋白。

1　生命科学研究院，江苏大学，镇江
2　医学科学与实验医学院，江苏大学，镇江
3　浙江省余姚市第二中学，余姚

Bombyx mori nucleopolyhedrovirus *BmP95* plays an essential role in budded virus production and nucleocapsid assembly

Xiang XW Shen YW Yang R
Chen L Hu XL Wu XF*

Abstract: *Bombyx mori* nucleopolyhedrovirus (*Bm*NPV) *BmP95* is a highly conserved gene that is found in all of the baculovirus genomes sequenced to date and is also found in nudiviruses. To investigate the role of *BmP95* in virus infection *in vitro*, a *BmP95* deletion virus (v*Bm*P95-De) was generated by homologous recombination in *Escherichia coli*. Fluorescence and light microscopy and titration analysis indicated that the *BmP95* deletion bacmid led to a defect in production of infectious budded virus (BV). However, deletion of *BmP95* did not affect viral DNA replication. Electron microscopy showed that masses of aberrant tubular structures were present in cells transfected with the *BmP95* deletion bacmid, indicating that deletion of *BmP95* affected assembly of the nucleocapsid. This defect could be rescued by insertion of full-length *BmP95* into the polyhedrin locus of the *BmP95*-knockout bacmid but not the N-terminal domain of *BmP95*. Together, these results showed that full-length *BmP95* is essential for BV production and is required for nucleocapsid assembly.

Published On: Journal of General Virology, 2013, 94(7), 1669-1679.

College of Animal Sciences, Zhejiang University, Hangzhou 310058, China.
*Corresponding author E-mail: wuxiaofeng@zju.edu.cn.

家蚕核型多角体病毒*BmP95*对于出芽型病毒粒子的产生和核衣壳的组装发挥重要作用

相兴伟　沈运旺　杨　锐
陈　琳　胡小龙　吴小锋*

摘要：家蚕核型多角体病毒（*Bm*NPV）*BmP95*存在于所有迄今为止已测序的杆状病毒基因组中，同时也发现存在于裸杆状病毒（nudiviruses）中，是一个高度保守的基因。为了在体外研究*BmP95*在病毒感染中的功能，在大肠杆菌中用同源重组的方法构建了*BmP95*敲除型病毒（v*Bm*P95-De）。荧光、光学显微镜和病毒滴度分析表明，*BmP95*敲除型杆状病毒导致有感染力的出芽型病毒粒子（BV）产生缺陷。然而，敲除*BmP95*并不影响病毒DNA复制。电子显微镜显示有大量异常的管状结构存在于转染*BmP95*敲除型杆状病毒的细胞中，表明敲除*BmP95*影响核衣壳的组装。将全长*BmP95*而非*BmP95* N端区域插入到*BmP95*敲除型杆粒的多角体位点，能够恢复这一缺陷。综上所述，这些结果表明全长的*BmP95*对于BV的产生和核衣壳的组装是必不可少的。

动物科学学院，浙江大学，杭州

The formation of occlusion-derived virus is affected by the expression level of ODV-E25

Chen L Yang R Hu XL
Xiang XW Yu SF Wu XF*

Abstract: *Odv-e25* is a core gene of baculoviruses and encodes a 25.5 kDa protein located on both budded virus (BV) and occlusion-derived virus (ODV). Our previous study demonstrated that ODV-E25 was required for the formation of intranuclear microvesicles and ODV, and an *odv-e25* deletion mutant could be rescued by re-expression of *odv-e25* under its native promoter. To investigate the functions of ODV-E25 expression level on ODV formation, the promoter of *ie-1* (*pIE1*), the *odv-e25* native promoter, and the *polyhedrin* promoter (*pPH*) were used to direct *odv-e25* expression. Our results showed that the production of ODVE25 under its native promoter was higher than that under *pIE1* but lower than that under *pPH*. Viral DNA replication and budded viruses (BVs) production showed that expression of *odv-e25* under *pIE1* and *pPH* could not completely repair the defects caused by the deletion of ODV-E25, while expression under its native promoter did. Electron microscopy showed that intranuclear microvesicles were found in all the constructs transfected cells except the *odv-e25*-null virus. However, mature ODVs only were detected in cells transfected with virus in which *odv-e25* was expressed under its native or *polyhedrin* promoter. These results indicated that the formation occlusion-derived virus was affected by the expression level of ODV-E25.

Published On: Virus Research, 2013, 173 (2), 404-414.

College of Animal Sciences, Zhejiang University, Hangzhou 310058, China.
*Corresponding author E-mail: wuxiaofeng@zju.edu.cn.

ODV-E25的表达水平影响包涵体衍生型病毒粒子的形成

陈　琳　杨　锐　胡小龙
相兴伟　于少芳　吴小锋*

摘要：*Odv-e25*是杆状病毒的核心基因，其编码一种位于出芽型病毒粒子（BV）和包涵体衍生型病毒粒子（ODV）中的大小为25.5 kDa的蛋白质。我们之前的研究中已经证明ODV-E25是核内微泡和ODV形成所必需的，并且敲除*odv-e25*的突变体可以被其自身启动子驱动表达的ODV-E25恢复。为了研究ODV-E25表达水平在ODV的形成过程中的作用，我们利用*ie-1*启动子（*pIE1*）、*odv-e25*自身启动子和多角体启动子（*pPH*）来启动*odv-e25*的表达。结果表明ODV-E25在自身启动子调控下的表达量高于*pIE1*的调控，却低于*pPH*的调控。此外，针对病毒DNA复制和出芽型病毒粒子（BV）产量，实验结果显示，*pIE1*和*pPH*表达*odv-e25*时均不能完全修复*odv-e25*缺失所带来的影响，但是它自身启动子却能完全修复缺陷。电子显微镜分析表明所有重组病毒转染的细胞中均有核内微泡，而*odv-e25*缺失型病毒转染的细胞却没有发现。此外，只有*odv-e25*自身启动子和多角体启动子下表达的ODV-E25才能在感染细胞中产生成熟的ODV。这些结果共同证明包涵体衍生型病毒粒子的形成受ODV-E25表达水平的影响。

动物科学学院，浙江大学，杭州

*Bm*91 is an envelope component of ODV but is dispensable for the propagation of *Bombyx mori* nucleopolyhedrovirus

Tang Q[1,2#] Li GH[2#] Yao Q[2] Chen L[2]
Lü P[2] Lian CQ[2] Chen KP[1,2*]

Abstract: *Orf91* (*Bm91*) of *Bombyx mori* nucleopolyhedrovirus (*Bm*NPV) is a highly conserved gene that encodes a predicted 105-amino-acid protein, but its function remains unknown. In the current study, 5'-RACE revealed that the transcription initiation site of *Bm91* was-12 nucleotides upstream of the start codon ATG, transcription of *Bm91* was detected from 12 to 96 h postinfection (p.i.) and *Bm*91 protein was detected from 24 to 96 h p.i. in *Bm*NPV-infected *BmN* cells. Furthermore, Western blot analysis revealed that *Bm*91 was in occlusion-derived virus (ODV) but not in budded virus (BV). To investigate the role of *Bm*91 in baculovirus life cycle, a *Bm91*-knockout virus was constructed by bacmid recombination in *E. coli.* Fluorescence and light microscopy showed that the production of BV and occlusion bodies (OBs) in *Bm91*-deficient-virus-infected *BmN* cells were similar to those in wild-type-virus-infected ones. Bioassay results showed that genetic deletion of *Bm91* did not significantly affect *Bm*NPV infectivity, but extended the median lethal time (LT_{50}). Taken together, these results indicate that *Bm91* is not essential for viral propagation *in vitro*, but absence of the gene may affect the virulence of ODVs in silkworm larvae.

Published On: Journal of Invertebrate Pathology, 2013, 113(1), 70-77.

1 School of Food and Biological Engineering, Jiangsu University, Zhenjiang 212013, China.
2 Institute of Life Sciences, Jiangsu University, Zhenjiang 212013, China.
#These authors contributed equally.
*Corresponding author E-mail: kpchen@ujs.edu.cn.

*Bm91*是ODV的包膜组件，但对于家蚕核型多角体病毒的繁殖是不必要的

唐　琦[1,2#]　李国辉[2#]　姚　勤[2]　陈　亮[2]
吕　鹏[2]　连超群[2]　陈克平[1,2*]

摘要：家蚕核型多角体病毒（*Bm*NPV）的*Orf91*（*Bm91*）是高度保守的基因，预测编码105个氨基酸的蛋白质，但其功能尚不清楚。在目前的研究中，5′-RACE显示*Bm91*的转录起始位点是起始密码子ATG上游的12个核苷酸，在*Bm*NPV感染的*BmN*细胞中，感染后12—96 h检测到*Bm91*的转录，24—96 h检测到*Bm*91蛋白。此外，Western blot分析显示*Bm*91存在于包涵体型的病毒（ODV）中，但不存在出芽病毒（BV）中。为了研究*Bm*91在杆状病毒生命周期中的作用，利用bacmid在大肠杆菌中构建了*Bm91*敲除病毒。荧光和光学显微镜显示，*Bm91*缺陷型病毒感染的*BmN*细胞中BV和包涵体（OB）的产生与野生型病毒感染的*BmN*细胞中的相似。生物测定结果表明，*Bm91*的遗传缺失对*Bm*NPV感染性没有显著影响，但延长了半致死时间（LT_{50}）。总之，这些结果表明*Bm91*对于体外病毒繁殖不是必需的，但缺失该基因可能会影响家蚕幼虫中ODV的毒力。

1　食品与生物工程学院，江苏大学，镇江
2　生命科学研究院，江苏大学，镇江

The *Bombyx mori* nucleopolyhedrovirus *Bm111* affects virulence but not virus replication

Han YY[1] Xia HC[1] Tang Q[2] Lü P[2] Ma SS[1]
Yang YH[1] Shao DD[1] Ma QB[1] Chen KP[1*]

Abstract: The *Bm111* of *Bombyx mori* nucleopolyhedrovirus (*Bm*NPV) encodes a small polypeptide (70 amino acids) of which the function remains unknown. To characterize its function, multiple sequence alignments were performed, and the predicted protein was found to share amazingly high (98%) sequence identity with the *Bombyx mandarina* nucleopolyhedrovirus ORF110 (*Boma110*) but negligible with proteins of other insect viruses, indicating the close relationship between these two NPVs with silkworm larvae. The transcription of *Bm111* was detected as early as 3 hpi in *Bm*NPV-infected *BmN* cells, suggesting it is an early gene. To investigate the role of *Bm111* in baculovirus life cycle, a *Bm111*-knockout virus was constructed by bacmid recombination in *Escherichia coli*. The results showed that knockout of the *Bm111* did not affect the replication of virus DNA, but significantly extended the death time of infected silkworm larvae compared to the wild-type or rescued viruses. We also successfully expressed the recombinant protein *Bm*111 in *E. coli* to provide sufficient material for subsequent studies. Taken together, our data indicate that *Bm*111 only affects the virulence of *Bm*NPV, but not its replication.

Published On: Current Microbiology, 2014, 69(1), 56-62.

1 Institute of Life Sciences, Jiangsu University, Zhenjiang 212013, China.
2 School of Food and Biological Engineering, Jiangsu University, Zhenjiang 212013, China.
*Corresponding author E-mail: kpchen@ujs.edu.cn.

家蚕核型多角体病毒*Bm111*影响毒力，但不影响病毒的复制

韩莹莹[1] 夏恒传[1] 唐 琦[2] 吕 鹏[2] 马上上[1]
杨艳华[1] 邵丹丹[1] 马全兵[1] 陈克平[1*]

摘要：家蚕核型多角体病毒*Bm111*编码一个功能未知的小肽（70个氨基酸）。为了分析其功能，进行了多重序列比对，并且所预测的蛋白质与野桑核型多角体ORF110（*Boma110*）有惊人（98%）的序列同一性，但在其他昆虫病毒蛋白中可以忽略不计，表明这两个家蚕幼虫病毒之间存在密切关系。早在*Bm*NPV感染家蚕细胞后3 h，就检测到*Bm111*的转录，这表明它是一个早期基因。为了探讨*Bm111*在杆状病毒生命周期中的作用，通过bacmid重组在大肠中构建了一个*Bm111*敲除病毒。结果表明，*Bm111*的敲除没有影响病毒DNA的复制，但相比野生型或获救病毒，它显著延长了被感染家蚕幼虫死亡时间。我们还在大肠杆菌中成功表达了重组蛋白*Bm*111，为后续研究提供了足够的材料。总之，我们的数据表明，*Bm*111仅影响*Bm*NPV的毒力，但不影响它的复制。

1 生命科学研究院，江苏大学，镇江
2 食品与生物工程学院，江苏大学，镇江

*Bm*NPV infection enhances ubiquitin-conjugating enzyme E2 expression in the midgut of *Bm*NPV susceptible silkworm strain

Gao L[2] Chen KP[1*] Yao Q[1] Chen HQ[1]

Abstract: The ubiquitin conjugating enzyme 2 (E2) is core component of ubiquitin proteasome pathway (UPP) which represents a selective mechanism for intracellular proteolysis in eukaryotic cells. The E2 has been implicated in the intracellular transfer of ubiquitin to target protein. We show here the involvement of E2 in antiviral immune of *Bombyx mori* to *Bombyx mori* nuclear polyhedrosis virus (*Bm*NPV). In this study, mRNA fluorescent differential display PCR(FDD-PCR) was performed with *Bm*NPV highly resistant silkworm strain *NB* and susceptible silkworm strain *306*. At 24 h post *Bm*NPV infection, FDD-PCR with the arbitrary primer AP34 showed that one cDNA band was downregulated in the midgut of resistant strain, but highly expressed in susceptible strain. The deduced amino acid sequence of this cDNA clone share 99% identity with the recently published *B. mori* ubiquitin conjugating enzyme E2 (Genbank NO: DQ311351). Fluorescent quantitative PCR corroborated downregulation of E2 in resistant strain. We there conclude that *Bm*NPV infection evokes strong response of susceptible strain including activation of UPP. *Bm*NPV may evolve escape mechanisms that manipulate the UPP in order to persist in the infected host. In addition, the identification of downregulation of E2 in resistant strain, as well as structure data, are essential to understanding how UPP operates in silkworm antiviral immune to *Bm*NPV disease.

Published On: International Journal of Indust Entomology, 2006, 13(1), 31-35.

1 Institute of Life Science, Jiangsu University, Zhenjiang 212013, China.

2 School of Medical Technology, Jiangsu University, Zhenjiang 212001, China.

*Corresponding author E-mail:kpchen@ujs.edu.cn.

*Bm*NPV 感染增强 *Bm*NPV 易感家蚕品系中肠内的泛素结合酶 E2 表达

高　路[2]　陈克平[1*]　姚　勤[1]　陈慧卿[1]

摘要：泛素结合酶 2（E2）是泛素蛋白酶体通路（UPP）的核心成分，代表了真核细胞中胞内蛋白水解的选择性机制。E2 涉及到泛素蛋白到靶蛋白的细胞内转移。本文中我们研究了 E2 参与家蚕对家蚕核型多角体病毒（*Bm*NPV）的抗病毒免疫。本研究用抗 *Bm*NPV 家蚕品系 *NB* 和易感 *Bm*NPV 家蚕品系 *306* 进行 mRNA 荧光差异显示 PCR（FDD-PCR）。在 *Bm*NPV 感染 24 h 后，任意引物 AP34 的 FDD-PCR 显示一条 cDNA 条带在抗性品系的中肠中被下调，但在易感品系中高度表达。该 cDNA 克隆的预测氨基酸序列与最近公布的家蚕泛素结合酶 E2（Genbank NO：DQ311351）有 99% 的同一性。荧光定量 PCR 证实了抗性株中 E2 的下调。我们得出结论，*Bm*NPV 感染引起易感品系的强烈反应，包括激活 UPP。*Bm*NPV 可能会进化出操纵 UPP 的逃逸机制，以便在受感染的宿主中持续存在。此外，抗性品系中 E2 下调以及结构数据，对于了解 UPP 如何在家蚕免疫 *Bm*NPV 疾病中发挥作用是至关重要的。

1　生命科学研究院，江苏大学，镇江
2　医学技术学院，江苏大学，镇江

Characterization of a protein tyrosine phosphatase as a host factor promoting baculovirus replication in silkworm, *Bombyx mori*

Wang F Xue RJ Li XY Hu CM Xia QY*

Abstract: The relevance of protein tyrosine phosphatase (PTP) to host-pathogen interaction is highlighted in mammalian studies, whereas less is known in insects. Here, we presented the categorization of the PTP complement of silkworm and characterized their homologous relationship with human and fruit fly PTPs. Among the 36 *PTP* genes, *ptp-h*, which was proposed to be the origin of baculovirus *ptp* belongs to atypical VH1-like dual-specific PTP subset and encodes a catalytic active protein. The maximum expression level of *Bmptp-h* was at 5th instar and in fat body. *Bombyx mori* nucleopolyhedrovirus (*Bm*NPV) infection potently induced its expression in silkworm larvae and in *BmE* cells. Knock-down of *Bmptp-h* by RNA interference significantly inhibited viral replication, and over-expression enhanced viral replication as determined by viral DNA abundance and *Bm*NPV-GFP positive cells. These results suggest that *Bm*PTP-h might be one of the host factors that is beneficial to baculovirus infection by promoting viral replication.

Published On: Development and Comparative Immunology, 2016, 57, 31-37.

State Key Laboratory of Silkworm Genome Biology, Southwest University, Chongqing 400716, China.
*Corresponding author E-mail: xiaqy@swu.edu.cn.

能促进家蚕杆状病毒复制的宿主因子蛋白酪氨酸磷酸酶的鉴定

王　菲　薛仁举　李显扬　胡翠美　夏庆友*

摘要：在哺乳动物中，关于蛋白酪氨酸磷酸酶（PTP）与宿主–病原体相互作用的报道较多，而在昆虫中较少。在本研究中，我们对家蚕所有PTP进行了鉴定和分类，并分析了其与人和果蝇PTPs的同源关系。在36个*PTP*基因中，*ptp-h*被认为来源于杆状病毒*ptp*，属于非典型VH1样双特异性PTP亚类，其编码一个具有催化活性的蛋白质。*Bmptp-h*在家蚕五龄脂肪体中表达量最高。感染家蚕核型多角体病毒（*Bm*NPV）能有效诱导其在家蚕幼虫和*BmE*细胞中的表达。通过RNAi降低*Bmptp-h*的表达可显著抑制病毒的复制，病毒DNA丰度和*Bm*NPV-GFP阳性细胞测定实验显示，过表达*Bmptp-h*能够增强病毒的复制。这些结果表明，*Bm*PTP-h可能是能促进病毒复制，有利于杆状病毒感染的宿主因子之一。

家蚕基因组生物学国家重点实验室，西南大学，重庆

Identification and characterization of the *Bombyx mori* myosin II essential light chain and its effect in *Bm*NPV infection

Hao LJ Lü P Gao L Zhou Y

Yao Q Yang YH Chen KP*

Abstract: Myosin, as a type of molecular motor, is mainly involved in muscle contraction. Recently, myosin research has made considerable progress. However, the function of *Bombyx mori* myosin remains unclear. In this study, we cloned the *BmMyosin II essential light chain* (*BmMyosin II ELC*) gene from a cDNA library of silkworm, which had an open reading frame (ORF) of 444 bp encoding 147 amino acids (about 16 kDa). After analyzing their sequences, *Bm*Myosin II ELC was similar to the ELCs of 27 other Myosin II types, which contained EFh domain that bound Ca^{2+}. In addition, 28 sequences had five motifs, motifs 1 and 3 were relatively conserved. We constructed two vectors with *BmMyosin* to transfect *MGC803* or *BmN*, monolayer wound healing of cells indicated they can promote cell migration successfully. For three fifth instar silkworms, *Bm306*, *BmNB*, *BmBC8*, we mainly analyzed the change of *BmMyosin II ELC* from transcription and translation after infecting with nucleopolyhedrovirus (*Bm*NPV). We found that gene expression of resistant strains were higher than susceptible strains at 12 h, while the result of the translation level was opposite that of the transcription level. Through *in vitro* protein interactions, we found *Bm*Myosin II ELC can interact with *Bm*NPV ubiquitin.

Published On: Invertebrate Survival Journal. 2015, 12, 38-45.

Institute of Life Sciences, Jiangsu University, Zhenjiang 212013, China.

*Corresponding author E-mail: kpchen@ujs.edu.cn.

家蚕肌球蛋白Ⅱ必需轻链基因的鉴定和表征及其在*Bm*NPV感染中的作用

郝丽娟　吕　鹏　高　力　周　阳
姚　勤　杨艳华　陈克平*

摘要：肌球蛋白作为一种分子马达，主要参与肌肉收缩。近年来，对肌球蛋白的研究取得了较好的进展。但是，家蚕肌球蛋白的功能尚不清楚。在本研究中，我们从蚕的cDNA文库中克隆了*BmMyosin II*必需的轻链(*BmMyosin II ELC*)基因，开放阅读框(ORF)长444 bp，可编码147个氨基酸(约16 kDa)。分析其序列后，*Bm*Myosin II ELC与其他27种肌球蛋白Ⅱ类型的ELC相似，都包含结合Ca^{2+}的EFh结构域。另外，这28个序列都有5个基序，并且基序1和3相对保守。我们用*BmMyosin*构建了两个载体，分别转染*MGC803*和*BmN*细胞，细胞单层伤口愈合实验表明它们可以促进细胞迁移。我们主要分析了3个品系的五龄蚕(*Bm306*, *BmNB*, *BmBC8*)在感染核型多角体病毒(*Bm*NPV)后，*BmMyosin II ELC*的转录和翻译的变化。发现在12 h时，*BmMyosin II ELC*的表达量在抗性品系中高于其在易感品系中，但是翻译水平的结果与转录水平相反。通过体外蛋白质相互作用，发现*Bm*Myosin II ELC可以与*Bm*NPV泛素发生相互作用。

生命科学研究院，江苏大学，镇江

Identification of a single-nucleocapsid baculovirus isolated from *Clanis bilineata tsingtauica* (Lepidoptera: Sphingidae)

Wang LQ[1,2] Yi JP[3] Zhu SY[1] Li B[4]
Chen Y[1] Shen WD[4] Wang WB[1*]

Abstract: A nucleopolyhedrovirus isolated from infected larvae of *Clanis bilineata tsingtauica* was characterized. Electron microscopical studies on the ultrastructure of *C. bilineata* nucleopolyhedrovirus (*ClbiS*NPV) occlusion bodies (OBs) showed several virions (up to 16) with a single nucleocapsid packaged with in a single viral envelope. The diameter of the OBs was 0.77-1.7 μm with a mean of (1.13±0.19)μm. The complete sequence of the *ClbiS*NPV polyhedrin (*polh*) gene contained 741 nucleotides, predicting a protein of 246 amino acids. Phylogenetic analyses using the complete sequence of the *polh* genes indicated that *ClbiS*NPV clusters with Group II NPVs. This is the first record of a baculovirus from *C. bilineata*.

Published On: Archives of Virology, 2008, 153(8), 1557-1561.

1 Institute of Life Sciences, Jiangsu University, Zhenjiang 212013, China.

2 Department of Chemical Engineering, Jiangsu Ploytechnic University, Changzhou 213016, China.

3 Shanghai Entry-Exit Quarantine and Inspection Bureau, Shanghai 200000, China.

4 School of Life Sciences, Soochow University, Suzhou 215123, China.

*Corresponding author E-mail: wenbingwang@ujs.edu.cn.

从豆天蛾中分离到的单粒包埋型杆状病毒的鉴定

王利群[1,2] 易建平[3] 朱姗颖[1] 李兵[4]
陈 言[1] 沈卫德[4] 王文兵[1*]

摘要：从被感染的豆天蛾幼虫体内分离到一种核型多角体病毒。利用电子显微镜观察超微结构发现豆天蛾核型多角体病毒（*ClbiS*NPV）的包涵体可以包被16个病毒粒子，其包埋型为单粒包埋。包涵体的直径大约在0.77—1.7 μm，平均为（1.13±0.19）μm。*ClbiS*NPV的多角体基因（*polh*）全长为741 bp，编码246个氨基酸。应用该序列进行进化分析，结果表明*ClbiS*NPV位于NPV II型分枝。这是首次从豆天蛾中发现的杆状病毒。一种核型多角体病毒从被感染的豆天蛾幼虫体内被分离出来。

1 生命科学研究院，江苏大学，镇江
2 化学工程学院，江苏工业大学，常州
3 上海出入境检验检疫局，上海
4 生命科学学院，苏州大学，苏州

Autographa californica multiple nucleopolyhedrovirus *odv-e25*(*Ac94*) is required for budded virus infectivity and occlusion derived virus formation

Chen L　Hu XL　Xiang XW

Yu SF　Yang R　Wu XF*

Abstract: *Autographa californica* multiple nucleopolyhedrovirus (*AcM*NPV) *odv-e25* is a core gene found in all lepidopteran baculoviruses, but its function is unknown. In this study, we generated an *odv-e25*-knockout *AcM*NPV and investigated the roles of ODV-E25 in the baculovirus life cycle. The *odv-e25* knockout was subsequently rescued by reinserting the *odv-e25* gene into the same virus genome. Fluorescence microscopy showed that transfection with the *odv-e25*-null bacmid vAcBac^{KO} was in sufficient for propagation in cell culture, whereas the 'repair' virus vAcBac^{RE} was able to function in a manner similar to that of the control vAcBac. We found that *odv-e25* was not essential for the release of budded viruses (BVs) into culture medium, although the absence of *odv-e25* resulted in a 100-fold lower viral titer at 24 h post-transfection (p. t.). Analysis of viral DNA replication in the absence of *odv-e25* showed that viral DNA replication was unaffected in the first 24 h p.t. Furthermore, electron microscopy revealed that polyhedra were found in the nucleus, while mature occlusion-derived viruses (ODVs) were not found in the nucleus or polyhedra in *odv-e25* null transfected cells, which indicated that ODV-E25 was required for the formation of ODV.

Published On: Archives of Virology, 2012, 157(4), 617-625.

College of Animal Sciences, Zhejiang University, Hangzhou 310058, China.

*Corresponding author E-mail: wuxiaofeng@zju.edu.cn.

苜蓿银纹夜蛾核型多角体病毒(*AcM*NPV)*odv-e25*(*Ac94*)是出芽型(BV)病毒感染和包涵体衍生型(ODV)病毒形成所必需

陈　琳　胡小龙　相兴伟
于少芳　杨　锐　吴小锋*

摘要：苜蓿银纹夜蛾核型多角体病毒(*AcM*NPV) *odv-e25*是一个存在于所有鳞翅目杆状病毒中的核心基因,但其功能并不清楚。在本研究中,我们构建了一个*odv-e25*敲除的*AcM*NPV,并探究了ODV-E25在杆状病毒生命周期中的作用。*odv-e25*敲除后可以通过重新插入*odv-e25*基因到该病毒基因组中进行挽救。荧光显微镜显示,转染*odv-e25*缺失的bacmid(vAcBac^{KO})不能在培养细胞中传播,而"修复型"病毒vAcBac^{RE}与对照组vAcBac的功能相似。我们发现*odv-e25*对BVs病毒释放到培养基中并不是必要的,尽管缺失*odv-e25*导致转染后24 h的病毒滴度仅为原来的1%。在缺失*odv-e25*的情况下分析病毒DNA复制情况,结果显示病毒DNA复制在转染后的前24 h不受影响。此外,电子显微镜显示在细胞核中发现多角体,而在缺失*odv-e25*转染细胞的核或多角体中未发现成熟的ODV,这表明ODV-E25是ODV形成所必需的。

动物科学学院,浙江大学,杭州

Characterization of *Antheraea pernyi* nucleopolyhedrovirus *p11* gene, a homologue of *Autographa californica* nucleopolyhedrovirus *orf108*

Shi SL[1,2,3,4] Pan MH[1] Lu C[1*]

Abstract: *Antheraea pernyi* nucleopolyhedrovirus (*Ap*NPV) *p11* gene is 309 bp long, potentially encoding 102 amino acids with a predicted molecular weight of 11.2 kDa. *Ap*NPV *p11* gene was cloned into the prokaryotic expression vector pQE-30 and *P11* was expressed in *E. coli* M15. Polyclonal antiserum was made against 6·His tagged P11 protein expressed in *E. coli* M15. *P11* gene transcription was detected as early as 36 h post-infection (p. i.) in tussah pupa and remain at high level up to 96 h p. i. Structural localization revealed that P11 protein was present in polyhedral inclusion bodies (PIB) dilute alkaline saline (DAS) pellet (P) fractions and occlusion-derived virus (ODV), but not in PIB DAS supernatant (S) fractions and budded virus (BV). These results indicated that P11 was associated with *Ap*NPV structure.

Published On: Virus Genes, 2007, 35(1), 97-101.

1 Key Sericultural Laboratory of Agricultural Ministry, College of Biotechnology, Southwest University, Chongqing 400716, China.

2 Institute of Applied Ecology, Chinese Academy of Sciences, Shenyang 110016, China.

3 College of Bioscience and Biotechnology, Shenyang Agricultural University, Shenyang 110161, China.

4 Graduate University of Chinese Academy of Sciences, Beijing 100049, China.

*Corresponding author E-mail: lucheng@swau.edu.cn.

苜蓿银纹夜蛾核型多角体病毒*orf108*的同系物，柞蚕核型多角体病毒*p11*基因表征研究

石生林[1,2,3,4*]　潘敏慧[1]　鲁　成[1*]

摘要：柞蚕核型多角体病毒（*Ap*NPV）*p11*基因长度为309 bp，可能编码102个氨基酸，预测分子质量为11.2 kDa。将*Ap*NPV *p11*基因克隆到原核表达载体pQE-30中，在大肠杆菌M15中表达*P11*。针对在大肠杆菌M15表达的带有6·His标签的P11蛋白制备了多克隆抗体。早在柞蚕蛹感染后36 h就可以检测到*p11*基因的转录，并且到感染后96 h一直保持高水平转录。结构定位显示P11蛋白存在于多面体包涵体（PIB）稀碱性盐水（DAS）沉淀（P）部分和包涵体来源病毒（ODV）中，但不存在于PIB DAS上清液（S）和芽殖病毒（BV）中。这些结果表明P11与核型多角体病毒结构相关。

1　农业部蚕桑学重点实验室，生物技术学院，西南大学，重庆
2　应用生态研究所，中国科学院，沈阳
3　生物科学技术学院，沈阳农业大学，沈阳
4　中国科学院研究生院，北京

Reclassification of the pathogen for empty-gut disease of Chinese oak silkworm, *Antheraea pernyi*

Wang LL[1] Qin F[1] Song C[2] Zhou ZY[1*]

Abstract: Empty-gut disease is a common disease of Chinese oak silkworm, *Antheraea pernyi* (Tussah). The study on the pathogen of the disease is of great help in prevention and cure of the disease. Previous studies showed that the pathogen was a new species in the genus *Streptococcus*, which was therefore named as *Streptococcus pernyi sp.nov.* In this study, with the purified pathogen, we performed some physiological and biochemical experiments, and also cloned and sequenced 16S rRNA gene of the pathogen. Subsequently, we analyzed its similarity and genetic distance with other related bacteria by using DNAstar software, and a NJ tree was then constructed for phylogenic analysis by MEGA 3.0. Unexpectedly, the results showed that the pathogen causing empty-gut disease of Tussah belongs to the *genus Enterococcus* with 99% bootstrap support. Therefore, *Streptococcus pernyi sp. nov* should be renamed as *Enterococcus pernyi*. Reclassification of the pathogen should keep pace with the development of modern taxonomy. In doing so, it will better our understanding of the pathogen and contribute to further study of the pathogen.

Published On: Journal of Food Agriculture & Enviroment, 2010, 8(2), 156-158.

1 College of Life Sciences, Chongqing Normal University, Chongqing 400047, China.

2 Liaoning Silkworm Scientific Research Institute, Dandong 118100, China.

*Corresponding author E-mail: zyzhou@cqnu.edu.cn.

柞蚕空肠病病原体重新分类

王林玲[1] 秦　凤[1] 宋　策[2] 周泽扬[1*]

摘要：空肠病是柞蚕的常见病，对该病病原体的研究有助于预防和治疗该疾病。以前的研究表明，该病病原体是链球菌属中的新物种，因此被命名为*sp.nov*柞蚕链球菌。在本研究中我们用纯化的病原体进行了一些生理生化实验，克隆并测序了病原体的16S rRNA基因。随后，我们使用DNAstar软件分析了与其他相关细菌的相似性和遗传距离，然后构建了一个进化树，采用MEGA 3.0进行了系统发育分析。出乎意料的是，结果表明，引起空肠病的病原体属于肠球菌属（bootstrap支持率99%）。因此，链球菌应更名为肠球菌。病原体的重新分类应跟上现代分类学的发展。这样做会更好地了解病原体，有助于进一步研究病原体。

1　生命科学学院，重庆师范大学，重庆
2　辽宁蚕业科学研究所，丹东

Bacillus bombysepticus α-toxin binding to G protein-coupled receptor kinase 2 regulates cAMP/PKA signaling pathway to induce host death

Lin P Cheng TC Ma SY Gao JP Jin SK Jiang L Xia QY*

Abstract: Bacterial pathogens and their toxins target host receptors, leading to aberrant behavior or host death by changing signaling events through subversion of host intracellular cAMP level. This is an efficient and widespread mechanism of microbial pathogenesis. Previous studies describe toxins that increase cAMP in host cells, resulting in death through G protein-coupled receptor (GPCR) signaling pathways by influencing adenylyl cyclase or G protein activity. G protein-coupled receptor kinase 2 (GRK2) has a central role in regulation of GPCR desensitization. However, little information is available about the pathogenic mechanisms of toxins associated with GRK2. Here, we reported a new bacterial toxin-*Bacillus bombysepticus* (*Bb*) α-toxin that was lethal to host. We showed that *Bb* α-toxin interacted with *Bm*GRK2. The data demonstrated that *Bb* α-toxin directly bound to *Bm*GRK2 to promote death by affecting GPCR signaling pathways. This mechanism involved stimulation of $G_{\alpha s}$, increase level of cAMP and activation of protein kinase A (PKA). Activated cAMP / PKA signal transduction altered downstream effectors that affected homeostasis and fundamental biological processes, disturbing the structural and functional integrity of cells, resulting in death. Preventing cAMP/PKA signaling transduction by inhibitions (NF449 or H-89) substantially reduced the pathogenicity of *Bb* α-toxin. The discovery of a toxin-induced host death specifically linked to GRK2 mediated signaling pathway suggested a new model for bacterial toxin action. Characterization of host genes whose expression and function are regulated by *Bb* α-toxin and GRK2 will offer a deeper understanding of the pathogenesis of infectious diseases caused by pathogens that elevate cAMP.

Published On: PLoS Pathogens, 2016, 12(3).

State Key Laboratory of Silkworm Genome Biology, Southwest University, Chongqing 400716, China.
*Corresponding author E-mail: xiaqy@swu.edu.cn.

黑胸败血芽孢杆菌α毒素结合到G蛋白偶联受体激酶2调节cAMP/PKA信号通路诱导宿主死亡

林 平 程廷才 马三垣 高军平 金盛凯 蒋 亮 夏庆友*

摘要：细菌病原体及其毒素可以特异性识别宿主受体，改变细胞信号通路，破坏胞内cAMP水平，从而导致宿主死亡或行为异常。这是一种有效且普遍的病原微生物致病机制。之前的研究证明了毒素能够影响腺苷酸环化酶或G蛋白活性，通过激活G蛋白偶联受体（GPCR）信号通路从而增加胞内cAMP水平，引起宿主死亡。G蛋白偶联受体激酶2（GRK2）主要参与G蛋白偶联受体的脱敏作用。但是关于GRK2与细菌毒素致病机制的相关报道很少。本研究报道了一种新的细菌毒素——黑胸败血芽孢杆菌（*Bb*）α-toxin。研究结果显示，*Bb* α-toxin与*Bm*GRK2可以直接相互作用，进而调控GPCR信号通路，导致宿主死亡。该致病过程包括$G_{\alpha s}$的激活，cAMP水平的升高和蛋白激酶A（PKA）的活化。激活的cAMP / PKA信号通路影响下游效应分子，进而影响细胞的稳态及生物过程，干扰细胞的结构和生理功能，进而导致宿主死亡。同时，通过抑制剂（NF449或H-89）阻碍cAMP/PKA信号通路可明显减少*Bb* α-toxin的致病性。*Bb* α-toxin通过GRK2介导的信号通路导致宿主死亡的致病机制是一种新的细菌毒素致病机理模型。研究*Bb* α-toxin和GRK2调控的宿主基因的表达特征和功能特性将有助于我们深入理解致病菌引起的cAMP增加的传染病的发病机制。

家蚕基因组生物学国家重点实验室，西南大学，重庆

PC, a novel oral insecticidal toxin from *Bacillus bombysepticus* involved in host lethality via APN and BtR-175

Lin P[#] Cheng TC[#] Jin SK Wu YQ
Fu BH Long RW Zhao P Xia QY[*]

Abstract: Insect pests have developed resistance to chemical insecticides, insecticidal toxins as bioinsecticides or genetic protection built into crops. Consequently, novel, orally active insecticidal toxins would be valuable biological alternatives for pest control. Here, we identified a novel insecticidal toxin, parasporal crystal toxin (PC), from *Bacillus bombysepticus* (*Bb*). PC shows oral pathogenic activity and lethality towards silkworms and Cry1Ac-resistant *Helicoverpa armigera* strains. *In vitro* assays, PC after activated by trypsin binds to *Bm*APN4 and BtR-175 by interacting with CR7 and CR12 fragments. Additionally, trypsin-activated PC demonstrates cytotoxicity against *Sf9* cells expressing *Bm*APN4, revealing that *Bm*APN4 serves as a functional receptor that participates in Bb and PC pathogenicity. *In vivo* assay, knocking out BtR-175 increased the resistance of silkworms to PC. These data suggest that PC is the first protein with insecticidal activity identified in *Bb* that is capable of causing silkworm death via receptor interactions, representing an important advance in our understanding of the toxicity of Bb and the contributions of interactions between microbial pathogens and insects to disease pathology. Furthermore, the potency of PC as an insecticidal protein makes it a good candidate for inclusion in integrated agricultural pest management systems.

Published On: Scientific Reports, 2015, 5.

State Key Laboratory of Silkworm Genome Biology, Southwest University, Chongqing 400716, China.
[#]These authors contributed equally.
[*]Corresponding author E-mail: xiaqy@swu.edu.cn.

黑胸败血芽孢杆菌新型口服杀虫毒素通过APN和BtR-175杀死宿主

林 平[#] 程廷才[#] 金盛凯 吴玉乾
傅博华 龙仁文 赵 萍 夏庆友[*]

摘要：害虫对化学杀虫剂、杀虫素毒（作为生物杀虫剂）和作物的遗传保护产生了抗性。因此，新型口服活性杀虫毒素将作为有害生物控制的有价值的生物替代品。本研究中，我们从黑胸败血芽孢杆菌（*Bb*）中鉴定出一种新型的杀虫毒素，为伴孢晶体毒素（PC）。PC对家蚕和Cry1Ac抗棉铃虫有经口致病活性和致死性。在体外实验中，胰蛋白酶活化后的PC通过与CR7和CR12片段相互作用结合*Bm*APN4和BtR-175。此外，胰蛋白酶激活的PC展现出对表达*Bm*APN4的*Sf9*细胞的毒性，揭示*Bm*APN4作为功能受体参与Bb和PC的致病性。在体内实验中，敲除BtR-175增强了家蚕对PC的抵抗力。以上研究表明，PC是第一种在*Bb*中鉴定出能够通过受体相互作用引起家蚕死亡的具有杀虫活性的蛋白质，代表着我们对Bb的毒性和微生物病原体与昆虫之间相互作用的理解更进了一步。此外，PC作为潜在的杀虫蛋白将成为农业害虫综合防治的良好选择。

家蚕基因组生物学国家重点实验室，西南大学，重庆

Expression of non-structural protein *NS3* gene of *Bombyx mori* densovirus (China isolate)

Yin HJ Yao Q* Guo ZJ Bao F

Yu W Li J Chen KP

Abstract: The invertebrate parvovirus *Bombyx mori* Densonucleosis Virus type 3 (China isolate), named *Bm*DNV-3, is a kind of bidensovirus. It is a new type of virus with unique replication mechanisms. To investigate the effects of the *NS3* gene during viral DNA replication, a pair of primers was designed for amplifying *NS3* gene of *Bombyx mori* densovirus (China isolate). Gene *NS3* amplified was cloned into a prokaryotic expression vector pET-30a and the donor plasmid pFastBacHTe, respectively. The NS3 protein was expressed in *Escherichia coli* BL21. The pFastBacHTe-NS3 was transformed to *E. coli* DH10Bac. The recombinant bacmid baculoviruses (rBacmid-EGFP-NS3) isolated from the white colonies were transfected into *BmN-4* cells using a transfection reagent. *BmN-4* cells were infected with recombinant virus to express fusion proteins. The expression of fusion protein around 30 kDa in *E. coli* BL21 was identified by SDS-PAGE, Western blot, and mass spectrometry. The expressed NS3 protein by *B. mori* nucleopolyhedrovirus bacmid system was confirmed by Western blot using an anti-NS3 polyclonal antibody. And about 45 kDa protein was found. The expressed fusion protein was smaller than the expected size of EGFP-NS3, 55 kDa. Western blot analysis indicated that EGFP-NS3 protein was expressed in infected larvae with smaller molecular size.

Published On: Journal of Genetics and Genomics, 2008, 35(4), 239-244.

Institute of Life Sciences, Jiangsu University, Zhenjiang 212013, China.

*Corresponding author E-mail:yaoqin@ujs.edu.cn.

家蚕二分浓核病毒(中国株)
非结构蛋白*NS3*基因的表达

尹慧娟　姚　勤*　郭忠建　包　方
余　蔚　李　军　陈克平

摘要：无脊椎动物细小病毒 *Bombyx mori* Densonocosis 病毒 3 型(中国株),命名为 *Bm*DNV-3,是一种二分浓核病毒。它是一种具有独特复制机制的新型病毒。为了研究 *NS3* 基因在病毒 DNA 复制过程中的作用,设计了一对引物,用于扩增家蚕二分浓核病毒(中国株)的 *NS3* 基因。将扩增的基因 *NS3* 分别克隆到原核表达载体 pET-30a 和供体质粒 pFastBacHTe 中。利用 *Escherichia coli* BL21 表达 NS3 蛋白。将 pFastBacHTe-NS3 转化到 *E. coli* DH10Bac 细胞中。利用转染试剂将从白色菌落分离的重组杆粒杆菌病毒(rBacmid-EGFP-NS3)转染 *BmN-4* 细胞以表达融合蛋白。通过 SDS-PAGE、Western blot 和质谱鉴定大肠杆菌 BL21 中融合蛋白(约 30 kDa)的表达。使用 NS3 多克隆抗体 Western blot 证实了家蚕核型多角体病毒杆粒系统表达的 NS3 蛋白(约 45 kDa)。表达的融合蛋白小于预期分子质量(55 kDa)。Western blot 分析表明,在感染幼虫中表达的 EGFP-NS3 蛋白分子质量较小。

生命科学研究院,江苏大学,镇江

Bmpkci is highly expressed in resistant strain of silkworm (Lepidoptera:Bombycidae): implication of its role in resistance to *Bm*DNV-Z

Chen KP[1*] Chen HQ[1] Tang XD[2] Yao Q[1] Wang LL[1] Han X[1]

Abstract: Using fluorescent differential display technique, a special band named *Bm*541 was identified by screening for gene expressed differentially among the resistant silkworm strain *Qiufeng*, the susceptible strain *Huaba35* and the near isogenic line BC_6, which carries the resistant gene to *Bombyx mori* densonucleosis virus(*Bm*DNV). After applying the 5'RACE technique with specially designed primers, a 1 148 bp cDNA clone containing a 387 bp open reading frame (ORF) was obtained. This gene was registered in GenBank under the accession number AY860950. The deduced amino acid sequence showed a 73.1% identity to the protein kinase C inhibitor(PKCI) of *Drosophila pseudoobscura*. In the deduced sequence of *Bm*PKCI, the histidine triad(HIT) motif, which is essential for PKCI function, and the α-helix region, which is conserved among the PKCI family, were present. The data from quantitative real-time PCR(qRT-PCR) suggested that the expression levels of *BmPKCI* in BC_6 and *Qiufeng* both treated with *Bm*DNV-Z are significantly higher than those in *Huaba35*, which indicated that *Bm*PKCI plays a role in resistance to *Bm*DNV-Z.

Published On: European Journal of Entomology, 2007, 104(3), 369-376.

1 Institute of Life Science, Jiangsu University, Zhenjiang 212013, China.

2 Institute of Life Science, Zhejiang University, Hangzhou 310058, China.

*Corresponding author E-mail: kpchen@ujs.edu.cn.

*Bmpkci*在抗*Bm*DNV品种中高水平表达预示着其在抗*Bm*DNV-Z中发挥着重要作用

陈克平[1*] 陈慧卿[1] 唐旭东[2] 姚 勤[1] 王林玲[1] 韩 旭[1]

摘要:采用荧光差异显示技术对*Bm*DNV抗性家蚕品种“秋风”,易感品系*Huaba35*和近等基因系BC_6进行抗性基因筛查,在抗性品种中获得一条特异的条带,命名为*Bm*541。设计特异引物进行5′末端快速扩增,获得一条包含387 bp开放阅读框(ORF)的总长1 148 bp的cDNA克隆。该基因已在GenBank进行注册,登录号为AY860950。该基因对应的氨基酸序列(*Bm*PKCI)与果蝇的PKCI有73.1%的同源性。在*Bm*PKCI的推导氨基酸序列中,存在PKCI功能必需的HIT基序和PKCI家族中保守的α-螺旋区。qRT-PCR数据分析显示,BC_6和“秋风”经*Bm*DNV-Z处理后,*BmPKCI*的表达水平显著高于其在*Huaba35*品系中的表达水平,这很可能预示*Bm*PKCI参与了家蚕对*Bm*DNV-Z病毒的抵抗。

1 生命科学研究院,江苏大学,镇江

2 生命科学研究院,浙江大学,杭州

Truncation of cytoplasmic tail of EIAV Env increases the pathogenic necrosis

Meng QL[1,2,4] Li SW[2] Liu LX[1,2,4] Xu JQ[2]
Liu Y[2] Zhang YZ[3] Zhang XY[2*] Shao YM[1,2*]

Abstract: Equine Infectious Anemia Virus (EIAV), like other lentiviruses, has a transmembrane glycoprotein with an unusually long cytoplasmic tail (CT). Viral envelope (Env) proteins having CT truncations just downstream the putative membrane-spanning domain (PMSD) are assumed to exist among all wild-type budded virions, and also in some cell-adapted strains. To determine whether CT-truncated Env proteins can cause particularly deleterious effects on the Env expressing cells and/or their neighboring cells, plasmids encoding codon-optimized *env* gene including full-length (pE863) or CT-truncated (pE686* and pE676*) were transiently transfected into 293T cells, respectively. Data from intracellular protein expression and cell death assays revealed that CT-truncated Env, compared to full-length Env, not only induced comparable apoptosis, but also caused much more intensive mitochondria-mediated necrosis that could simultaneously induce significant decrease of intracellular protein expression in the Env expressing cells. Moreover, results from flow cytometric analysis showed that mitochondrial depolarization preceded the caspase activation in cells no matter which *env* construct was delivered, and indicated that both full-length and CT-truncated Env proteins share a common intrinsic mitochondrial pathway to induce apoptosis. Our results partially elucidate the mechanisms underlying cell death resulting from EIAV pathogenesis.

Published On: Virus Research, 2008, 133(2), 201-210.

1 State Key Laboratory of Virology, Wuhan Institute of Virology, Chinese Academy of Sciences, Wuhan 430071, China.
2 State Key Laboratory for Infectious Disease Prevention and Control, National Center for AIDS/STD Control and Prevention, Chinese Center for Disease Control and Prevention, Beijing 100050, China.
3 Institute of Biochemistry, College of Life Sciences, Zhejiang Sci-Tech University, Hangzhou 310018, China.
4 Graduate School of Chinese Academy of Sciences, Beijing 100039, China.
*Corresponding author E-mail: zhangxy@chinaaids.cn; yshao@bbn.cn.

贫血病毒囊膜蛋白胞内区截短增加病原性坏死

孟庆来[1,2,4] 李深伟[2] 刘连兴[1,2,4] 徐建青[2]
刘 颖[2] 张耀洲[3] 张晓燕[2*] 邵一鸣[1,2*]

摘要：与其他慢病毒一样，马传染性贫血病毒（EIAV）的跨膜糖蛋白具有异常长的胞内区。推测在假定跨膜结构域（PMSD）下游具有胞内区截短的病毒囊膜（Env）蛋白质，并存在于所有野生型出芽病毒体以及一些细胞适应株中。为了确定跨膜蛋白胞内区截短的病毒囊膜蛋白质是否可以对病毒囊膜表达细胞及其相邻细胞产生特别有害的影响，将编码密码子优化的病毒囊膜基因的质粒，包括全长（pE863）及截短的（pE686 *和 pE676 *）质粒，分别瞬时转染293T细胞。细胞内蛋白质表达和细胞死亡测定的数据显示，与全长Env相比，CT截短的Env不仅诱导相似的凋亡，而且会加剧线粒体介导的坏死，同时还能引起Env蛋白质表达量的显著降低。此外，流式细胞检测分析结果表明，无论是哪种*env*质粒，线粒体去极化都先于细胞中的半胱天冬酶活化，并且表明全长和截短的Env蛋白质都具有共同的内在线粒体途径以诱导细胞凋亡。我们的结果部分阐明了EIAV发病机制造成细胞死亡的作用机理。

1 病毒学国家重点实验室，武汉病毒研究所，中国科学院，武汉
2 传染病预防控制国家重点实验室，性病艾滋病预防控制中心，中国疾病预防与控制中心，北京
3 生物化学研究所，生命科学学院，浙江理工大学，杭州
4 研究生院，中国科学院，北京

Transcriptional profiling of midgut immunity response and degeneration in the wandering silkworm, *Bombyx mori*

Xu QY[1#] Lu AR[1#] Xiao GH[1] Yang B[1] Zhang J[1]
Li XQ[1] Guan JM[1] Shao QM[1] Brenda T. Beerntsen[2]
Zhang P[3] Wang CS[1] Ling EJ[1*]

Abstract:

Backgrounds: Lepidoptera insects have a novel development process comprising several metamorphic stages during their life cycle compared with vertebrate animals. Unlike most Lepidoptera insects that live on nectar during the adult stage, the *Bombyx mori* silkworm adults do not eat anything and die after egg-laying. In addition, the midguts of Lepidoptera insects produce antimicrobial proteins during the wandering stage when the larval tissues undergo numerous changes. The exact mechanisms responsible for these phenomena remain unclear.

Principal Findings: We used the silkworm as a model and performed genome-wide transcriptional profiling of the midgut between the feeding stage and the wandering stage. Many genes concerned with metabolism, digestion and ion and small molecule transportation were down-regulated during the wandering stage, indicating that the wandering stage midgut loses its normal functions. Microarray profiling, qRT-PCR and Western blot proved the production of antimicrobial proteins (peptides) in the midgut during the wandering stage. Different genes of the immune deficiency (IMD) pathway were up-regulated during the wandering stage. However, some key genes belonging to the Toll pathway showed no change in their transcription levels. Unlike butterfly (*Pachliopta aristolochiae*), the midgut of silkworm moth has a layer of cells, indicating that the development of midgut since the wandering stage is not usual. Cell division in the midgut was observed only for a short time during the wandering stage. However, there was extensive cell apoptosis before pupation. The imbalance of cell division and apoptosis probably drives the continuous degeneration of the midgut in the silkworm since the wandering stage.

Conclusions: This study provided an insight into the mechanism of the degeneration of the silkworm midgut and the production of innate immunity-related proteins during the wandering stage. The imbalance of cell division and apoptosis induces irreversible degeneration of the midgut. The IMD pathway probably regulates the production of antimicrobial peptides in the midgut during the wandering stage.

Published On: PLoS ONE, 2012, 7(8).

1 Key Laboratory of Insect Developmental and Evolutionary Biology, Institute of Plant Physiology and Ecology, Shanghai Institutes for Biological Sciences, Chinese Academy of Sciences, Shanghai 200032, China;

2 Department of Veterinary Pathobiology, University of Missouri, Columbia, Missouri, USA;

3 National Laboratory of Plant Molecular Genetics, Institute of Plant Physiology and Ecology, Shanghai Institutes for Biological Sciences, Chinese Academy of Sciences, Shanghai 200032, China.

[#]These authors contributed equally.

[*]Corresponding author E-mail: erjunling@sippe.ac.cn.

上蔟期家蚕中肠免疫反应与解离的转录组图谱

徐秋云[1#] 路岸瑞[1#] 肖国华[1] 杨 兵[1] 张 洁[1]
李旭全[1] 管京敏[1] 邵奇妙[1] Brenda T.Beerntsen[2]
张 鹏[3] 王成树[1] 凌尔军[1*]

摘要：背景——相比脊椎动物，鳞翅目昆虫具有一个新的发育过程，其生命周期包括几个变态阶段。不同于在成虫期取食花蜜的大多数鳞翅目昆虫，家蚕成虫不吃任何东西且在产卵后即死亡。此外，鳞翅目昆虫的中肠在上蔟期幼虫组织经历很多变化时会产生抗菌蛋白。这些现象的确切机制目前仍不清楚。主要研究结果——我们以家蚕为模型，在摄食期和上蔟期进行了中肠全基因组转录组分析。在上蔟期，许多与代谢、消化、离子和小分子转运相关的基因表达下调，表明上蔟期中肠失去其正常功能。芯片分析、qRT-PCR和Western blot证实在上蔟期中肠会产生抗菌蛋白（肽）。上蔟期时免疫缺陷途径（IMD）的不同基因表达上调。然而，Toll途径的一些关键基因的转录水平却没有变化。与蝴蝶（*Pachliopta aristolochiae*）不同，家蚕蚕蛾的中肠有一层细胞，表明上蔟期的中肠发育不同寻常。中肠细胞分裂仅在上蔟期很短的时间被观察到。然而，化蛹前有大量的细胞凋亡。细胞分裂和凋亡失衡可能是导致家蚕中肠自上蔟时期以来持续解离的原因。结论——本研究为深入了解上蔟期家蚕中肠解离和产生先天免疫相关蛋白的机制提供了参考。细胞分裂和凋亡失衡导致中肠不可逆解离。IMD通路可能在上蔟期调节中肠抗菌肽的产生。

1 昆虫发育与进化生物学重点实验室，植物生理生态研究所，上海生命科学研究院，中国科学院，上海
2 兽医病理学系，密苏里大学，哥伦比亚，美国
3 植物分子遗传国家重点实验室，植物生理生态研究所，上海生命科学研究院，中国科学院，上海

Detection and characterization of *Wolbachia* infection in silkworm

Zha XF[#] Zhang WJ[#] Zhou CY

Zhang LY Xiang ZH Xia QY[*]

Abstract: *Wolbachia* naturally infects a wide variety of arthropods, where it plays important roles in host reproduction. It was previously reported that *Wolbachia* did not infect silkworm. By means of PCR and sequencing, we found in this study that *Wolbachia* is indeed present in silkworm. Phylogenetic analysis indicates that *Wolbachia* infection in silkworm may have occurred via transfer from parasitic wasps. Furthermore, Southern blot results suggest a lateral transfer of the *wsp* gene into the genomes of some wild silkworms. By antibiotic treatments, we found that tetracycline and ciprofloxacin can eliminate *Wolbachia* in the silkworm and *Wolbachia* is important to ovary development of silkworm. These results provide clues towards a more comprehensive understanding of the interaction between *Wolbachia* and silkworm and possibly other lepidopteran insects.

Published On: Genetics and Molecular Biology, 2014, 37(3), 573-580.

State Key Laboratory of Silkworm Genome Biology, Southwest University, Chongqing 400716, China.

[#]These authors contributed equally.

[*]Corresponding author E-mail: xiaqy@swu.edu.cn.

沃尔巴克氏体感染的家蚕检测与表征

查幸福[#]　张文姬[#]　周春燕　张李颖　向仲怀　夏庆友[*]

摘要：沃尔巴克氏体能自然地感染各种节肢动物，并在宿主繁殖中起重要作用。在以前的报道中，沃尔巴克氏体不感染家蚕。通过PCR和测序，我们在本研究中发现沃尔巴克氏体确实存在于家蚕体内。系统发生分析表明，沃尔巴克氏体感染家蚕可能是通过寄生蜂传递发生的。此外，Southern印迹结果表明侧向转移的*wsp*基因进入一些野生蚕的基因组。通过抗生素处理，我们发现四环素和环丙沙星可以消除家蚕体内的沃尔巴克氏体，并且沃尔巴克氏体对家蚕的卵巢发育很重要。本研究为更全面地了解沃尔巴克氏体和家蚕及其他可能的鳞翅目昆虫之间的相互作用提供了线索。

家蚕基因组生物学国家重点实验室，西南大学，重庆

Immune signaling pathways activated in response to different pathogenic micro-organisms in *Bombyx mori*

Liu W[1#] Liu JB[1#] Lu YH[1] Gong YC[1] Zhu M[1] Chen F[1] Liang Z[1]
Zhu LY[1] Kuang SL[1] Hu XL[1,2] Cao GL[1] Xue RY[1,2] Gong CL[1,2*]

Abstract: The JAK/STAT, Toll, IMD and RNAi pathways are the major signaling pathways associated with insect innate immunity. To explore the different immune signaling pathways triggered in response to pathogenic micro-organism infections in the silkworm, *Bombyx mori*, the expression levels of the signal transducer and activator of transcription (*Bm*STAT), spatzle-1 (*Bm*spz-1), peptidoglycan-recognition protein LB (*Bm*PGRP-LB), peptidoglycan-recognition protein LE (*Bm*PGRP-LE), argonaute 2 (*Bm*ago2), and dicer-2 (*Bm*dcr2) genes after challenge with *Escherichia coli* (*E. coli*), *Serratiamarcescens* (*Sm*), *Bacillus bombyseptieus* (*Bab*), *Beauveriabassiana* (*Beb*), *nucleopolyhedrovirus* (*Bm*NPV), *cypovirus* (*Bm*CPV), *bidensovirus* (*Bm*BDV), or *Nosemabombycis* (*Nb*) were determined using real-time PCR. We found that the JAK/STAT pathway could be activated by challenge with *Bm*NPV and *Bm*BDV, the Toll pathway could be most robustly induced by challenge with *Beb*, the IMD pathway was mainly activated in response to infection by *E. coli* and *Sm*, and the RNAi pathway was not activated by viral infection, but could be triggered by some bacterial infections. These findings yield insights into the immune signaling pathways activated in response to different pathogenic micro-organisms in the silkworm.

Published On: Molecular Immunology, 2015, 65(2), 391-397

1 School of Biology and Basic Medical Sciences, Soochow University, Suzhou 215123, China.
2 National Engineering Laboratory for Modern Silk, Soochow University, Suzhou 215123, China.
#These authors contributed equally.
*Corresponding author E-mail: gongcl@suda.edu.cn.

家蚕免疫信号通路对不同病原微生物的应答研究

刘　伟[1#]　刘佳宾[1#]　陆亚红[1]　龚永昌[1]　朱　敏[1]　陈　菲[1]　梁　子[1]
朱丽媛[1]　匡苏兰[1]　胡小龙[1,2]　曹广力[1]　薛仁宇[1,2]　贡成良[1,2*]

摘要：JAK / STAT，Toll，IMD 和 RNAi 信号通路是昆虫先天免疫相关的主要信号通路。为探讨家蚕天然免疫相关通路对不同微生物感染后的应答差异，我们采用实时定量 PCR 检测了经大肠杆菌（*E. coli*），灵杆菌（*Sm*），黑胸败血芽孢杆菌（*Bab*），白僵菌（*Beb*），家蚕核型多角体病毒（*Bm*NPV），家蚕质型多角体病毒（*Bm*CPV），家蚕浓核病毒（*Bm*BDV）以及家蚕微孢子虫（*Nb*）诱导对信号转导和转录激活因子 *Bm*STAT，*Bm*spz-1，肽聚糖识别蛋白 *Bm*PGRP-LB，肽聚糖识别蛋白 *Bm*PGRP-LE、*Bm*ago2 和 *Bm*dcr2 基因表达水平的影响。结果发现 JAK / STAT 通路可以被 *Bm*NPV 和 *Bm*BDV 激活，Toll 通路在 *Beb* 感染后被诱导激活，IMD 途径主要在大肠杆菌和 *Sm* 感染后被激活，RNAi 途径不能被病毒感染激活，但在细菌感染后会被激活。本研究为家蚕不同免疫信号通路对不同病原微生物的应答研究提供了基础。

1　基础医学与生物科学学院，苏州大学，苏州
2　现代丝绸国家工程实验室，苏州大学，苏州

Resistance comparison of domesticated silkworm (*Bombyx mori* L.) and wild silkworm (*Bombyx mandarina* M.) to phoxim insecticide

Li B[1,2] Wang YH[2] Liu HT[2] Xu YX[1,2]
Wei ZG[1,2] Chen YH[1,2] Shen WD[1,2*]

Abstract: In this study, the resistance difference to phoxim between *Bombyx mori* L. and *Bombyx mandarina* M was investigated. For the both silkworm species, the whole body of each larval were collected, and on the third day of the 5th instar, the brain, midgut, fat bodies, and silk gland were collected for enzymatic activity assay of acetylcholinesterase (AChE). Our results showed that in the early larval stages, the resistance difference to phoxim was not significant between the two species. However, in the 4th and 5th instar, the resistance differences showed significant increase. When compared to *B. mori* L., the LC_{50} of *B. mandarina* was 4.43 and 4.02-fold higher in the 4th and 5th instar, respectively. From the 1st to 5th instar, the enzymatic activities of AChE of *B. mandarina* were 1.60, 1.65, 1.81, 1.93 and 2.28-fold higher than that of *B. mori*, respectively. For the brain, midgut, fat body, and silk gland on the third day of the 5th instar, the enzymatic activity ratios of *B. mandarina* to *B. mori* were 1.90, 2.23, 2.76, and 2.78, respectively. The AChE-I_{50} values of *B. mori* and *B. mandarina* detected by eserine method were 5.02×10^{-7} and 5.23×10^{-7} mol/L, respectively. Thus, our results indicate that the higher enzymatic activities of AChE and the insensitivity to specific inhibitor of the enzyme might be the underlying mechanisms for higher phoxim resistance in *B. mandarina*.

Published On: African Journal of Biotechnology, 2010, 9(12), 1771-1775.

1 National Engineering Laboratory for Modern Silk, Soochow University 215123, Suzhou, China.
2 School of Basic Medicine and Biological Sciences, Soochow University 215123, Suzhou, China.
*Corresponding author E-mail:shenwd@suda.edu.cn.

家蚕和野桑蚕对辛硫磷杀虫剂的抗性比较

李　兵[1,2]　王燕红[2]　刘海涛[2]　许雅香[1,2]
卫正国[1,2]　陈玉华[1,2]　沈卫德[1,2*]

摘要:本研究调查了家蚕和野桑蚕对辛硫磷抗性的差异。分别收集了家蚕和野桑蚕不同龄期的个体和五龄第3天的脑、中肠、脂肪体和丝腺,用于检测乙酰胆碱酯酶的活性。结果表明,小蚕期家蚕和野桑蚕对辛硫磷的敏感性差异相对较小,四龄和五龄的抗性差异显著。和家蚕相比,四龄和五龄野桑蚕的LC_{50}分别是家蚕的4.43倍和4.02倍。从一龄至五龄,野桑蚕的AChE的活性分别是家蚕的1.60倍、1.65倍、1.81倍、1.93倍和2.28倍。野桑蚕和家蚕五龄第3天的脑、中肠、脂肪体和丝腺的AChE的活性比分别是为1.90、2.23、2.76和2.78。用毒扁豆碱法分析测定了家蚕和野桑蚕的AChE-I_{50}分别是5.02×10^{-7} mol/L和5.23×10^{-7} mol/L。因此,我们的结果表明,较高的AChE活性和对专一性抑制剂的不敏感,可能是野桑蚕对辛硫磷抗性高的内在机制。

1　现代丝绸国家工程实验室,苏州大学,苏州
2　基础医学与生物科学学院,苏州大学,苏州

Effects of *Bm*CPV infection on silkworm *Bombyx mori* intestinal bacteria

Sun ZL[1#] Lu YH[1#] Zhang H[1] Kumar Dhiraj[1] Liu B[1]
Gong YC[1] Zhu M[1] Zhu LY[1] Liang Z[1] Kuang SL[1]
Chen F[1] Hu XL[1,2] Cao GL[1,2] Xue RY[1,2] Gong CL[1,2*]

Abstract: The gut microbiota has a crucial role in the growth, development and environmental adaptation in the host insect. The objective of our work was to investigate the microbiota of the healthy silkworm *Bombyx mori* gut and changes after the infection of *B. mori* cypovirus (*Bm*CPV). Intestinal contents of the infected and healthy larvae of *B. mori* of fifth instar were collected at 24, 72 and 144 h post infection with *Bm*CPV. The gut bacteria were analyzed by pyrosequencing of the 16S rRNA gene. 147(135) and 113(103) genera were found in the gut content of the healthy control female (male) larvae and *Bm*CPV-infected female (male) larvae, respectively. In general, the microbial communities in the gut content of healthy larvae were dominated by *Enterococcus*, *Delftia*, *Pelomonas*, *Ralstonia* and *Staphylococcus*, however the abundance change of each genus was depended on the developmental stage and gender. Microbial diversity reached minimum at 144 h of fifth instar larvae. The abundance of *Enterococcus* in the females was substantially lower and the abundance of *Delftia*, *Aurantimonas* and *Staphylococcus* was substantially higher compared to the males. Bacterial diversity in the intestinal contents decreased after post infection with *Bm*CPV, whereas the abundance of both *Enterococcus* and *Staphylococcus* which belongs to Gram-positive were increased. Therefore, our findings suggested that observed changes in relative abundance was related to the immune response of silkworm to *Bm*CPV infection. Relevance analysis of plenty of the predominant genera showed the abundance of the *Enterococcus* genus was in negative correlation with the abundance of the most predominant genera. These results provided insight into the relationship between the gut microbiota and development of the *Bm*CPV-infected silkworm.

Published On: PLoS ONE, 2015, 11(1).

1 School of Biology and Basic Medical Sciences ,Soochow University, Suzhou 215123, China.
2 National Engineering Laboratory for Modern Silk, Soochow University, Suzhou 215123, China.
#These authors contributed equally.
*Corresponding author E-mail: gongcl@suda.edu.cn.

*Bm*CPV感染对家蚕肠道菌群的影响

孙振丽[1#] 陆亚红[1#] 张 皓[1] Kumar Dhiraj 刘 波[1]
龚永昌[1] 朱 敏[1] 朱丽媛[1] 梁 子[1] 匡苏兰[1] 陈 菲[1]
胡小龙[1,2] 曹广力[1,2] 薛仁宇[1,2] 贡成良[1,2*]

摘要:肠道菌群在宿主昆虫的生长,发育和环境适应中起着至关重要的作用。本研究调查了健康家蚕的肠道菌群及感染*Bm*CPV后肠道菌群的变化。分别收集*Bm*CPV感染后24 h、72 h、144 h小时及健康五龄家蚕的肠道内容物。通过16S rRNA基因的焦磷酸测序分析肠道菌群。在健康对照的雌性(雄性)幼虫和*Bm*CPV感染的雌性(雄性)幼虫的肠道内容物中分别发现了147(135)个和113(103)个属。健康幼虫的肠道微生物群落主要是*Enterococcus*,*Delftia*,*Pelomonas*,*Ralstonia*和*Staphylococcus*,但每个属的丰度变化取决于发育阶段和性别。在144 h时,微生物种类最少。与雄性相比,雌性*Enterococcus*的丰度显著降低,*Delftia*,*Aurantimonas*和*Staphylococcus*的丰度明显升高。*Bm*CPV感染后,肠道微生物多样性降低,然而同属于革兰氏阳性菌的*Enterococcus*和*Staphylococcus*的丰度有所增加。因此,我们的研究结果表明,肠道微生物丰度的变化与家蚕对*Bm*CPV感染的免疫反应有关。优势菌属的丰度相关性分析结果显示,肠球菌属的丰度与其他的优势菌属成负相关。这些结果为研究肠道微生物菌群与*Bm*CPV感染家蚕后的发育两者之间的关系提供了启示。

1 基础医学与生物科学学院,苏州大学,苏州
2 现代丝绸国家工程实验室,苏州大学,苏州

cDNA cloning and characterization of *LASP1* from silkworm, *Bombyx mori*, involved in cytoplasmic polyhedrosis virus infection

Gao K[1,2,3] Deng XY[3] Qian HY[1,2] Wu P[1,2]
Qin GX[1,2] Liu T[1,2] Guo XJ[1,2*]

Abstract: Full-length cDNA of a LIM and SH3 contained protein 1 (named *Bm*LASP1) was identified from the silkworm, *Bombyx mori*, for the first time by rapid amplification of cDNA ends. The full-length cDNA of *BmLASP1* is 2 094 bp, consisting of a 5′-terminal untranslated region (UTR) of 117 bp, and a 3′-UTR of 610 bp with two poly-adenylation signal sequence AATAAA and a poly (A) tail. The *BmLASP1* cDNA encodes a polypeptide comprising 455 amino acids, including a LIM domain, two nebulin domains and an SH3 domain. The theoretical isoelectric point is 7.07 and the predicted molecular weight is 51.8 kDa. *Bm*LASP1 has no signal peptide but three potential N-glycosylation sites. Sequence similarity and phylogenic analyses indicated that *Bm*LASP1 belonged to the group of insect LASP1 with a longer linker region which is different from vertebrate LASP1. The *LASP1* in silkworm contained eight exons in its coding regions, and the last exon-intron boundary was conserved the same as in mammalian and *Ciona intestinalis LASP1* genes. By fluorescent quantitative real-time polymerase chain reaction, the mRNA transcripts of *BmLASP1* were mainly detected in the gonad, head, and spiracle, and slightly in the silk gland, vasa mucosa, midgut, fat body, and hemocytes. After silkworm larvae were infected by *B. mori* cytoplasmic polyhedrosis virus (*Bm*CPV), the relative expression level of *Bm*LASP1 was down-regulated in the midgut. This result suggested that *Bm*LASP1 may play an important role in the response of silkworm to *Bm*CPV infection.

Published On: Gene, 2012, 511(2), 389-397.

1 Sericultural Research Institute, Jiangsu University of Science and Technology, Zhenjiang 212018, China.
2 Sericultural Research Institute, Chinese Academy of Agricultural Sciences, Zhenjiang 212018, China.
3 College of Biotechnology and Chemical Engineering, Jiangsu University of Science and Technology, Zhenjiang 212003, China.
*Corresponding author E-mail: guoxijie@126.com.

家蚕*LASP1*基因全长cDNA的克隆及其在*Bm*CPV感染过程中的特征分析

高　坤[1,2,3]　邓祥元[3]　钱荷英[1,2]　吴　萍[1,2]
覃光星[1,2]　刘　挺[1,2]　郭锡杰[1,2*]

摘要：通过CDNA末端快速扩增(RACE)，本文首次从家蚕中鉴定出了一种含有LIM和SH3的蛋白质1(*Bm*LASP1)的全长CDNA。该cDNA全长为2 094 bp，包含一个117 bp 5′非翻译区和一个610 bp的3′非翻译区，3′非翻译区包含2个加尾信号AATAAA和poly(A)尾；其开放阅读框(ORF)编码一个455个氨基酸组成的蛋白质，预测蛋白质分子质量为51.8 kDa，等电点为7.07；没有信号肽和糖基化位点，包含一个LIM结构域、两个Nebulin结构域和一个SH3结构域。多序列比对和系统进化树显示，家蚕*Bm*LASP1属于昆虫类LASP1，其连接体区域较长，与脊椎动物LASP1不同。*BmLASP1*的编码区域有8个外显子，最后一个外显子和内含子之间的剪切位点在哺乳动物和海鞘动物中都是非常保守的。实时荧光定量PCR(qRT-PCR)分析表明*BmLASP1*主要在家蚕的生殖腺、头部和气管组织表达，在丝腺、vasa mucosa、脂肪体和血细胞中仅有少量表达。且家蚕在感染质型多角体病毒(*Bombyx mori* cytoplasmic polyhedrosis virus，*Bm*CPV)后，*Bm*LASP在中肠的表达量明显下调。综上所述，*Bm*LASP1在家蚕感染*Bm*CPV过程中发挥着重要的作用。

1　蚕业研究所，江苏科技大学，镇江
2　蚕业研究所，中国农业科学院，镇江
3　生物与化学工程学院，江苏科技大学，镇江

Microarray analysis of the gene expression profile in the midgut of silkworm infected with cytoplasmic polyhedrosis virus

Wu P[1,2] Wang X[1] Qin GX[2] Liu T[2]
Jiang YF[1] Li MW[2] Guo XJ[1,2*]

Abstract: In order to obtain an overall view on silkworm response to *Bombyx mori* cytoplasmic polyhedrosis virus(*Bm*CPV) infection, a microarray system comprising 22 987 oligonucluotide 70-mer probes was employed to compare differentially expressed genes in the midguts of *Bm*CPV-infected and normal silkworm larvae. At 72 h post-inoculation, 258 genes exhibited at least 2.0-fold differences in expression level. Out of these, 135 genes were up-regulated, while 123 genes were down-regulated. According to gene ontology (GO), 140 genes were classified into GO categories. Kyoto Encyclopedia of Genes and Genomes (KEGG) pathway analysis indicates that 35 genes were involved in 10 significant (P<0.05) KEGG pathways. The expressions of genes related to valine, leucine,and isoleucine degradation, retinol metabolism, and vitamin B_6 metabolism were all down-regulated. The expressions of genes involved in ribosome and proteasome pathway were all up-regulated. Quantitative real-time polymerase chain reaction was performed to validate the expression patterns of 13 selected genes of interest. The results suggest that *Bm*CPV infection resulted in the disturbance of protein and amino acid metabolism and a series of major physiological and pathological changes in silkworm. Our results provide new insights into the molecular mechanism of *Bm*CPV infection and host cell response.

Published On: Molecular Biology Reports, 2011, 38(1), 333-341.

1 School of Biotechnology and Environmental Engineering, Jiangsu University of Science and Technology, Zhenjiang 212018, China.
2 Key Laboratory of Silkworm and Mulberry Genetic Improvement, Ministry of Agriculture of The People's Republic of China, Sericultural Research Institute, Chinese Academy of Agricultural Sciences, Zhenjiang 212018, China.
*Corresponding author E-mail: guoxijie@126.com.

家蚕感染质型多角体病毒后中肠基因表达谱的微阵列分析

吴　萍[1,2]　王　秀[1]　覃光星[2]　刘　挺[2]
蒋云峰[1]　李木旺[2]　郭锡杰[1,2*]

摘要：为了研究家蚕在质型多角体病毒(*Bm*CPV)感染后的整体情况，利用包含22 987个70-mer寡核苷酸探针的微阵列系统来比较*Bm*CPV感染和正常家蚕幼虫中肠中的差异表达基因。在接种72 h后，258个基因在表达水平上表现出至少2.0倍的差异。其中，135个基因表达上调，123个基因表达下调。根据基因本体论(GO)，将140个基因GO分类。KEGG分析表明，35个基因参与10个显著性($P<0.05$)的KEGG通路。与缬氨酸、亮氨酸、异亮氨酸降解，视黄醇代谢和维生素B_6代谢相关的基因都下调表达。核糖体和蛋白酶体通路相关基因表达均上调，通过实时定量PCR对挑选的13个有趣基因的表达模式进行验证。结果表明*Bm*CPV感染家蚕后导致蛋白质和氨基酸代谢紊乱，并引起一系列生理病理变化。我们的结果为了解*Bm*CPV感染和宿主细胞应答的分子机制提供了更新的见解。

1　生物技术与环境工程学院，江苏科技大学，镇江
2　农业部蚕桑遗传改良重点实验室，蚕业研究所，中国农业科学院，镇江

A novel protease inhibitor in *Bombyx mori* is involved in defense against *Beauveria bassiana*

Li YS[#] Zhao P[#] Liu SP Dong ZM
Chen JP Xiang ZH Xia QY[*]

Abstract: Entomopathogenic fungi, such as *Beauveria bassiana*, penetrate the insect cuticle using a plethora of hydrolytic enzymes including cuticle-degrading proteases and chitinases, which are important virulence factors. The insect integument and hemolymph contains a relatively high concentration of protease inhibitors, which are closely involved with defense against pathogenic microorganisms. To elucidate the molecular mechanism underlying resistance against entomopathogenic fungi and to identify a new molecular target for improving fungal resistance in the silkworm, *Bombyx mori*, we cloned and expressed a novel silkworm TIL-type protease inhibitor *Bm*SPI38, which was very stable over a wide range of temperatures and pH values. An activity assay suggested that *Bm*SPI38 potently inactivated the insecticidal cuticle-degrading enzyme (CDEP-1) produced by *B. bassiana* and subtilisin A produced by *Bacillus licheniformis*. The melanization of silkworm induced by CDEP-1 protease could also be blocked by *Bm*SPI38. These results provided new insights into the molecular mechanisms whereby insect protease inhibitors provide resistance against entomopathogenic fungi, suggesting the possibility of using fungal biopesticides in sericulture.

Published On: Insect Biochemistry and Molecular Biology, 2012, 42(10), 766-775.

State Key Laboratory of Silkworm Genome Biology, Southwest University, Chongqing 400716, China.
[#]These authors contributed equally.
[*]Corresponding author E-mail: xiaqy@swu.edu.cn.

一个新的家蚕蛋白酶抑制剂参与抵御球孢白僵菌

李游山[#] 赵 萍[#] 刘仕平 董照明
陈建平 向仲怀 夏庆友[*]

摘要：昆虫致病性真菌，如球孢白僵菌，利用大量的表皮降解蛋白酶和几丁质酶水解昆虫表皮，从而达到侵染昆虫的目的。昆虫的表皮和血淋巴含有一种浓度相当高的蛋白酶抑制剂，这种抑制剂与抵抗病原微生物侵染密切相关。为了阐明家蚕抵抗昆虫致病性真菌的分子机制，以及筛选一种新的可以提高真菌抵抗力的分子靶标，我们克隆并表达了一个新的家蚕TIL类蛋白酶抑制剂*Bm*SPI38。该蛋白酶抑制剂在很广泛的温度和pH范围内都非常稳定。活性分析表明，*Bm*SPI38能够有效地使球孢白僵菌产生的可杀虫的表皮降解酶（CDEP-1）和枯草杆菌产生的枯草杆菌蛋白酶A失活。家蚕中由CDEP-1所诱导的黑化作用也受到*Bm*SPI38的抑制。这些结果为解析昆虫蛋白酶抑制剂抵抗昆虫致病性真菌的分子机制提供了一些新的见解，也为在蚕桑生产中使用真菌生物农药提供了新的参考。

家蚕基因组生物学国家重点实验室，西南大学，重庆

Hindgut innate immunity and regulation of fecal microbiota through melanization in insects

Shao QM[1] Yang B[1#] Xu QY[1#] Li XQ[1] Lü ZQ[2]
Wang CS[1] Huang YP[1] Kenneth Söderhäll[3*] Ling EJ[1*]

Abstract: Many insects eat the green leaves of plants but excrete black feces in a yet-unknown mechanism. Insects cannot avoid ingesting pathogens with food that will be specifically detected by the midgut immune system. However, just as in mammals, many pathogens can still escape the insect midgut immune system and arrive in the hindgut, where they are excreted out with the feces. Here we show that the melanization of hindgut content induced by prophenoloxidase, a key enzyme that induces the production of melanin around invaders and at wound sites, is the last line of immunity to clear bacteria before feces excretion. We used the silkworm *Bombyx mori* as a model and found that prophenoloxidase produced by hindgut cells and is secreted into the hindgut contents. Several experiments were done to clearly demonstrate that the blackening of the insect feces was due to activated phenoloxidase, which served to regulate the number of bacteria in the hindgut. Our analysis of the silkworm hindgut prophenoloxidase disclose the natural secret why the insect feces is black and provides insight into the hindgut innate immunity that is still rather unclear in mammals.

Published On: Journal of Biological Chemistry, 2012, 287(17), 14270-14279.

1 Key Laboratory of Insect Developmental and Evolutionary Biology, Shanghai Institute of Plant Physiology and Ecology, Shanghai Institutes for Biological Sciences, Chinese Academy of Sciences, Shanghai 200032, China.

2 College of Plant Protection, Northwest A&F University, Yangling 712100, China.

3 Department of Comparative Physiology, Uppsala University, Uppsala 75218, Sweden.

[#]These authors contributed equally.

[*]Corresponding author E-mail: kenneth.soderhall@ebc.uu.se; erjunling@sippe.ac.cn.

昆虫通过黑化作用调节后肠先天免疫以及粪便菌群

邵奇妙[1]　杨　兵[1#]　徐秋云[1#]　李旭全[1]　吕志强[2]
王成树[1]　黄勇平[1]　Kenneth Söderhäll[3]　凌尔军[1*]

摘要：许多昆虫会取食植物的绿色叶片，但是排出的却是黑色的粪便，这种机制目前尚不明确。昆虫无法避免在取食时摄入病原菌，但这些病原菌却能被中肠免疫系统特异性地识别。然而就如在哺乳动物中一样，许多病原菌仍然能够逃逸昆虫中肠免疫系统到达后肠，并随粪便排出体外。本研究发现后肠内容物的黑化反应是由多酚氧化酶（PPO）催化引起的，PPO是在病原周围和伤口处诱导产生黑色素的关键酶，是在粪便排出前免疫系统清除细菌的最后一道防线。我们使用家蚕作为模型发现PPO由后肠细胞产生并分泌到后肠内容物中。多个实验清楚地证实了昆虫粪便黑化是由PPO的活化所致，这是为了调节后肠细菌数量。我们对家蚕后肠PPO的分析揭示了为什么昆虫的粪便是黑色的这一自然奥秘，并为目前尚不清楚的哺乳动物后肠先天免疫提供了参考。

1　昆虫发育与进化生物学重点实验室，上海植物生理生态研究所，上海生命科学研究院，中国科学院，上海
2　植物保护学院，西北农林科技大学，杨凌
3　比较生理学系，乌普萨拉大学，乌普萨拉

第四章

家蚕性别调控机制研究

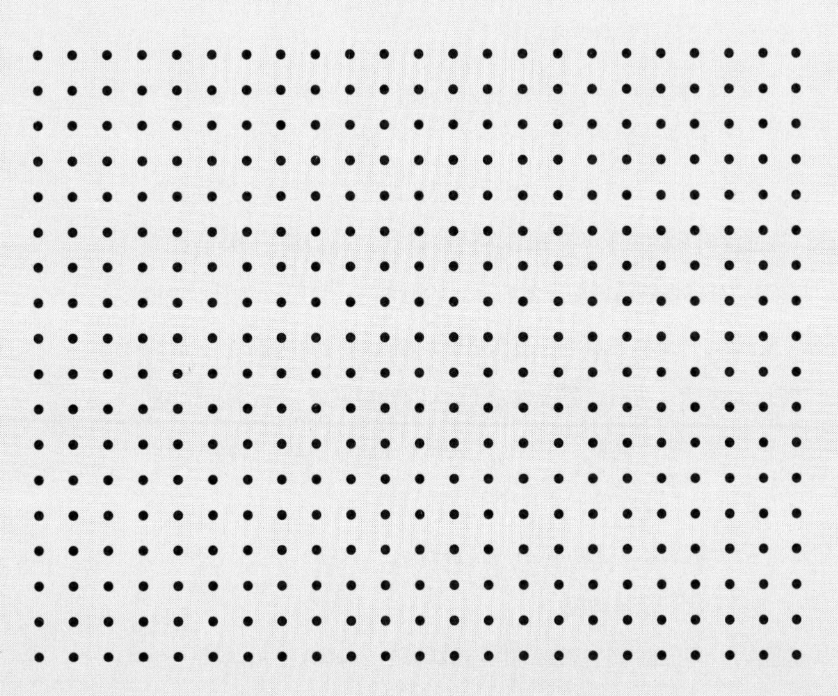

Chapter 4

家蚕雌雄个体在产业经济价值方面具有差异。其性别决定既是基础的科学问题，也是蚕业的重要课题。近年来，随着家蚕性别决定途径末端关键的双性基因（Bmdsx）被鉴定报道以来，其调控机制和网络一直是此领域的研究热点。项目组对Bmdsx的自身调控及上游调节因子进行了深入研究，取得了一系列的重要进展。

1. 家蚕双性基因（*Bmdsx*）自身剪接调控及功能

项目组系统分析了家蚕双性基因*Bmdsx*的转录，发现*Bmdsx*以前未报道过的新外显子，对*Bmdsx*的基因结构和表达进行了重新认知。*Bmdsx*转录本经剪接可产生17个选择性剪接体和11个推定的反式剪接体。发现新的多种形式剪接体可产生多种形式Dsx蛋白，且Dsx蛋白都具有调控下游靶基因表达的能力。项目组进一步研究反式剪接，发现反式剪接体可调节家蚕雌特异转录本*Bmdsx*^{F}的表达。利用转基因方法将*BmDSX*^{F}在雄蚕中异位表达，可抑制信息素结合蛋白基因和精巢特异表达基因*A2*的表达，并激活储存蛋白1基因的表达。即反式剪接是*Bmdsx*的剪接方式，通过*BmDSX*^{F}的表达，参与了雌蚕的性别发育。项目组通过转基因方法异位表达双性基因雄特异剪接体*Bmdsx*^{M}，结果发现*Bmdsx*可通过调节EGFR信号通路活性和*Abd-B*的表达来调控第八腹节的发育命运，从而控制家蚕雌雄腹部性特征差异。

2. 家蚕*Bmdsx*的上游调节因子及其调控

家蚕性别决定基因*Bmdsx*通过选择性剪接产生雌雄特异的mRNA形式，从而产生雌雄特异的*Bm*DSX蛋白。项目组基于同源性，在家蚕中克隆了果蝇性别决定基因*dsx*前mRNA的剪接因子RBP1的同源体（*Bmrbp1*）。*Bmrbp1*为SR家族剪接因子的成员，是家蚕*Bmdsx*前mRNA重要的候选剪接因子。另一方面，项目组根据*Bmdsx*第四外显子上的选择性剪接顺式元件CE1，通过凝胶阻滞和LC-MS / MS等实验，鉴定了*Bmdsx*上游调控蛋白*Bm*Hrp28。克隆和过量表达*Bm*Hrp28蛋白及EMSA实验发现*Bm*HRP28蛋白特异性结合*Bmdsx*的CE1+6元件上。项目组进一步分析了*Bm*Hrp28与*Bmdsx*另一个上游调控蛋白*Bm*PSI的相互关系，酵母双杂交（Y2H）结果显示*Bm*HRP28和*Bm*PSI蛋白之间几乎没有或没有直接结合；免疫共沉淀（Co-IP）实验发现*Bm*HRP28和*Bm*PSI共存于多蛋白复合物中。为了寻找多蛋白复合物中的其他调节因子，项目组构建了一个雄性家蚕早期胚胎的Y2H cDNA文库。通过Y2H筛查，鉴定了一个RNA结合蛋白*Bm*SPX，其能够与*Bm*PSI结合。这些结果为阐明*Bmdsx*上游的调控机制提供了很好的线索。

3. 家蚕生殖腺发育相关基因的鉴定和表达谱

生殖腺是雌雄分化最为明显的组织器官。项目组克隆和分析了原始生殖细胞迁移与分化所必需的基因*nanos*。结果发现Nanos和Vasa蛋白在卵母细胞中有表达，在精原细胞中高表达，后期阶段的精母细胞中表达相对较弱。为了系统地了解生殖腺蛋白表达的性别二态性，项目组对家蚕幼虫生殖腺蛋白质组进行鉴定。从精巢和卵巢中分别鉴定到286个和205个非冗余蛋白，其中绝大部分蛋白是新鉴定的。通过蛋白表达谱和生物信息学分析方法发现，胞内维甲酸结合蛋白（CRABP）在精巢中极高丰度地表达，微管蛋白在卵巢中高丰度表达，以及一些果蝇配子形成关键蛋白的同源蛋白在家蚕中也被鉴定。这些结果为系统理解性别二态性和配子形成提供了依据。

查幸福

*Bm*TGIF, a *Bombyx mori* homolog of *Drosophila Dm*TGIF, regulates progression of spermatogenesis

Zhang PJ[1#] Cao GL[1,2#] Sheng J[1] Xue RY[1,2] Gong CL[1,2*]

Abstract: TG-interacting factor (TGIF) in *Drosophila* consists of two tandemly-repeated genes, *achintya* (*Dmachi*) and *vismay* (*Dmvis*), which act as transcriptional activators in *Drosophila* spermatogenesis. In contrast, TGIF in humans is a transcriptional repressor that binds directly to DNA or interacts with corepressors to repress the transcription of target genes. In this study, we investigated the characteristics and functions of *Bm*TGIF, a *Bombyx mori* homolog of *Dm*TGIF. Like *DmTGIF, BmTGIF* is predominantly expressed in the testes and ovaries. Four alternatively spliced isoforms could be isolated from testes, and two isoforms from ovaries. Quantitative polymerase chain reaction indicated *BmTGIF* was abundantly expressed in the testis of 3rd instar larvae, when the testis is almost full of primary spermatocytes. The results of luciferase assays indicated that *Bm*TGIF contains two adjacent acidic domains that activate the transcription of reporter genes. Immunofluorescence assay in *BmN* cells showed that the *Bm*TGIF protein was located mainly in the nucleus, and paraffin sections of testis showed *Bm*TGIF was grossly expressed in primary spermatocytes and mature sperms. Consistent with the role of *Dm*Vis in *Drosophila* development, *Bm*TGIF significantly affected spermatid differentiation, as indicated by hematoxylin-eosin staining of paraffin sections of testis from *BmTGIF*-small interfering RNA (siRNA) - injected male silkworms. Co-immunoprecipitation experiments suggested that *Bm*TGIF interacted with *Bm*Aly, and that they may recruit other factors to form a complex to regulate the genes required for meiotic divisions and spermatid differentiation. The results of this analysis of *Bm*TGIF will improve our understanding of the mechanism of spermatid differentiation in *B. mori*, with potential applications for pest control.

Published On: PLoS ONE, 2012, 7(11).

1 School of Biology and Basic Medical Science, Soochow University, Suzhou 215123, China.

2 National Engineering Laboratory for Modern Silk, Soochow University, Suzhou 215123, China.

[#]These authors contributed equally.

[*]Corresponding author E-mail: gongcl@suda.edu.cn

果蝇*Dm*TGIF同族体家蚕*Bm*TGIF对精子形成过程的调节

张鹏杰[1#] 曹广力[1,2#] 盛 洁[1] 薛仁宇[1,2] 贡成良[1,2*]

摘要：果蝇中TG互作因子(TGIF)包含两个串联重复基因，*Dmachi*和*Dmvis*，它们在果蝇精子形成过程中起转录激活剂作用。相反，人体中的TGIF是转录抑制因子，通过直接结合DNA或与阻遏辅助物互作，进而抑制靶基因的转录。本文中我们研究了*Bm*TGIF——*Dm*TGIF的家蚕同源物的特征和功能和*DmTGIF*一样，*BmTGIF*主要在家蚕精巢和卵巢中表达。我们从家蚕精巢中分离到了4种剪接亚型，从卵巢中分离到了2种剪接亚型。定量PCR结果显示*BmTGIF*在三龄家蚕幼虫的精巢中大量表达，该时期精巢中充满大量初级精母细胞。荧光素酶实验结果表明，家蚕*Bm*TGIF包含两个相邻酸性域，能激活报告基因的转录。免疫荧光实验表明，*Bm*TGIF蛋白主要定位于家蚕卵巢细胞(*BmN*)的细胞核中，家蚕精巢的石蜡切片结果显示，*Bm*TGIF主要在初级精母细胞和精子成熟时期大量表达。与*Dm*Vis在果蝇发育阶段扮演的角色类似，用siRNA沉默*BmTGIF*基因表达后，家蚕的精巢石蜡切片染色结果显示：*Bm*TGIF能显著影响精子细胞分化过程。免疫共沉淀实验结果表明，家蚕*Bm*TGIF与*Bm*Aly相互作用，它们可能“募集”其他的因子形成一个复合物，该复合物调控减数分裂和精细胞分化过程中所需基因的表达。对家蚕TGIF的研究将有助于我们对家蚕精细胞分化机制的认识以及挖掘TGIF在昆虫防治中潜在的应用价值。

1 基础医学与生物科学学院，苏州大学，苏州
2 现代丝绸国家工程实验室，苏州大学，苏州

Female qualities in males: vitellogenin synthesis induced by ovary transplants into the male silkworm, *Bombyx mori*

Yang CW[#] Lin Y[#] Shen GW Chen EX Wang YX
Luo J Zhang HY Xing RM Xia QY[*]

Abstract: Female qualities in males are common in vertebrates but have not been extensively reported in insects. Vitellogenin (Vg) is highly expressed in the female fat body and is generally required for the formation of yolk proteins in the insect egg. Vg upregulation is generally regarded as a female quality in female oviparous animals. In this study, we found that *Bombyx mori* Vg (*Bm*Vg) is especially highly expressed in the female pupa. Downregulation of the *BmVg* gene in the female pupa by RNA interference (RNAi) interfered with egg formation and embryonic development, showing the importance of *Bm*Vg in these processes. So, we used *Bm*Vg as a biomarker for female qualities in the silkworm. Hematoxylin-eosin staining and immunofluorescence histochemistry showed that ovary transplants induced *Bm*Vg synthesis in the male pupa fat body. Ovaries transplanted into male silkworms produced only a few eggs with deformed yolk granules. These results suggested that the amount of *Bm*Vg in the male silkworm was insufficient for eggs to undergo complete embryonic development. After 17-beta-estradiol was used to treat male pupae and male pupal fat bodies, *BmVg* was upregulated *in vivo* and *in vitro*. These findings indicated that the male silkworm has innate female qualities that were induced by a transplanted ovary and 17β - estradiol. However, in silkworms, female qualities in males are not as complete as in females.

Published On: Biochemical and Biophysical Research Communications, 2014, 453(1), 31-36.

State Key Laboratory of Silkworm Genome Biology, Southwest University, Chongqing 400716, China.
[#]These authors contributed equally.
[*]Corresponding author E-mail: xiaqy@swu.edu.cn.

雄性的雌性特征：卵巢移植诱导雄性家蚕合成卵黄原蛋白

杨从文[#] 林 英[#] 沈关望 陈恩祥 王艳霞
罗 娟 张海燕 邢润苗 夏庆友[*]

摘要：雄性的雌性特征在脊椎动物中普遍存在，但在昆虫中并未见报道。昆虫中卵黄原蛋白(Vg)在雌性脂肪体中大量表达，且是卵中卵黄蛋白颗粒形成的必需物质。Vg的上调表达一般被认为是雌性卵生动物的雌性特征。本研究发现家蚕Vg(*Bm*Vg)在雌蛹中大量表达，RNAi下调雌蛹中的*BmVg*基因发现卵的发育及胚胎的形成受阻，表明*Bm*Vg在这些过程中的重要性。因此，可将*Bm*Vg作为家蚕雌性特征的生物标记物。HE染色和免疫组化结果显示，移植卵巢的雄蚕，蛹期脂肪体被诱导合成了Vg，且能产生少量卵黄颗粒残缺的卵，但雄蚕体内被诱导合成的Vg不足以支撑卵完成整个胚胎发育过程。在体内和体外实验中分别用雌激素处理雄蚕蛹和其脂肪体，发现*Bm*Vg的表达上调。本研究结果证实卵巢移植和17β雌激素处理能诱导雄性家蚕产生先天的雌性特征。然而，家蚕中这种雄性的雌性特征并不如雌性中的完整。

家蚕基因组生物学国家重点实验室，西南大学，重庆

Vertebrate estrogen regulates the development of female characteristics in silkworm, *Bombyx mori*

Shen GW Lin Y Yang CW Xing RM Zhang HY
Chen EX Han CS Liu HL Zhang WW Xia QY*

Abstract: The vertebrate estrogens include 17-β-estradiol (E2), which has an analog in silkworm ovaries. In this study, the *Bombyx mori* vitellogenin gene (*BmVg*) was used as a biomarker to analyze the function of the E2 in silkworm. In most oviparous animals, *Vg* has female-specific expression. However, *BmVg* expression was also detected in *B. mori* males. Stage specific fluctuation of *BmVg* expression was similar in males and females, but expression levels in males were lower than in females. E2 treatment by injection or feeding of male larvae in the final instar stage induced and stimulated male *BmVg* transcription and protein synthesis. When silkworm ovary primordia were transplanted into males, *BmVg* was induced in male fat bodies. Transplanted ovaries primordia were also able to develop into ovaries and produce mature eggs. When females were treated with E2 promoted *Bm*Vg/*Bm*Vn protein accumulation in hemolymph, ovaries and eggs. However, *BmVg* transcription was decreased in female fat bodies. An E2 analog was identified in the hemolymph of day 3 wandering silkworms using high-performance liquid chromatography. Estradiol titers from fifth late-instar larvae to pupal stage were determined by enzyme-linked immunosorbent assay. The results suggested that silkworms synthesized a vertebrate E2 analog. This study found that E2 promoted the synthesis of *Bm*Vg, a female typical protein in silkworms.

Published On: General and Comparative Endocrinology, 2015, 210, 30-37.

State Key Laboratory of Silkworm Genome Biology, Southwest University, Chongqing 400716, China.
*Corresponding author E-mail: xiaqy@swu.edu.cn.

脊椎动物雌激素调控家蚕雌性特征的发生

沈关望　林　英　杨从文　邢润苗　张海燕
陈恩祥　韩超珊　刘红玲　张薇薇　夏庆友*

摘要：脊椎动物雌激素包括17-β-雌二醇（E2），家蚕卵巢中具有其类似物。在本研究中，将家蚕卵黄原蛋白基因（*BmVg*）用作生物标志物来分析家蚕E2的功能。在大多数卵生动物中，*Vg*具有雌性特异性表达特征。然而，*BmVg*表达也在雄性家蚕中检测到。*BmVg*表达的时期特异性波动在雄性和雌性中是相似的，但雄性的表达水平低于雌性。对末龄的雄性幼虫注射或添食E2能够诱导和刺激雄性*BmVg*转录和蛋白质合成。将卵巢移植到雄性中时，*BmVg*在雄性脂肪体中被诱导，且移植的卵巢还能够发育并产生成熟的卵。用E2处理雌性家蚕则能促进*Bm*Vg/*Bm*Vn蛋白在血淋巴、卵巢、卵中积聚。然而，*BmVg*转录在雌性脂肪体中减少。使用高效液相色谱法在上蔟第3天家蚕血淋巴中鉴定到E2类似物。通过酶联免疫吸附测定法从第五龄晚期幼虫到蛹期测定到了雌二醇滴度。以上结果表明，家蚕合成了一种脊椎动物E2类似物。本研究发现E2促进了家蚕中雌性特征性蛋白（*Bm*Vg）的合成。

家蚕基因组生物学国家重点实验室，西南大学，重庆

Proteomics analysis of adult testis from *Bombyx mori*

Zhang Y[1] Dong ZM[1] Gu PM[2] Zhang WW[1]
Wang DD[1] Guo XM[1] Zhao P[1*] Xia QY[1]

Abstract: The development of the testis involves a large number of tissue-specific proteins, possibly because the sperms in it are the most divergent of all cell types. In this study, LC-MS/MS was employed to investigate the protein compositions of the adult testis of silkworm. A total of 14 431 peptides were identified in the adult testis of *Bombyx mori*, which were matched to 2 292 proteins. Thirty-two HSPs constitute a group of most abundant proteins in the adult testis, suggesting that they are critical for the development, differentiation, and survival of germ cells. Other proteins in this analysis were also involved in testis-specific processes mainly including sperm motility, meiosis, germ cell development, and spermatogenesis.

Published On: Proteomics, 2014, 14(20), 2345-2349.

1 State Key Laboratory of Silkworm Genome Biology, Southwest University, Chongqing 400716, China.
2 Thermo Fisher, Shanghai 201206, China.
*Corresponding author E-mail: zhaop@swu.edu.cn.

家蚕成虫精巢蛋白组学分析

张　艳[1]　董照明[1]　顾培明[2]　张薇薇[1]
王丹丹[1]　郭晓朦[1]　赵　萍[1*]　夏庆友[1]

摘要：精巢的发育涉及大量的组织特异性蛋白质，可能是因为其中的精子是所有细胞类型中分化最多的。在本研究中，利用LC-MS / MS技术研究家蚕成虫精巢的蛋白质组成。在家蚕的成虫精巢中共鉴定到14 431个肽段，与2 292个蛋白质匹配。蚕蛾精巢中32个HSP构成一组最丰富的蛋白质，表明它们对生殖细胞的发育、分化和存活至关重要。该分析中的其他蛋白质也参与精巢特异性过程，主要包括精子活动、减数分裂、生殖细胞发育和精子发育。

1　家蚕基因组生物学国家重点实验室，西南大学，重庆
2　赛默飞，上海

The *nanos* gene of *Bombyx mori* and its expression patterns in developmental embryos and larvae tissues

Zhao GL[1] Chen KP[1*] Yao Q[1] Wang WH[2] Wang Y[3]
Mu RH[4] Chen HQ[1] Yang HJ[1] Zhou HL[1]

Abstract: The *nanos* gene encodes a zinc-finger protein which is required for the migration and differentiation of primordial germ cells as well as for their fate maintenance. In this study, a 1 913 bp *nanos* gene was cloned and characterized in silkworm (*Bombyx mori*). RT-PCR and Western blot analysis showed that the *nanos* was expressed in developing embryos and various silkworm larval tissues. The expression patterns of *Nanos* and *Vasa* in silkworm larval gonads were analyzed using immunohistochemistry. It was found that, in silkworm larval ovaries, the Nanos and Vasa proteins were expressed in oocytes. While in testes, high expression of Nanos and Vasa was detected in spermatogonia and relatively weaker expression was found in spermatocytes at latter stages.

Published On: Gene Expression Patterns, 2008, 8(4), 254-260.

1 Institute of Life Sciences, Jiangsu University, Zhenjiang 212013, China.
2 College of Life Science, Henan Normal University, Xinxiang 453007, China.
3 Department of Biotechnology, Faculty of Food and Biological Engineering, Jiangsu University, Zhenjiang 212013, China.
4 School of Medicine, Beihua University, Jilin 132013, China.
*Corresponding author E-mail:kpchen@ujs.edu.cn.

家蚕的*nanos*基因及其在胚胎和幼虫组织中的表达模式分析

赵国力[1] 陈克平[1*] 姚 勤[1] 王伟华[2] 王 勇[3]
母润红[4] 陈慧卿[1] 杨华军[1] 周红亮[1]

摘要：*nanos*基因编码锌指蛋白，其是原始生殖细胞迁移、分化以及命运维系所必需的。在本研究中，我们克隆了一个家蚕(*Bombyx mori*)*nanos*基因，长1 913 bp。RT-PCR和蛋白质免疫印迹分析显示*nanos*在发育胚胎和各种幼虫组织中都表达。利用免疫组织化学分析了家蚕幼虫性腺中Nanos和Vasa的表达模式。结果发现，在家蚕幼虫卵巢中，Nanos和Vasa蛋白在卵母细胞中表达。在精巢中，Nanos和Vasa主要在精原细胞中高表达，在发育后期的精母细胞中表达量相对较低。

1 生命科学研究院，江苏大学，镇江
2 生命科学研究院，河南师范大学，新乡
3 生物技术系，食品与生物工程学院，江苏大学，镇江
4 医学院，北华大学，吉林

*Bm*Hrp28 is a RNA-binding protein that binds to the female-specific exon 4 of *Bombyx mori dsx* pre-mRNA

Wang ZL Zhao M Li D Zha XF Xia QY Xiang ZH He NJ*

Abstract: The *Bombyx mori* sex determination gene *Bmdsx* is alternatively spliced in the male and female to produce the male- and female-specific proteins. In an effort to better understand the mechanism of the alternative splicing regulation of *Bmdsx*, we conducted agel-shift assay followed by LC-MS/MS analysis toidentify the putative proteins bound to the cis-element CE1+6 in the exon 4 of *Bmdsx*. A protein named as *Bm*Hrp28 is homologous to the *Drosophila* Hrp48, a member of the hnRNPA/B family, was identified and expressed in *Escherichia coli* for testing RNA-protein binding *in vitro*. All of the results showed that *Bm*Hrp28 specifically bound to the CE1+6 RNA probe. *Bm*Hrp28 has two RNA recognition motifs atthe N-terminal and a glycine-rich motif at the C-terminal.It might be one of the factors involved in the male-specific splicing of *Bmdsx*.

Published On: Insect Molecular Biology, 2009, 18(6), 795-803.

State Key Laboratory of Silkworm Genome Biology, Southwest University, Chongqing 400716, China.
*Corresponding author E-mail: hejia@swu.edu.cn.

*Bm*Hrp28是一个能与家蚕雌特异基因*dsx* mRNA前体第4外显子结合的RNA结合蛋白

王子龙　赵　敏　李　东　查幸福　夏庆友　向仲怀　何宁佳*

摘要：雌蛹和雄蛹性别决定基因*Bmdsx*可选择地剪接，产生雄性和雌性特异性蛋白质。为了更好地了解*Bmdsx*的选择性剪接机制，我们进行了凝胶阻滞实验，然后进行LC-MS / MS分析，鉴定到与*Bmdsx*的第4外显子中的顺式元件CE1+6结合的蛋白质。利用体外重组表达的方法，我们在大肠杆菌中表达了*Bm*Hrp28，该蛋白质与果蝇Hrp48同源，属于hnRNPA / B家族。在体外，我们检测了重组蛋白与RNA的结合。所有结果表明*Bm*Hrp28特异性结合CE1+6 RNA探针。*Bm*Hrp28的N-末端具有两个RNA识别基序，C-末端具有富含甘氨酸的基序，这可能是*Bmdsx*雄性特异性剪接的原因之一。

家蚕基因组生物学国家重点实验室，西南大学，重庆

New insights into the genomic organization and splicing of the *Doublesex* gene, a terminal regulator of sexual differentiation in the silkworm *Bombyx mori*

Duan JP[1,2#] Xu HF[1#] Guo HZ[1] David A. O'Brochta[3] Wang F[1]
Ma SY[1] Zhang LY[1] Zha XF[1] Xiang ZH[1] Xia QY[1*]

Abstract: Sex-determination mechanisms differ among organisms. The primary mechanism is diverse, whereas the terminal regulator is relatively-conserved. We analyzed the transcripts of the *Bombyx mori doublesex gene* (*Bmdsx*), and reported novel results concerning the genomic organization and expression of *Bmdsx*. *Bmdsx* consists of nine exons and eight introns, of which two exons are novel and have not been reported previously. *Bmdsx* transcripts are spliced to generate seventeen alternatively-spliced forms and eleven putative trans-spliced variants. Thirteen of the alternatively-spliced forms and five of the putative trans-spliced forms are reported here for the first time. Sequence analysis predicts that ten female-specific, six male-specific splice forms and one splice form found in males and females will result in four female-specific, two male-specific Dsx proteins and one Dsx protein common to males and females. The Dsx proteins are expected to be functional and regulate downstream target genes. Some of the predicted Dsx proteins are described here for the first time. Therefore, the expression of the *dsx* gene in *B. mori* results in a variety of cis- and trans-spliced transcripts and multiple Dsx proteins. These findings show that in *B. mori* there is a complicated pattern of *dsx* splicing, and that the regulation of splicing and sex-specific functions of lepidopteran *dsx* have evolved complexity.

Published On: PLoS ONE, 2013, 8(11).

1 State Key Laboratory of Silkworm Genome Biology, Southwest University, Chongqing 400716, China.
2 Henan Provincial Key Laboratory of Funiu Mountain Insect Biology, Nanyang Normal University, Nanyang 473061, China.
3 Department of Entomology, University of Maryland, College Park MD 20742, USA.
#These authors contributed equally.
*Corresponding author E-mail: xiaqy@swu.edu.cn.

深入分析家蚕性别决定终端调控基因*Doublesex*的基因组结构和剪接形式

段建平[1,2#]　徐汉福[1#]　郭慧珍[1]　David A. O′Brochta[3]　王　峰[1]
马三垣[1]　张李颖[1]　查幸福[1]　向仲怀[1]　夏庆友[1*]

摘要：不同生物性别决定机制各不相同。在性别决定中，基本机制多种多样，但末端调控机制相对保守。我们系统分析了家蚕*Bmdsx*的转录，对*Bmdsx*的基因结构和表达有了新的认识。*Bmdsx*由9个外显子和8个内含子组成，其中2个外显子为首次报道。*Bmdsx*转录本经剪接可产生17个选择性剪接体和11个可能的反式剪接体，其中13种选择性剪接体和5种可能的反式剪接体为首次报道。序列分析结果显示，发现的10个雌特异、6个雄特异和1个雌雄共有剪接体可翻译合成4个雌特异、2个雄特异和1个雌雄共有Dsx蛋白，且Dsx蛋白都具有调控下游靶基因表达的能力。一些潜在的Dsx蛋白也是首次报道。因此，*Bmdsx*的表达可产生多种顺式和反式剪接体，并翻译成多种Dsx蛋白。这些研究结果暗示，家蚕*Bmdsx*剪接形式复杂，且鳞翅目昆虫*dsx*的剪接和性别特异性调控方式也很复杂。

1　家蚕基因组生物学国家重点实验室，西南大学，重庆
2　河南省伏牛山昆虫生物学重点实验室，南阳师范大学，南阳
3　昆虫系，马里兰大学帕克分校，马里兰，美国

Novel female-specific *trans*-spliced and alternative splice forms of *dsx* in the silkworm *Bombyx mori*

Duan JP Xu HF Wang F Ma SY Zha XF
Guo HZ Zhao P Xia QY*

Abstract: The *Bombyx mori doublesex gene* (*Bmdsx*) plays an important role in somatic sexual development. Its pre-mRNA splices in a sex-specific manner to generate two female-specific and one male-specific splice forms. The present study investigated six novel *dsx* variants generated by *trans*-splicing between female *dsx* transcripts and two additional novel genes, *dsr1* and *dsr2*. Expression analysis indicated that *Bmdsx-dsr1* represented splicing noise, whereas *dsr2*, which *trans*-spliced with dsx to generate five variants, regulated the expression of the female-specific *B. mori dsx* transcript *Bmdsx*^{F}s. We unexpectedly found a novel exon *2n* insertion during *Bmdsx* transcription, which did not influence the validity of the novel protein, *Bm*DSX^{F3}. Ectopic expression of *Bm*DSX^{F3} repressed the pheromone-binding protein gene and the testis-specific gene *A2* in males, and activated of the storage protein 1 gene. Our findings suggest that *trans*-splicing is a novel regulatory function of *Bmdsx*, which participates in female sexual development by regulating the expression of three *Bm*DSX^{F3} proteins..

Published On: Biochemical and Biophysical Research Communications, 2013, 431(3), 630-635.

State Key Laboratory of Silkworm Genome Biology, Southwest University, Chongqing 400716, China.
*Corresponding author E-mail: xiaqy@swu.edu.cn.

家蚕双性基因*dsx*全新的雌特异反式剪接体和选择性剪接体的鉴定

段建平 徐汉福 王 峰 马三垣
查幸福 郭慧珍 赵 萍 夏庆友*

摘要：家蚕双性基因*doublesex*(*Bmdsx*)在体细胞性别发育中具有重要的作用。其mRNA前体以性别特异性方式剪接产生2种雌特异性和1种雄特异性剪接体。本文分析了在雌性dsx转录本和2个新基因drs1和drs2之间通过反式剪接产生的6个新dsx。表达分析显示，*Bmdsx-dsr1*为剪接噪音，而*dsr2*与*dsx*反式剪接产生的5个反式剪接体可调节家蚕雌特异转录本*Bmdsx*^{F}*s*的表达。发现新外显子*2n*的插入，并不影响蛋白*Bm*DSX^{F3}的调控功能。*Bm*DSX^{F3}在雄蚕中的异位表达，可抑制信息素结合蛋白基因和精巢特异表达基因*A2*的表达，并激活储存蛋白1基因的表达。研究结果暗示，反式剪接是*Bmdsx*的一种全新的剪接方式，通过调节3种*Bm*DSX^{F3}蛋白的表达参与雌蚕的性别发育。

家蚕基因组生物学国家重点实验室，西南大学，重庆

Proteome analysis of silkworm, *Bombyx mori*, larval gonads: characterization of proteins involved in sexual dimorphism and gametogenesis

Chen JE[1,2] Li JY[1,3] You ZY[1] Liu LL[4] Liang JS[5]
Ma YY[6] Chen M[4] Zhang HR[6] Jiang ZD[1] Zhong BX[1*]

Abstract: Sexual dimorphism is initialed by the components of the sex determination pathway and is most evident in gonads and germ cells. Although striking dimorphic expressions have been detected at the transcriptional level between the silkworm larval testis and the ovary, the sex-dimorphic expressions at the protein level have not yet been well characterized. The proteome of silkworm larval gonads was investigated using a shotgun-based identification. A total of 286 and 205 non-redundant proteins were identified from the silkworm testis and ovary, respectively, with a false discovery rate (FDR) lower than 1%. Only 40 and 16 proteins were previously identified, and 246 and 189 proteins were newly identified in the silkworm testis and the ovary, respectively. The gametogenesis mechanism of silkworm was demonstrated using the protein expression profile and bioinformatics analysis. Cellular retinoic acid binding protein (CRABP) showed to be highly abundant in testis, while tubulins were abundant in ovary. Several homologies of *Drosophila* essential proteins for gametogenesis were identified in silkworm, such as male meiotic arrest gene product ALY and VISMAY in testis, and maternal mRNA localization protein exuperantia and SQUID in ovary. The gene ontology (GO) annotation and pathway analysis provide system-level insights into the sexual dimorphism and gametogenesis.

Published On: Journal of Proteome Research, 2013, 12(6), 2422-2438.

1 College of Animal Sciences, Zhejiang University, Hangzhou 310058, China.
2 Institute of Sericultural Research, Zhejiang Academy of Agricultural Sciences, Hangzhou 310021, China.
3 Institute of Developmental and Regenerative Biology, Hangzhou Normal University, Hangzhou 310036, China.
4 College of Life Sciences, Zhejiang University, Hangzhou 310058, China.
5 College of Environmental and Resource Sciences, Zhejiang University, Hangzhou 310058, China.
6 Zhejiang California International NanoSystems Institute (ZCNI), Zhejiang University, Hangzhou 310029, China.
*Corresponding author E-mail: bxzhong@zju.edu.cn

家蚕幼虫生殖腺蛋白质组分析：对性别二态性和配子形成相关蛋白的鉴定

陈金娥[1,2] 李建营[1,3] 尤征英[1] 刘丽丽[4] 梁建设[5]
马莹莹[6] 陈 铭[4] 张华蓉[6] 蒋振东[1] 钟伯雄[1*]

摘要：性别二态性由性别决定通路元件激活，在性腺和生殖细胞中表现最明显。尽管转录水平研究已在家蚕幼虫精巢和卵巢间检测到明显的二态性表达，然而蛋白质水平的鉴定和研究还未有报道。本研究利用鸟枪法对家蚕幼虫生殖腺蛋白质组进行鉴定，从家蚕精巢和卵巢中分别鉴定到286个和205个非冗余蛋白，假阳性率低于1%。精巢和卵巢中分别只有40个和16个蛋白质是以前鉴定过的，分别有246个和189个蛋白质是本研究新鉴定的。通过蛋白质表达谱和生物信息学分析方法对家蚕配子形成机制进行了阐述。发现胞内维甲酸结合蛋白（CRABP）在精巢中非常高丰度地表达，微管蛋白在卵巢中高丰度表达。一些果蝇配子形成关键蛋白的同源蛋白在家蚕中也有鉴定到，包括在家蚕精巢中鉴定的雄性减数分裂阻滞基因蛋白ALY和VISMAY，家蚕卵巢中的母源mRNA定位蛋白exuperantia和SQUID。GO注释和通路分析为系统理解性别二态性和配子形成提供了依据。

1 动物科学学院，浙江大学，杭州
2 蚕桑研究所，浙江省农业科学院，杭州
3 发育与再生生物学研究所，杭州师范大学，杭州
4 生命科学学院，浙江大学，杭州
5 环境与资源学院，浙江大学，杭州
6 浙江加州国际纳米技术研究院，浙江大学，杭州

Analysis of interaction between *Bm*hrp28 and *Bm*PSI in sex-specific splicing of *Bombyx mori Bmdsx* gene

Zha XF# Zhao M# Zhou CY Guo HZ
Zhao P Xiang ZH Xia QY*

Abstract: *Bombyx mori Bm*HRP28 and *Bm*PSI, which belong to the family of RNA-binding proteins, have been identified binding to the female-specific exon 4 of the sex-determining gene *Bmdsx* pre-mRNA. However, the relationships between *Bm*HRP28 and *Bm*PSI still remain unclear. In this study, we carried out yeast two-hybrid (Y2H) and co-immunoprecipitation (Co-IP) analyses to address them. Y2H analysis showed that there was little or no direct binding between the *Bm*HRP28 and *Bm*PSI proteins. Also, the Co-IP experiments revealed that *Bm*HRP28 and *Bm*PSI coexisted in a multiprotein complex. Our results suggested that *Bm*HRP28 and *Bm*PSI form a muliprotein complex to regulate the splicing of *Bmdsx* pre-mRNA, but are not directly bound to each other. In an effort to find other regulatory factors in the multiprotein complex, we constructed a silkworm Y2H cDNA library of male early embryo. By Y2H screening, we identified an RNA-binding protein *Bm*SPX, a putative component of the spliceosome, binding to *Bm*PSI. These results indicated that *Bm*HRP28 and *Bm*PSI make up a spliceosome complex to regulate *Bmdsx* splicing and that *Bm*SPX is another potential protein involved in this process. Our study provides some clues to better understand the mechanism of sex determination in the silkworm.

Published On: Genetics and Molecular Research, 2014. 13(3), 5452-5462.

State Key Laboratory of Silkworm Genome Biology, Southwest University, Chongqing 400716, China.
#These authors contributed equally.
*Corresponding author E-mail: xiaqy@swu.edu.cn.

*Bm*hrp28 和 *Bm*PSI 在家蚕 *Bmdsx* 基因的性别特异性剪接中的相互作用分析

查幸福[#] 赵 敏[#] 周春燕 郭慧珍
赵 萍 向仲怀 夏庆友[*]

摘要：*Bm*HRP28 和 *Bm*PSI 属于 RNA 结合蛋白家族，它们已被鉴定为与性别决定基因 *Bmdsx* 前体 mRNA 上的雌性特异性外显子 4 结合。然而，*Bm*HRP28 和 *Bm*PSI 之间的关系仍然不清楚。在本研究中，我们采用酵母双杂交（Y2H）和免疫共沉淀（Co-IP）实验来验证两者关系。Y2H 分析显示 *Bm*HRP28 和 *Bm*PSI 蛋白之间几乎没有结合或没有直接结合。此外，Co-IP 实验显示 *Bm*HRP28 和 *Bm*PSI 共存于多蛋白复合物中。我们的研究结果表明，*Bm*HRP28 和 *Bm*PSI 形成一个多蛋白复合物，以调节 *Bmdsx* 前体 mRNA 的剪接，但它们不直接相互结合。为了寻找多蛋白复合物中的其他调节因子，我们构建了一个雄性家蚕早期胚胎的 Y2H cDNA 文库。通过 Y2H 筛查，确定了一种 RNA 结合蛋白 *Bm*SPX，它是一种剪接体的假定成分，能够与 *Bm*PSI 结合。这些结果表明，*Bm*HRP28 和 *Bm*PSI 组成剪接体复合物以调节 *Bmdsx* 剪接，而 *Bm*SPX 是参与此过程的另一种潜在蛋白质。本研究为更好地了解家蚕性别决定的机制提供了一些线索。

家蚕基因组生物学国家重点实验室，西南大学，重庆

Ectopic expression of the male *BmDSX* affects formation of the chitin plate in female *Bombyx mori*

Duan JP[1,2] Xu HF[1] Ma SY[1] Guo HZ[1] Wang F[1]
Zhang LY[1] Zha XF[1] Zhao P[1] Xia QY[1*]

Abstract: Mating structures are involved in successful copulation, intromission, and / or insemination. These structures enable tight coupling between external genitalia of two sexes. During *Bombyx mori* copulation, the double harpagones in the external genitalia of males clasp the female chitin plate, which is derived from the larval eighth abdominal segment; abnormal development of the female chitin plate affects copulation. We report that ERK phosphorylation (p-ERK) and expression of *Abdominal-B* (*Abd-B*) in the posterior abdomen of the female adult is lower than in the male. Ectopic expression of the male-specific spliced form of *B. mori doublesex* (*BmdsxM*) in females, however, up-regulates *Abd-B* and *spitz* (*spi*) expression, increasing EGFR signaling activity, and thus forming an abnormal chitin plate and reduced female copulation. These findings indicate that *Bmdsx* affects the development of the eighth abdominal segment by regulating the activity of EGFR signaling and the expression of *Abd-B*, resulting in an extra eighth abdominal segment (A8) in males versus the loss of this segment in adult females.

Published On: Molecular Reproduction and Development, 2014, 81(3), 240-247.

1 State Key Laboratory of Silkworm Genome Biology, Southwest University, Chongqing 400716, China.
2 China-UK Nanyang Normal University-Rothamsted Research Joint Laboratory of Insect Biology, Henan Provincial Key Laboratory of Funiu Mountain Insect Biology, Nanyang Normal University, Nanyang 473061, China.
*Corresponding author E-mail: xiaqy@swu.edu.cn.

雄蚕*BmDSX*基因异常表达影响雌蚕几丁质板形成

段建平[1,2] 徐汉福[1] 马三垣[1] 郭慧珍[1] 王峰[1]
张李颖[1] 查幸福[1] 赵萍[1] 夏庆友[1*]

摘要：昆虫交配器官参与正常求偶、插入或授精等过程，交配中雌雄外部生殖器必须紧密结合。家蚕交配时，雄蚕外部生殖器上的一对抱器需紧握由雌蚕幼虫第8腹节分化而来的几丁质板。雌蚕几丁质板的异常发育会影响交配行为。本研究发现雌蛾后腹部ERK蛋白的磷酸化水平和*Abdominal-B*(*Abd-B*)的表达水平较雄蛾低。双性基因雄特异剪接体*BmdsxM*在雌蚕中异常表达可导致*Abd-B*和*spitz*(*spi*)的表达量上调，EGFR信号通路活性提高，进而形成异常的几丁质板，最终雌蛾不能交配。暗示*Bmdsx*可通过调节EGFR信号通路活性和*Abd-B*的表达来调控第8腹节的发育，使雌蛾较雄蛾少1个腹节。

1 家蚕基因组生物学国家重点实验室，西南大学，重庆
2 中英-南阳洛桑昆虫生物学联合实验室，河南省伏牛山昆虫生物学重点实验室，南阳师范学院，南阳

Genetic marking of sex using a W chromosome-linked transgene

Ma SY[#] Wang XG[#] Fei JT Liu YY

Duan JP Wang F Xu HF Zhao P Xia QY[*]

Abstract: Many species belonging to the order Lepidoptera are major pests in agriculture and arboriculture. The sterile insect technique (SIT) is an eco-friendly and highly efficient genetically targeted pest management approach. In many cases, it is preferable to release only sterile males in an SIT program, and efficient sexing strategies are crucial to the successful large-scale implementation of SIT. In the present study, we established 160 transgenic silkworm (*Bombyx mori*) lines to test the possibility of genetic sexing using a W chromosome-linked transgene, which is thought to be the best sexing strategy for lepidopteran species. One transgenic line with a female-specific expression pattern of reporter gene was obtained. The expression level of the W-linked transgene was comparable with autosomal insertions and was stable for 17 continuous generations. Molecular characterization showed this line contained a single copy of the reporter gene on the W chromosome, and the integration site was TTAG in contig W-BAC-522N19-C9. The feasibility of using a W chromosome-linked transgene demonstrated here and the possible improvements discussed will provide valuable information for other lepidopteran pests. The novel W chromosome-linked transgenic line established in this study will serve as an important resource for fundamental research with the silkworm *B. mori*.

Published On: Insect Biochemistry and Molecular Biology, 2013, 43(12), 1079-1086.

State Key Laboratory of Silkworm Genome Biology, Southwest University, Chongqing 400716, China.

[#]These authors contributed equally.

[*]Corresponding author E-mail: xiaqy@swu.edu.cn.

使用W染色体连锁转基因进行性别遗传标记

马三垣[#] 汪小刚[#] 费纪涛 刘园园
段建平 王 峰 徐汉福 赵 萍 夏庆友[*]

摘要：鳞翅目中的大部分物种都属于农业和树木栽培的主要害虫。昆虫不育技术(SIT)是一种环保高效的基因靶向害虫管理方法。在许多情况下，昆虫不育技术更偏好将不育的雄性释放，有效的性别策略对SIT大规模成功实施至关重要。在本研究中，我们建立了160个转基因家蚕（*Bombyx mori*）系，以检测使用W染色体连锁转基因进行遗传性育的可能性，这被认为是鳞翅目物种的最佳性别选择策略。我们获得了一个雌特异性表达报告基因的转基因品系。W连锁转基因的表达水平与常染色体插入转基因的表达水平相当，并且可以连续稳定地表达17个世代。分子鉴定显示，该品系在W染色体上的报告基因为单拷贝，插入位点在W-BAC-522N19-C9中的TTAG。本研究证明了使用W染色体连锁转基因的可行性，所提到的潜在改善也为其他鳞翅目害虫防治提供了有价值的信息。本研究建立的新型W染色体连锁转基因品系将是桑蚕基础研究的重要资源。

家蚕基因组生物学国家重点实验室，西南大学，重庆

Proteome analysis on lethal effect of l_2 in the sex-linked balanced lethal strains of silkworm, *Bombyx mori*

Chen JE[1,2]　Niu BL[2]　Wang YQ[2*]　Liu Y[2]
Liu PG[2]　Meng ZQ[2]　Zhong BX[1*]

Abstract: The sex-linked balanced lethal (SLBL) strains of silkworm serve as an effective system for sex-control in silkworm. To gain comprehensive insight into the effect of one *sex-linked balanced lethal* gene l_2, comparative proteomic analysis was carried out between the survival embryos ($W^{+l1}Z^{l1+l2}$) and lethal embryos ($W^{+l1}Z^{+l1l2}$) stage. The lethal stage of l_2 was confirmed by observing the typical dead embryo morphology. The two genotype embryos before lethal stage were distinguished using polymorphic simple sequence repeats (SSR) markers closely linked to l_2 on the sex chromosome. Finally, 11 differentially expressed protein spots were successfully identified by MALDI-TOF / TOF mass spectrometry (MS). Among them, only 1 protein identified as heat shock protein 20.4 (HSP20.4) was up-regulated in the lethal embryos, while the other 10 were down-regulated. The up-regulation of HSP20.4 suggests that there may be abnormal polypeptides produced in the lethal embryos. The gene ontology (GO) annotation indicated those down-regulated proteins are involved in important biological processes including embryo development, nucleoside metabolism, tRNA splicing, translation and protein folding. The biological pathway analysis showed that those down-regulated proteins are mainly involved in spindle assemblage and morphogenesis. Based on our results, we suggest that the l_2 may be the mutant expressing abnormal polypeptides. Its expression has a negative effect on mitosis and morphogenesis processes. The death of the embryos may be caused by the accumulation of abnormal polypeptides and the handicap of cell proliferation and morphogenesis.

Published On: Biotechnology and Bioprocess Engineering, 2012, 17(2), 298-308.

1　College of Animal Sciences, Zhejiang University, Hangzhou 310029, China.
2　State Key Laboratory Breeding Base for Zhejiang Sustainable Pest and Disease Control, Institute of Sericultural Research, Zhejiang Academy of Agricultural Sciences, Hangzhou 310021, China.
*Corresponding author E-mail: bxzhong@zju.edu.cn; wangyq@mail.zaas.ac.cn.

家蚕平衡致死系致死基因l_2致死效应的蛋白质组分析

陈金娥[1,2] 牛宝龙[2] 王永强[2*] 刘 岩[2]
刘培刚[2] 孟智启[2] 钟伯雄[1*]

摘要：性连锁平衡致死系是实现家蚕性别控制的有效方法。为了全面地了解性连锁平衡致死系致死基因l_2的致死效应，我们用蛋白质组学的方法比较分析了存活型基因($W^{+l1}Z^{l1+l2}$)胚胎和致死型基因($W^{+l1}Z^{+l1l2}$)胚胎在致死期的蛋白表达差异。致死时期我们用观察胚胎发育期的典型特征来确定。致死期前两种胚胎的基因型通过性染色体上与致死基因l_2紧密关联的SSR标记确定。最终通过MALDI-TOF/TOF鉴定到11个差异表达的蛋白质点。11个差异表达蛋白质中，只有一个蛋白质热激蛋白HSP20.4在致死型胚胎中表达上调，而其他10个蛋白质都下调。热激蛋白HSP20.4的表达上调暗示在致死胚胎中可能存在异常多肽。通过GO功能注释发现这些下调的蛋白质参与重要的生物学过程，包括胚胎发育、核苷酸代谢、tRNA剪切、翻译和蛋白质折叠。生物学通路分析揭示，这些下调蛋白质参与纺锤体装配和形态发生。基于我们的结果，我们推测l_2导致异常多肽的产生。这些异常多肽的表达对有丝分裂和形态发育过程产生负面影响。胚胎的死亡可能是异常多肽累积、细胞增殖和形态发生受阻的结果。

1 动物科学学院，浙江大学，杭州
2 浙江省植物有害生物防控重点实验室——省部共建国家重点实验室培育基地，蚕桑研究所，浙江省农业科学院，杭州

Proteome profiling reveals tissue-specfic protein expression in male and female accessory glands of the silkworm, *Bombyx mori*

Dong ZM Wang XH Zhang Y Zhang LP
Chen QM Zhang XL Zhao P* Xia QY

Abstract: Male accessory gland (MAG) and female accessory gland (FAG) of the reproductive system are, respectively, responsible for producing seminal proteins and adhesive proteins during copulation and ovulation. Seminal proteins are ejaculated to female along withsperms, whereas adhesive proteins are excreted along with eggs. Proteins from the male and female reproductive organs are usually indicative of rapid adaptive evolution. Understanding the reproductive isolation and species divergence requires identifying reproduction-related proteins from many different species. Here, we present our proteomic analyses of male and female accessory glands of the silkworm, *Bombyx mori*. Using LC/MS - MS, we identifed 2 133 MAG proteins and 1 872 FAG proteins. In total, 652 proteins were signifcant more abundant in the MAG than in the FAG, including growth factors, odorant-binding proteins, enzymes, and proteins of unknown function. Growth factors and odorant-binding proteins are potentialsignaling molecules, whereas most of proteins of unknown function were found to be Lepidoptera-specfic proteins with high evolutionary rates. Microarray experiments andsemi-quantitative RT-PCR validated that MAG-specfic proteins were expressed exclusively in male moths. Totally 192 proteins were considered as FAG-specfic proteins, including protease inhibitors, enzymes, and other proteins. Protease inhibitors were found to be the most abundant FAG-specfic proteins, which may protect eggs from infection by inhibiting pathogen-derived proteases. These results provide comprehensive insights into copulation and oviposition. Moreover, the newly identifed Lepidoptera-specfic MAG proteins provide useful data for future research on the evolution of reproductive proteins in insects.

Published On: Amino Acids, 2016, 48(5), 1173-1183.

State Key Laboratory of Silkworm Genome Biology, Southwest University, Chongqing 400716, China.
*Corresponding author E-mail: zhaop@swu.edu.cn.

家蚕蛹期雌、雄附腺特异表达蛋白的分析

董照明　王晓欢　张　艳　张利平
陈全梅　张晓璐　赵　萍*　夏庆友

摘要：雄性附腺（MAG）和雌性附腺（FAG）分别负责在交配和排卵期间产生精液蛋白和黏合蛋白。精液蛋白质与精液一起排出，而黏合蛋白与卵一起排出。精巢与卵巢蛋白通常表现为快速适应性进化。了解生殖分离和物种分歧需要鉴定许多不同物种的繁殖相关蛋白。我们利用LC / MS-MS，对家蚕的雄性和雌性附腺的蛋白质组学进行分析，共鉴定到2 133个MAG蛋白和1 872个FAG蛋白。分析发现652个蛋白在MAG中比在FAG中丰度高，包括生长因子、气味结合蛋白、酶和未知功能蛋白质。生长因子和气味结合蛋白是潜在的信号分子，而大多数未知功能的蛋白质被发现是具有高进化速率的鳞翅目特异性蛋白质。芯片数据和RT-PCR验证了MAG特异性蛋白质仅在雄蛾中表达。共有192个蛋白质在FAG中特异表达，包括蛋白酶抑制剂、酶和其他蛋白。蛋白酶抑制剂是最丰富的FAG特异性蛋白质，其可以通过抑制病原体来源的蛋白酶来保护蚕卵免受感染。这些结果使我们对交配和产卵有了更全面的了解。此外，新鉴定的鳞翅目MAG特异表达蛋白为未来对昆虫繁殖蛋白进化的研究提供了有用的数据。

家蚕基因组生物学国家重点实验室，西南大学，重庆

Dosage analysis of Z chromosome genes using microarray in silkworm, *Bombyx mori*

Zha XF[1] Xia QY[1,2*] Duan J[1,2]
Wang CY[1] He NJ[1] Xiang ZH[1]

Abstract: In many organisms, dosage compensation is needed to equalize sex-chromosome gene expression in males and females. Several genes on silkworm Z chromosome were previously detected to show a higher expression level in males and lacked dosage compensation. Whether silkworm lacks global dosage compensation still remains poorly known. Here, we analyzed male: female (M: F) ratios of expression of chromosome-wide Z-linked genes in the silkworm using microarray data. The expression levels of genes on Z chromosome in each tissue were significantly higher in males compared to females, which indicates no global dosage compensation in silkworm. Interestingly, we also found some genes with no bias (M∶F ratio: 0.8-1.2) on the Z chromosome. Comparison of male-biased (M∶F ratio more than 1.5) and unbiased genes indicated that the two sets of the genes have functional differences. Analysis of gene expression by sex showed that M:F ratios were, to some extent, associated with their expression levels. These results provide useful clues to further understanding roles of dosage of Z chromosome and some Z-linked sexual differences in silkworms.

Published On: Insect Biochemistry and Molecular Biology, 2009, 39(5-6), 315-321.

1 Key Sericultural Laboratory of Agricultural Ministry, College of Biotechology, Southwest University, Chongqing 400716, China.
2 The Institute of Agronomy and Life Sciences, Chongqing University, Chongqing 400030, China.
*Corresponding author E-mail: xiaqy@swu.edu.cn.

利用芯片技术对家蚕Z染色体基因进行剂量分析

查幸福[1] 夏庆友[1,2*] 段 军[1,2]
汪春云[1] 何宁佳[1] 向仲怀[1]

摘要：在许多生物体中，雄性和雌性都需要剂量补偿来平衡性染色体基因的表达。先前检测到的家蚕Z染色体上的几个基因在雄性家蚕中表达水平较高而且缺乏剂量补偿。然而家蚕是否缺乏全面性的剂量补偿依然不甚清楚。在这里，我们利用芯片数据分析家蚕染色体范围内的Z-连锁基因在雄性和雌性(M:F)家蚕中的表达比例。雄性家蚕各组织中Z染色体上基因的表达水平显著高于雌性家蚕，这表明家蚕无全面性的剂量补偿。有趣的是，我们还发现家蚕体内Z染色体上一些基因没有性别偏好(M:F比率:0.8—1.2)。雄性偏好(M:F比率大于1.5)基因和无偏好基因的比较表明两组基因具有功能差异。对性别差异表达基因分析结果表明，M:F比率在一定程度上与其表达水平相关。这些结果为进一步了解Z染色体的剂量补偿和一些Z染色体相关的性别差异提供了有用的线索。

1 农业部蚕桑学重点实验室，生物技术学院，西南大学，重庆
2 农学及生命科学研究院，重庆大学，重庆

Molecular characterization and functional analysis of serine/threonine protein phosphatase of *Toxocara canis*

Ma GX[1] Zhou RQ[1*] Hu SJ[1] Huang HC[1] Zhu T[1] Xia QY[2*]

Abstract: *Toxocara canis* (*T. canis*) is a widely prevalent zoonotic parasite that infects a wide range of mammalian hosts, including humans. We generated the full-length complementary DNA (cDNA) of the serine/threonine phosphatase gene of *T. canis* (*Tc stp*) using 5' rapid amplification of the cDNA ends. The 1 192 bp sequence contained a continuous 942 nucleotide open reading frame, encoding a 313 amino acid polypeptide. The *Tc*STP polypeptide shares a high level of amino-acid sequence identity with the predicted STPs of *Loa loa* (89%), *Brugia malayi* (86%), *Oesophagostomum columbianum* (76%), and *Oesophagostomum dentatum* (76%). The *Tc*STP contains GDXHG, GDXVDRG, GNHE motifs, which are characteristic of members of the phosphoprotein phosphatase family. Our quantitative real-time polymerase chain reaction analysis showed that the *Tc*STP was expressed in six different tissues in the adult male, with high-level expression in the spermary, vas deferens, and musculature, but was not expressed in the adult female, suggesting that *Tc*STP might be involved in spermatogenesis and mating behavior. Thus, STP might represent a potential molecular target for controlling *T. canis* reproduction.

Published On: Experimental Parasitology, 2014, 141, 55-61.

1 Department of Veterinary Medicine, Rongchang Campus, Southwest University, Chongqing 402460, China.

2 State Key Laboratory of Silkworm Genome Biology, Southwest University, Chongqing 400716, China.

*Corresponding author E-mail: rongqiongzhou@126.com; xiaqy@swu.edu.cn.

犬弓蛔虫丝氨酸/苏氨酸蛋白磷酸酶的分子特性和功能分析

马光旭[1] 周荣琼[1*] 胡世军[1] 黄汉成[1] 朱 涛[1] 夏庆友[2*]

摘要：犬弓蛔虫是一种普遍的人畜共患的动物传染病寄生虫，它们能够感染不同种类的哺乳动物，包括人类。我们通过5′端快速扩增获得犬弓蛔虫丝氨酸/苏氨酸蛋白磷酸酶(*Tc*stp)的全长cDNA。该片段全长1 192 bp，包含一个942个核苷酸的开放阅读框，编码313个氨基酸残基。*Tc*STP的氨基酸序列同罗阿丝虫(89%)、马来丝虫(86%)、哥伦比亚节线虫(76%)、口线虫(76%)预测的STPs序列具有较高的相似性。*Tc*STP包含蛋白磷酸酶家族特征性的GDXHG、GDXVDRG和GNHE结构域。实时定量聚合酶链式反应(qRT-PCR)分析显示*Tc*STP在雄性成虫的6种组织中有表达，并且在精巢、输精管和肌肉组织中表达水平较高，但是，在雌性成虫中无表达，这表明*Tc*STP可能参与精子的形成及交配行为。因此，STP可能作为控制犬弓蛔虫繁殖的潜在分子靶标。

1 动物医学院，荣昌校区，西南大学，重庆
2 家蚕基因组生物学国家重点实验室，西南大学，重庆

第五章

蚕的遗传操作与基因编辑

Chapter 5

遗传操作是功能基因研究和品种素材创新的重要手段，尤其是转座子介导的转基因技术和最近发展起来的基因组编辑技术已经成为许多物种遗传改造的优选技术。作为重要的模式生物和经济昆虫，家蚕的遗传操作技术发展相对比较滞后，为此项目组开展了深入研究，建立和研究了基于转座子、重组酶、整合酶等一系列元件的家蚕转基因技术体系，突破了以锌指核酸酶（ZFN）、类转录激活效应因子核酸酶（TALEN）、成簇规律间隔的短回文重复序列（CRISPR）的家蚕基因组编辑技术。

建立和研究了以转座子为主的转基因技术。基于全基因组分析和*piggy*Bac转座子元件分析，系统地鉴定了家蚕中*piggy*Bac-like元件，分析了其结构特征；比较了*piggy*Bac元件不同结构域在家蚕细胞系（*BmN*）和胚胎中的活性，在此基础上建立了高效的家蚕转基因技术体系；进一步比较分析了*piggy*Bac在不同家蚕品系中的转基因效率；利用W染色体特异的转基因标记揭示了*piggy*Bac转座子在家蚕中的稳定性，并采用多重复末端组合的策略实现了整合位点的固定化；利用TALE蛋白与*piggy*Bac结合的方式提高了转基因整合的效率。此外，还建立了基于Cre/LoxP的家蚕基因表达激活、基于整合酶φC31的家蚕基因整合等遗传操作方法。

突破了家蚕基因组编辑技术。基因组编辑技术对家蚕基础理论研究和蚕桑产业的可持续发展都具有革命性和开拓性意义。项目组紧跟基因组编辑前沿技术的发展，相继建立了基于ZFN、TALEN和CRISPR的家蚕基因组编辑技术体系；首次在昆虫细胞中实现了多个基因的同时编辑，发现了如何制备和筛选获得没有明显表型的家蚕突变体；利用TALEN或CRISPR实现了可遗传的长片段基因组删除和长达8.9Mb的片段的删除、翻转和重复等基因组结构变异。

遗传操作技术促进家蚕功能研究和素材创制。随着遗传操作技术在家蚕中的逐步完善、发展和应用，家蚕功能基因研究的广度和深度都发生了显而易见的改变。比如利用Cas9敲除了DNA修复途径中NHEJ通路中关键作用酶*Bm*Ku70，并发现得到的突变体中同源重组的效率得到了一定的提升；利用ZFN敲除了*BmFib-H*基因，发现丝腺变小、变中空，茧层薄而脆，茧丝只含丝胶蛋白，同时也在*BmFib-H*基因的遗传特性、丝蛹平衡、蛋白分泌等方面有一些新的发现。

马三垣

Identification of a functional element in the promoter of the silkworm (*Bombyx mori*) fat body-specific gene *Bmlp3*

Xu HF[1] Deng DJ[1,2] Yuan L[1]
Wang YC[1] Wang F[1] Xia QY[1*]

Abstract: 30K proteins are a group of structurally related proteins that play important roles in the life cycle of the silkworm *Bombyx mori* and are largely synthesized and regulated in a time-dependent manner in the fat body. Little is known about the upstream regulatory elements associated with the genes encoding these proteins. In the present study, the promoter of *Bmlp3*, a fat body-specific gene encoding a 30K protein family member, was characterized by joining sequences containing the *Bmlp3* promoter with various amounts of 5' upstream sequences to a *luciferase* reporter gene. The results indicated that the sequences from -150 to -250 bp and -597 to -675 bp upstream of the *Bmlp3* transcription start site were necessary for high levels of luciferase activity. Further analysis showed that a 21 bp sequence located between - 230 to - 250 was specifically recognized by nuclear factors from silkworm fat bodies and *BmE* cells, and could enhance luciferase reporter-gene expression 2.8-fold in *BmE* cells. This study provides new insights into the *Bmlp3* promoter and contributes to the further clarification of the function and developmental regulation of *Bmlp3*.

Published On: Gene, 2014, 546(1), 129-134.

1 State Key Laboratory of Silkworm Genome Biology, Southwest University, Chongqing 400716, China.
2 The Institute of Forensic Science, Chongqing Public Security Bureau, Chongqing 401147, China.
*Corresponding author E-mail: xiaqy@swu.edu.cn.

家蚕脂肪体特异基因*Bmlp3*启动子功能元件的鉴定

徐汉福[1]　邓党军[1,2]　袁　林[1]
王元成[1]　王　峰[1]　夏庆友[1*]

摘要：30K蛋白是家蚕脂肪体在特定时期大量合成的一类结构蛋白，在家蚕的生命活动过程中扮演着重要角色，但目前有关30K蛋白基因上游调控元件的信息知之甚少。本研究利用荧光素酶报告基因系统，研究了30K蛋白家族基因*Bmlp3*启动子的序列元件特征。结果显示，*Bmlp3*转录起始位点上游-150 bp至 -250 bp和-597 bp至-675 bp区域的启动子序列对荧光素酶基因高量表达至关重要。进一步分析表明，位于-230 bp至-250 bp区域的一段21 bp的序列能够特异识别家蚕脂肪体和*BmE*细胞中的核因子；在*BmE*细胞中，该21 bp序列能够使荧光素酶基因的表达量提高2.8倍。本研究提供了*Bmlp3*启动子的新信息，有助于进一步阐释*Bmlp3*基因的功能与发育调控。

1　家蚕基因组生物学国家重点实验室，西南大学，重庆
2　物证鉴定所，重庆市公安局，重庆

Elementary research into the transformation *BmN* cells mediated by the *piggy*Bac transposon vector

Xue RY[#] Li X[#] Zhao Y Pan XL

Zhu XX Cao GL Gong CL[*]

Abstract: To generate stable transformants of *Bombyx mori* silkworm *BmN* cells continuously expressing a target gene from a *piggy*Bac-derived vector, *BmN* cells were transfected with a *piggy*Bac vector containing a neomycin-resistance gene, green fluorescent protein gene, and human insulin-like growth factor *I* gene (*hIGF-I*) and a helper plasmid containing the *piggy*Bac transposase sequence under the control of the *B. mori actin 3* (*A3*) promoter. With the antibiotic G418, we selected stably transformed *BmN* cells expressing hIGF-*I* from the *piggy*Bac-derived vector containing a neomycin-resistance gene driven by the *ie-1* promoter from the *B. mori* nucleopolyhedrovirus. However, no stably transformed *BmN* cells transformed with the *piggy*Bac element vector containing an SV40-promoter-driven neomycin-resistance gene were isolated. Determined with an enzyme-linked immunosorbent assay, the expression level of hIGF1 was about 7.8 ng in 5×10^5 cells in which the *hIGF-I* gene was driven by the *sericin-1* promoter, and 147.5 ng in 5×10^5 cells in which the *hIGF-I* was under control of *B. mori* fibroin heavy chain gene (*fib-H*) promoter with its downstream signal peptide sequence. Analysis of the chromosomal insertion site by inverse PCR showed that the exogenous DNA was inserted into the cell genome randomly or at a TTAA target sequence, characteristic of *piggy*Bac element transposition. These results are particularly important because *piggy*Bac has been suggested for use in the transgenesis of silkworm cells.

Published On: Journal of Biotechnology, 2009, 144(4), 272-278.

School of Medicine, Soochow University, Suzhou 215123, China.

[#]These authors contributed equally.

[*]Corresponding author E-mail: gongcl@suda.edu.cn.

*piggy*Bac转座子介导的家蚕*BmN*细胞转基因研究初探

薛仁宇[#] 李 曦[#] 赵 越 潘兴亮
诸戌娴 曹广力 贡成良[*]

摘要：为了利用*piggy*Bac衍生的载体筛选稳定表达靶基因的家蚕*BmN*细胞系，用包含新霉素抗性基因、绿色荧光蛋白基因、人胰岛素生长因子Ⅰ基因(*hIGF-1*)的*piggy*Bac的载体以及含有家蚕*actin 3*(*A3*)启动子控制的*piggy*Bac转座子序列的helper质粒转染*BmN*细胞。在G418抗生素的作用下，利用含有家蚕核型多角体病毒的*ie-1*启动子启动新霉素抗性基因的*piggy*Bac衍生载体筛选获得稳定表达hIGF-Ⅰ的细胞系。然而，转染SV40启动子启动抗性基因的*piggy*Bac载体并未筛选出稳定转化的细胞系。酶联免疫吸附实验发现，*sericin-1*启动子启动时，5×10^5个细胞中hIGF1表达量为7.8 ng；*fib-H*启动子和其下游信号肽序列启动时，5×10^5个细胞中hIGF1表达量为147.5 ng。通过反向PCR分析表明，在转化细胞中外源DNA可通过随机整合或按照*piggy*Bac特定的靶位点序列TTAA插入细胞基因组。这些结果说明*piggy*Bac可以被用来筛选家蚕转基因细胞系。

医学院，苏州大学，苏州

Comparison of transformation efficiency of *piggy*Bac transposon among three different silkworm *Bombyx mori* strains

Zhong BX[1*] Li JY[1] Chen JE[1,2] Ye J[1] Yu SD[1]

Abstract: The transformation rate of three different strains of silkworm *Bombyx mori* was compared after the introduction of enhanced green fluorescence protein (EGFP)-encoding genes into the silkworm eggs by microinjection of a mixture of *piggy*Bac vector and helper plasmid containing a transposase-encoding sequence. Although there were no significant differences among the three strains in the percentages of fertile moths in microinjected eggs (P=0.125 8), the percentages of G_0 transformed moths in fertile moths and injected eggs were both significantly different (P=0.013 68 and P=0.023 98, respectively). The transformation rate of the *Nistari* strain (Indian strain) was significantly higher than that of the other two strains, *Golden-yellow-cocoon* (Vietnamese strain) and *Jiaqiu* (Chinese strain), which had similar rate. These results indicate that the transformation efficiency of the *piggy*Bac-based system might vary with silkworm strains with different genetic backgrounds. The presence of endogenous *piggy*Bac-like elements might be an important factor influencing the transformation efficiency of introduced *piggy*Bac-derived vectors, and the diverse amount and activation in different silkworm strains might account for the significant differences.

Published On: Acta Biochimica et Biophysica Sinica, 2007, 39(2), 117-122.

1 College of Animal Sciences, Zhejiang University, Hangzhou 310029, China.

2 Institute of Sericultural Research, Zhejiang Academy of Agricultural Sciences, Hangzhou 310021, China.

*Corresponding author E-mail: bxzhong@zju.edu.cn.

三个不同家蚕品种中*piggy*Bac转座子转座效率的比较分析

钟伯雄[1*] 李建营[1] 陈金娥[1,2] 叶 健[1] 余颂东[1]

摘要：本研究采用显微注射技术，将编码增强型绿色荧光蛋白的*piggy*Bac转基因载体和编码转座酶的辅助质粒一起注射到卵中，比较了3个不同家蚕品种的转基因效率。虽然注射的卵产生的蛾子的百分比没有明显的差异（P=0.125 8），但是，G_0代转基因的蛾子和注射的卵之间具有显著的差异（分别为P=0.013 68和P=0.023 98）。*Nistari*（印度品种）转座效率明显地高于其他两个品种，即金黄茧（越南品种）和*Jiaqiu*（中国品种），在这两个品种中具有类似的效率。结果说明利用*piggy*Bac转座子系统获得的转基因效率在不同遗传背景的家蚕品种中存在差异。内源的*piggy*Bac类似转座元件的存在有可能是影响*piggy*Bac载体转座效率的一个重要因素，在不同的家蚕品种中的数量和活性有可能是造成这种差异的原因。

动物科学学院，浙江大学，杭州
蚕桑研究所，浙江省农业科学院，杭州

The relationship between internal domain sequences of *piggy*Bac and its transposition efficiency in *BmN* cells and *Bombyx mori*

Zhuang LF[1] Wei H[1] Lu CD[2] Zhong BX[1*]

Abstract: The *piggy*Bac transposon, which includes terminal inverted repeat sequences and internal domain (ID) sequences, is widely used as a tool for insect transformation. To optimize this system for transgenic research on *Bombyx mori*, we examined the effects of the amount of the transposase plasmid and its ID sequences on the expression of green fluorescent protein (GFP). Four kinds of transposon plasmids, pB [A3GFP]-1 with the full length of ID sequences, pB[A3GFP]-2 having only the 3' ID sequence, pB[A3GFP]-3 without ID sequences, and pB[A3GFP]-4 containing 333 bp of the 5' ID sequence, and 179 bp of the 3' ID sequence were constructed with *GFP* as the marker. After transfecting these four plasmids into *BmN* cells, we analyzed the transfecting efficiency by comparing the GFP positive to negative cell ratio. Our results indicated that plasmid pB-4 got the highest ratio on the 22nd day. Moreover, the GFP positive to negative cell ratio increased with higher amount of transposase plasmid without overproduction inhibition. Furthermore, we injected three *piggy*Bac transposon plasmids, pB[A3GFP]-1, pB[33P3GFP]-3, and pB[33P3RFP]-4 harboring different markers into preblastoderm stage eggs of *B. mori*, and found that the transformation efficiency of pB [33P3RFP] - 4 was 3.8 folds higher than pB[A3GFP] - 1, whereas pB[33P3GFP] - 3 failed to produce transformants. Our results suggested that pB-4 may be one of the best *piggy*Bac transposon plasmids currently available for germline transformation in *B. mori*.

Published On: Acta Biochimica et Biophysica Sinica, 2010, 42(6), 426-413.

1 College of Animal Sciences, Zhejiang University, Hangzhou 310029, China.

2 State Key Laboratory of Molecular Biology, Institute of Biochemistry and Cell Biology, Shanghai Institutes for Biological Sciences, Chinese Academy of Sciences, Shanghai 200032, China.

*Corresponding author E-mail: bxzhong@zju.edu.cn.

*piggy*Bac转座子内部功能域及其在*BmN*细胞和家蚕中的转座效率的相关分析

庄兰芳[1]　危　浩[1]　陆长德[2]　钟伯雄[1*]

摘要：包含末端反向重复序列和内部区域(ID)的*piggy*Bac转座子，被广泛用作昆虫转化的工具。为了优化这个系统以便更好地服务于家蚕的转基因研究，我们检测了转座酶质粒的量和其ID序列对GFP蛋白表达的影响。4种转座质粒被构建，它们分别是包含全长ID序列的pB[A3GFP]-1，包含3′ID序列的pB[A3GFP]-2，不包含ID序列的pB[A3GFP]-3，包含333 bp的5′ID序列和179 bp的3′ID序列的pB[A3GFP]-4，其中*GFP*作为标志基因。转染这4个载体到家蚕*BmN*细胞中，我们通过比较GFP阳性细胞与阴性细胞的比例来分析转染效率。结果表明pB-4质粒在第22天获得了最高的效率。此外，随着转座酶质粒的增加，GFP阳性细胞相对于阴性细胞增加，没有出现过量的抑制现象。此外，我们注射了3个包含不同标记基因的pB[A3GFP]-1、pB[33P3GFP]-3和pB[33P3RFP]-4载体到家蚕胚胎中，发现pB[33P3RFP]-4的转基因效率比pB[A3GFP]-1高3.8倍，而pB[33P3GFP]-3没能获得转基因家蚕。结果表明pB-4有可能是目前最好的适合于家蚕的转基因载体。

1　动物科学学院，浙江大学，杭州

2　分子生物学国家重点实验室，生物化学与细胞生物学研究所，上海生命科学研究院，中国科学院，上海

Postintegration stability of the silkworm *piggy*Bac transposon

Jiang L[1,2#] Sun Q[1,2#] Liu WQ[1,2] Guo HZ[1,2] Peng ZW[1,2]
Dang YH[1,2] Huang CL[1,2] Zhao P[1,2] Xia QY[1,2*]

Abstract: The *piggy*Bac transposon is the most widely used vector for generating transgenic silkworms. The silkworm genome contains multiple *piggy*Bac-like sequences that might influence the genetic stability of transgenic lines. To investigate the postintegration stability of *piggy*Bac in silkworms, we used random insertion of the *piggy*Bac[3×p3 EGFP afm] vector to generate a W chromosome-linked transgenic silkworm, named *W-T*. Results of Southern blot and inverse PCR revealed the insertion of a single copy in the W chromosome of *W-T* at a standard TTAA insertion site. Investigation of 11 successive generations showed that all *W-T* females were EGFP positive and all males were EGFP negative; PCR revealed that the insertion site was unchanged in *W-T* offspring. These results suggested that endogenous *piggy*Bac-like elements did not affect the stability of *piggy*Bac inserted into the silkworm genome.

Published On: Insect Biochemistry and Molecular Biology, 2014, 50, 18-23.

1 State Key Laboratory of Silkworm Genome Biology, Southwest University, Chongqing 400716, China.
2 College of Biotechnology, Southwest University, Chongqing 400716, China.
#These authors contributed equally.
*Corresponding author E-mail: xiaqy@swu.edu.cn.

*piggy*Bac转座子整合进家蚕基因组后的稳定性分析

蒋　亮[1,2#]　孙　强[1,2#]　刘纬强[1,2]　郭慧珍[1,2]　彭正文[1,2]
党颖慧[1,2]　黄春林[1,2]　赵　萍[1,2]　夏庆友[1,2*]

摘要：在制备转基因家蚕时，*piggy*Bac转座子是最广泛使用的载体。家蚕基因组包含多个*piggy*Bac-like序列，它们可能会影响转基因家蚕的遗传稳定性。为了调查*piggy*Bac整合进家蚕基因组后的稳定性，我们利用*piggy*Bac[3×p3 EGFP afm]载体的随机插入特性制备了一个雌特异的转基因家蚕品系*W-T*。Southern blot和反向PCR结果显示外源片段以单拷贝方式插入W染色体，插入位点为标准的TTAA。连续11代的遗传调查结果发现*W-T*的所有雌个体都发出绿色荧光而所有雄个体都没有绿色荧光，PCR结果显示，*W-T*后代的插入位点没有发生改变。这些结果表明，内源的*piggy*Bac-like元件不会影响*piggy*Bac整合进家蚕基因组后的稳定性。

1　家蚕基因组生物学国家重点实验室，西南大学，重庆
2　生物技术学院，西南大学，重庆

Remobilizing deleted *piggy*Bac vector post-integration for transgene stability in silkworm

Wang F[1,2] Wang RY[1] Wang YC[1,2] Xu HF[1] Yuan L[1]
Ding H[1] Ma SY[1] Zhou Y[2] Zhao P[1] Xia QY[1,2*]

Abstract: Deletion of transposable elements post-genomic integration holds great promise for stability of the transgene in the host genome and has an essential role for the practical application of transgenic animals. In this study, a modified *piggy*Bac vector that mediated deletion of the transposon sequence post-integration for transgene stability in the economically important silkworm *Bombyx mori* was constructed. The *piggy*Bac vector architecture contains inversed terminal repeat sequences L1, L2 and R1, which can form L1/R1 and L2/R1 types of transposition cassettes. hsp70-PIG as the *piggy*Bac transposase expression cassette for initial transposition, further remobilization and transgene stabilization test was transiently expressed in a helper vector or integrated into the modified vector to produce a transgenic silkworm. Shortening L2 increased the transformation frequency of L1/R1 into the silkworm genome compared to L2/R1. After the integration of L1/R1 into the genome, the remobilization of L2/R1 impaired the transposon structure and the resulting transgene linked with an impaired transposon was stable in the genome even in the presence of exogenously introduced transposase, whereas those flanked by the intact transposon were highly mobile in the genome. Our results demonstrated the feasibility of post-integration deletion of transposable elements to guarantee true transgene stabilization in silkworm. We suggest that the modified vector will be a useful resource for studies of transgenic silkworms and other *piggy*Bac-transformed organisms.

Published On: Molecular Genetics and Genomics, 2015, 290(3), 1181-1189.

1 State Key Laboratory of Silkworm Genome Biology, Southwest University, Chongqing 400716, China.
2 College of Biotechnology, Southwest University, Chongqing 400716, China.
*Corresponding author E-mail: xiaqy@swu.edu.cn.

再转座删除*piggy*Bac载体提高家蚕转基因稳定性研究

王　峰[1,2]　王日远[1]　王元成[1,2]　徐汉福[1]　袁　林[1]
丁　欢[1]　马三垣[1]　周　游[2]　赵　萍[1]　夏庆友[1,2*]

摘要：删除宿主基因组中的外源转座元件对维持转基因稳定具有广阔的前景，同时在转基因动物的实际应用中具有重要的作用。在本研究中，为了提高重要经济昆虫家蚕的转基因稳定性，构建了一个改造后的*piggy*Bac载体，可以介导转座子序引整合后的删除。*piggy*Bac载体带有反向末端重复序列L1、L2和R1，它们能形成L1/R1、L2/R1两种转座框。为了得到转基因家蚕，辅助载体或插入到修饰载体中的*piggy*Bac转座酶表达框hsp70-PIG瞬时表达用于起始转座、再转座和转基因稳定。与L2/R1相比，缩短的L2提高了L1/R1转座到家蚕基因组的效率。L1/R1整合到基因组后L2/R1再转座破坏转座子结构，即使有外源转座酶存在，与受损转座子相连的转基因在基因组中也都是稳定的，然而带有完整转座子的转基因很容易移动。我们的结果证明转座元件整合后的删除可确保家蚕转基因稳定。我们的改造载体对于转基因家蚕和其他*piggy*Bac转座生物的研究都大有帮助。

1　家蚕基因组生物学国家重点实验室，西南大学，重庆
2　生物技术学院，西南大学，重庆

Advanced technologies for genetically manipulating the silkworm *Bombyx mori*, a model Lepidopteran insect

Xu HF[1#*] David A. O'Brochta[2#*]

Abstract: Genetic technologies based on transposon-mediated transgenesis along with several recently developed genome-editing technologies have become the preferred methods of choice for genetically manipulating many organisms. The silkworm, *Bombyx mori*, is a Lepidopteran insect of great economic importance because of its use in silk production and because it is a valuable model insect that has greatly enhanced our understanding of the biology of insects, including many agricultural pests. In the past ten years, great advances have been achieved in the development of genetic technologies in *B. mori*, including transposon-based technologies that rely on *piggy*Bac-mediated transgenesis and genome-editing technologies that rely on protein - or RNA-guided modification of chromosomes. The successful development and application of these technologies has not only facilitated a better understanding of *B. mori* and its use as a silk production system, but also provided valuable experiences that have contributed to the development of similar technologies in non-model insects. This review summarizes the technologies currently available for use in *B. mori*, their application to the study of gene function and their use in genetically modifying *B. mori* for biotechnology applications. The challenges, solutions and future prospects associated with the development and application of genetic technologies in *B. mori* are also discussed.

Published On: Proceedings of the Royal Sciety B: Biological Sciences, 2015, 282(1810).

1 State Key Laboratory of Silkworm Genome Biology, Southwest University, Chongqing 400716, China.

2 Department of Entomology, The Institute for Bioscience and Biotechnology Research, University of Maryland, College Park, Rockville, MD 20850, USA.

#These authors contributed equally.

*Corresponding author E-mail: xuhf@swu.edu.cn; dobrocht@umd.edu.

鳞翅目模式昆虫——家蚕的遗传改造技术前沿

徐汉福[1#*]　David A.O′Brochta[2#*]

摘要：基于转座子介导的转基因技术和最近发展起来的基因组编辑技术已经成为许多物种遗传改造的优选方法。鳞翅目昆虫家蚕是一个能生产蚕丝的重要经济昆虫，同时也是一个重要的模式昆虫，以家蚕为模式的研究促进了我们对包括许多农业害虫在内的昆虫生物学的理解。在过去的10年里，家蚕遗传技术研究取得了巨大进展，分别建立了以*piggy*Bac转座子为介导的转基因操作技术和依赖于蛋白质或RNA引导进行染色体修饰的基因组编辑技术。这些遗传技术的成功建立和应用不仅促进了家蚕研究及其在蚕丝生产中的应用，而且还为在非模式昆虫中建立类似的遗传操作技术提供了宝贵经验。本综述总结了目前在家蚕中建立的遗传操作技术及其在基因功能研究和生物技术领域的应用进展，同时还讨论了当前家蚕遗传操作技术研究与应用所面临的挑战、解决方案和未来前景。

1　家蚕基因组生物学国家重点实验室，西南大学，重庆

2　昆虫系，生物科学与生物技术研究所，马里兰大学，马里兰，美国

Cre-mediated targeted gene activation in the middle silk glands of transgenic silkworms (*Bombyx mori*)

Duan JP[#] Xu HF[#] Ma SY Guo HZ
Wang F Zhao P Xia QY[*]

Abstract: Cre-mediated recombination is widely used to manipulate defined genes spatiotemporally *in vivo*. The present study evaluated the *Cre/loxP* system in *Bombyx mori* by establishing two transgenic lines. One line contained a Cre recombinase gene controlled by a *sericin-1* gene (*Ser1*) promoter. The other line contained a loxP-Stop-loxP-DsRed cassette driven by the same *Ser1* promoter. The precise deletion of the Stop fragment was found to be triggered by Cre-mediated site-specific excision, and led to the expression of DsRed fluorescence protein in the middle silk glands of all double-transgenic hybrids. This result was also confirmed by phenotypical analysis. Hence, the current study demonstrated the feasibility of Cre-mediated site-specific recombination in *B. mori*, and opened a new window for further refining genetic tools in silkworms.

Published On: Transgenic Research, 2013, 22(3), 607-619.

State Key Laboratory of Silkworm Genome Biology, Southwest University, Chongqing 400716, China.
[#]These authors contributed equally.
[*]Corresponding author E-mail: xiaqy@swu.edu.cn.

Cre介导的转基因家蚕中部丝腺中的靶基因激活

段建平# 徐汉福# 马三垣 郭慧珍
王 峰 赵 萍 夏庆友*

摘要： Cre介导的重组被广泛用于体内操控靶标基因时空特异性表达。本研究通过建立两个转基因品系，对家蚕中*Cre* / *loxP*系统进行了评估。一个转基因品系含有丝胶1基因（*Ser1*）启动子驱动的Cre重组酶基因。另一个转基因品系含有同一*Ser1*启动子驱动的loxP-Stop-loxP-DsRed表达框。杂交结果显示，由Cre介导的位点特异性切除可以精确删除Stop片段，导致DsRed荧光蛋白在所有双转基因杂交体中部丝腺表达。表型分析也证实了该结果。因此，本研究证实了Cre介导位点特异性重组在家蚕中是可行的，为进一步改良家蚕遗传工具提供了新思路。

家蚕基因组生物学国家重点实验室，西南大学，重庆

Construction of transformed, cultured silkworm cells and transgenic silkworm using the site-specific integrase system from phage φC31

Yin YJ[1#] Cao GL[1,2#] Xue RY[1,2] Gong CL[1,2*]

Abstract: The *Streptomyces* bacteriophage, φC31, uses a site-specific integrase enzyme to perform efficient recombination. The recombination system uses specific sequences to integrate exogenous DNA from the phage into a host. The sequences are known as the *att*P site in the phage and the *att*B site in the host. The system can be used as a genetic manipulation tool. In this study it has been applied to the transformation of cultured *BmN* cells and the construction of transgenic *Bombyx mori* individuals. A plasmid, pSK-attB/Pie1-EGFP/Zeo-PASV40, containing a cassette designed to express a *egfp-zeocin* fusion gene, was co-transfected into cultured *BmN* cells with a helper plasmid, pSK-Pie1/NLS-Int/NSL. Expression of the *egfp-zeocin* fusion gene was driven by an *ie-1* promoter, downstream of a φC31 *att*B site. The helper plasmid encoded the φC31 integrase enzyme, which was flanked by two nuclear localization signals. Expression of the *egfp-zeocin* fusion gene could be observed in transformed cells. The two plasmids were also transferred into silkworm eggs to obtain transgenic silkworms. Successful integration of the fusion gene was indicated by the detection of green fluorescence, which was emitted by the silkworms. Nucleotide sequence analysis demonstrated that the *att*B site had been cut, to allow recombination between the *att*B and endogenous pseudo *att*P sites in the cultured silkworm cells and silkworm individuals.

Published On: Molecular Biology Reports, 2014, 41(10), 6449-6456.

1 School of Biology and Basic Medical Sciences ,Soochow University, Suzhou 215123, China.

2 National Engineering Laboratory for Modern Silk, Soochow University, Suzhou 215123, China.

[#]These authors contributed equally.

[*]Corresponding author E-mail: gongcl@suda.edu.cn.

利用噬菌体φC31位点特异性整合酶系统构建转化培养的家蚕细胞及转基因家蚕

殷亚娟[1#]　曹广力[1,2#]　薛仁宇[1,2]　贡成良[1,2*]

摘要： φC31是一种可以利用位点特异性整合进行有效重组的链霉菌噬菌体。重组系统使用噬菌体*att*P位点和宿主*att*B位点将外源DNA从噬菌体整合到宿主中。该系统可作为一种基因操作的工具。本研究利用该系统构建转基因家蚕细胞和转基因家蚕。本研究构建了由*ie-1*启动子驱动φC31 *att*B位点下游的*egfp-zeocin*融合基因表达盒pSK-attB/Pie1-EGFP/Zeo-PASV40载体；同时构建了一个包含φC31整合酶基因瞬时表达盒的辅助质粒pSK-Pie1/NLS-Int/NSL，其中两侧均带有核定位信号序列的φC31整合酶基因由*Bm*NPV的*ie-1*启动子驱动。在共转染供体质粒和辅助质粒的家蚕培养细胞中可检测到*egfp-zeocin*融合基因的表达，同样，将供体质粒和辅助质粒通过精子介导法转入家蚕的卵中，通过绿色荧光筛选获得转基因家蚕。反向PCR产物测序结果显示，φC31整合酶可以准确地识别和切割供体质粒上的*att*B位点，并介导*att*B位点和细胞内源性假*att*P位点之间的重组。

1　基础医学与生物科学学院，苏州大学，苏州

2　现代丝绸国家工程实验室，苏州大学，苏州

TAL effectors mediate high-efficiency transposition of the *piggy*Bac transposon in silkworm *Bombyx mori* L.

Ye LP You ZY Qian QJ Zhang YY
Che JQ Song J Zhong BX*

Abstract: The *piggy*Bac (*PB*) transposon is one of the most useful transposable elements, and has been successfully used for genetic manipulation in more than a dozen species. However, the efficiency of *PB*-mediated transposition is still insufficient for many purposes. Here, we present a strategy to enhance transposition efficiency using a fusion of transcription activator-like effector (TALE) and the *PB* transposase (*PBase*). The results demonstrate that the TALE-*PBase* fusion protein which is engineered in this study can produce a significantly improved stable transposition efficiency of up to 63.9%, which is at least 7 times higher than the current transposition efficiency in silkworm. Moreover, the average number of transgene-positive individuals increased up to 5.7-fold, with each positive brood containing an average of 18.1 transgenic silkworms. Finally, we demonstrate that TALE-*PBase* fusion–mediated *PB* transposition presents a new insertional preference compared with original insertional preference. This method shows a great potential and value for insertional therapy of many genetic diseases. In conclusion, this new and powerful transposition technology will efficiently promote genetic manipulation studies in both invertebrates and vertebrates.

Published On: Scientific Reports, 2015, 5.

College of Animal Sciences, Zhejiang University, Hangzhou 310058, China.
*Corresponding author E-mail: bxzhong@zju.edu.cn.

类转录激活因子效应蛋白介导*piggy*Bac转座子的高效转座

叶露鹏　尤征英　钱秋杰　张玉玉
车家倩　宋　佳　钟伯雄*

摘要： *piggy*Bac转座子是用途非常广泛的一种转座子，已成功应用到10多种模式生物的遗传操作中。然而，*piggy*Bac介导的转基因效率仍然有待提高。本研究创建了一种有效提高转座效率的新方法，该方法的核心是构建一个类转录激活因子效应蛋白（transcription activator-like effector，TALE）和*piggy*Bac转座酶（*PBase*）的融合蛋白（TALE-*PBase*）。实验证明TALE-*PBase*融合蛋白可以显著地将转座效率提高到63.9%，这比目前家蚕的转座效率至少高7倍。而且新方法的转基因阳性蛾区中的阳性个体数每区平均达到近18.1个，比目前家蚕转基因实验的平均水平提高了5.7倍。TALE-*PBase*融合蛋白介导的转基因技术不仅提高了转基因阳性蛾区数，而且也明显提高了阳性蛾区中的阳性个体数。最后，我们还发现TALE-*PBase*融合蛋白介导的转基因与传统的转基因相比出现了新的插入偏好性。这些结果表明该技术在今后遗传性疾病治疗上有一定的应用价值。总之，这种高效的转座系统可以有效地推动脊椎和非脊椎动物的遗传操作研究。

动物科学学院，浙江大学，杭州

The clustered regularly interspaced short palindromic repeats/associated proteins system for the induction of gene mutations and phenotypic changes in *Bombyx mori*

Song J Che JQ You ZY Ye XG Li JS

Ye LP Zhang YY Qian QJ Zhong BX*

Abstract: To probe the general phenomena of gene mutations, *Bombyx mori*, the lepidopterous model organism, was chosen as the experimental model. To easily detect phenotypic variations, the *piggy*Bac system was utilized to introduce two marker genes into the silkworm, and 23.4% transposition efficiency aided in easily breeding a new strain for the entire experiment. Then, the clustered regularly interspaced short palindromic repeats/an associated protein (Cas9) system was utilized. The results showed that the Cas9 system can induce efficient gene mutations and the base changes could be detected since the G_0 individuals in *B.mori*; and that the mutation rates on different target sites were diverse. Next, the gRNA2-targeted site that generated higher mutation rate was chosen, and the experimental results were enumerated. First, the mutation proportion in G_1 generation was 30.1%, and some gene mutations were not inherited from the G_0 generation; second, occasionally, base substitutions did not lead to variation in the amino-acid sequence, which decreased the efficiency of phenotypic changes compared with that of genotypic changes. These results laid the foundation for better use of the Cas9 system in silkworm gene editing.

Published On: Acta Biochimica et Biophysica Sinica, 2016, 48(12),1112-1119.

College of Animal Sciences, Zhejiang University, Hangzhou 310058, China.

*Corresponding author E-mail: bxzhong@zju.edu.cn.

CRISPR/Cas9系统诱导家蚕基因突变及表型改变

宋 佳 车家倩 尤征英 叶小刚 李季生
叶露鹏 张玉玉 钱秋杰 钟伯雄*

摘要：为了探讨基因突变现象，家蚕作为鳞翅目的模式昆虫被选择作为实验模型。为了容易地检测表型突变，利用*piggy*Bac系统引入两个标记基因到家蚕，在整个实验过程中，23.4%的转基因效率有利于选育新的家蚕品系。本研究采用CRISPR/Cas9系统，结果表明Cas9系统可有效地诱导基因突变，碱基变化可以在G_0代个体中被检测到。同时突变在不同的靶位点具有不同的效率。随后，选择具有较高突变效率的gRNA2-靶向位点用于后续实验并对实验数据进行统计。首先，G_1代的突变率是30.1%，其中一些突变没有从G_0代被遗传到后代。其次，碱基突变有时候并不会导致氨基酸的改变，因此，较基因型而言表型改变的效率降低。这些结果为更好地利用Cas9系统编辑家蚕基因奠定了基础。

动物科学学院，浙江大学，杭州

Multiplex genomic structure variation mediated by TALEN and ssODN

Ma SY Wang XG Liu YY Gao J Zhang SL
Shi R Chang JS Zhao P Xia QY*

Abstract: Background—Genomic structure variation (GSV) is widely distributed in various organisms and is an important contributor to human diversity and disease susceptibility. Efficient approaches to induce targeted genomic structure variation are crucial for both analytic and therapeutic studies of GSV. Here, we presented an efficient strategy to induce targeted GSV including chromosomal deletions, duplications and inversions in a precise manner. Results—Utilizing Transcription Activator-Like Effector Nucleases (TALEN) designed to target two distinct sites, we demonstrated targeted deletions, duplications and inversions of an 8.9 Mb chromosomal segment, which is about one third of the entire chromosome. We developed a novel method by combining TALEN - induced GSV and single stranded oligodeoxynucleotide (ssODN) mediated gene modifications to reduce unwanted mutations occurring during the targeted GSV using TALEN or Zinc finger nuclease (ZFN). Furthermore, we showed that co-introduction of TALEN and ssODN generated unwanted complex structure variation other than the expected chromosomal deletion. Conclusions—We demonstrated the ability of TALEN to induce targeted GSV and provided an efficient strategy to perform GSV precisely. Furthermore, it is the first time to show that co-introduction of TALEN and ssODN generated unwanted complex structure variation. It is plausible to believe that the strategies developed in this study can be applied to other organisms, and will help understand the biological roles of GSV and therapeutic applications of TALEN and ssODN.

Published On: BMC Genomics, 2014, 15.

State Key Laboratory of Silkworm Genome Biology, Southwest University, Chongqing 400716, China.
*Corresponding author E-mail: xiaqy@swu.edu.cn.

位点特异性核酸酶TALEN和ssODN介导的多重基因组结构变异

马三垣　汪小刚　刘园园　高　杰　张圣棂
施　润　常珈菘　赵　萍　夏庆友*

摘要：背景——基因组结构变异(GSV)广泛分布在各种生物体内，是人类多样性和疾病易感性的重要因素。诱导靶向基因组结构变异的有效方法对GSV的分析和治疗研究至关重要。本研究中，作者提出了一种有效的策略，可以精确的方式诱导靶向GSV，包括染色体缺失、重复和反转。结果——我们展示了利用设计的两个不同靶位点的转录激活因子样效应核酸酶(TALEN)，对占染色体长度约三分之一的片段(8.9 Mb)进行定向缺失、重复和反转。作者开发了一种新的方法，通过联合运用TALEN诱导的GSV和单链寡脱氧核苷酸(ssODN)介导的基因修饰，可以减少在用TALEN或锌指核酸酶(ZFN)在靶向GSV时发生不想要的突变。此外，作者发现TALEN和ssODN的共同引入可以产生除染色体缺失等预期之外的复杂结构变异。结论—— 作者展示了TALEN诱导靶向GSV的能力，并提供了一种精确执行GSV的有效策略。此外，本文第一次展示了TALEN和ssODN共同引入可以产生预期之外的复杂结构变化。相信本研究中开发的策略可以应用于其他生物体，这将有助于了解GSV的生物学作用以及TALEN和ssODN在治疗上的应用。

家蚕基因组生物学国家重点实验室，西南大学，重庆

Highly efficient and specific genome editing in silkworm using custom TALENs

Ma SY　Zhang SL　Wang F　Liu Y　Liu YY

Xu HF　Liu C　Lin Y　Zhao P　Xia QY*

Abstract: Establishment of efficient genome editing tools is essential for fundamental research, genetic engineering, and gene therapy. Successful construction and application of transcription activator-like effector nucleases (TALENs) in several organisms herald an exciting new era for genome editing. We describe the production of two active TALENs and their successful application in the targeted mutagenesis of silkworm, *Bombyx mori*, whose genetic manipulation methods are parallel to those of *Drosophila* and other insects. We will also show that the simultaneous expression of two pairs of TALENs generates heritable large chromosomal deletion. Our results demonstrate that (i) TALENs can be used in silkworm and (ii) heritable large chromosomal deletions can be induced by two pairs of TALENs in whole organisms. The generation and the high frequency of TALENs-induced targeted mutagenesis in silkworm will promote the genetic modification of silkworm and other insect species.

Published On: PLoS ONE, 2012, 7(9).

State Key Laboratory of Silkworm Genome Biology, Southwest University, Chongqing 400716, China.

*Corresponding author E-mail: xiaqy@swu.edu.cn.

TALENs介导的家蚕高效特异基因组编辑

马三垣　张圣棂　王　峰　刘　勇　刘园园
徐汉福　刘　春　林　英　赵　萍　夏庆友*

摘要：建立有效的基因组编辑工具对于基础研究、基因工程和基因治疗至关重要。转录激活子样效应因子核酸酶(TALENs)在多种生物中的成功构建和应用预示着一个激动人心的基因组编辑新时代的到来。作者描述了两种活性TALEN的构建及其在家蚕的定点突变中的成功应用,其遗传操作方法与果蝇和其他昆虫的遗传操作方法类似。结果显示将两对TALEN同时在家蚕中表达可产生能够遗传的染色体大片段缺失。结果表明:(i)TALENs可以用于家蚕基因组编辑,(ii)两对TALEN同时起作用可以在全部生物体中产生可遗传的染色体大片段缺失。TALENs介导的高效可遗传定点突变可以促进家蚕和其他昆虫的遗传修饰。

家蚕基因组生物学国家重点实验室,西南大学,重庆

High-effciency system for construction and evaluation of customized TALENs for silkworm genome editing

Wang F[#] Ma SY[#] Xu HF Duan JP Wang YC
Ding H Liu YY Wang XG Zhao P Xia QY[*]

Abstract: Transcription activator-like effector nuclease(TALEN) possesses the characteristics of ease design and precise DNA targeting. In the silkworm *Bombyx mori*, TALEN has been successfully used to knockout an endogenous *Bombyx* gene, and shown the huge potential in functional genes research and improvement of the economical characteristics of silkworm. Thus, there is an urgent need to develop an applicable system that permits the efficient construction of customized TALEN with high activity that could efficiently induce the hereditable mutagenesis in the silkworm. In this study, we constructed an efficient assembly and evaluation system of the customized TALEN especially for silkworm genome editing by combination of a modified Golden Gate ligation strategy, a luciferase(LUC) reporter system in insect cell culture for binding activity and a surveyor nuclease assay system in silkworm embryos for cleavage efficiency. We showed the reliability of this system by assembling a pair of TALENs targeting a silkworm genome locus and assaying their binding and cleavage activities. The assembly strategy was convenient and efficient which allows the rapid construction of customized TALEN in less than 1 week, and the evaluation system was reliable and necessary for screening of the customized TALEN pair with high binding and cleavage activities. The results showed this system is a reliable and efficient tool for the construction of customized TALEN with high activity for gene targeting of silkworm, and will contribute to the wide application of TALEN technology in the functional gene research of silkworm.

Published On: Molecular Genetics and Genomics. 2013, 288(12), 683-690.

State Key Laboratory of Silkworm Genome Biology, Southwest University, Chongqing 400716, China.
[#]These authors contributed equally.
[*]Corresponding author E-mail: xiaqy@swu.edu.cn.

适用于家蚕基因组编辑的TALENs高效系统的构建和评估

王　峰[#]　马三垣[#]　徐汉福　段建平　王元成
丁　欢　刘园园　汪小刚　赵　萍　夏庆友[*]

摘要：转录激活因子样效应因子核酸酶(TALEN)具有易于设计和精确靶向DNA的特征。在家蚕中,TALEN已经成功地用于敲除内源性基因,在功能基因研究和改良家蚕经济性状方面显示出巨大潜力。因此,迫切需要开发一种可用于高效构建具有高活性的TALEN系统,从而在家蚕中产生可遗传变异。通过改良的Golden Gate连接策略,在昆虫细胞培养中用荧光素酶(LUC)报告系统检测TALEN结合效率,用核酸酶测定法检测家蚕胚胎系统的切割效率,作者构建了一个有针对性的TALEN组装和评估系统,特别适用于家蚕基因组编辑。作者组装了一对靶向蚕基因组的TALEN,并测定其结合和切割活性,结果显示该系统非常可靠。该组装策略方便有效,可在1周内快速构建特定的TALEN系统,该评估系统对于筛选具有高结合和切割活性的特定TALEN对是非常可靠和必要的。结果显示,该系统是一种可靠且有效的工具,可以用于构建家蚕基因打靶高活性的TALEN对,有助于促进TALEN技术在家蚕功能基因研究中的广泛应用。

家蚕基因组生物学国家重点实验室,西南大学,重庆

Highly efficient multiplex targeted mutagenesis and genomic structure variation in *Bombyx mori* cells using CRISPR/Cas9

Liu YY[#] Ma SY[#] Wang XG Chang JS Gao J

Shi R Zhang JD Lu W Liu Y Zhao P Xia QY[*]

Abstract: *Bombyx mori* is an important economical insect and a model organism for studying lepidopteran and arthropod biology. Using a highly efficient CRISPR / Cas9 system, we showed that this system could mediate highly efficient targeted genome editing of a single gene locus, large chromosomal deletion or inversion, and also multiplex genome editing of 6 genes simultaneously in *BmNs* cell line derived from *B. mori*. The simplicity and high efficiency of our system provide unprecedented possibilities for researchers to implement precise and sophisticated manipulation of a chosen *B. mori* gene in *BmNs* cells easily in a limited time course, and perhaps new opportunities for functional genomics of *B. mori* and other lepidopteran insects.

Published On: Insect Biochemistry and Molecular Biology, 2014, 49, 35-42.

State Key Laboratory of Silkworm Genome Biology, Southwest University, Chongqing 400716, China.

[#]These authors contributed equally.

[*]Corresponding author E-mail: xiaqy@swu.edu.cn.

运用CRISPR/Cas9实现家蚕细胞高效的多重定向诱变和基因组结构变异

刘园园# 马三垣# 汪小刚 常珈菘 高 杰
施 润 张建铎 陆 卫 刘 越 赵 萍 夏庆友*

摘要：家蚕是一种重要的经济昆虫，也是研究鳞翅目昆虫和节肢动物生物学的模式生物。我们发现使用高效的CRISPR / Cas9系统，可以介导单个基因座的高效靶向基因组编辑，染色体大片段缺失或倒位，而且还可以在家蚕*BmNs*细胞系中同时对6个基因进行基因组编辑。我们构建的基因编辑系统简单、高效，为研究人员提供了前所未有的可能性，可以在有限的时间内轻松地在*BmNs*细胞中实现精确和复杂的基因组编辑，该系统也可能是研究家蚕和其他鳞翅目昆虫功能基因组学的新方法。

家蚕基因组生物学国家重点实验室，西南大学，重庆

Heritable genome editing with CRISPR/Cas9 in the silkworm, *Bombyx mori*

Wei W[1#] Xin HH[2#] Bhaskar Roy[1]

Dai JB[1] Miao YG[2*] Gao GJ[1*]

Abstract: We report the establishment of an efficient and heritable gene mutagenesis method in the silkworm *Bombyx mori* using modified type II clustered regularly interspaced short palindromic repeats (CRISPR) with an associated protein (Cas9) system. Using four loci *Bm-ok*, *BmKMO*, *BmTH*, and *Bmtan* as candidates, we proved that genome alterations at specific sites could be induced by direct microinjection of specific guide RNA and Cas9-mRNA into silkworm embryos. Mutation frequencies of 16.7% – 35.0% were observed in the injected generation, and DNA fragments deletions were also noted. *Bm-ok* mosaic mutants were used to test for mutant heritability due to the easily determined translucent epidermal phenotype of *Bm-ok* disrupted cells. Two crossing strategies were used. In the first, injected *Bm-ok* moths were crossed with wild-type moths, and a 28.6% frequency of germline mutation transmission was observed. In the second strategy, two *Bm-ok* mosaic mutant moths were crossed with each other, and 93.6% of the offsprings appeared mutations in both alleles of *Bm-ok* gene (compound heterozygous). In summary, the CRISPR/Cas9 system can act as a highly specific and heritable gene-editing tool in *Bombyx mori*.

Published On: PLoS ONE. 2014, 9(7).

1 School of Life Sciences, Tsinghua University, Beijing 100084, China.

2 College of Animal Sciences, Zhejiang University, Hangzhou 310058, China.

#These authors contributed equally.

*Corresponding author E-mail: miaoyg@zju.edu.cn; gaogu@mail.tsinghua.edu.cn.

应用CRISPR/Cas9系统对家蚕基因组进行可遗传编辑

魏　薇[1#]　辛虎虎[2#]　Bhaskar Roy[1]
戴俊彪[1]　缪云根[2*]　高冠军[1*]

摘要： 本文报道了CRISPR/Cas9系统可在家蚕中建立有效和可遗传的基因编辑方法。我们选取了4个与家蚕体壁相关的基因座*Bm-ok*、*BmKMO*、*BmTH*和*Bmtan*作为靶基因，通过特异性gRNA和Cas9-mRNA直接显微注射到家蚕胚胎中，诱导特定部位的基因组变异。在注射当代中观察到突变频率为16.7%—35.0%，并且记录了DNA片段的缺失。因为*Bm-ok*嵌合突变体容易确定，所以*Bm-ok*破坏细胞的半透明表皮表型被用于测试突变遗传力。使用两种交叉杂交策略，当*Bm-ok*嵌合突变体与野生型蛾杂交，观察到28.6%的种系突变频率；当两个*Bm-ok*嵌合突变体蛾相互交叉，93.6%的后代在*Bm-ok*基因的两个等位基因中均出现突变（复合型杂合子）。总之，CRISPR/Cas9系统可以作为家蚕高度特异性和遗传性的基因编辑工具。

1　生命科学学院，清华大学，北京
2　动物科学学院，浙江大学，杭州

CRISPR/Cas9 mediated multiplex genome editing and heritable mutagenesis of *BmKu70* in *Bombyx mori*

Ma SY[#] Chang JS[#] Wang XG Liu YY Zhang JD
Lu W Gao J Shi R Zhao P Xia QY[*]

Abstract: CRISPR / Cas9, a bacterial adaptive immune system derived genome-editing technique, has become to be one of the most compelling topics in biotechnology. *Bombyx mori* is an economically important insect and a model organism for studying lepidopteran and arthropod biology. Here we reported highly efficient and multiplex genome editing in *B. mori* cell line and heritable site-directed mutagenesis of *Bmku70*, which is required for NHEJ pathway and also related to antigen diversity, telomere length maintenance and subtelomeric gene silencing, using CRISPR/Cas9 system. We established a simple and practicable method and obtained several *Bmku70* knockout *B. mori* lines, and showed that the frequency of HR was increased in embryos of the *Bmku70* knockout *B. mori*. The mutant lines obtained in this study could be a candidate genetic resource for efficient knock-in and fundamental research of DNA repair in *B. mori*. We also provided a strategy and procedure to perform heritable genome editing of target genes with no significant phenotype effect.

Published On: Scientific Reports, 2013, 4, 4489.

State Key Laboratory of Silkworm Genome Biology, Southwest University, Chongqing 400716, China.
[#]These authors contributed equally.
[*]Corresponding author E-mail: xiaqy@swu.edu.cn.

CRISPR/Cas9介导的家蚕多重基因组编辑与*BmKu70*的可遗传突变

马三垣[#] 常珈菘[#] 汪小刚 刘园园 张建铎
陆卫 高杰 施润 赵萍 夏庆友[*]

摘要：CRISPR/Cas9是一种细菌的后天免疫系统衍生出的基因组编辑技术，是目前最受关注的生物技术之一。家蚕是一种重要的经济昆虫，也是研究鳞翅目昆虫和节肢动物的模式生物。本研究中，我们报道了家蚕细胞系高效基因编辑技术——CRISPR/Cas9系统，并且建立了被多重编辑的可遗传的家蚕*Bmku70*突变细胞系。*Bmku70*是家蚕重要基因，参与家蚕DNA双链断裂(DSBs)修复通路非同源末端连接(NHEJ)，和抗原多样性、端粒长度维护、亚端粒基因沉默等重要功能相关。作者建立了一个简单的方法，并且获得了多个*Bmku70*突变品系。研究结果显示，*Bmku70*突变品系的胚胎中DSBs修复通路同源重组(HR)效率显著提高。*Bmku70*突变品系不仅可以用于高效基因敲入(knock-in)，而且可以用于研究家蚕DNA损伤修复机制。同时作者提供了一种筛选敲除基因后无表型的突变体的策略。

家蚕基因组生物学国家重点实验室，西南大学，重庆

Effects of $BmKIT_3^R$ gene transfer on pupal development of *Bombyx mori* linnaeus using a *Gal4/UAS* binary transgenic system

Zhang HK[1#] Cao GL[1,2#] Zhang XR[1]
Wang XJ[1] Xue RY[1,2] Gong CL[1,2*]

Abstract: The pupal stage of the silkworm *Bombyx mori* Linnaeus lasts for approximately two weeks. However, prolongation of pupal duration would reduce the labor required to process and dry fresh cocoons. This study investigated the effects of $BmKIT_3^R$ gene (from the Chinese scorpion *Buthus martensii* Karsch) transfer on the pupal development of *B. mori* using a *Gal4/UAS* binary transgenic system. *Gal4* driven by a pupa-specific promoter *BmWCP4* (from a *B. mori* wing-cuticle protein gene) or *PDP* (from a *B. mori* cocoonase gene), and $BmKIT_3^R$ driven by a *UAS cis*-acting element were used to construct novel *piggy*Bac-derived plasmids containing a neomycin-resistance gene (*neo*) controlled by the *Bombyx mori* nucleopolyhedrovirus (*Bm*NPV) *ie-1* (immediate-early gene) promoter and a green fluorescent protein gene (*gfp*) under the control of the *B. mori* actin 3 (*A3*) promoter. The vector was transferred into silkworm eggs by sperm-mediated gene transfer. Transgenic silkworms were produced after screening for *neo* and *gfp* genes, and gene transfer was verified by polymerase chain reaction and dot-blot hybridization. The larval development of the hybrid progeny of *Gal4*- and *UAS*-transgenic silkworms was similar to that of normal silkworms, but some pupae failed to metamorphose into moths, and the development of surviving pupae was arrested as a result of $BmKIT_3^R$ expression. Moreover, *Gal4* driven by the *BmWCP4* promoter delayed pupal development more effectively than that driven by the *PDP* promoter in the *Gal4/UAS* binary transgenic system. Pupal durations of hybrid transgenic silkworm progeny with *BmWCP4* and *PDP* promoters were approximately 5, 2, and 4 days longer, respectively, compared to corresponding normal silkworms, *BmWCP4/Gal4*, and $UAS/BmKIT_3^R$ transgenic silkworms, respectively. These results suggest new avenues of research for prolonging the pupal duration of silkworms.

Published On: Journal of Agricultural and Food Chemistry, 2012, 60(12), 3173-3179.

1 School of Biology & Basic Medical Sciences, Soochow University, Suzhou 215123, China.
2 National Engineering Laboratory for Modern Silk, Soochow University, Suzhou 215123, China.
[#]These authors contributed equally.
[*]Corresponding author E-mail: gongcl@suda.edu.cn.

Gal4/UAS 转基因系统表达 $BmKIT_3^R$ 对家蚕蛹期发育的影响

张昊堃[1#] 曹广力[1,2#] 张晓荣[1]
王晓娟[1] 薛仁宇[1,2] 贡成良[1,2*]

摘要：家蚕的蛹期一般维持在两周左右，但是通过延长蛹期，可以节约大量花费在鲜茧加工过程中的人力物力。本文旨在利用 *Gal4/UAS* 双元转基因系统，研究东亚钳蝎毒素基因（$BmKIT_3^R$）对家蚕蛹发育的影响。*Gal4* 基因由蛹期特异性表达启动子 *BmWCP4* 或家蚕溶茧酶 *PDP* 所控制，$BmKIT_3^R$ 基因由可以被 Gal4 蛋白特异性识别的 *UAS* 序列控制表达，双元系统的激活系统和效应系统都装载在 *piggy*Bac 转座子载体上。另外，*piggy*Bac 载体中均含有由家蚕核型多角体病毒 *Bm*NPV *ie-1* 极早期基因启动子控制的新霉素抗性基因 *neo*，以及家蚕 *A3* 基因启动子控制的绿色荧光蛋白基因 *gfp*。载体通过精子介导法导入家蚕卵中。经过 *neo* 和 *gfp* 初步筛选，再运用 PCR 和点杂交进一步确定所筛选的转基因家蚕。*Gal4*-和 *UAS*-转基因家蚕的杂交种的幼虫发育与正常家蚕类似，但 $BmKIT_3^R$ 基因的表达导致有些家蚕在蛹期死亡，而存活的蛹发育停滞。并且，在 *Gal4/UAS* 双元系统中，*BmWCP4* 启动子对延迟蛹发育的效果优于 *PDP* 启动子。利用 *BmWCP4* 和 *PDP* 启动子的转基因家蚕杂交后代的蛹期较正常家蚕、*BmWCP4/Gal4* 和 *UAS/*$BmKIT_3^R$ 转基因家蚕而言，分别延长 5 d、2 d 和 4 d。本研究提供了一种延长家蚕蛹期的新方法。

1 基础医学与生物科学学院，苏州大学，苏州
2 现代丝绸国家工程实验室，苏州大学，苏州

Effects of *egt* gene transfer on the development of *Bombyx mori*

Zhang X[1#] Xue RY[1,2#] Cao GL[1,2] Hu XL[1] Wang XJ[1]
Pan ZH[1,2] Xie M[1] Yu XH[1,2] Gong CL[1,2*]

Abstract: This study investigated the effects of gain of ecdysteroid UDP-glucosyltransferase (*egt*) gene function mutation on the development of the silkworm, *Bombyx mori*. A novel *piggy*Bac-derived plasmid containing the *egt* gene from *B. mori* nucleopolyhedrovirus (*Bm*NPV) driven by a heat-shock protein (*hsp*) 23.7 promoter, with a neomycin-resistance gene (*neo*) controlled by the *Bm*NPV *ie-1* promoter and a green fluorescent protein gene (*gfp*) under the control of the *B. mori actin 3 (A3)* promoter was constructed. The vector was transferred into silkworm eggs by sperm-mediated gene transfer. Transgenic silkworms were produced after screening for *neo* and *gfp* genes and gene transfer was verified by polymerase chain reaction, dot-blot hybridization and Western blot. The hatching rate of G_1 generation silkworm eggs was about 60% lower than that of normal silkworm eggs. The duration of the G_1 generation larval period was extended, and the G_2 generation pupal stage lasted four days longer than that in non-transgenic silkworms. The ecdysone blood level in G_2 silkworms in the third instar molting stage was reduced by up to 90%. These results show that *egt* suppressed transgenic silkworm molting, and that *egt* expression in *egt*-transgenic silkworms resulted in arrest of metamorphosis from pupae to moths.

Published On: Gene, 2012, 491(2), 272-277.

1 School of Biology & Basic Medical Sciences, Soochow University, Suzhou 215123, China.
2 National Engineering Laboratory for Modern Silk, Soochow University, Suzhou 215123, China.
#These authors contributed equally.
*Corresponding author E-mail: gongcl@suda.edu.cn.

转 *egt* 基因对家蚕发育的影响

张　星[1#]　薛仁宇[1,2#]　曹广力[1,2]　胡小龙[1]　王晓娟[1]
潘中华[1,2]　谢　敏[1]　虞晓华[1,2]　贡成良[1,2*]

摘要：本研究探讨了蜕皮类固醇UDP-葡萄糖基转移酶（*egt*）基因功能突变对家蚕发育的影响。我们构建了一种新型*piggy*Bac质粒，该质粒包含热休克蛋白（*hsp*）23.7启动子控制的家蚕核型多角体病毒（*Bm*NPV）*egt*基因，*Bm*NPV *ie-1*启动子控制的新霉素耐药性基因*neo*和由家蚕*A3*启动子控制的绿色荧光蛋白基因*gfp*。利用精子介导法将该质粒导入蚕卵中。通过*neo*和*gfp*基因筛选转基因家蚕，利用PCR、点杂交和蛋白质免疫印迹法鉴定转基因家蚕。结果显示：第一代转基因蚕卵（G_1）的孵化率比对照组的低60%。G_1代家蚕幼虫龄期延长，G_2代蛹期比非转基因家蚕延长4 d。在三龄眠期的G_2代转基因家蚕的血液中，脱皮激素水平降低90%以上。以上结果表明*egt*能够抑制转基因家蚕的蜕皮，并且在*egt*转基因家蚕中*egt*基因的过表达导致家蚕从蛹到蛾的变态过程停滞。

1　基础医学与生物科学学院，苏州大学，苏州
2　现代丝绸国家工程实验室，苏州大学，苏州

附图

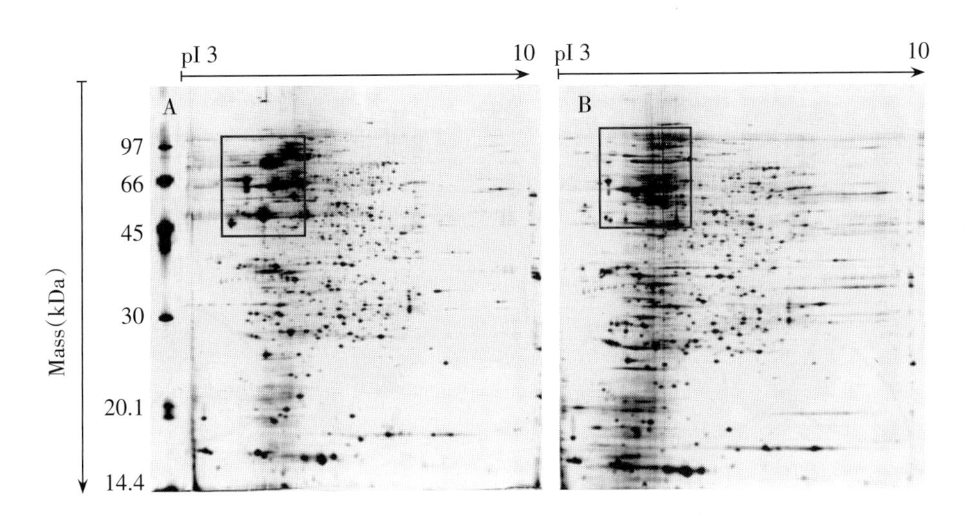

Figure 1. 2-DE map of proteins extracted from MSG and PSG. Proteins (0.15 mg) from MSG (A) and PSG(B)were placed on pH 3—10, 18 cm ReadyStrip IPG strips for electrophoresis (10 h) followed by SDS-PAGE(12.5% gel, 5 h), and silver staining. Significant differences were noted between MSG and PSG (square). [选自:*Studies on middle and posterior silk glands of silkworm (Bombyx mori) using two-dimensional electrophoresis and mass spectrometry*,参见本书第6页。]

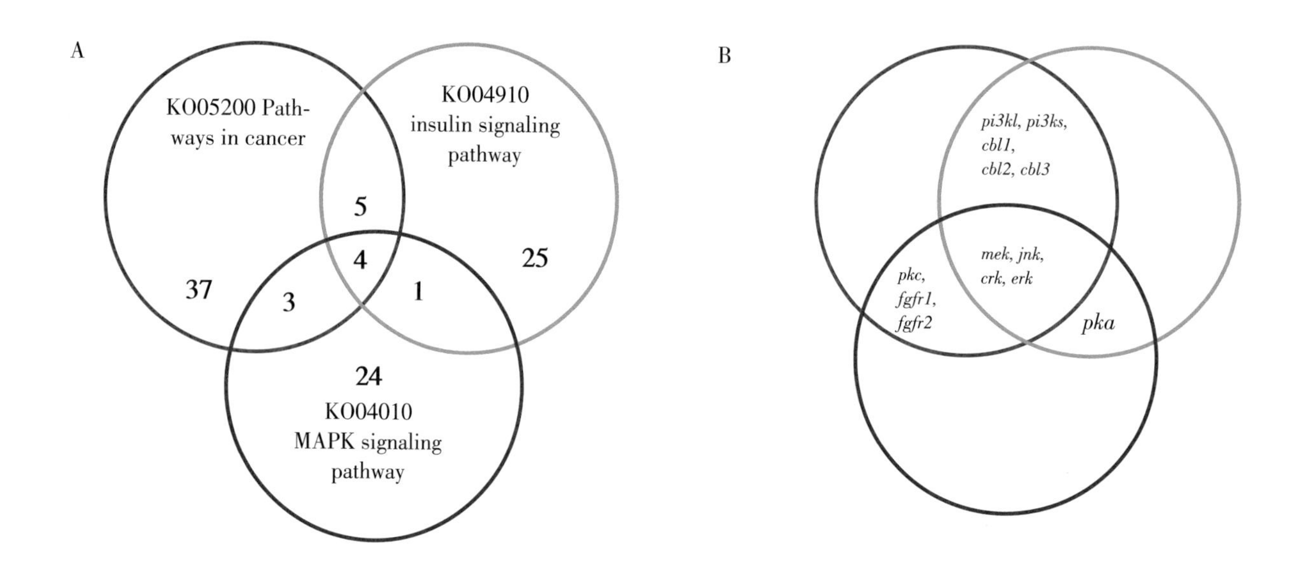

Figure 2. DEGs distribution in "pathways in cancer", "insulin signaling pathway" and "MAPK signaling pathway". (A) Showing the number of DEGs. (B) Showing the common DEGs.（选自：*Transcriptomic analysis of differentially expressed genes in the Ras1^{CA}-overexpressed and wildtype posterior silk glands*，参见本书第16页。）

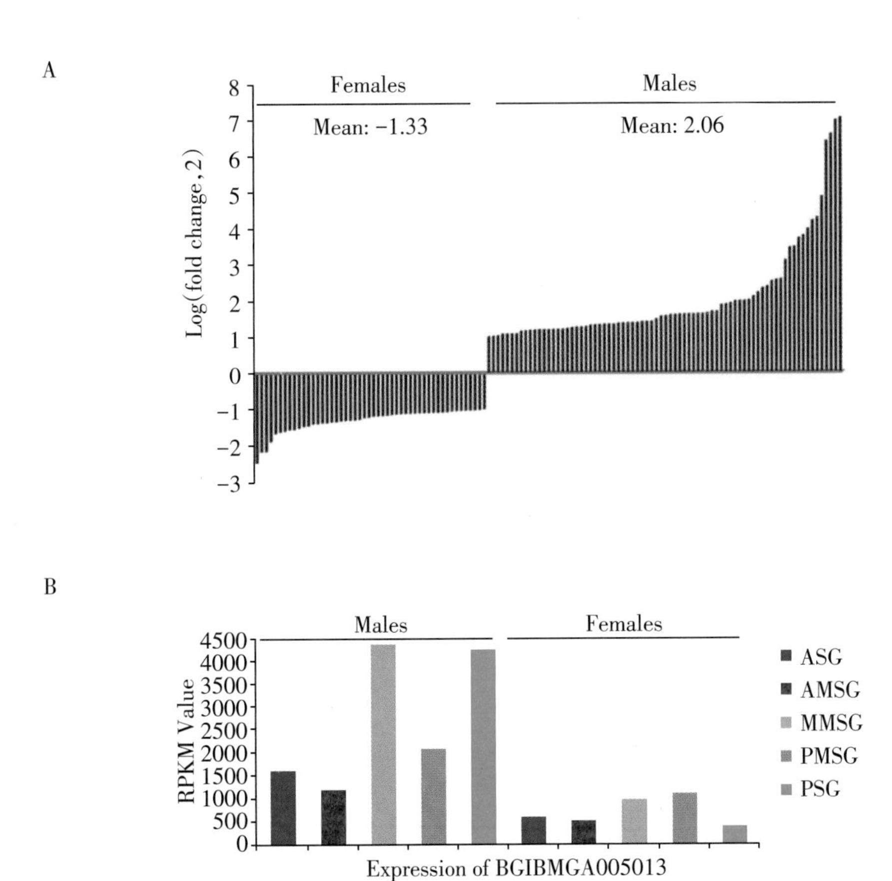

Figure 3. Expression patterns of up-regulated genes in males and females. (A) PSG as an example of the gene expression pattern based on the fold-change between males and females. There are more up-regulated genes in the PSG of males than of females. (B) Expression pattern of the gene *BGIBMGA005013*.[选自:*Transcriptomic analysis of the anterior silk gland in the domestic silkworm (Bombyx mori) - Insight into the mechanism of silk formation and spinning*,参见本书第18页。]

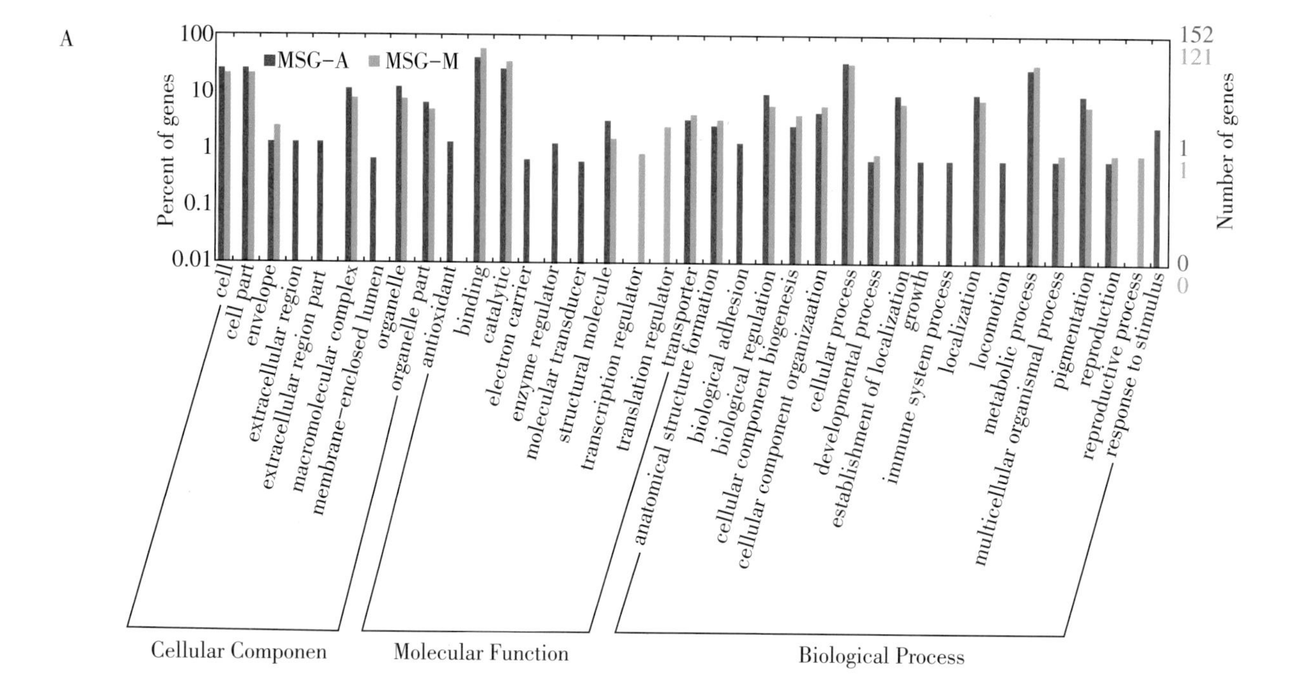
A
MSG-A
MSG-M
Percent of genes
Number of genes
100
10
1
0.1
0.01
152
121
1
1
0
0
cell
cell part
envelope
extracellular region
extracellular region part
macromolecular complex
membrane-enclosed lumen
organelle
organelle part
antioxidant
binding
catalytic
electron carrier
enzyme regulator
molecular transducer
structural molecule
transcription regulator
translation regulator
transporter
anatomical structure formation
biological adhesion
biological regulation
cellular component biogenesis
cellular component organization
cellular process
developmental process
establishment of localization
growth
immune system process
localization
locomotion
metabolic process
multicellular organismal process
pigmentation
reproduction
reproductive process
response to stimulus
Cellular Componen
Molecular Function
Biological Process

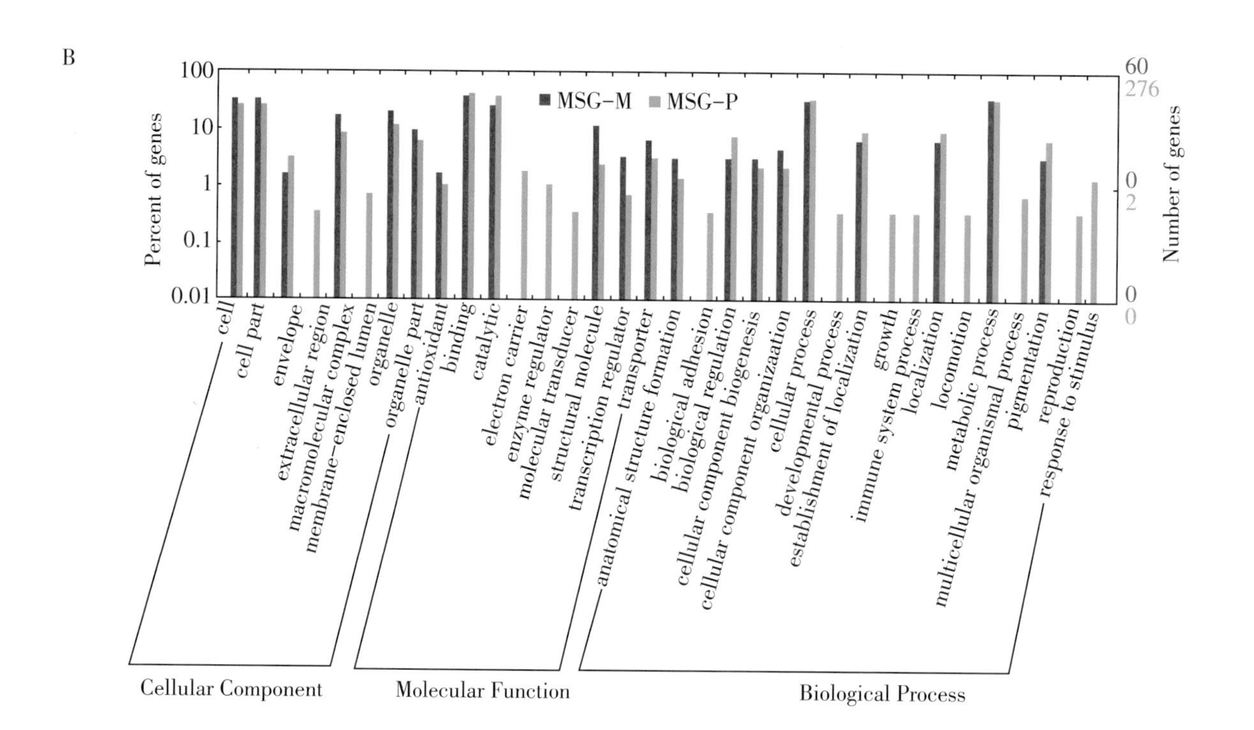
B
MSG-M
MSG-P
Percent of genes
Number of genes
100
10
1
0.1
0.01
60
276
0
2
0
0
cell
cell part
envelope
extracellular region
macromolecular complex
membrane-enclosed lumen
organelle
organelle part
antioxidant
binding
catalytic
electron carrier
enzyme regulator
molecular transducer
structural molecule
transcription regulator
transporter
anatomical structure formation
biological adhesion
biological regulation
cellular component biogenesis
cellular component organization
cellular process
developmental process
establishment of localization
growth
immune system process
localization
locomotion
metabolic process
multicellular organismal process
pigmentation
reproduction
response to stimulus
Cellular Component
Molecular Function
Biological Process

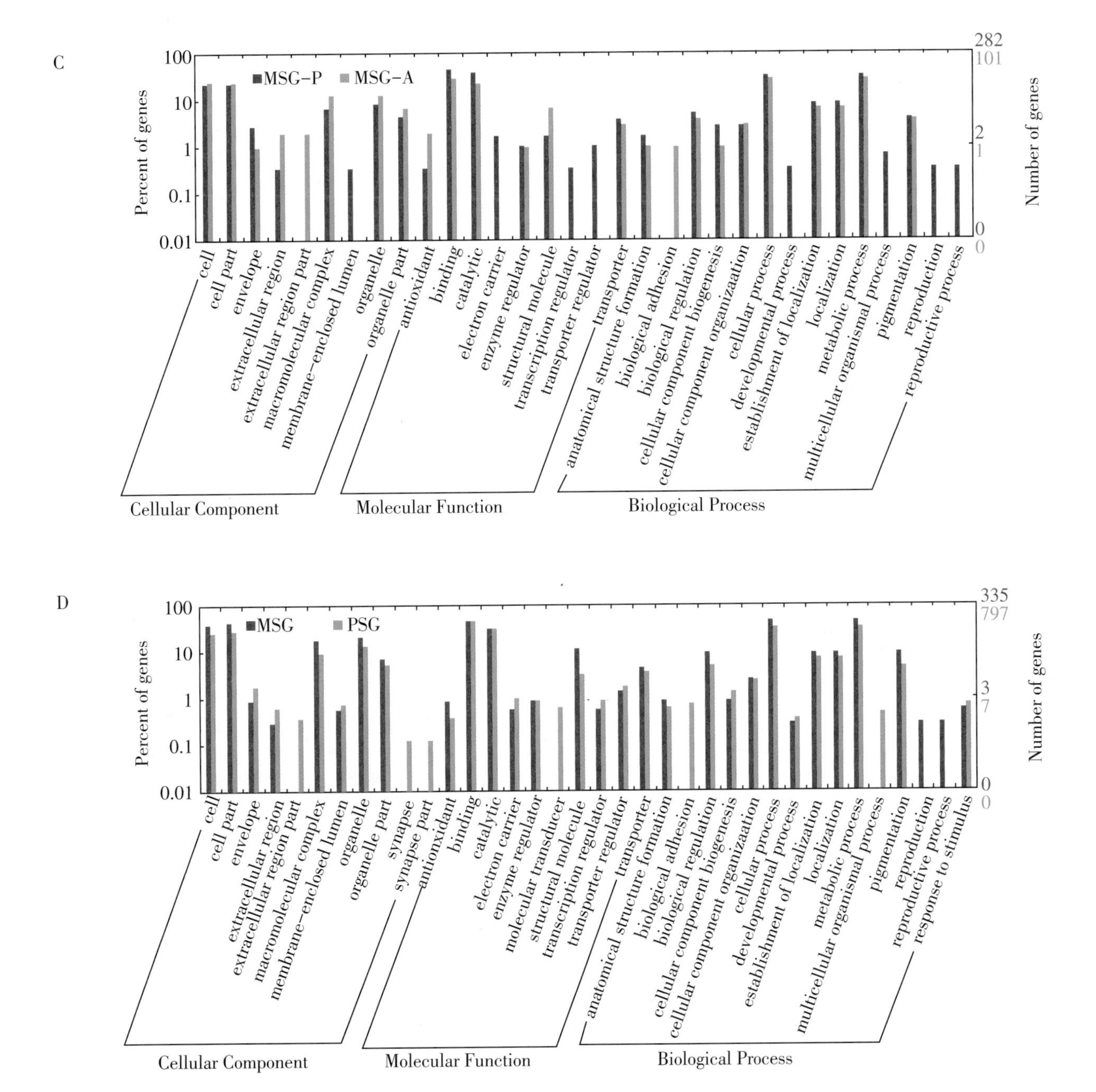

Figure 4. GO categories of the differentially expressed proteins. A, the GO categories of proteins with specific and significantly higher expression in the MSG-A and MSG-M; B, the GO categories of proteins with specific and significantly higher expression in the MSG-M and MSG-P; C, the GO categories of proteins with specific and significantly higher expression in the MSG-A and MSG-P; D, the GO categories of proteins with specific and significantly higher expression in the MSG and PSG.（选自：*Comparative proteomic analysis of the silkworm middle silk gland reveals the importance of ribosome biogenesis in silk protein production*，参见本书第20页。）

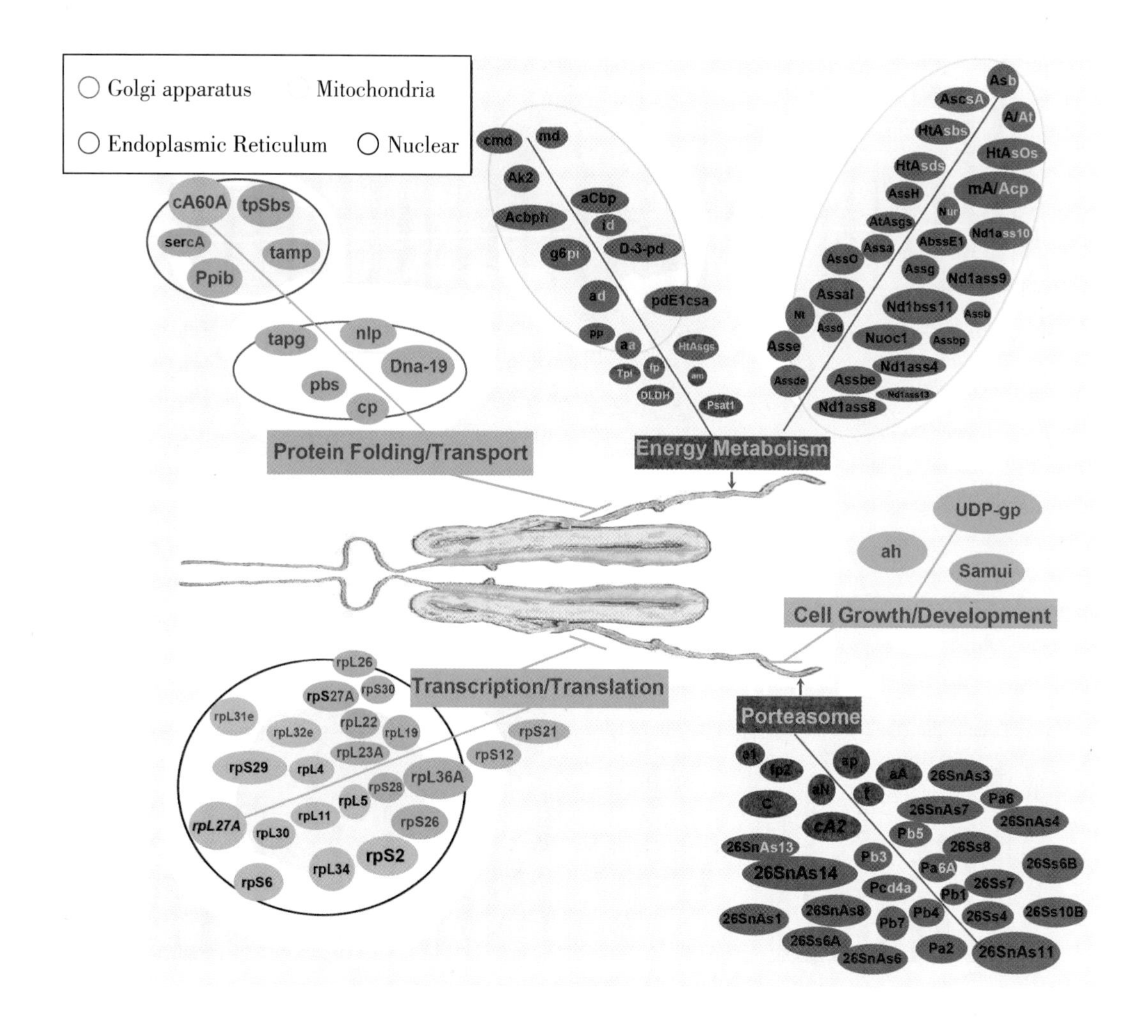

Figure 5. Whole view of five-way regulated functional unit involved in molecular mechanism of low silk production. Color coding is as follows for items from the present study: black letters and red background represents the up-regulated transcripts, yellow letters and red background represents increased proteins, black letters and green background represents the down-regulated transcripts, and blue letters and green background represents decreased proteins. Color coding is as follows for items from the previous study: black letters and pink background represents the up-regulated transcripts, yellow letters and pink background represents increased proteins, black letters and lake blue background represent the down-regulated transcripts, and blue letters and lake blue background represent decreased proteins. The half pink/half red background represents genes that were up-regulated in both studies. The half black/half yellow letters represent the genes from both proteins and transcripts. (选自：*Analyses of the molecular mechanisms associated with silk production in silkworm by iTRAQ-based proteomics and RNA-sequencing-based transcriptomics*，参见本书第22页。)

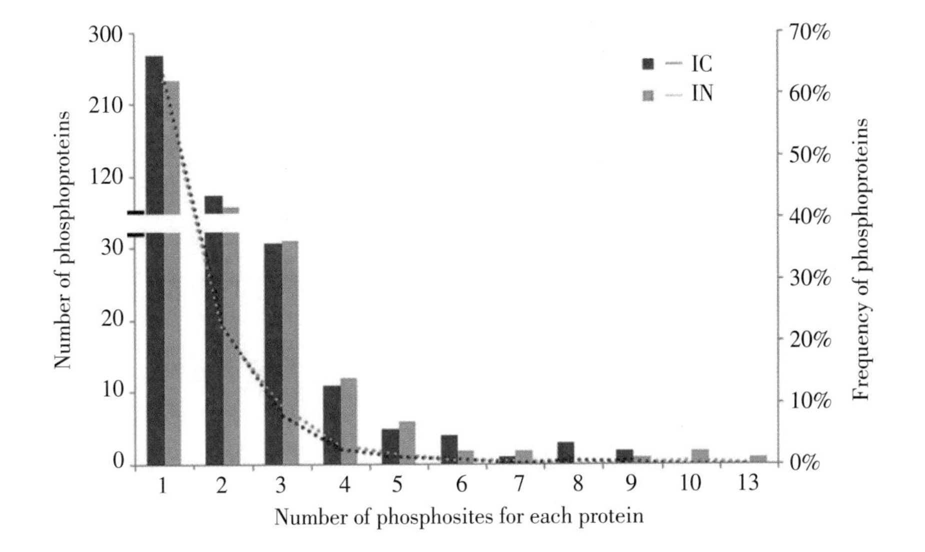

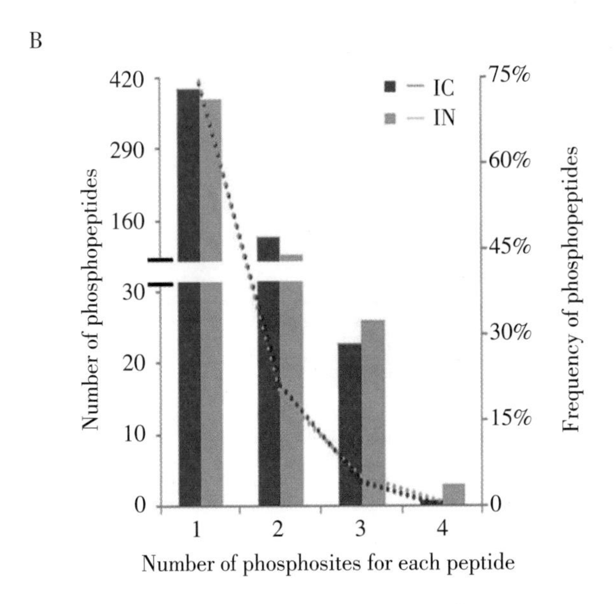

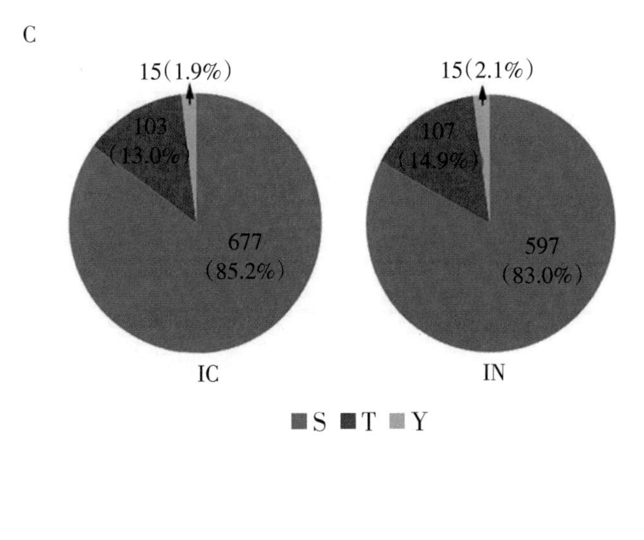

Figure 6. Overview of the phosphorylation sites in the PSG of silkworm. Bar graphs show the number of phosphosites for each protein (A) and each peptide (B) detected in the distinct strains. The lines represent the corresponding percent. Pie chart shows the distribution of the phosphorylated amino acids in the IC strain (upper) and the IN strain (lower) (C).（选自：*Phosphoproteomic analysis of the posterior silk gland of Bombyx mori provides novel insight into phosphorylation regulating the silk production*，参见本书第24页。）

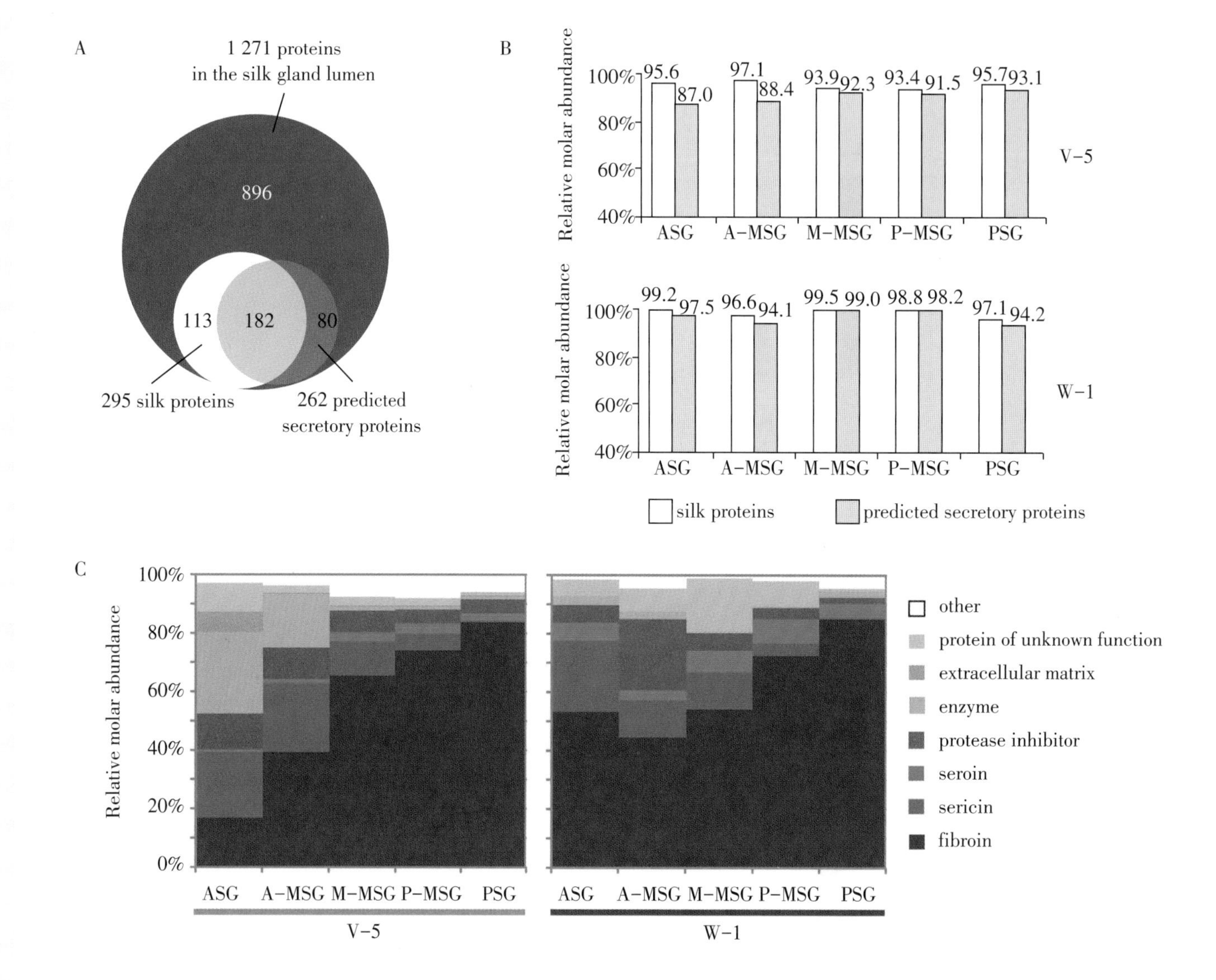

Figure 7. Identification, annotation and classification of proteins in the silk gland lumen. (A) LC-MS/MS identified 1 271 proteins in the silk gland lumen, 295 proteins of them was identified as silk proteins (Dong *et al*. 2013; Zhang *et al*. 2015), and 262 proteins were predicted as secretory proteins by Phobius website. The secretory proteins contain signal peptides but no transmembrane regions. (B) The relative abundances of silk proteins and predicted secretory proteins in the lumen of silk gland. (C) The relative abundances of the proteins in the silk gland lumen according to their functional categories. The protein abundances were calculated with intensity-based absolute quantification (iBAQ).（选自：*Analysis of proteome dynamics inside the silk gland lumen of Bombyx mori*，参见本书第26页。）

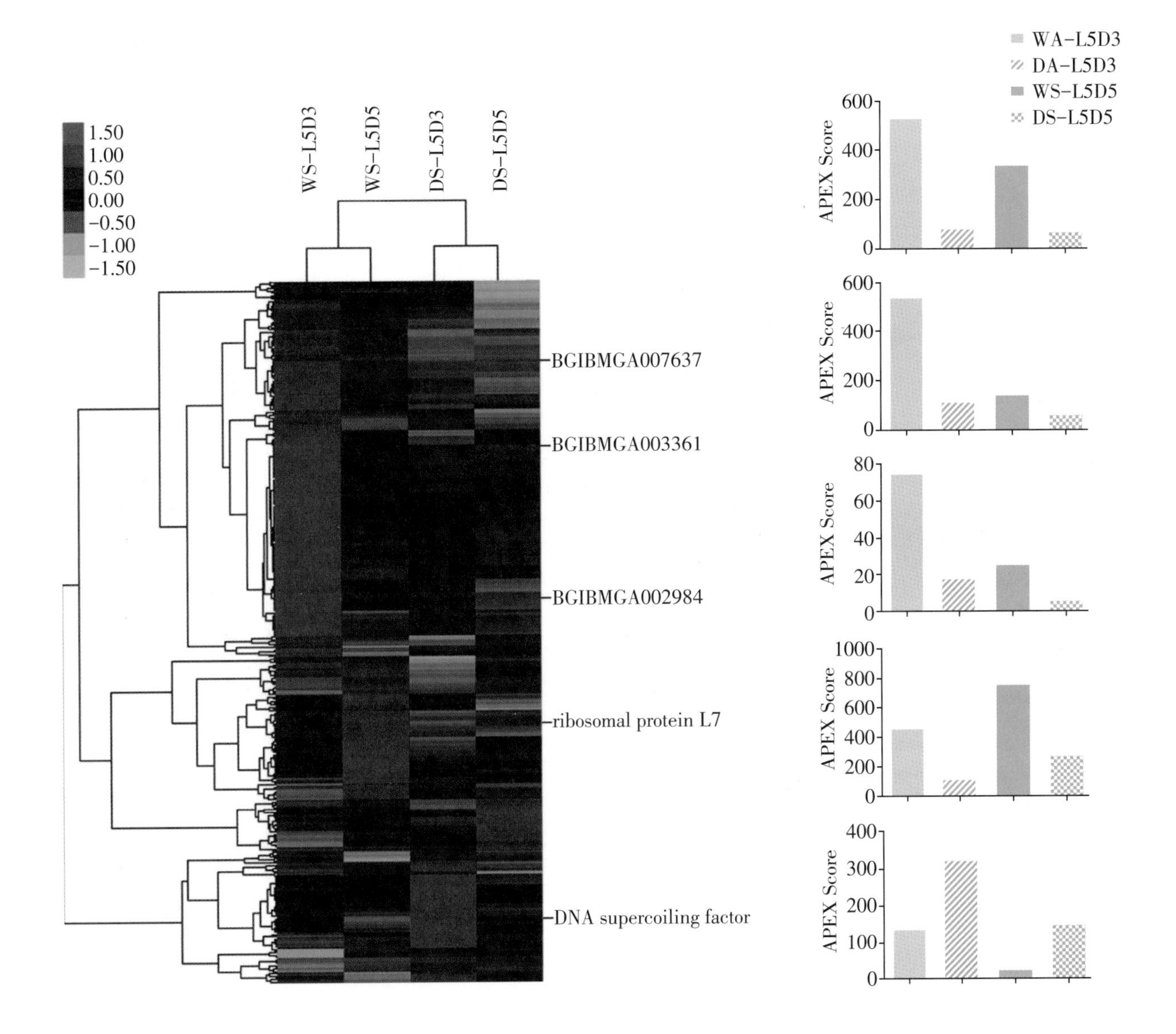

Figure 8. TreeView analysis of the differentially expressed proteins between wild and domesticated silkworm PSGs. Differentially expressed proteins in pairwise comparisons of wild and domesticated silkworms at L5D3 and L5D5 were clustered with Gene Cluster 3.0 and viewed with Java TreeView software. The heat map on the upper left indicates the differential expression levels. The relative expression of representative proteins evaluated with APEX scores is shown in the right panel.（选自：*Comparative proteomic analysis of posterior silk glands of wild and domesticated silkworms reveals functional evolution during domestication*，参见本书第28页。）

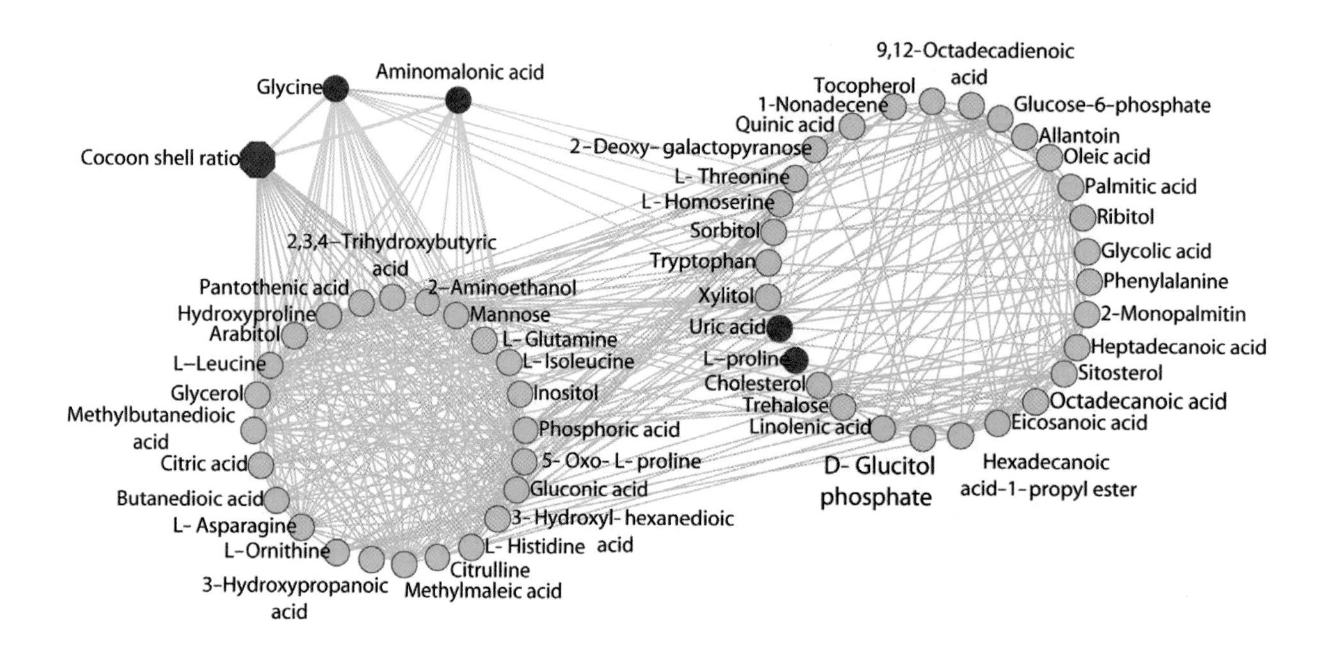

Figure 9. The correlation networks diagram of the cocoon shell ratio and differential metabolites from FGKO group. Highly correlated metabolites (|Cij|> 0.65) were connected with a line. FGKO means *fib-H* gene knocked-out, red dots mean up-regulated in FGKO samples, green dots mean down-regulated in FGKO samples. Blue line indicates negative relationship, pink line represents positive relationship. (For interpretation of the references to color in this figure legend, the reader is referred to the web version of this article.)（选自：*GC/MS-based metabolomic studies reveal key roles of glycine in regulating silk synthesis in silkworm, Bombyx mori*，参见本书第30页。）

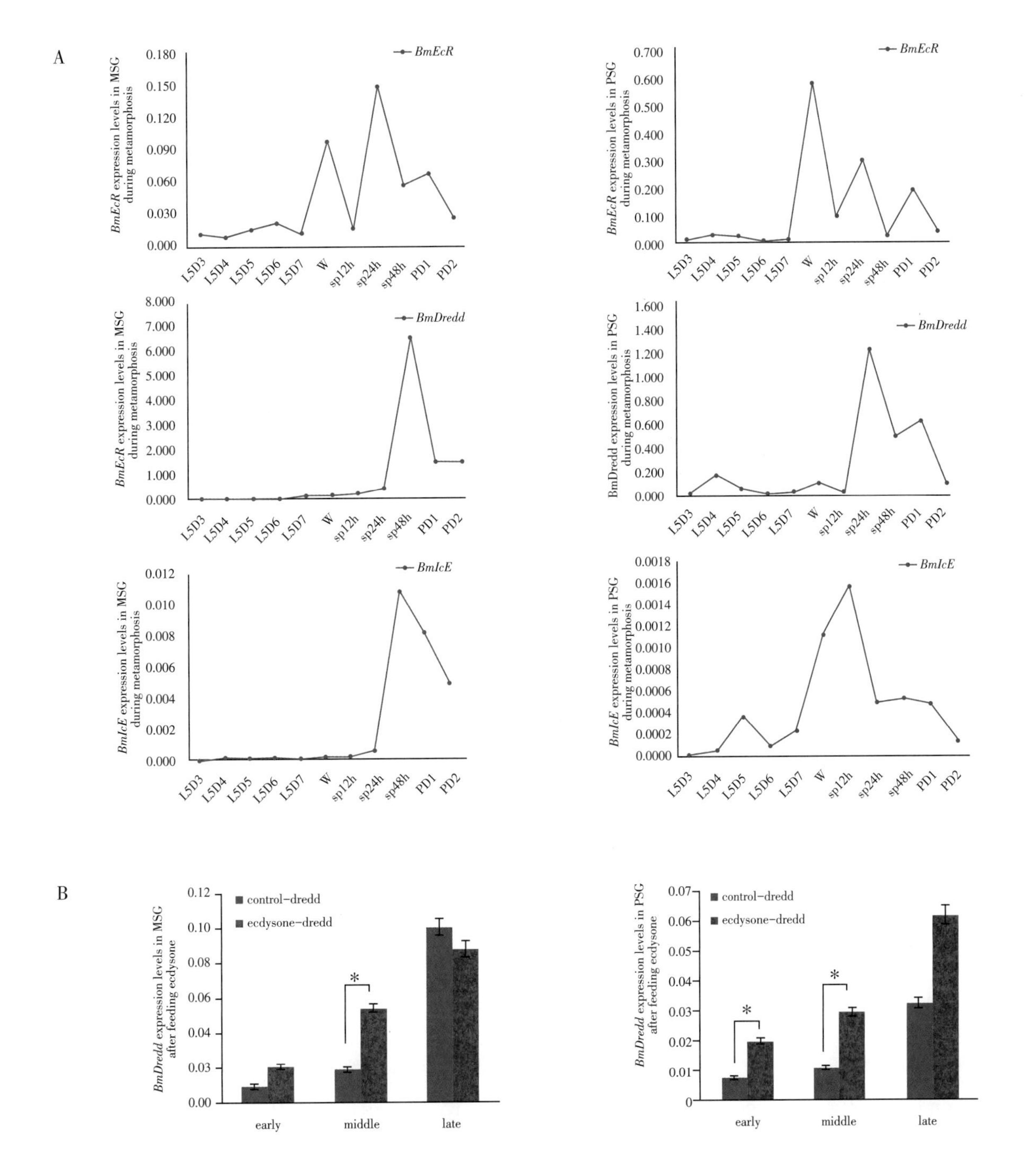

Figure 10. Expression level interactions between *BmDredd* and other two apoptosis-related genes in MSG and PSG during metamorphosis. (A) Expression levels in MSG and in PSG. L5D3-L5D7: day 3 to day 7 of the fifth larval instar. W: wandering stage. sp12h: 12 h of spinning. PD1: 1-day-old pupa. PD2: 2-day-old pupa. (B) The increase of *BmDredd* expression induced by ecdysone in MSG and PSG, P<0.5. Early, middle and late means the time of feeding ecdysone at L5D3, L5D5 to L5D6 and L5D7 to wandering respectively.（选自：*Role of BmDredd during apoptosis of silk gland in silkworm, Bombyx mori*，参见本书第38页。）

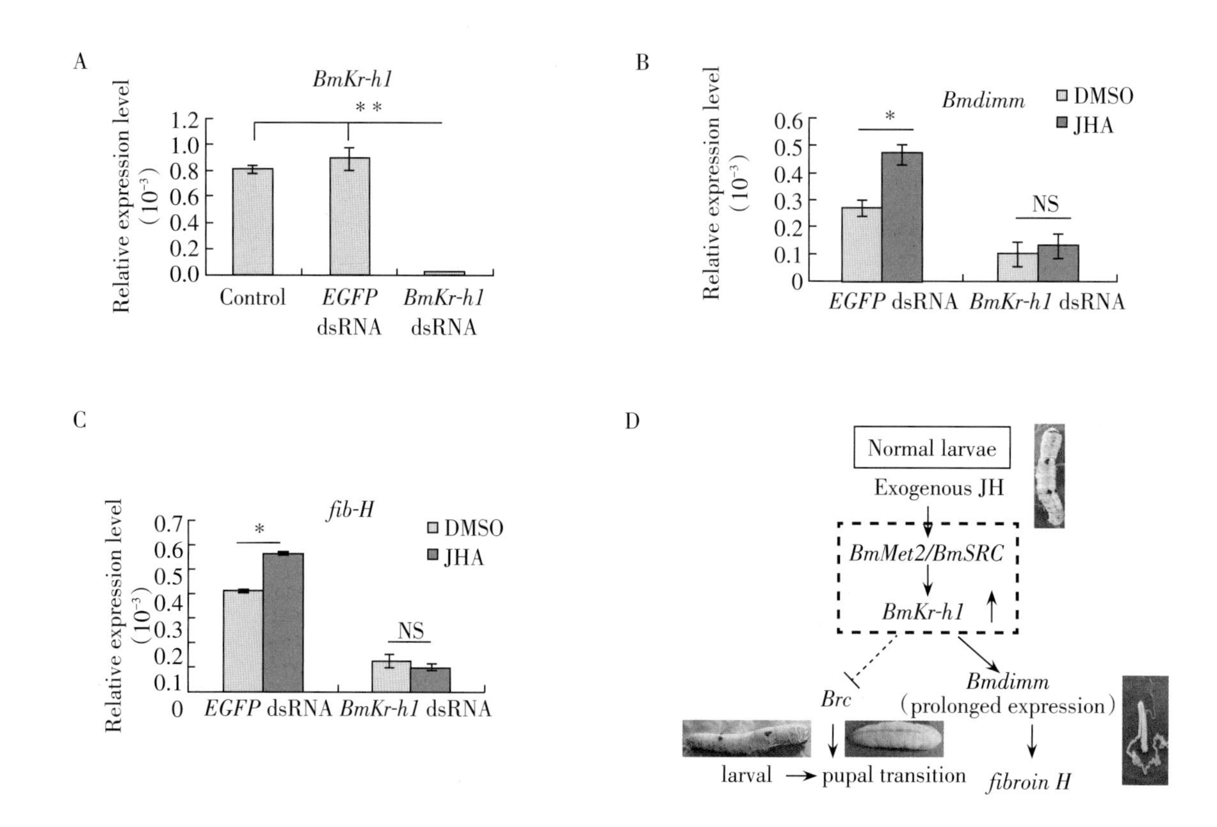

Figure 11. Knockdown of *BmKr-h1* by RNAi in *BmE* cells and schematic description of hypothesized regulation of *fib-H*. A, knockdown of *BmKr-h1* by RNAi assayed by qRT-PCR. B, expression of *Bmdimm* after knocking down *BmKr-h1* assayed by qRT-PCR. C, expression of *fib-H* after knocking down *BmKr-h1* assayed by qRT-PCR. *BmE* cells were transfected with 5 μg of *BmKr-h1* dsRNA or *EGFP* dsRNA for 12 h and incubated in medium containing 10 μM JHA or DMSO for 12 h. *BmRpl3* expression is shown as a control. Results are expressed as means ±S.D. of three independent experiments; *, $P<0.05$; **, $P<0.01$. D, schematic representation of the hypothesis for molecular regulation of *fib-H*. NS, no significance.（选自：*A juvenile hormone transcription factor Bmdimm-fibroin H chain pathway is involved in the synthesis of silk protein in silkworm, Bombyx mori*，参见本书第42页。）

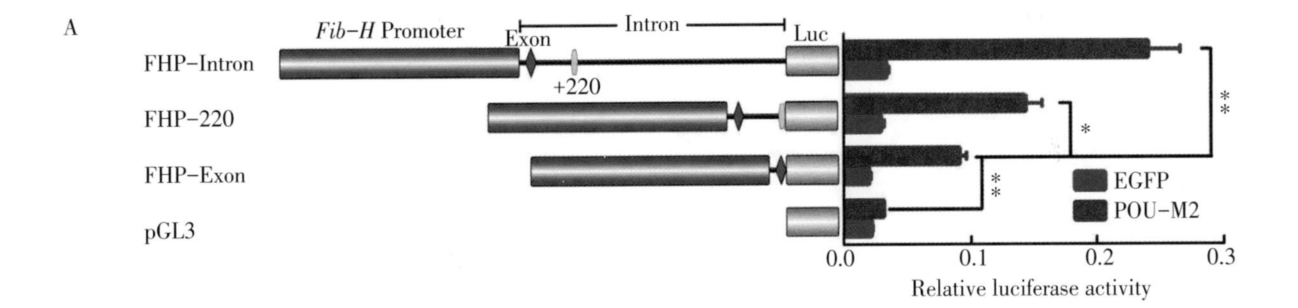

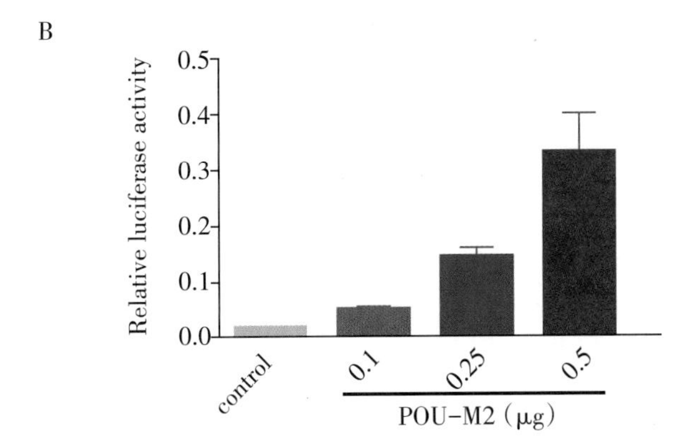

Figure 12. Effects of POU-M2 on the promoter activity of *fib-H*. A. Different promoters of *fib-H* are constructed and transfected into *BmN* cells for luciferase activity analysis. B. Effects of the amount of POU-M2 on the promoter activity of FHP-Intron. The result is repeated three times independently. All data are analyzed statistically by independent *t*-test. Asterisk indicates the value is significantly different from control (**P*<0.05, ***P*<0.01, ****P*<0.001).（选自：*Biochemical characterization and functional analysis of the POU transcription factor POU-M2 of Bombyx mori*，参见本书第48页。）

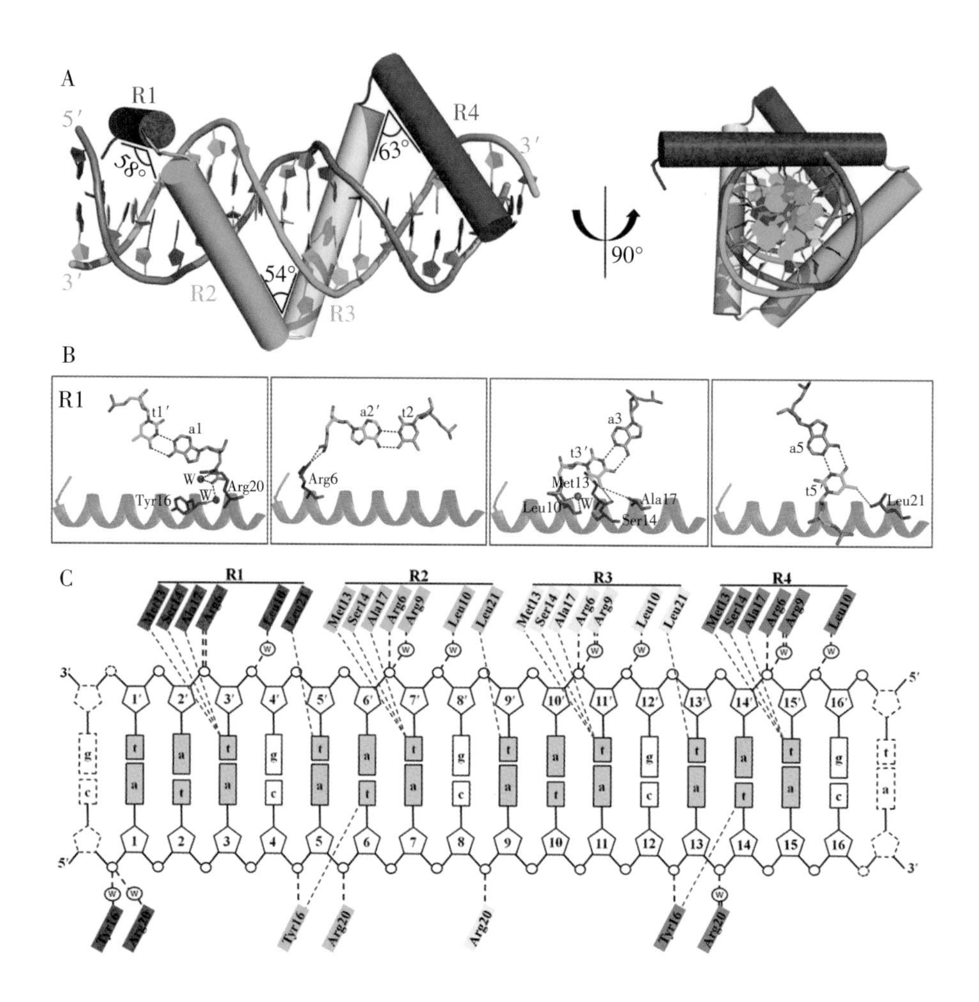

Figure 13. The structure of *Bm*STPR in complex with 18-bp DNA containing four repeats of 5′-atac-3′. (A) Cartoon representation of *Bm*STPR in complex with 18-bp DNA. The DNA strands and repeats of *Bm*STPR adopt the same colour coding as Figure 1C, in addition to R1 coloured in red. (B) Cartoon representation of the contacts between R1 and corresponding nucleotides in the 18-bp DNA complexed structure. The involved nucleotides and residues are labelled and shown as sticks. The water molecules are indicated as red spheres and marked with the letter 'W'. (C) A diagram of the interactions between *Bm*STPR and 18-bp DNA. (选自:*Structures of an all-α protein running along the DNA major groove*,参见本书第50页。)

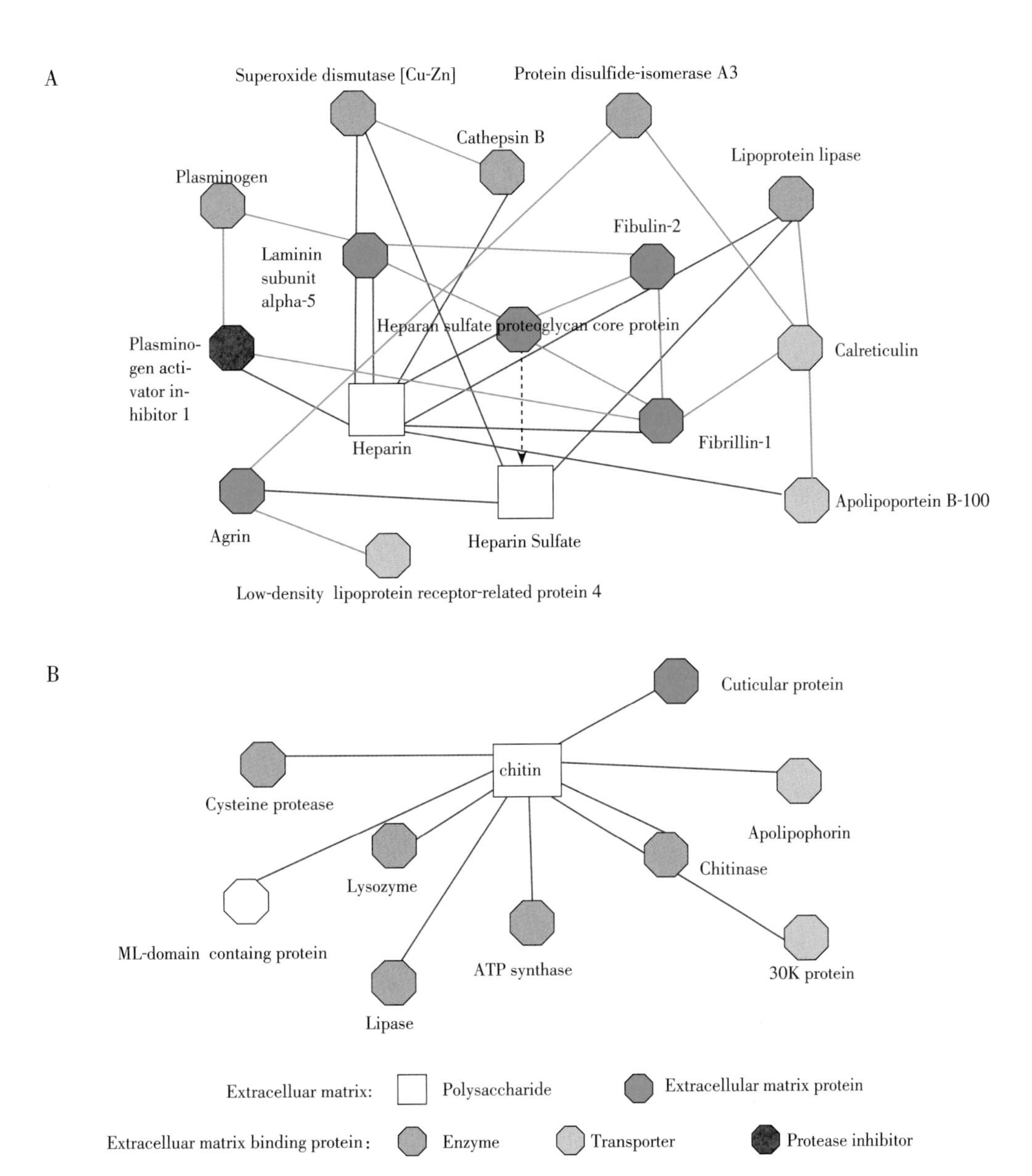

Figure 14. Interaction network model of extracellular-matrix-associated proteins found in *B. mori* silks. The extracellular-matrix-associated proteins in *B. mori* silks include homologues of human extracellular-matrix-associated proteins (A) and chitin-binding proteins (B). Interaction networks of homologues of human extracellular matrix proteins were built by using the tools provided on the website of MatrixDB. Chitin-binding proteins were reported by Tang *et al.* (2010). Hexagons represent proteins and squares represent polysaccharides. Colors show the protein category: extracellular matrix proteins, blue; protease inhibitors, red; enzymes, green; binding and transport proteins, yellow.（选自：*Comparative proteomics reveal diverse functions and dynamic changes of Bombyx mori silk proteins spun from different development stages*，参见本书第54页。）

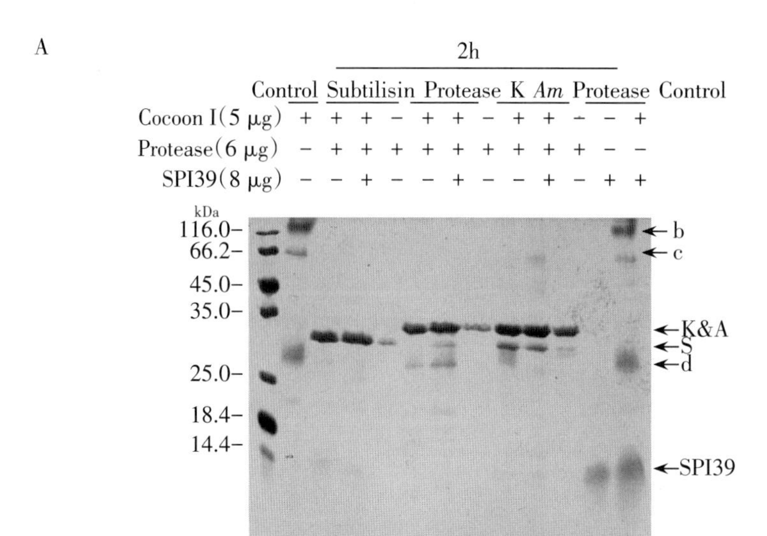

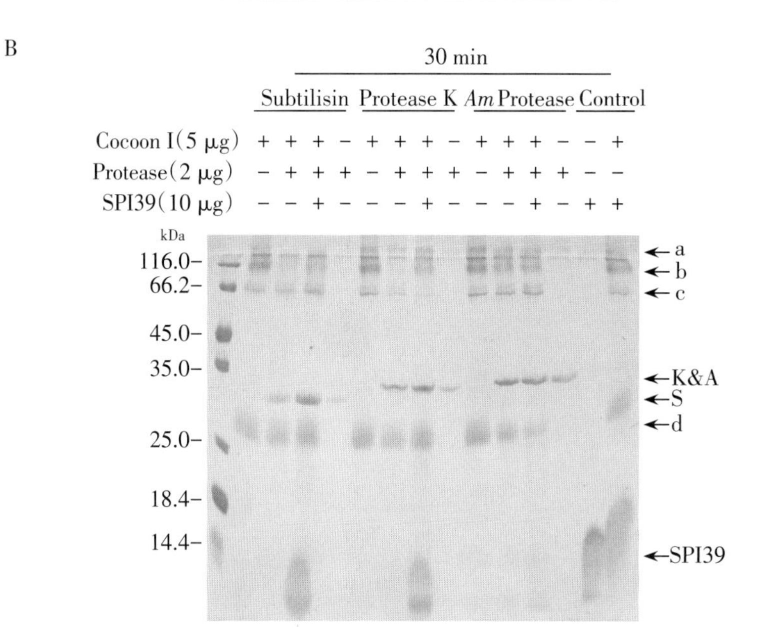

Figure 15. The hydrolysis of cocoon proteins caused by microbial proteases can be partially suppressed by *Bm*SPI39. (A) Degradation of cocoon proteins by microbial proteases. *In vitro* enzymatic hydrolysis of cocoon proteins was performed using an excess of microbial proteases, including subtilisin A from *Bacillus licheniformis*, protease K from *Engyodontium album*, and proteases from *Aspergillus melleus*. (B) The inhibition of enzymatic hydrolysis of cocoon proteins by *Bm*SPI39. *Am*Protease, protease from *A. melleus*; Cocoon I, Cocoon Proteins I; SPI39, recombinant His_6-*Bm*SPI39 protein. The letters ‘a’, ‘b’, ‘c’ and ‘d’ indicate four predominant protein bands in cocoon lysates under test conditions. The letters ‘S’, ‘K’ and ‘A’ indicate subtilisin A, protease K, and proteases from *A. melleus*, respectively. The symbols ‘+’ or ‘ − ’ indicate with or without material, respectively. Corresponding proteases and *Bm*SPI39 were used as blank controls. Cocoon Proteins I alone or in combination with *Bm*SPI39 were used as negative controls.（选自：*Protease inhibitors in Bombyx mori silk might participate in protecting the pupating larva from microbial infection*，参见本书第58页。）

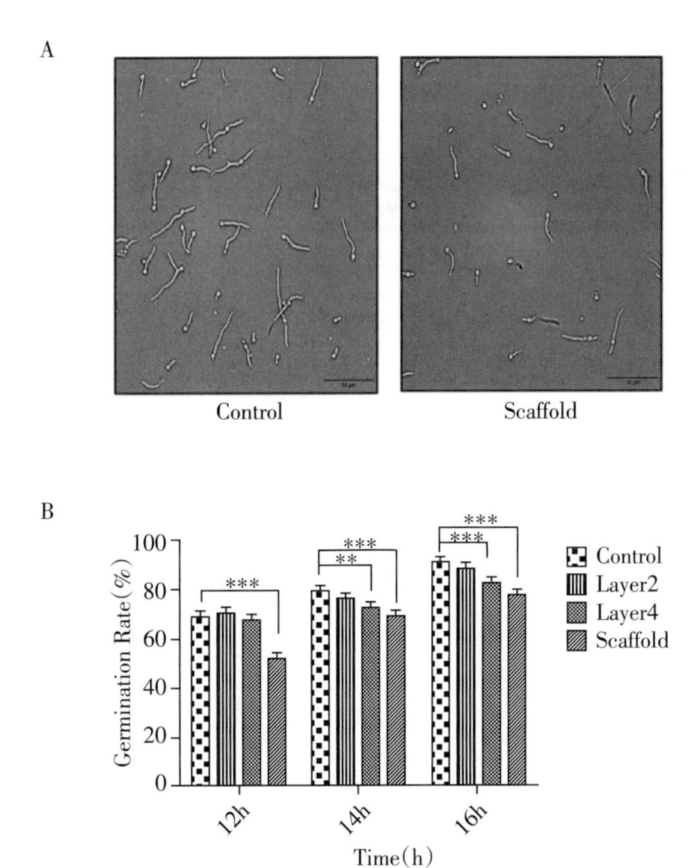

Figure 16. Inhibition of *B. Bassiana* conidia by proteins extracted from different cocoon layers. (A)Photographs showing the inhibition on germination of *B. bassiana* spores by the scaffold proteins. (B) The inhibition of *B. Bassiana* conidia by proteins from different cocoon layers. Equal amounts (65 μg) of proteins from different cocoon layers were incubated with the *B. bassiana* conidia. ***P<0.001 and **P<0.01 versus the control. Error bars indicate the standard error of the mean (n = 3).（选自：*Proteins in the cocoon of silkworm inhibit the growth of Beauveria bassiana*，参见本书第60页。）

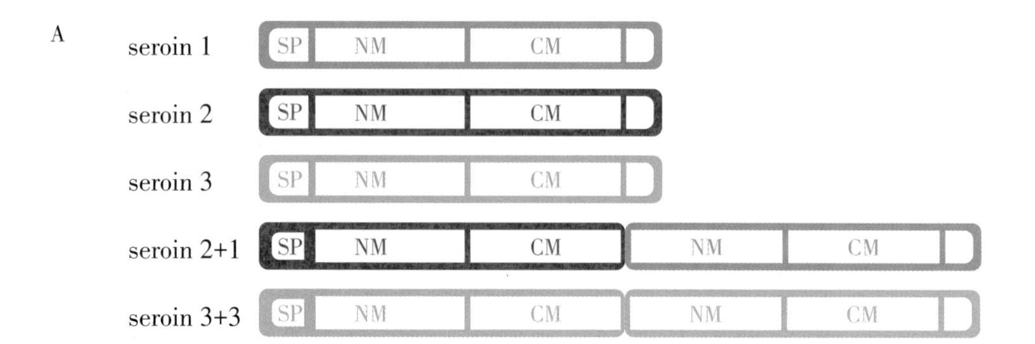

Figure 17. Structure and classification of lepidopteran seroins. (A) Classification of seroins. Seroins were divided into five subfamilies: seroin 1, seroin 2, seroin 3, seroin 2+1, and seroin 3+3. Boxes in different colors represent seroin 1 domain (blue), seroin 2 domain (red), and seroin 3 domain (green), respectively. Each seroin domain contains a signal peptide (SP), a proline-rich N-terminal motif (NM), a conserved C-terminal motif (CM), and an extension after CM. (B) WebLogos for CMs from different seroin domains. WebLogos were constructed at http://weblogo.berkeley.edu/logo.cgi. Gray boxes represent seroin 1 and 3 specific sequences in the CM. (For interpretation of the references to colour in this figure legend, the reader is referred to the web version of this article.)（选自：*Structure, evolution, and expression of antimicrobial silk proteins, seroins in Lepidoptera*，参见本书第62页。）

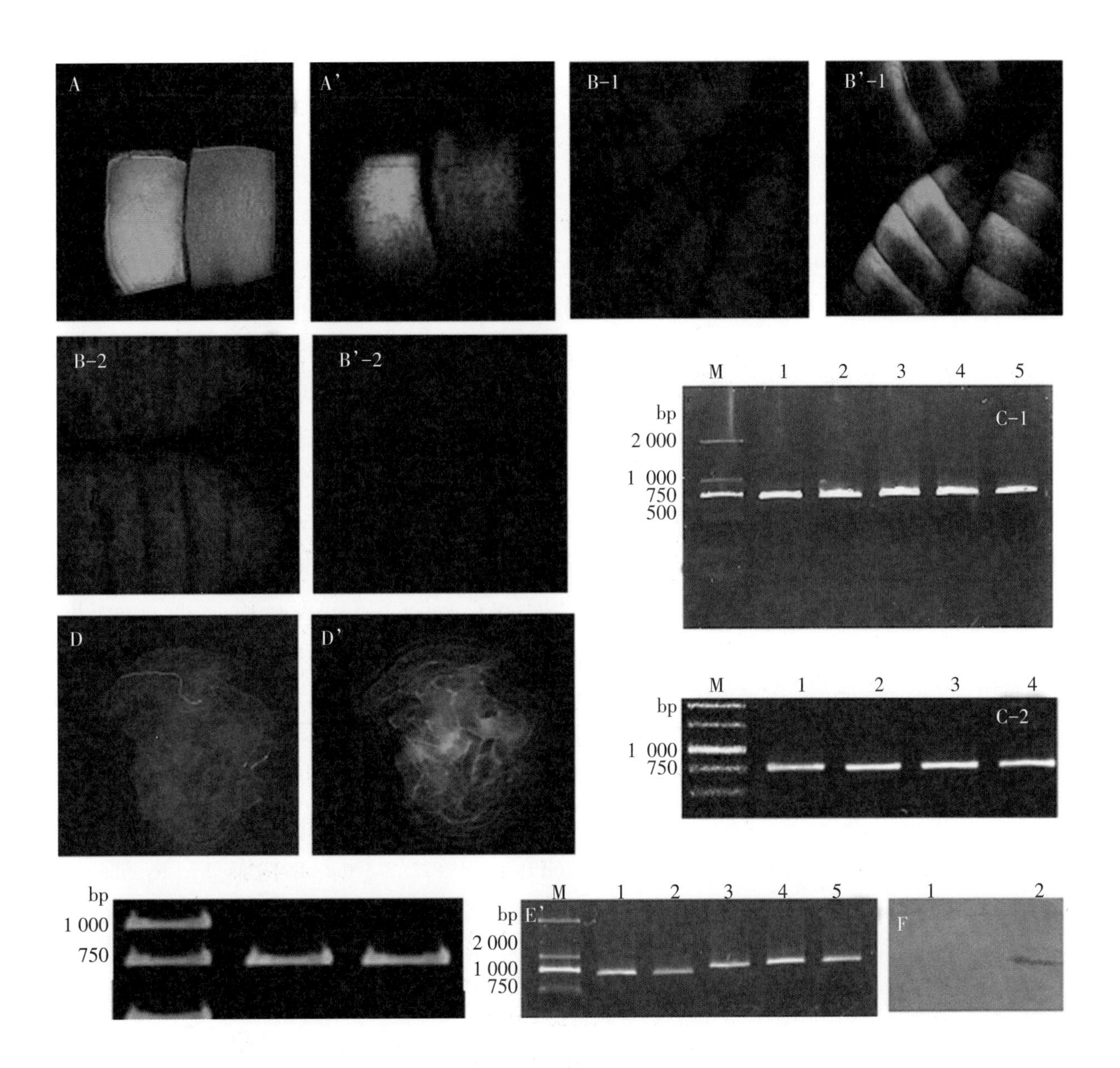
A
A'
B-1
B'-1
B-2
B'-2
M 1 2 3 4 5
bp
C-1
2 000
1 000
750
500
M 1 2 3 4
bp
C-2
1 000
750
D
D'
bp
1 000
750
M 1 2 3 4 5
bp E'
2 000
1 000
750
1 2
F

Figure 18. Screening and identification of transgenic silkworms A and A', a cocoon piece of the G_0 generation transgenic silkworm under normal light and fluorescent light (480 nm), respectively. Left transgenic; right non-transgenic; B, the observation of the pupa of the G_0 generation transgenic silkworm under different light sources. B-1 and B'-1, the pupa under normal light and fluorescent light (480 nm), respectively. B-2 and B'-2, the pupa under bright light and fluorescent light (554 nm); C-1, identification of the G_0-G_3 generations of the transgenic silkworm by PCR with primers FGFP-1/FGFP-2. Lanes M DNA ladder; 1 the template was the targeting vector; 2-5 the template is the genomic DNA extracted from G0-G3 generation silkworm. C-2, identification of the G_4-G_6 generations of the transgenic silkworm by PCR with primers FGFP-1/FGFP-2. Lanes M DNA ladder; 1 the template was the targeting vector; 2-4 the template is the genomic DNA extracted from G_4-G_6 generation silkworm. D and D', The silk glands of G_3 generation silkworm under the normal light and fluorescent light (480 nm), respectively., E identification of the G_2-G_3 generations of the transgenic silkworm by RT-PCR with primers FGFP-1/FGFP-2; lanes M DNA marker; 1 and 2 the cDNA used as template was from the silk glands of G_2 and G_3 generation transgenic silkworm, respectively. E' The identification of G_4-G_6 generation transgenic silkworm by RT-PCR with primers FGFP-1/EGFP-2 or FGFP-1/cec-2; lanes M DNA marker; 1 and 2 the cDNA used as template was from silk glands of the G_4 and G_5 generations of the transgenic silkworm, respectively, the primers were FGFP-1/EGFP-2; 3-5 the cDNA used as template was from silk glands of the G_4-G_6 generations of the transgenic silkworm, respectively, the primers were FGFP-1/cec-2. F identification of the G_6 generation of the transgenic silkworm by western blot. Lanes 1 the protein extracted from the PSG of the non-transgenic silkworm; 2 the protein extracted from the PSG of the transgenic silkworm. The primary antibody was rabbit antibody against GFP, the second antibody was HRP-conjugated goat antibody against rabbit IgG（选自：*Construction of transgenic silkworm spinning antibacterial silk with fluorescence*，参见本书第66页。）

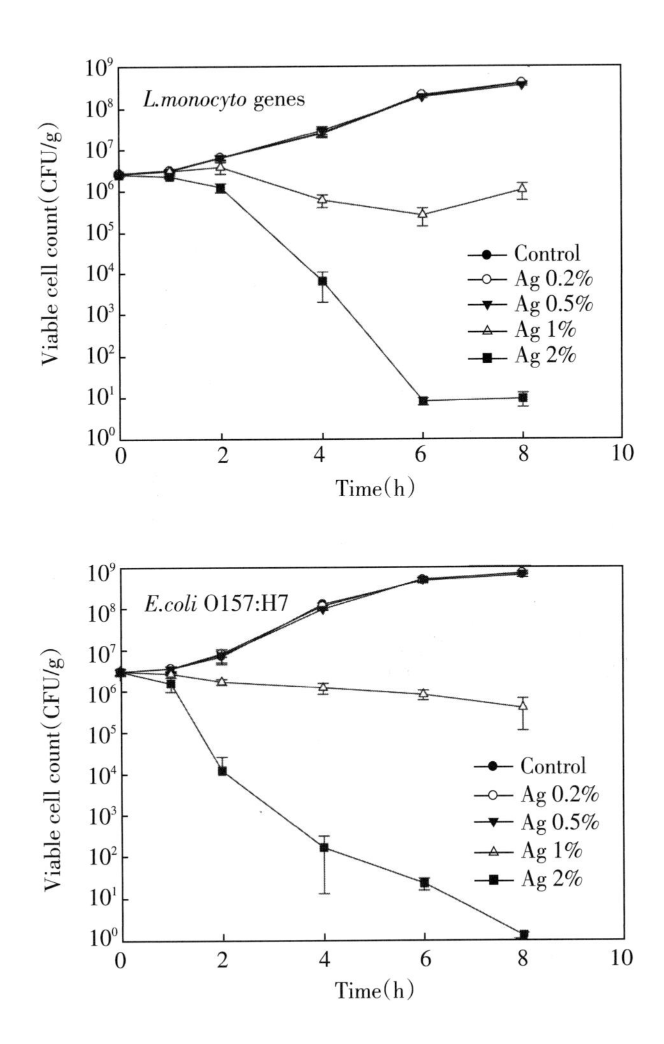

Figure 19. Antimicrobial activity of agar and agar/AgNPs composite films against pathogenic Gram^{+} and Gram^{-} bacteria.（选自：*Preparation and characterization of silver nanoparticles composited on polyelectrolyte film coated sericin gel for enhanced antibacterial application*，参见本书第80页。）

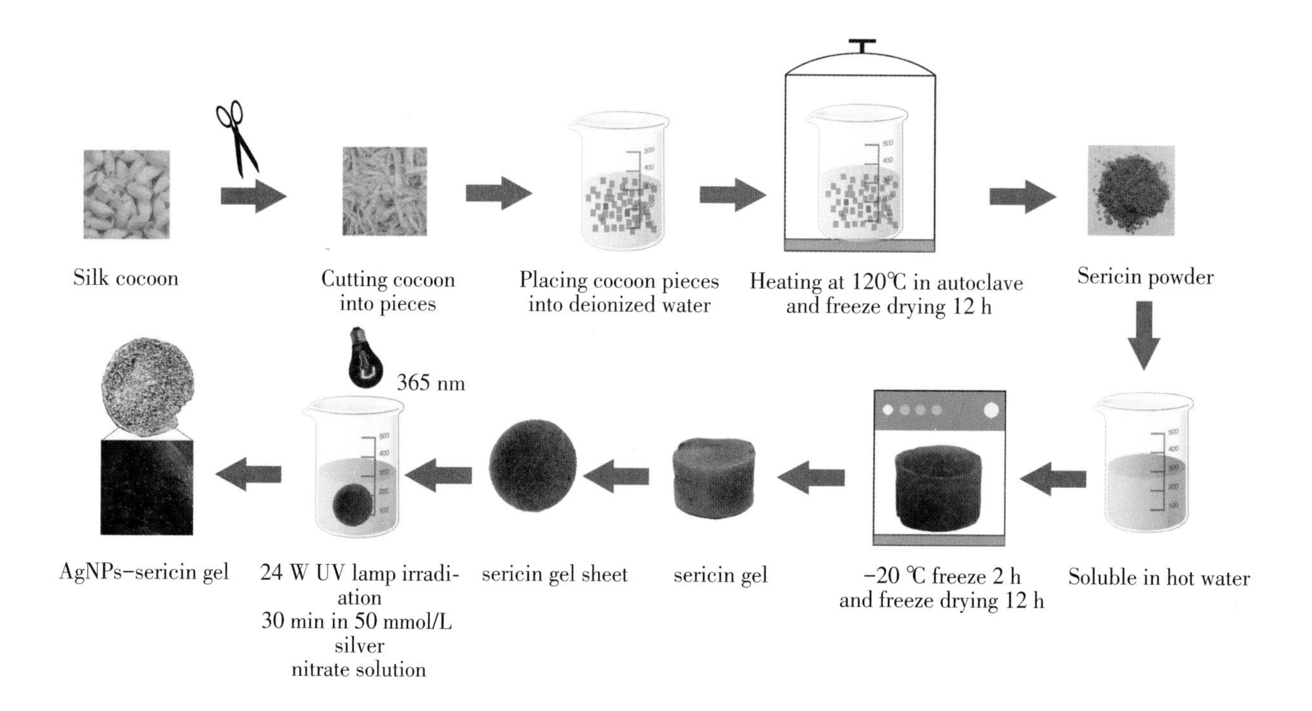

Figure 20. A flowchart illustrates the procedure of the extraction and fabrication of sericin, sericin hydrogel, sericin gel, and AgNPs-sericin gel.（选自：*Characterization of silver nanoparticle in situ synthesis on porous sericin gel for antibacterial application*，参考本书第82页。）

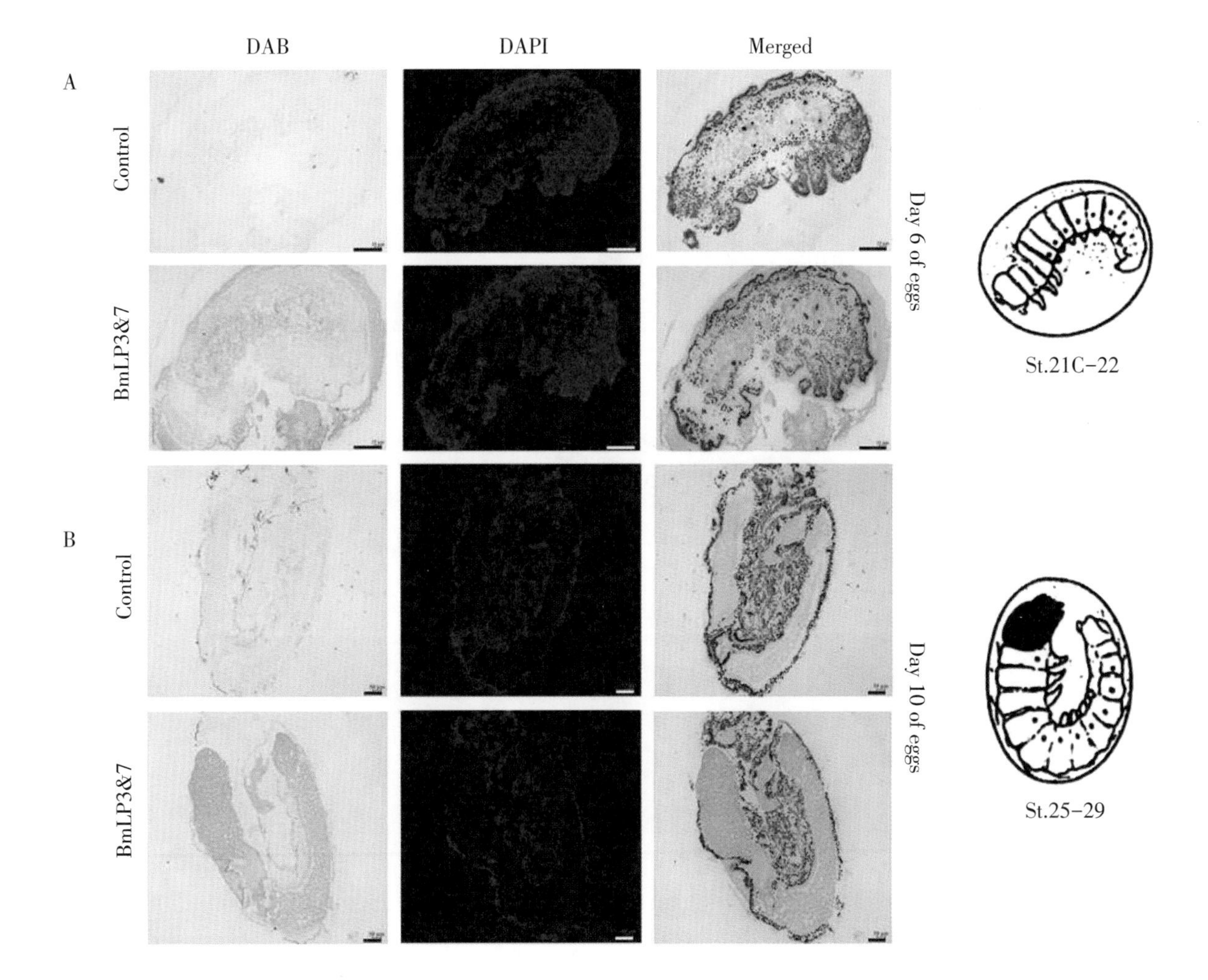

Figure 21. Immunolocalization of *Bm*LP3 and *Bm*LP7 in eggs. *Bm*LP3 and *Bm*LP7 were located by the appropriate antibodies in day 6 (A) and day 10 eggs (B) with sections treated with normal serum as the control. The scale bar represents 50 μm. The *Bm*LP3 and *Bm*LP7 signals were shown by staining with DAB (tan) and the morphology of the embryo was shown by counterstaining with DAPI (blue). The diagrams represent two stages of embryogenesis (Kuwana and Takami, 1968; Ueno *et al.*, 1995). (For interpretation of the references to colour in this figure legend, the reader is referred to the web version of this article.)(选自：*The synthesis, transportation and degradation of BmLP3 and BmLP7, two highly homologous Bombyx mori 30K proteins*，参见本书第88页。)

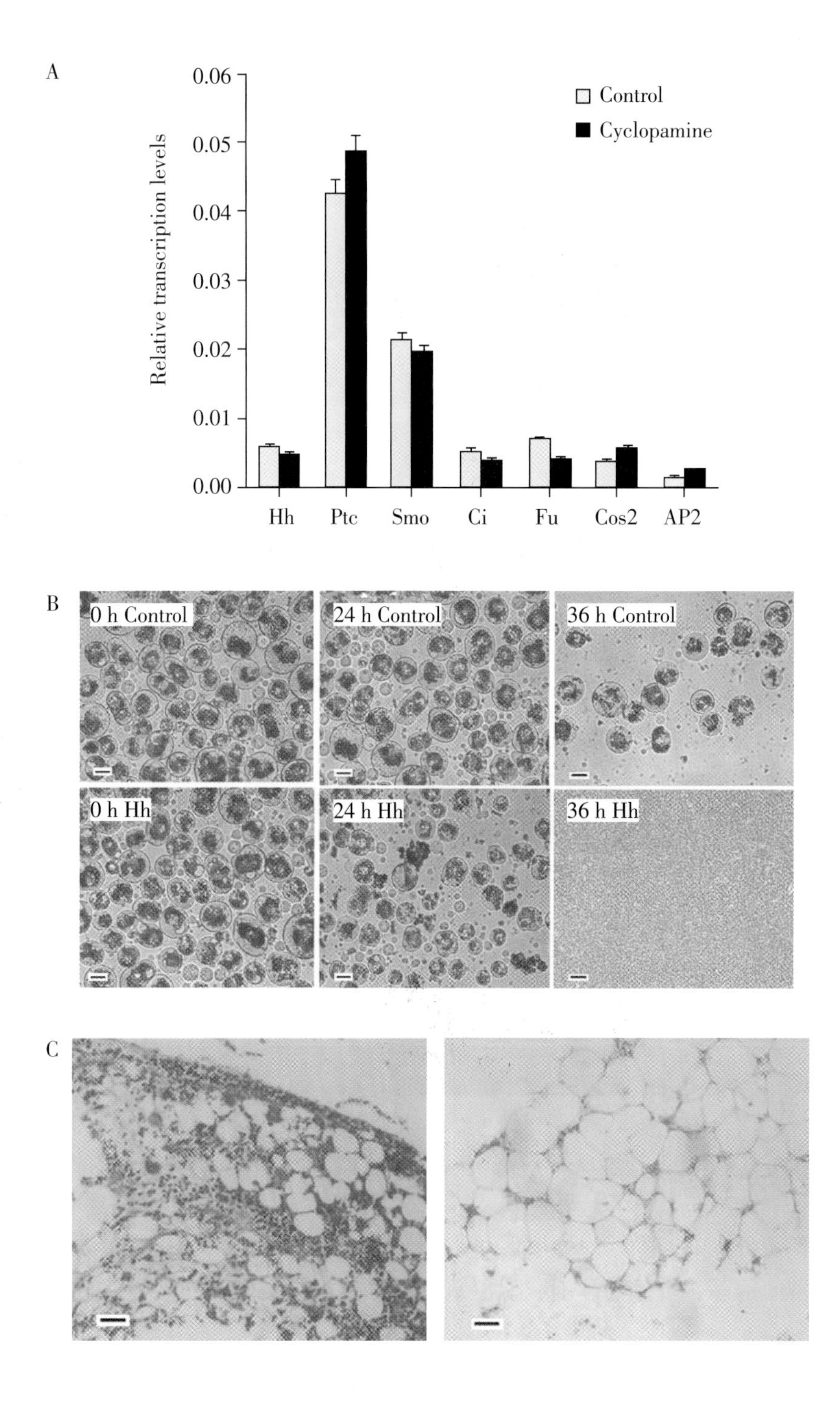

Figure 22. Cyclopamine inhibits Hedgehog (Hh) pathway stimulates *Bombyx mori* fat formation. A: The expression levels of Hh pathway and adipocyte marker gene *AP2* after cyclopamine treatment in silkworm adipose cells. B: Adipose cells were incubated in TC100 media in the presence or absence (control) of 10 μmol/L cyclopamine, and lipid formation was assessed based upon morphology (B). C: Hematoxylin-eosin stained sections of adipose tissues from cyclopamine-treated (i.e., Hh inhibited) larvae and control larvae. Scale=20 μm.（选自：*Hedgehog signaling pathway regulated the target genes for adipogenesis in silkworm Bombyx mori*，参见本书第98页。）

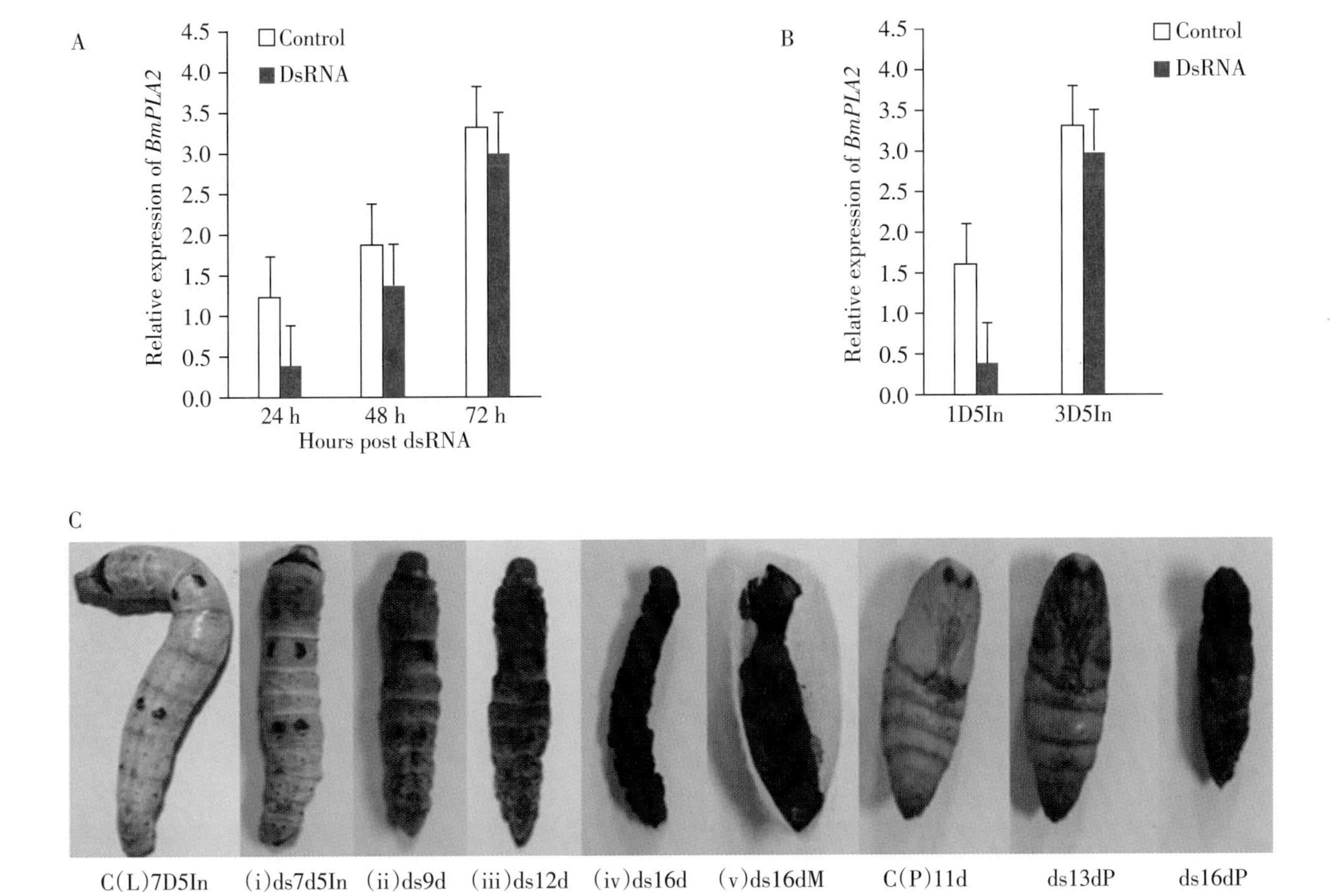

Figure 23. Transcript level of RNAi mediated knockdown of *BmPLA2* in silkworm *Bombyx mori.* (A) *BmPLA2* expression in fat body tissue after knockdown in post-24 h, 48 h and 72 h of double-stranded DNA (dsRNA) treatment. (B) *BmPLA2* expression in fat body tissue-specific after knockdown in fifth instar larvae (1D5In, first day fifth instar; 3D5In, third day fifth instar). (C) dsRNAi mediated knockdown of *BmPLA2* larvae and pupae leading to pharate adult lethality and underdevelopment of the larvae fat body and pharate pupa. C(L) 7D5In, 7-day-old fifth instar larva control; (i) ds7d5In, 7-day-old fifth instar dsRNAi knockdown larva; (ii) ds9d, 9-day-old knockdown larva; (iii) ds12d, 12-day-old knockdown larva; (iv) ds16d, 16-day-old knockdown dead and dried larva; (v) ds16dM, 16-dayold knockdown and melted larva; and C(P)11d, 11-day-old pupa control; ds13dP, 13-day-old knockdown pupa; and ds16dP, 16-day-old knockdown dead and dried pupa), respectively. ds-*EGFP* of 5 μL (1 μg/μL dissolved in DEPC water) was injected as control. All experiments were performed in three independent biological replications and reactions of each sample were carried out in triplicate.（选自：*BmPLA2 containing conserved domain WD40 affects the metabolic functions of fat body tissue in silkworm, Bombyx mori*，参见本书第100页。）

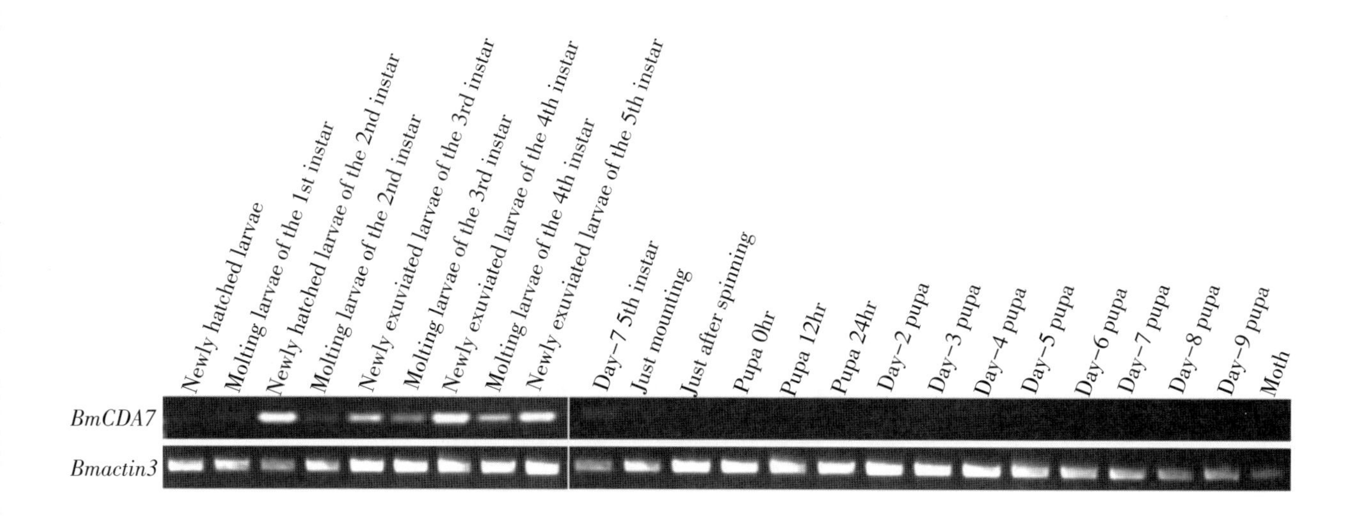

Figure 24. The temporal expression profiles of *BmCDA7*. To determine the temporal specificity of expression, three whole silkworm bodies (except for the midgut contents) were collected at the times of the first-fourth instar larvae, middle-stage silkworm and molting silkworm. Total RNA was extracted and was used as template for first-strand cDNA synthesis. *Bmactin 3* transcript with the same cDNA template served as an internal control. (选自:*Identification and molecular characterization of a chitin deacetylase from Bombyx mori peritrophic membrane*,参考本书第104页。)

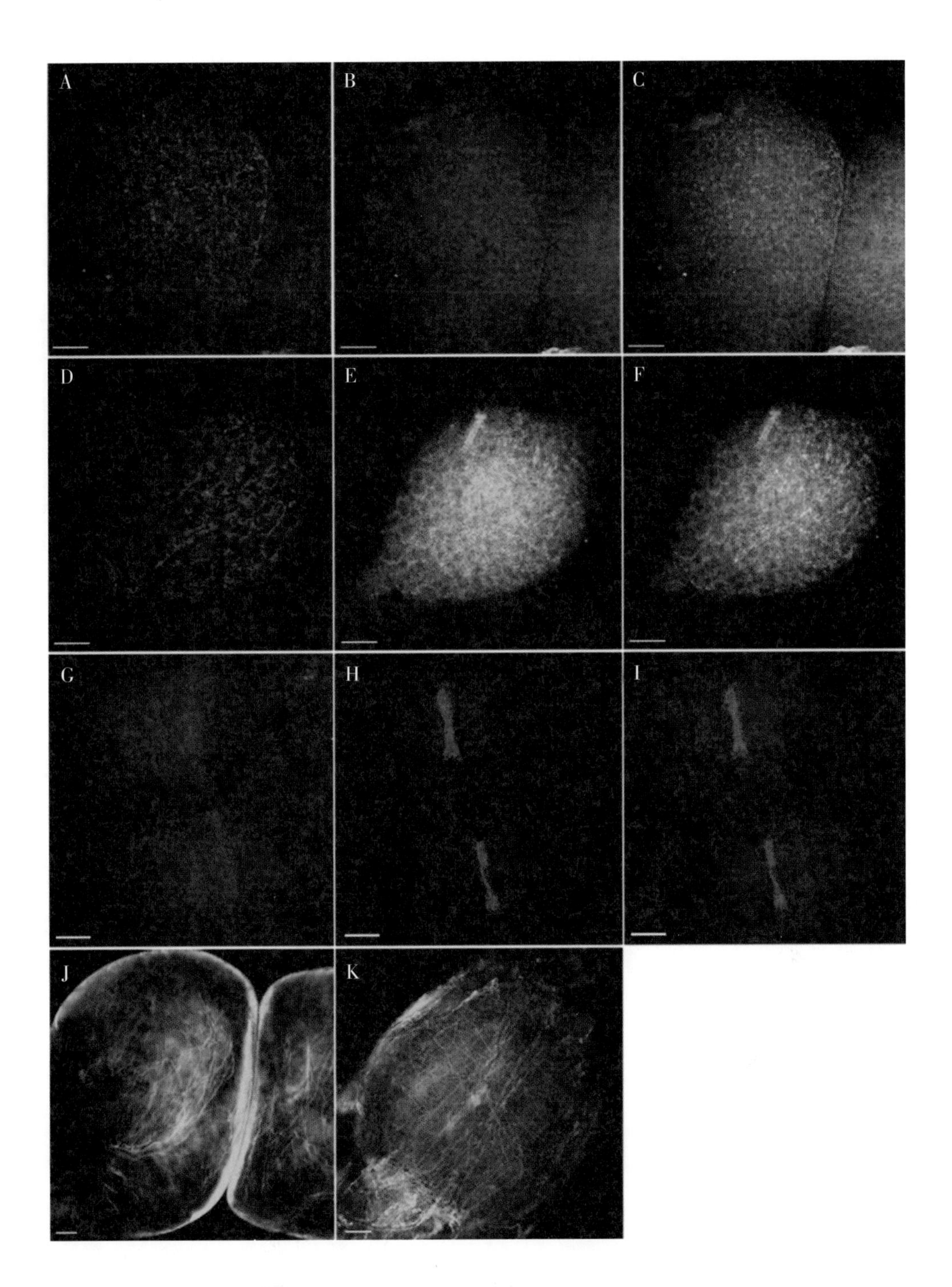

Figure 25. Expression patterns of *Bmrobo2/3* in the nervous system. (A, D) *Bmrobo2/3* mRNA was expressed in the brain (A) and the ganglion (D) of day 2 first instar larvae. (B, E) The central nervous system was revealed by anti-HRP-FITC antibody. (C, F) The merged image of (A) and (B), (D) and (E), respectively. (G-I) Localization of *Bm*Robo2/3 (G) and *Bm*Slit (H) in the CNS of the silkworm embryo. (I) The merged image of (G) and (H). (J, K) Localization of *Bm*Robo2/3 in the brain and ganglion of day 3 fifth instar larva. Scale bars: 200 μm in (A-I); 50 μm in (J, K).（选自：*BmRobo2/3 is required for axon guidance in the silkworm Bombyx mori*，参见本书第106页。）

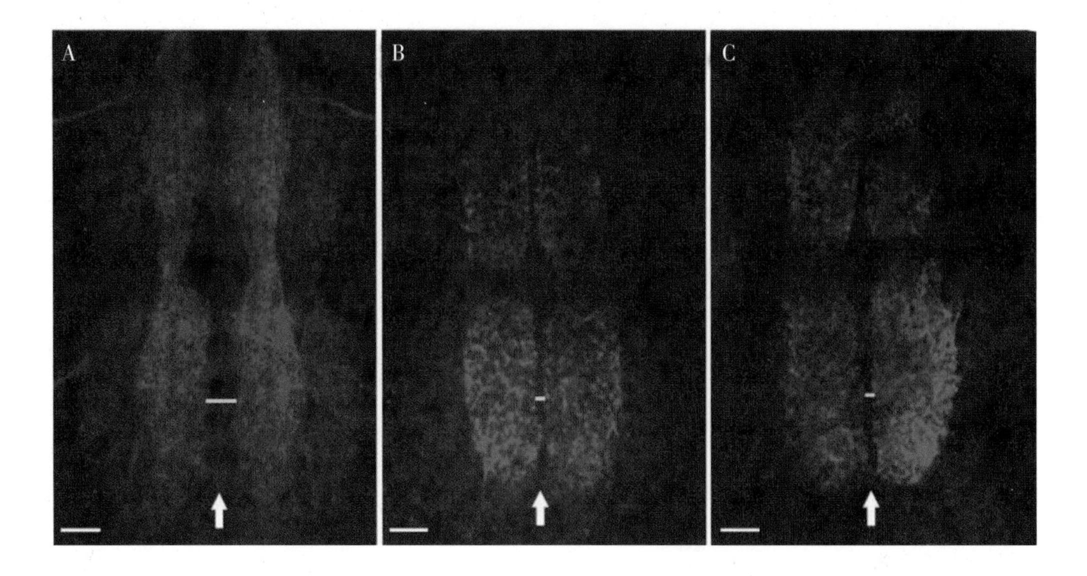

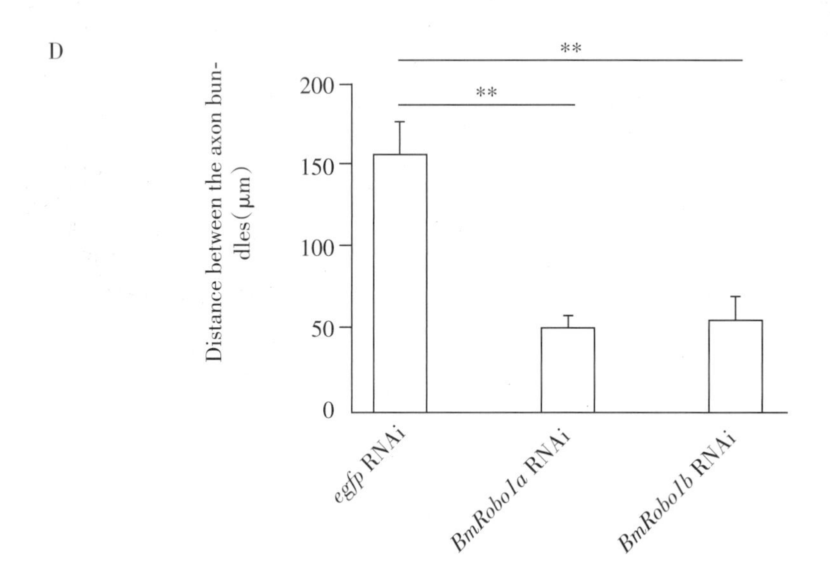

Figure 26. The knock-down phenotypes of *Bmrobo1a* and *Bmrobo1b*. (A) The control embryos were injected with *egfp* dsRNA. (B and C) The phenotype of embryos injected with *Bmrobo1a* (82.7%, *n* = 75) and Bmrobo1b (83.3%, *n* = 78) dsRNAs, respectively. Axons of embryos injected with *Bmrobo1a* and *Bmrobo1b* dsRNAs were closer to the midline compared with the control silkworm. The embryos were stained by 22C10 antibody (A-C). Arrow indicates the midline of CNS. Green line indicates the distance between the two axon bundles. (D) Quantification of the distances between the axon bundles in the embryos injected dsRNAs of *egfp* (A), *Bmrobo1a* (B) and *Bmrobo1b* (C). Error bars, SD. Student's *t*-tests, $^{**}P < 0.01$. Scale bars: 200 μm.（选自：*BmRobo1a and BmRobo1b control axon repulsion in the silkworm Bombyx mori*，参见本书第108页。）

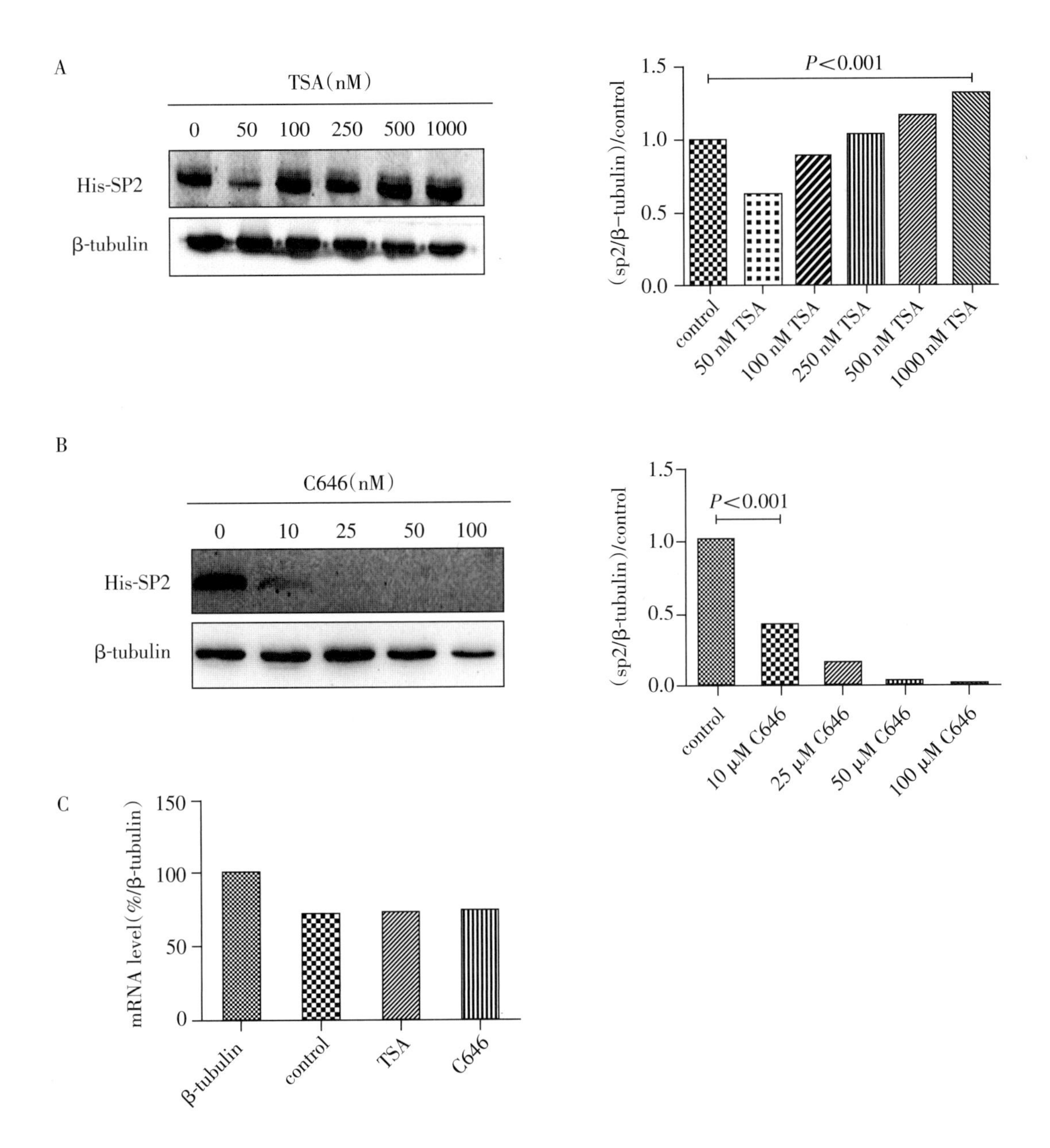

Figure 27. Expression of SP2 regulated by TSA and C646 in *BmN* cells. The protein level of His-fused SP2 protein was up-regulated by TSA (A) and down-regulated by C646 (B) with different concentrations; however, the mRNA level of SP2 was not changed (C), suggesting that lysine acetylation could up-regulate the protein level of SP2 not at the transcriptional or translational level but at the post-translational level. The expression of SP2 declined quickly and could not be detected when the concentration of C646 was greater than 25 mM.(选自:*Lysine acetylation stabilizes SP2 protein in the silkworm, Bombyx mori*,参见本书第112页。)

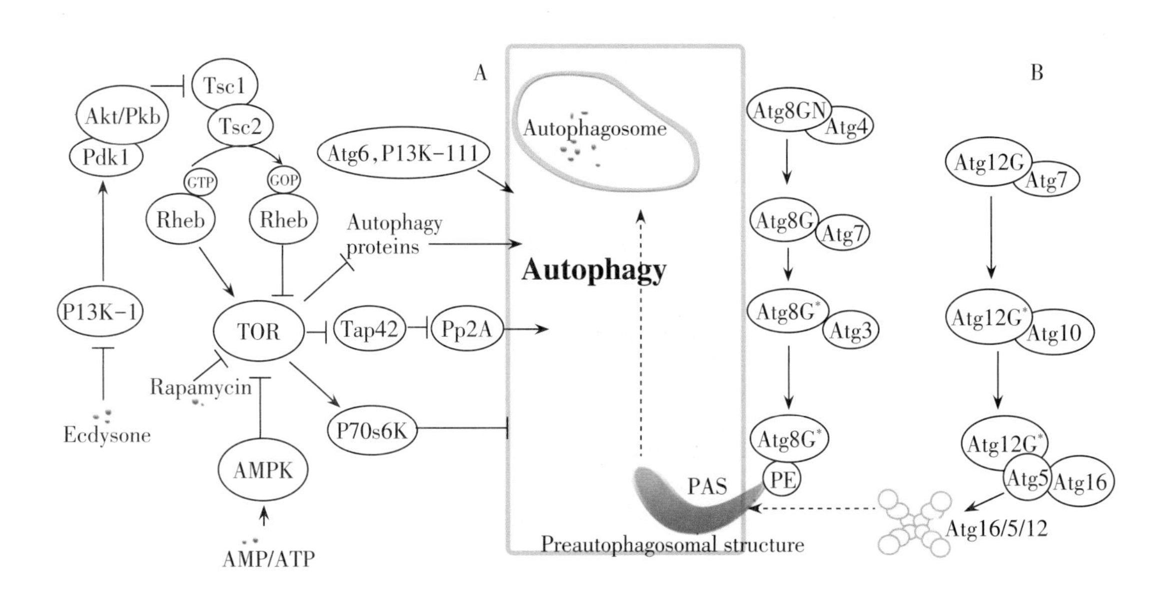

Figure 28. Autophagy pathway in *Bombyx mori*. (A) Signal transduction pathway. (B) Two ubiquitin-like protein conjugation systems. The absent Atg10 is colored in grey. (选自：*Systematic cloning and analysis of autophagy-related genes from the silkworm Bombyx mori*，参见本书第120页。)

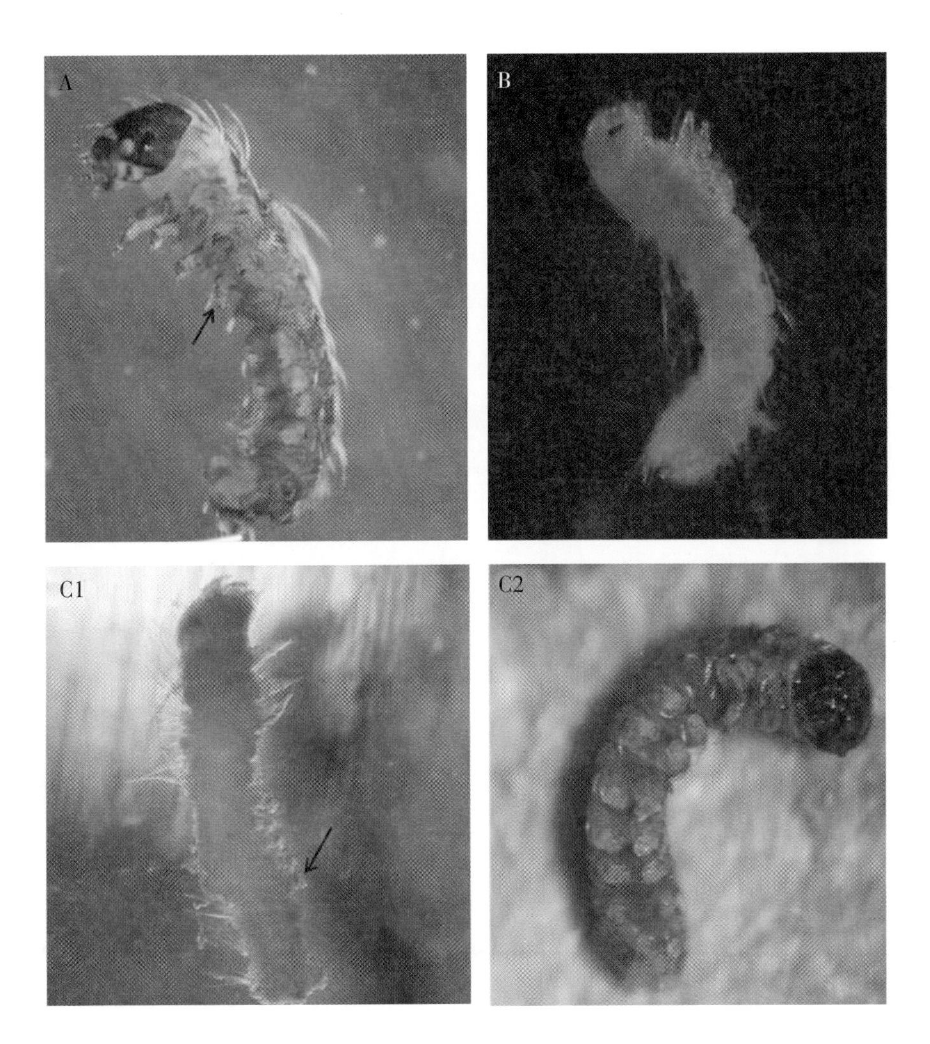

Figure 29. RNAi phenotypes of the silkworm E-complex genes. (A) The *Ubx* RNAi phenotype. A pair of supernumerary legs grow in A1 segment. (B) The *abd-A* RNAi phenotype. Prolegs from A3 to A6 segments are all severely defected. These affected embryos cannot complete the blastokinesis and died when removed from the chorion. (C) The *Abd-B* RNAi phenotype. Dose-dependent effect is detected. (C1) Mild affected embryos have a pair of supernumerary prolegs in A7 segment. (C2) Strong affected embryos have supernumerary prolegs in more posterior segments.(选自:*Influence of RNAi knockdown for E-complex genes on the silkworm proleg development*,参见本书第122页。)

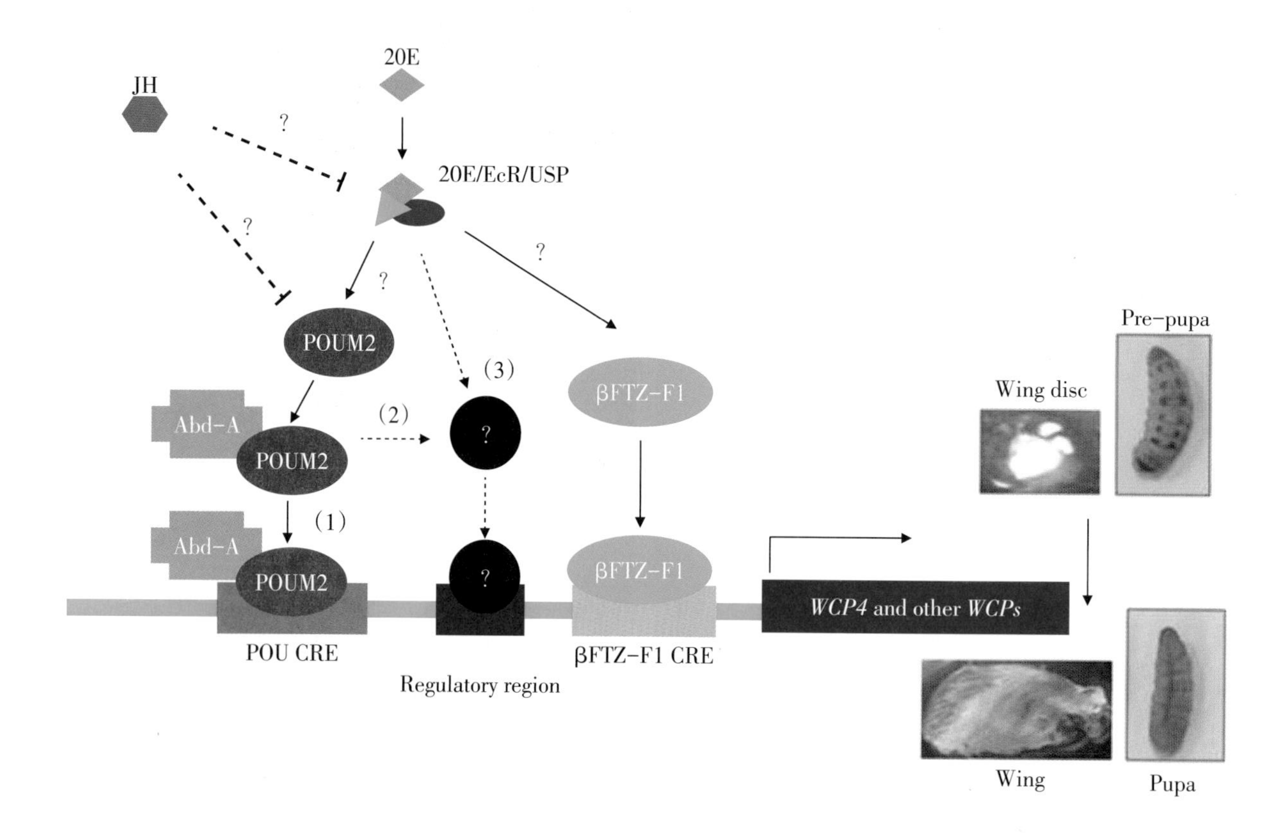

Figure 30. Schematic representation of the model of molecular regulation of the expression of *BmWCP4* as well as other *BmWCP* genes by nuclear transcription factors in response to 20-E and JH during metamorphosis in B. mori. 20E binds to the heterodimer of the USP/ecdysone receptor, which either directly or indirectly activates the expression of the transcription factors *Bm*POUM2 and *Bm*βFTZ-F1. While *Bm*βFTZ-F1 binds to the corresponding *Bm*βFTZ-F1 CRE, the nuclear homeodomain transcription factor *Bm*POUM2 interacts with another homeodomain transcription factor, *Bm*Abd-A, to form a complex, which binds to the corresponding POU CRE in the *BmWCP4* regulatory region to activate the expression of *BmWCP4*, as well as other genes coding for pupal wing disc cuticle proteins (pathway 1). *Bm*POUM2 may also indirectly activate expression of the genes coding for other *Bm*WCP proteins by activating other unidentified transcription factor(s) that interact with CREs in the promoter of these genes (pathway 2). Activation of other *BmWCP* genes may not act through the direct (pathway 1) or indirect (pathway 2) action of *Bm*POUM2 but through unknown transcription factor (s) in the regulation of the expression of *BmWCP* genes (pathway 3). All these pathways work together, leading to the post-larval development and differentiation of the wing disc during metamorphosis.（选自：*Homeodomain POU and Abd-A proteins regulate the transcription of pupal genes during metamorphosis of the silkworm, Bombyx mori*，参见本书第136页。）

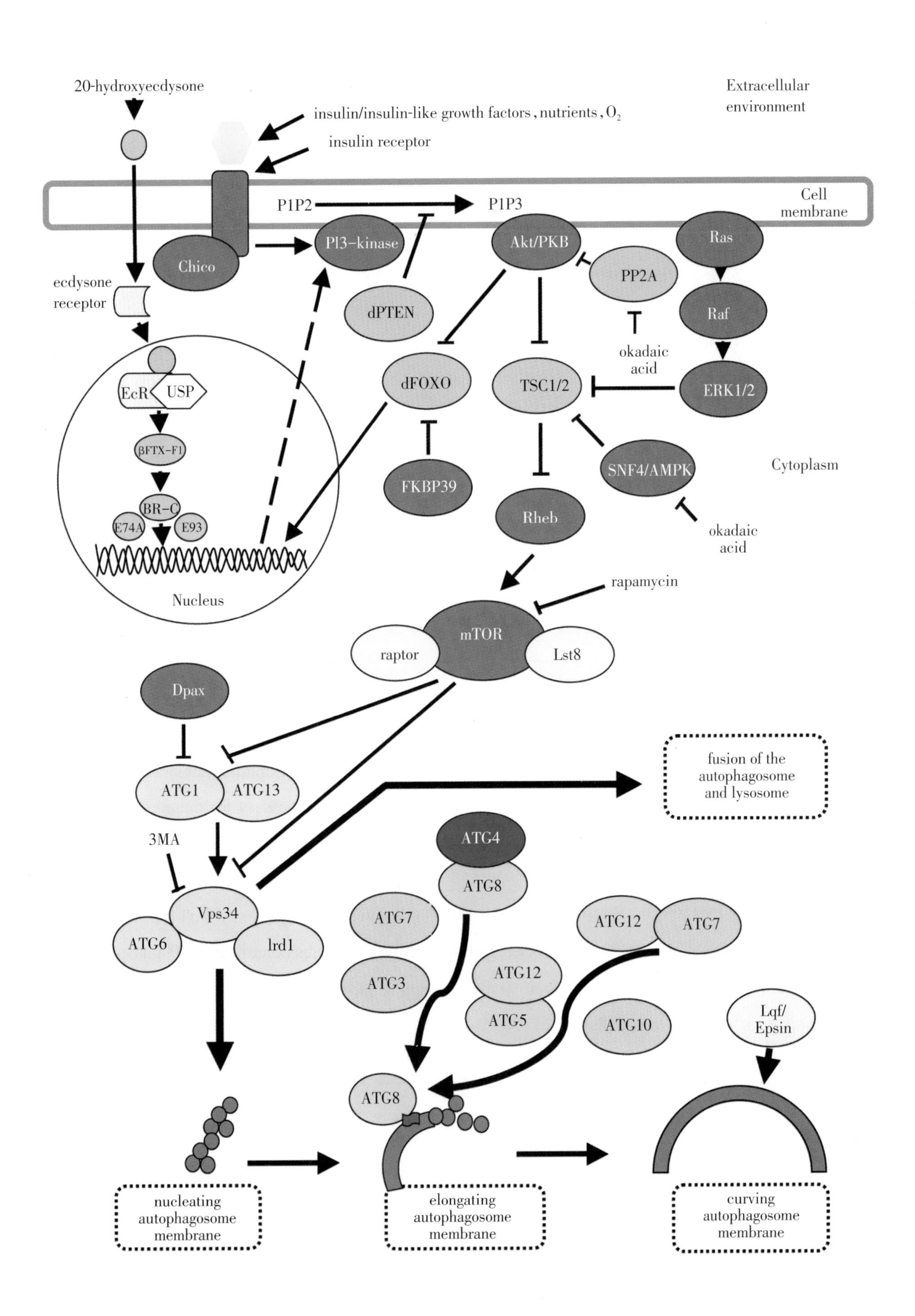
20-hydroxyecdysone
Extracellular
environment
insulin/insulin-like growth factors, nutrients, O2
insulin receptor
P1P2
P1P3
Cell
membrane
Chico
Pl3-kinase
Akt/PKB
Ras
PP2A
ecdysone
receptor
dPTEN
Raf
okadaic
acid
EcR
USP
dFOXO
TSC1/2
ERK1/2
βFTX-F1
SNF4/AMPK
Cytoplasm
FKBP39
BR-C
E74A
E93
Rheb
okadaic
acid
Nucleus
rapamycin
mTOR
raptor
Lst8
Dpax
fusion of the
autophagosome
and lysosome
ATG1
ATG13
3MA
ATG4
ATG8
Vps34
ATG7
ATG12
ATG7
ATG6
lrd1
ATG3
ATG12
ATG5
ATG10
Lqf/
Epsin
ATG8
nucleating
autophagosome
membrane
elongating
autophagosome
membrane
curving
autophagosome
membrane

Figure 31. The backbone of the regulation of autophagy in insects is the evolutionarily highly conserved insulin receptor-Pt-dIns3K-Akt/PKB-TSC1/2-Rheb-mTOR complex pathway. The activity of this pathway is influenced by 20-hydroxyecdysone (20E) at the level of PtdIns3K (most probably via activation of steroid-dependent genes). 20E induces the expression of transcription factors βFTZ-F1, BR-C, E74A and E93, thus activating the autophagic pathway. The PIP3 level is regulated by dPTEN. PP2A inhibits the activity of Akt/PKB (Bánréti and Sass unpublished results), whereas Akt/PKB and FKBP downregulate the transcription factor dFOXO. TSC1/2 is regulated by the Ras-Raf-ERK1/2 pathway and by SNF4/AMPK. The activity of the mTOR complex depends on the vectorial effect of all of the above-mentioned members of the regulatory network. The mTOR complex suppresses autophagy under normal conditions by inhibiting the activity of ATG1 and ATG13. At the same time, the mTOR complex prevents the recruitment of the Vps34 complex to the phagophore membrane. When the concentration of the insulin-like growth factors, nutrients, and oxygen declines, the activity of the mTOR complex decreases. Due to the low level of mTOR activity, the ATG1-ATG13 and the ATG6-Vps34-Ird1 complexes can be formed and the nucleation of the autophagosome membrane begins. The activity of the serine/threonine protein kinase ATG1 is regulated by Dpax. In the elongation of the phagophore membrane, two ubiquitin-like protein conjugation systems play important roles. Due to their functions, the ATG8-PE and the ATG12-ATG5 complex become integral parts of the phagophore membrane in insects, as in other organisms. The protein product of the Lqf/Epsin gene might be necessary for the curving of the autophagosome membrane. Vps34 plays a regulatory role in the fusion of autophagosomes and lysosomes. Codes: → activation; ⊣ inhibition; ▬ activates autophagy; ▬ inhibits autophagy; ▭ necessary for the membrane of the autophagosome; ▬ necessary for the elongation of the phagophore; ▬ enzymes of the ubiquitin-like protein conjugation systems; ▭ necessary for the curvature of the phagophore; ⚑ phosphatidylethanolamine (PE); 3MA, 3-methyladenine.(选自:*Autophagy and its physiological relevance in arthropods: Current knowledge and perspectives*,参见本书第138页。)

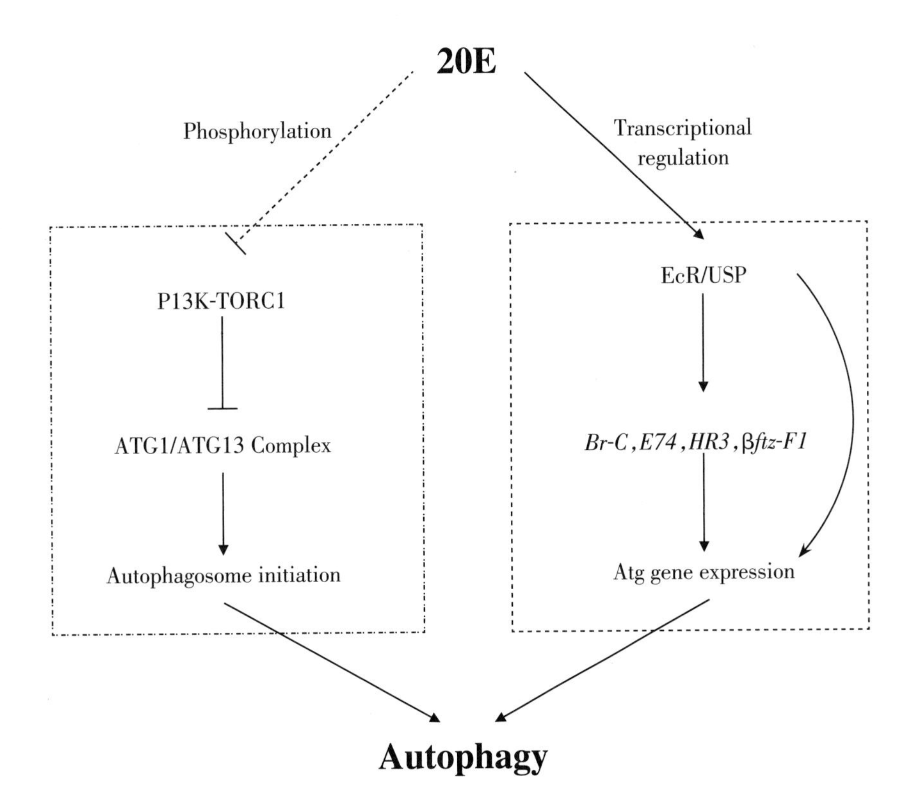

Figure 32. A model for the 20-E signal to induce autophagy by two indispensible means. 20-E (top: the upstream power source) blocks torC1 activity to induce autophagosome initiation (left: to break the gate) by phosphorylation and upregulates the *Atg* genes (right: to provide the flow) by transcriptional regulation resulting in autophagy induction (bottom: the downstream effect).（选自：*20-hydroxyecdysone upregulates Atg genes to induce autophagy in the Bombyx fat body*，参见本书第140页。）

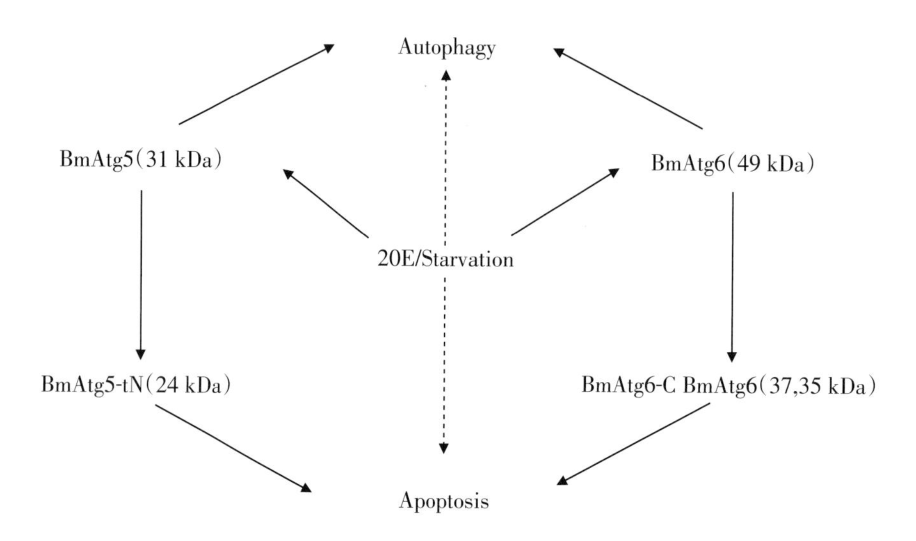

Figure 33. A model for 20-E and starvation signals to induce autophagy and apoptosis in *Bombyx*. Treatments with 20-E or starvation (middle: the upstream stimuli) activate *Bm*ATG5 and *Bm*ATG6 to facilitate autophagy and mediate the autophagy-induced apoptosis.(选自:*BmATG5 and BmATG6 mediate apoptosis following autophagy induced by 20-hydroxyecdysone or starvation*,参见本书第142页。)

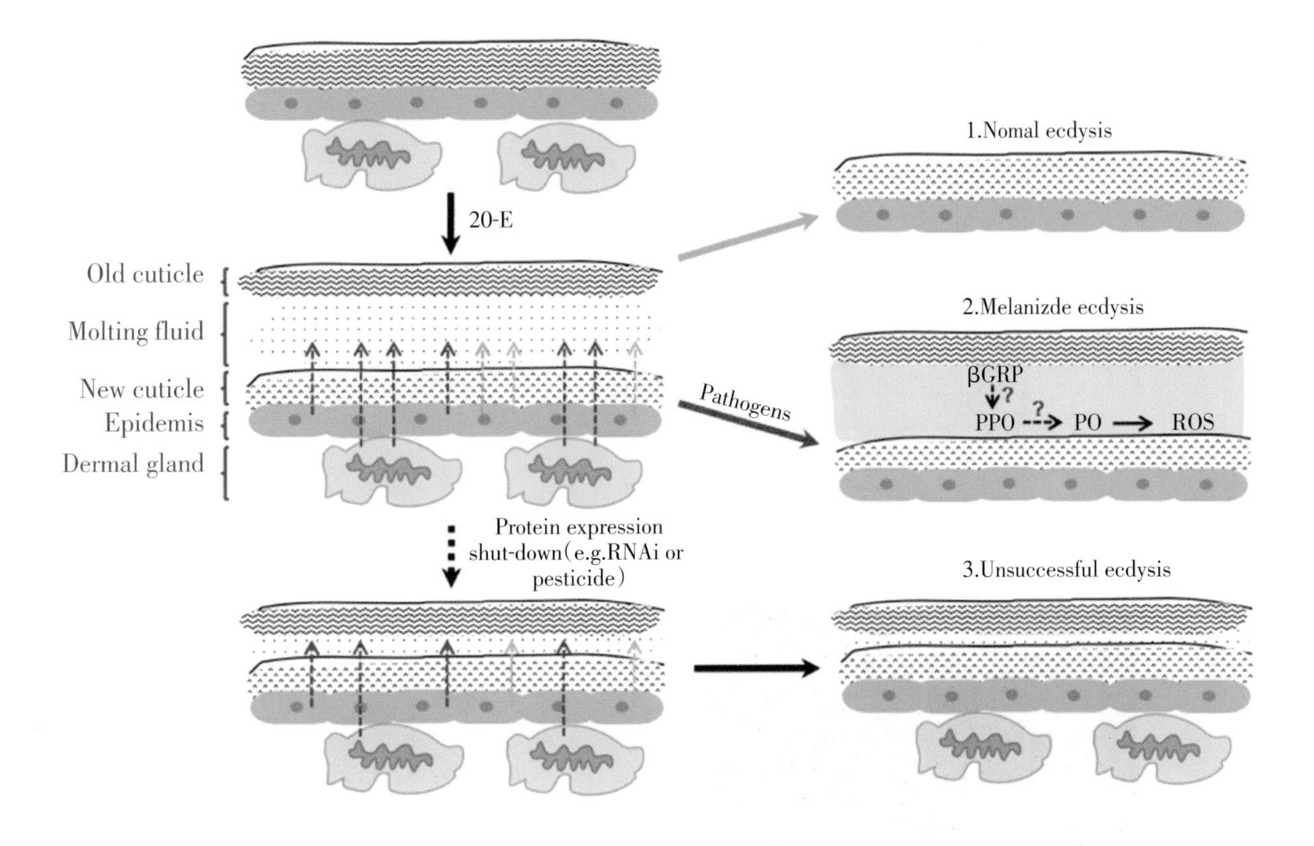

Figure 34. Regulation of insect ecdysis. Insects undergo several types of ecdysis during a life cycle: larval-larval, larval-pupal, and pupal-adult ecdysis. During ecdysis, the increasing concentration of 20E promotes protein expression and the production of new cuticles. Many proteins in the molting fluids are indispensable and must work in coordination to induce normal ecdysis. When insects were infected by pathogens before ecdysis, PPO was activated in molting fluids, and melanization occurred. Immune responses indirectly regulated ecdysis in some insects, according to decreased chitinase activities after melanizaton. The pharate instar insects under the melanized cuticle were intact. This melanized ecdysis is a result of immune responses to pathogens. When molting protein expression was interrupted, the insect could not shed the old cuticle on time. Therefore, the insect cannot undergo successful ecdysis unless all molting proteins are expressed and work in coordination to remove the old cuticle.(选自:*Functional analysis of insect molting fluid proteins on the protection and regulation of ecdysis*,参见本书第144页。)

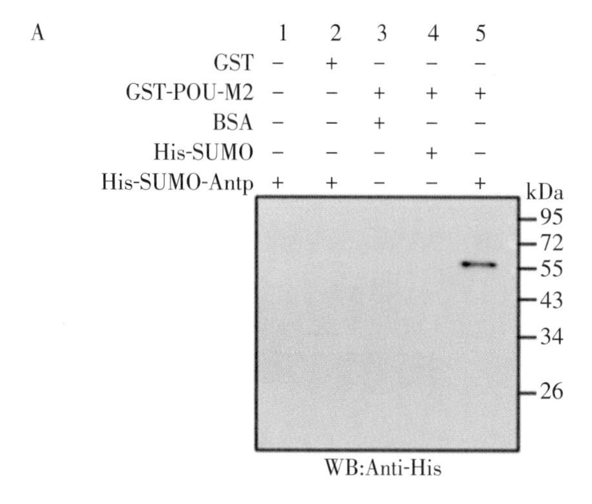
A
1 2 3 4 5
GST − + − − −
GST-POU-M2 − − + + +
BSA − − + − −
His-SUMO − − − + −
His-SUMO-Antp + + − − +
kDa
95
72
55
43
34
26
WB:Anti-His

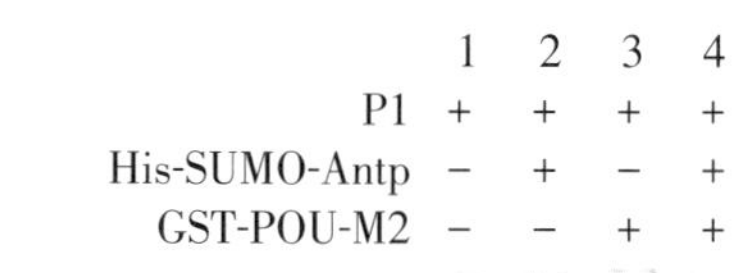
B
1 2 3 4
P1 + + + +
His-SUMO-Antp − + − +
GST-POU-M2 − − + +

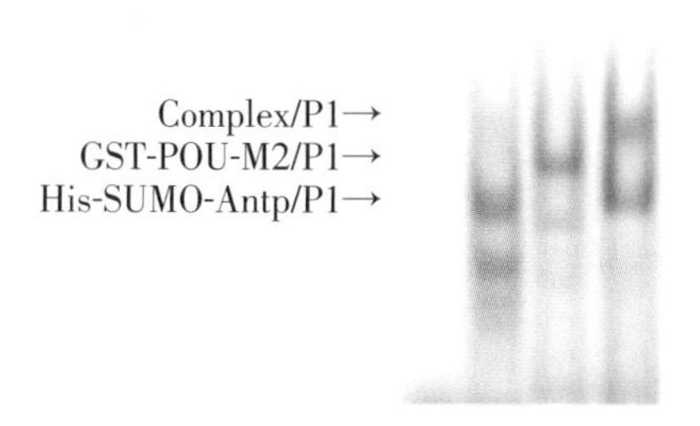
Complex/P1→
GST-POU-M2/P1→
His-SUMO-Antp/P1→

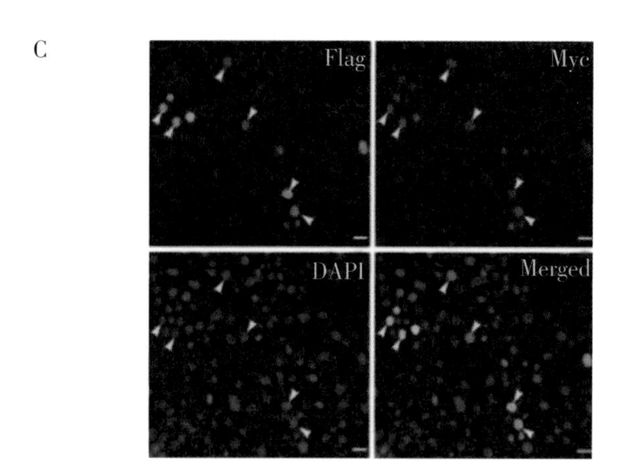
C
Flag
Myc
DAPI
Merged

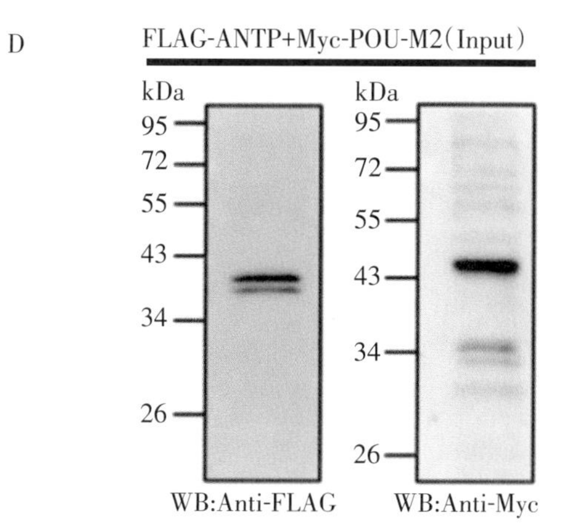
D
FLAG-ANTP+Myc-POU-M2(Input)
kDa
95
72
55
43
34
26
WB:Anti-FLAG
kDa
95
72
55
43
34
26
WB:Anti-Myc

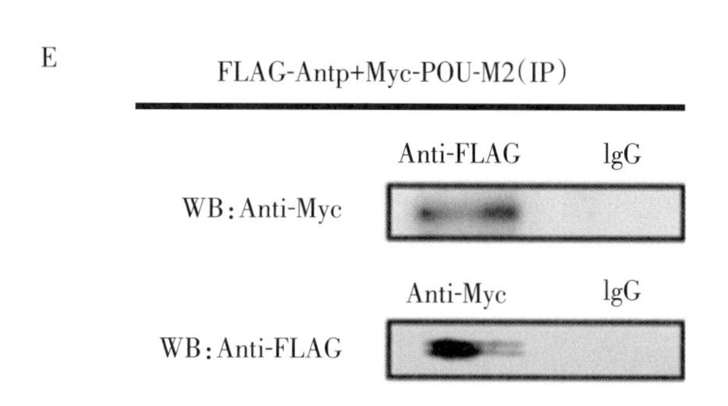
E
FLAG-Antp+Myc-POU-M2(IP)
Anti-FLAG
IgG
WB:Anti-Myc
Anti-Myc
IgG
WB:Anti-FLAG

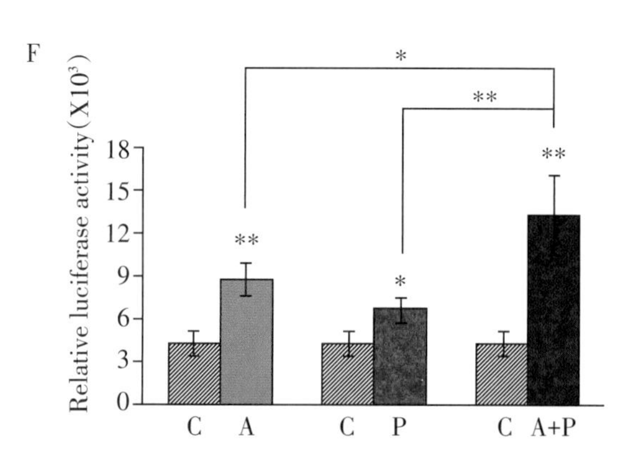
F
Relative luciferase activity(X10³)
18
15
12
9
6
3
0
*
**
**
*
**
C A
C P
C A+P

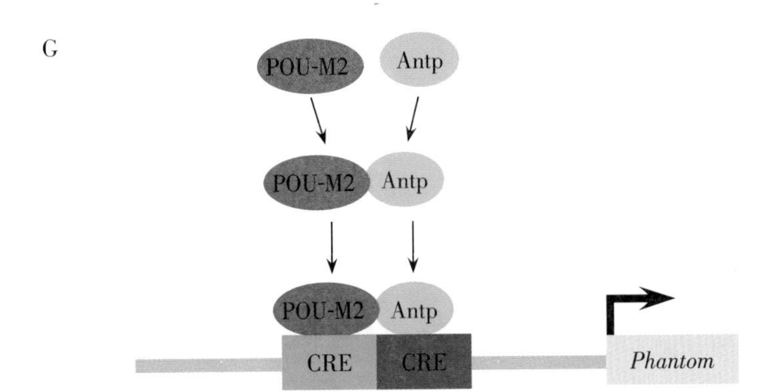

Figure 35. Interaction between Antp and POU-M2. A, GST pulldown assay to determine the interaction between Antp and POU-M2. Recombinant His-SUMO Antp was incubated with recombinant GST-POU-M2 and then detected by Western blot using a monoclonal antibody against the His tag. B, EMSA showed that recombinant His-SUMO-Antp and GST-POU-M2 formed a complex that bound to overlapping CREs in the silkworm *Phm* promoter. C, immunohistochemical analysis of the co-localization of the recombinant FLAG-tagged Antp and Myc-tagged POU-M2 in *BmE* cells by using monoclonal antibodies against the FLAG and Myc tags. Scale bar, 20 μm. D, Western blot analysis of the nucleoproteins isolated from silkworm *BmE* cells overexpressing *FLAG*-tagged *Antp* and *Myc*-tagged *POU-M2* by using monoclonal antibodies against the FLAG and Myc tags. E, co-IP assays of the interaction between Antp and POU-M2. The nucleoproteins isolated from the *BmE* cells overexpressing *FLAG*-tagged *Antp* and *Myc*-tagged *POU-M2* were immunoprecipitated with a monoclonal antibody against the FLAG tag or the Myc tag followed by Western blot analysis using a monoclonal antibody against the Myc tag or the FLAG tag. F, effects of *Antp* and *POU-M2* co-expression on the activity of the silkworm *Phm* promoter in *BmE* cells. Constructs overexpressing *Antp*, *POU-M2*, or both were separately cotransfected into *BmE* cells with a construct containing the luciferase gene driven by the *Phm* promoter. A construct overexpressing the *EGFP* gene was used as the control. The cells were collected for luciferase activity analysis at 48 h after transfection. The experiments were independently repeated three times, and the data represent the mean ± S.E. (n = 3). *, $P<0.05$; **, $P<0.01$, compared with the control. A, *Antp* overexpression; P, *POU-M2* overexpression; C, *EGFP* overexpression. G, proposed model for the transcriptional regulation of the silkworm *Phm* gene. The homeodomain transcription factors Antp and POU-M2 form a heterodimer via a protein interaction, and this complex subsequently binds specifically to different motifs in the overlapping CREs for Antp and POU-M2 in the *Phm* promoter to regulate the transcription of the *Phm* gene.（选自：*The homeodomain transcription factors Antennapedia and POU-M2 regulate the transcription of the steroidogenic enzyme gene Phantom in the silkworm*，参见本书第146页。）

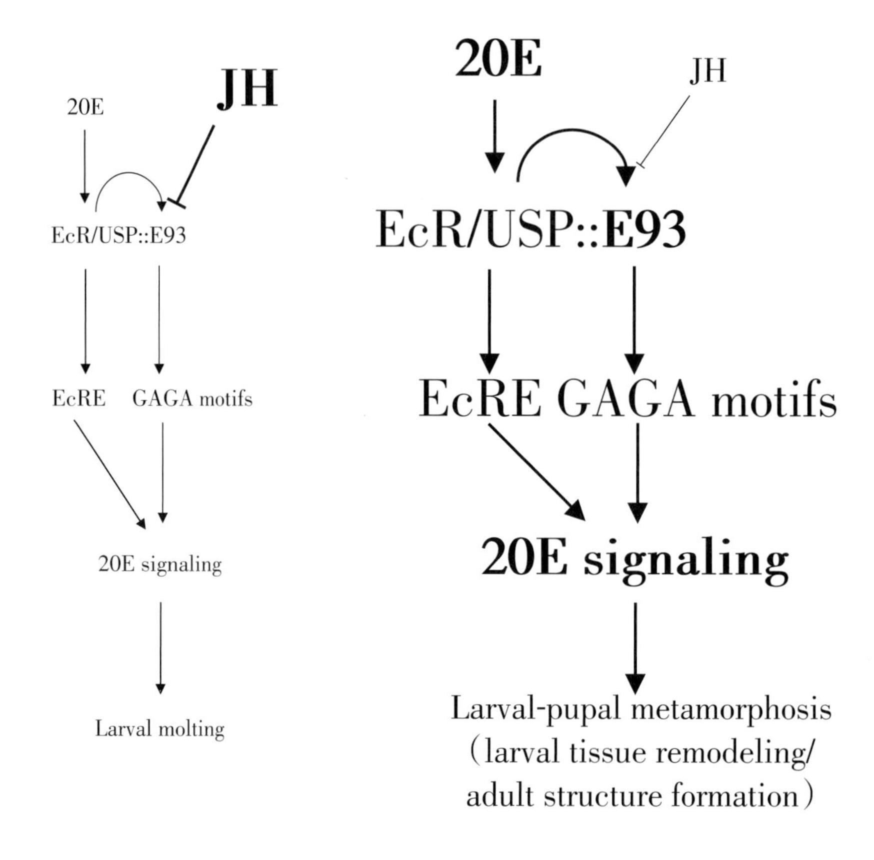

Figure 36. A model: E93 transcriptionally modulates 20-E signaling to promote *Bombyx* larval-pupal metamorphosis. Induced by 20-E and suppressed by JH, *E93* is abundantly expressed during the larval-pupal metamorphosis. At this stage, E93 acts through GAGA-containing motifs to induce expression of a subset of 20-E-response genes, positively affects 20-E signaling, and promotes larval tissue remodeling and adult tissue formation. The expression level of *E93* is induced by 20-E-EcR-USP, but the transcriptional activity of *E93* is attenuated by its physical association with USP. Notably, the E93 action to modulate 20E signaling is both dependent and independent on 20-E-EcR-USP. *Text and arrow sizes* convey magnitude of signal transduction.[选自：*20-hydroxyecdysone (20E) primary response gene E93 modulates 20E signaling to promote Bombyx larval-pupal metamorphosis*，参见本书第148页。]

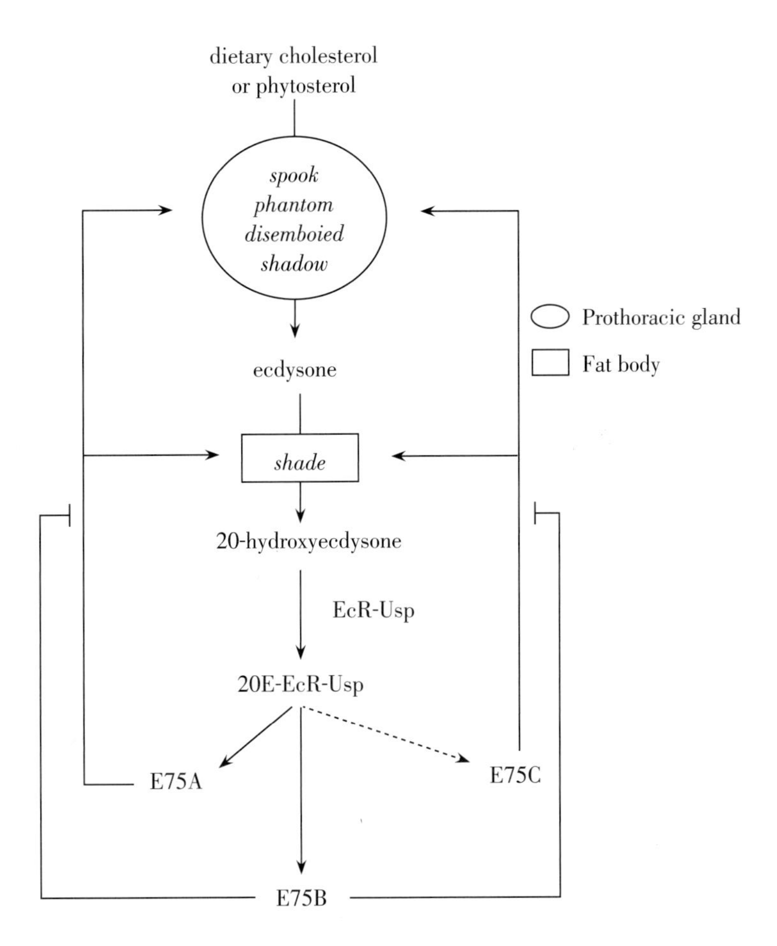

Figure 37. A model, E75 isoforms mediate a fine regulatory loop between ecdysteroid biosynthesis and 20-E signaling. 20-E rapidly induces the expression of *E75A* and *E75B*, whereas its induction of *E75C* expression is slow. E75A/C induces the Halloween gene expression responsible for ecdysone biosynthesis in the prothoracic glands and the conversion from ecdysone to 20-E in the fat body, whereas E75B antagonizes the transactivation ability of E75A/C. This model supports the central role of the 20-E-response gene *E75* in regulating ecdysteroid biosynthesis. The E75-mediated regulatory loop represents a fine autoregulation of steroidogenesis which contributes to the precise control of developmental timing.［选自：*20-Hydroxyecdysone (20-E) primary response gene E75 isoforms mediate steroidogenesis autoregulation and regulate developmental timing in Bombyx*，参见本书第150页。］

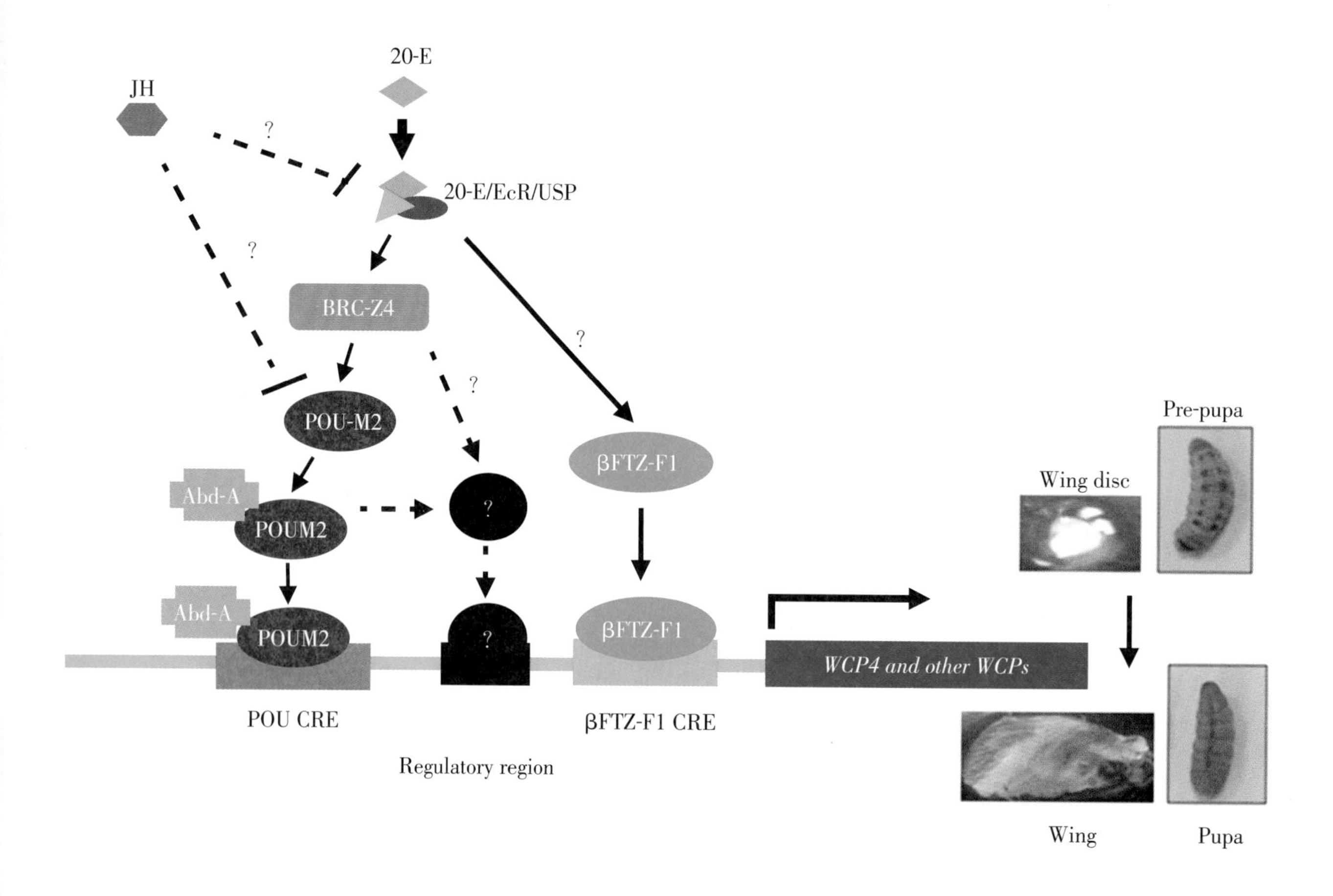

Figure 38. Schematic representation of the molecular regulation of *BmWCP4* expression by nuclear transcription factors in response to 20-E and JH during metamorphosis in *B. mori*. (This scheme is modified based on the original model in Deng et al., 2012). 20-E binds to its heterodimer receptor (EcR and USP complex), which activates the expression of the transcription factor *Bm*BRC-Z4, as well as βFTZ-F1. *BmBRC-Z4* expression generates *Bm*BRC-Z4 protein that initiates the expression of *BmPOUM2* by binding to the *Bm*BRC-Z4 CRE of *BmPOUM2*. *Bm*POUM2 then binds with the *Bm*Abd-A. The complex of the two transcription factors subsequently binds to the corresponding POU CRE of the *BmWCP4* gene and activates the expression of *Bm*-WCP4, as well as other pupal wing disc cuticle proteins, leading to the post larval development and differentiation of wing disc during the larval-pupal transformation.（选自：*BmBR-C Z4 is an upstream regulatory factor of BmPOUM2 controlling the pupal specific expression of BmWCP4 in the silkworm, Bombyx mori*，参见本书第152页。）

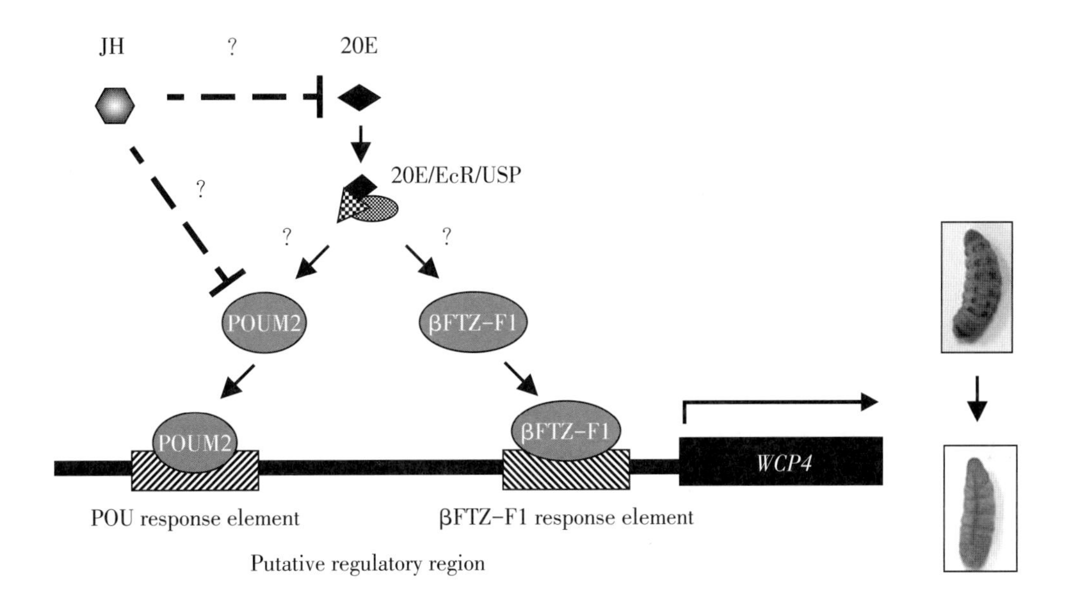

Figure 39. Schematic description of the hypothesis of the molecular regulation of *Bombyx mori* wing cuticle protein 4 (*Bm*WCP4) expression. 20-hydroxyecdysone (20-E) binds to the heterodimer of the ecdysone receptor, which either directly or indirectly activates the expression of the transcription factors POUM2 and βFTZ-F1. The transcription factors bind to their corresponding POUM2 and βFTZ-F1 cis-regulatory elements (CREs) in the *Bm*WCP4 putative regulatory region to activate the expression of the gene, resulting in the synthesis of the pupal wing discs cuticle protein during the larval to pupal metamorphosis. Juvenile hormone (JH) suppresses the action of 20-E in unidentified mechanisms. EcR, ecdysone receptor; USP, ultraspiracle.（选自：*Transcription factors BmPOUM2 and BmβFTZ-F1 are involved in regulation of the expression of the wing cuticle protein gene BmWCP4 in the silkworm, Bombyx mori*，参见本书第154页。）

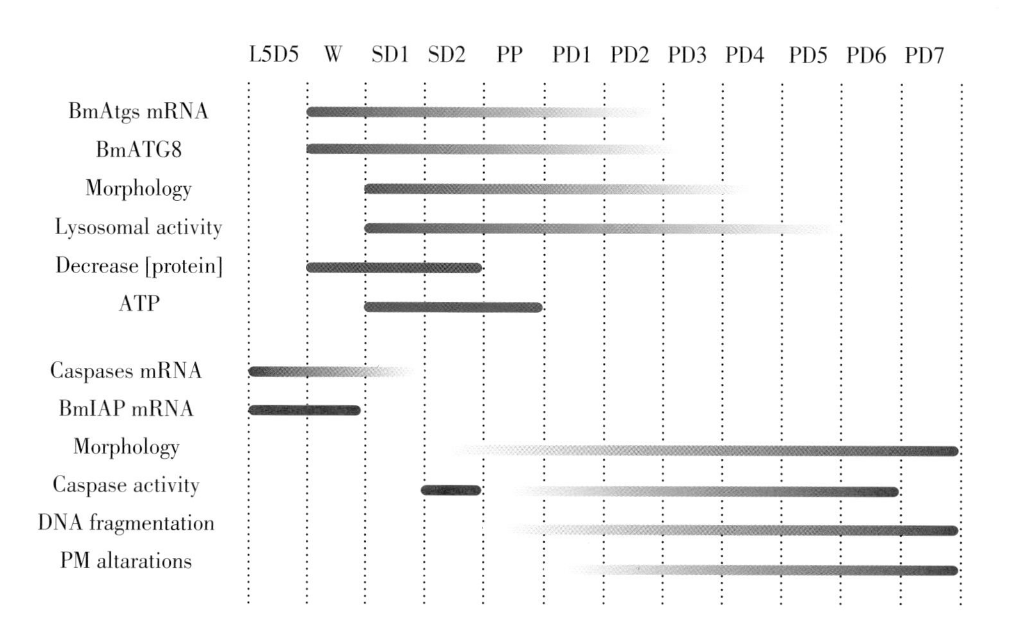

Figure 40. Summary scheme of the most representative cellular, biochemical, and molecular events that occur in silkworm midgut during the period L5D5-PD7. Processes related to autophagy are represented by blue bars (top), while those referring to apoptosis are represented by red bars (bottom) (Color figure online).（选自：*Autophagy precedes apoptosis during the remodeling of silkworm larval midgut*，参见本书第160页。）

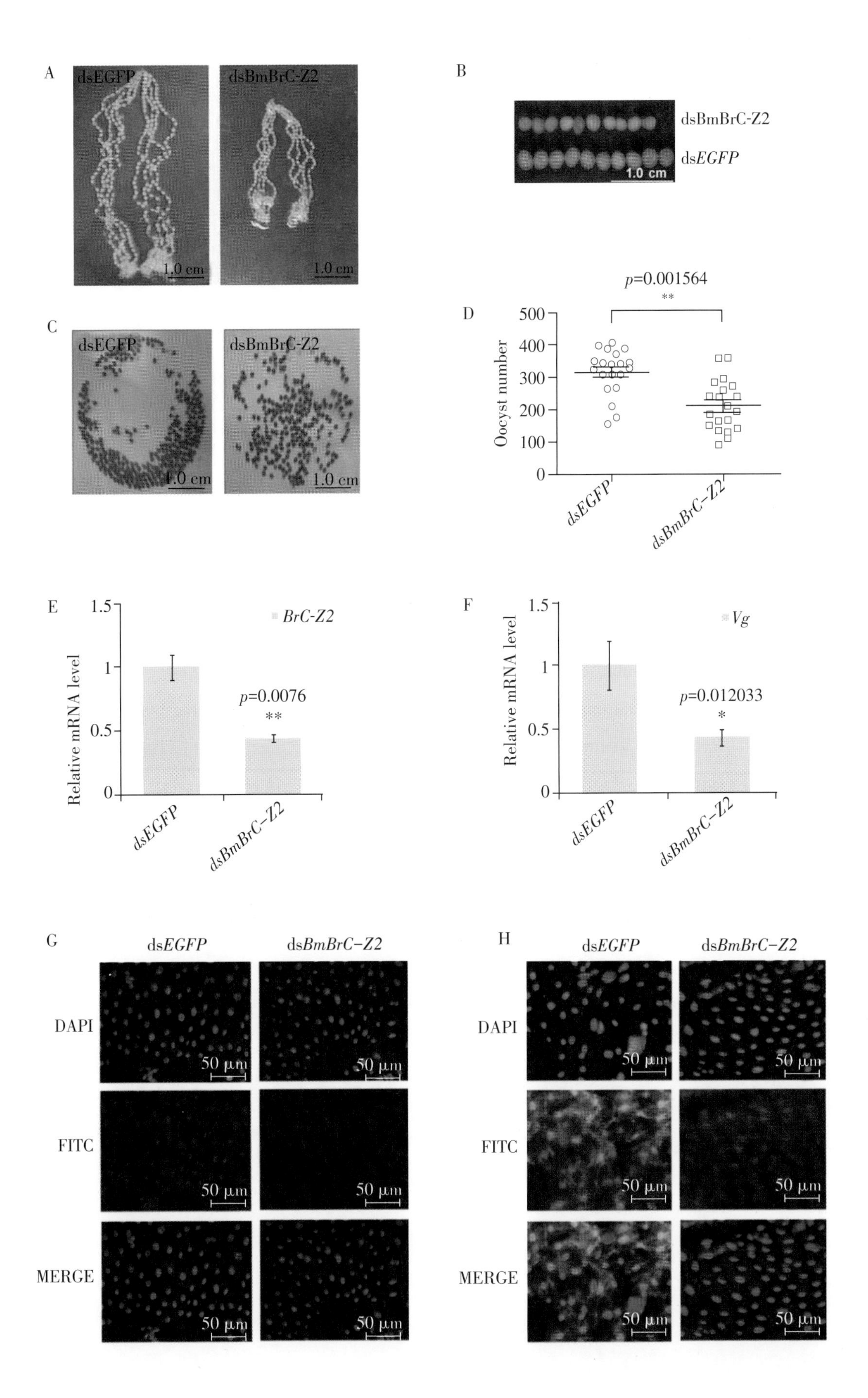
A
dsEGFP
dsBmBrC-Z2
1.0 cm
1.0 cm
B
dsBmBrC-Z2
dsEGFP
1.0 cm
C
dsEGFP
dsBmBrC-Z2
1.0 cm
1.0 cm
D
p=0.001564
**
Oocyst number
500
400
300
200
100
0
dsEGFP
dsBmBrC-Z2
E
BrC-Z2
Relative mRNA level
1.5
1
0.5
0
p=0.0076
**
dsEGFP
dsBmBrC-Z2
F
Vg
Relative mRNA level
1.5
1
0.5
0
p=0.012033
*
dsEGFP
dsBmBrC-Z2
G
dsEGFP
dsBmBrC-Z2
DAPI
FITC
MERGE
50 μm
H
dsEGFP
dsBmBrC-Z2
DAPI
FITC
MERGE
50 μm

Figure 41. Effect of RNAi *BmBrC-Z2* in the female silkworm. (A) Differentiation of wild type and dsRNA-treated silkworm ovaries. (B) Eggs from *dsBrC-Z2*-treated adults were smaller and whiter compared to *dsEGFP*-treated adults. (C) Eggs from wild type and *dsBmBrC-Z2*-treated female adults. (D) Number of eggs from *dsBmBrC-Z2*-treated adults after fertilization. Each group had 9 samples. (E) qRT-PCR analysis of *BmBrC-Z2* transcription in dsRNA-treated female pupal fat body at 2 days after treatment with dsRNA. (F) *BmVg* transcription in dsRNA-treated female pupal fat body at 2 days after treatment with dsRNA. (G) Protein synthesis of *Bm*BrC-Z2 in treatment with dsRNA of the female pupal fat body at 2 days after treatment with dsRNA detected by immunofluorescence histochemistry. (H) Protein synthesis of *Bm*Vg in dsRNA-treated female pupal fat body at 2 days after treatment of female pupae with dsRNA detected by immunofluorescence histochemistry. (*dsBmBrC-Z2* treatment vs. *dsEGFP* treatment control; *t*-test).［选自：*The Broad Complex isoform 2 (BrC-Z2) transcriptional factor plays a critical role in vitellogenin transcription in the silkworm Bombyx mori*，参见本书第162页。］

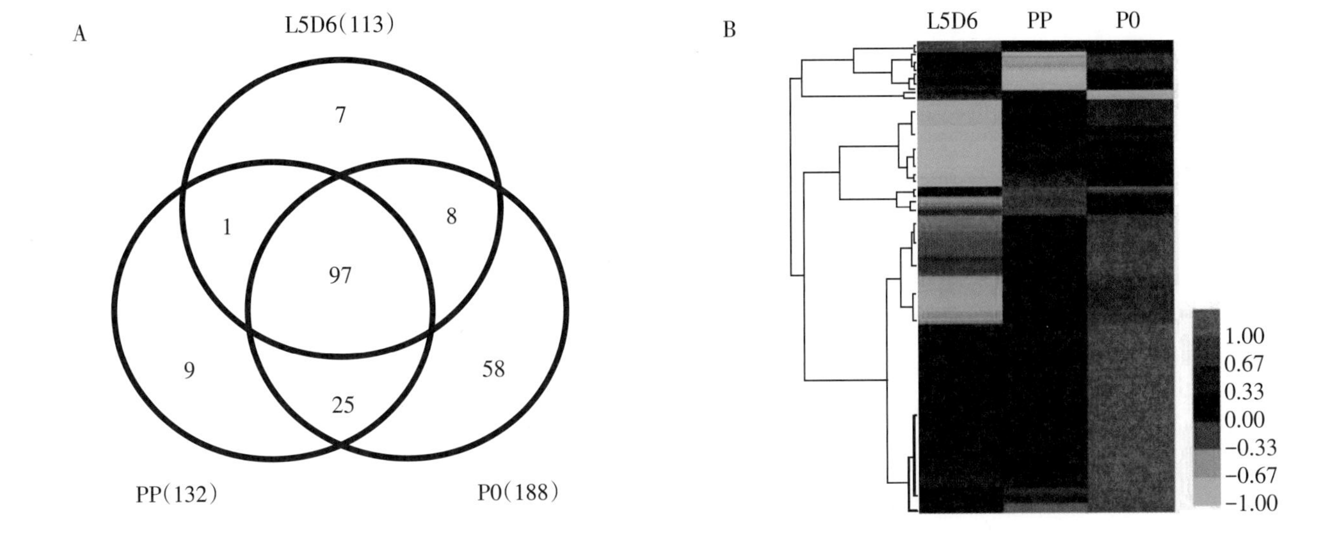

Figure 42. Expression analysis of cuticular or cuticle protein coding genes. The RPKM values of undetected genes are viewed as 0.001. Cluster 3.0 software was used to plot the heat map (Similarity Metric: Euclidean distance, Clustering method: Complete linkage), and Treeview software was used to generate the heat map. The colors in the map display the relative values at the given 3 developmental stages. Green indicates the lowest level of expression, black indicates the intermediate level of expression, and red indicates the highest level of expression. (A) Distribution of cuticular or cuticle protein coding genes at the three stages. (B) Heat map of hierarchical clustering of 148 differentially expressed cuticle protein genes.（选自：*Transcriptomic analysis of developmental features of Bombyx mori wing disc during metamorphosis*，参见本书第164页。）

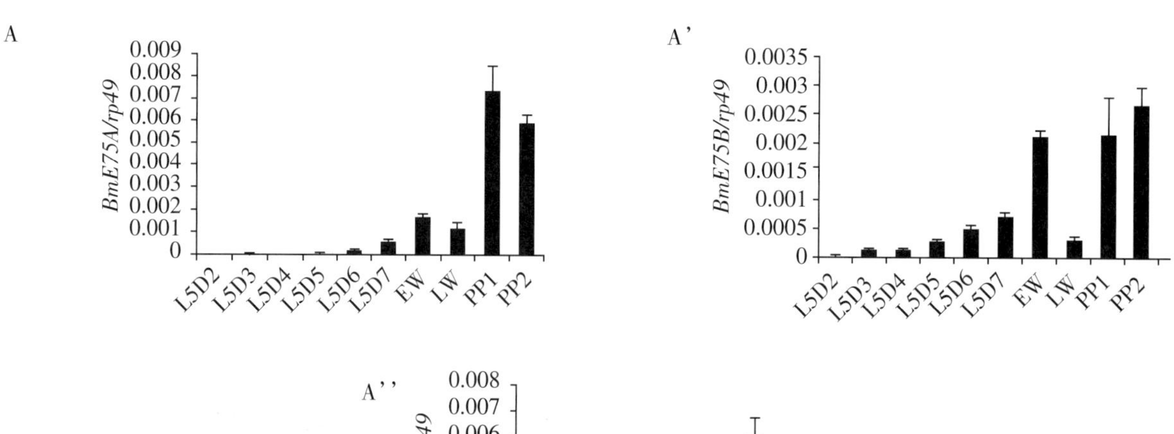
A
BmE75A/rp49
0.009
0.008
0.007
0.006
0.005
0.004
0.003
0.002
0.001
0
L5D2 L5D3 L5D4 L5D5 L5D6 L5D7 EW LW PP1 PP2
A'
BmE75B/rp49
0.0035
0.003
0.0025
0.002
0.0015
0.001
0.0005
0
L5D2 L5D3 L5D4 L5D5 L5D6 L5D7 EW LW PP1 PP2

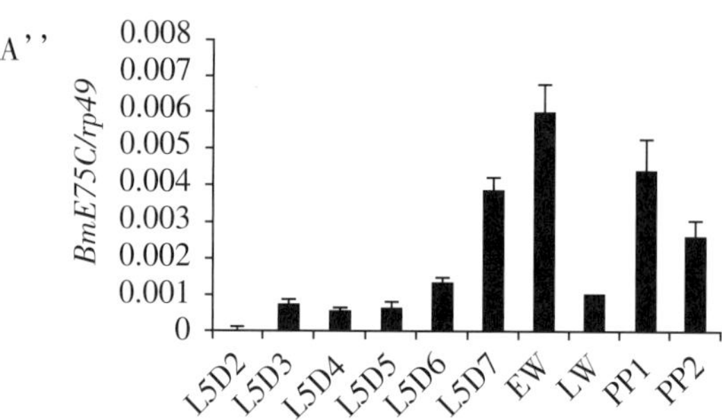
A''
BmE75C/rp49
0.008
0.007
0.006
0.005
0.004
0.003
0.002
0.001
0
L5D2 L5D3 L5D4 L5D5 L5D6 L5D7 EW LW PP1 PP2

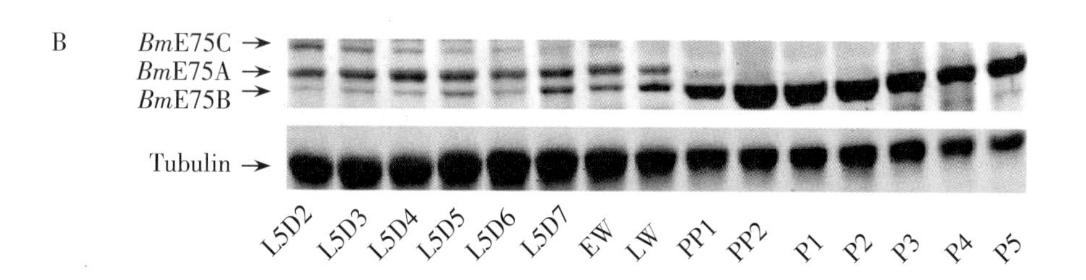
B
BmE75C →
BmE75A →
BmE75B →
Tubulin →
L5D2 L5D3 L5D4 L5D5 L5D6 L5D7 EW LW PP1 PP2 P1 P2 P3 P4 P5

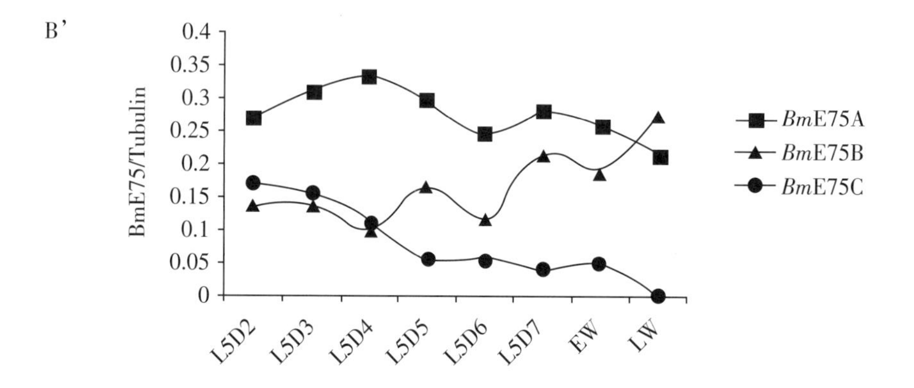
B'
BmE75/Tubulin
0.4
0.35
0.3
0.25
0.2
0.15
0.1
0.05
0
L5D2 L5D3 L5D4 L5D5 L5D6 L5D7 EW LW
BmE75A
BmE75B
BmE75C

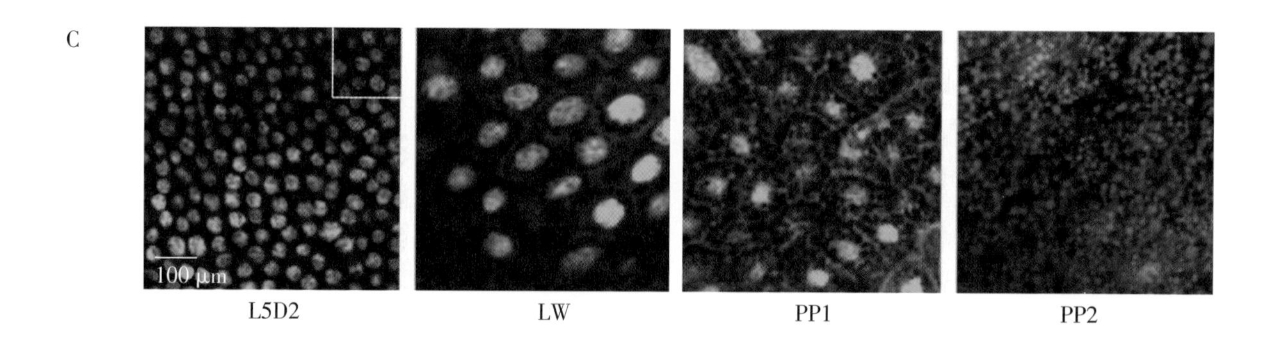
C
100 μm
L5D2
LW
PP1
PP2

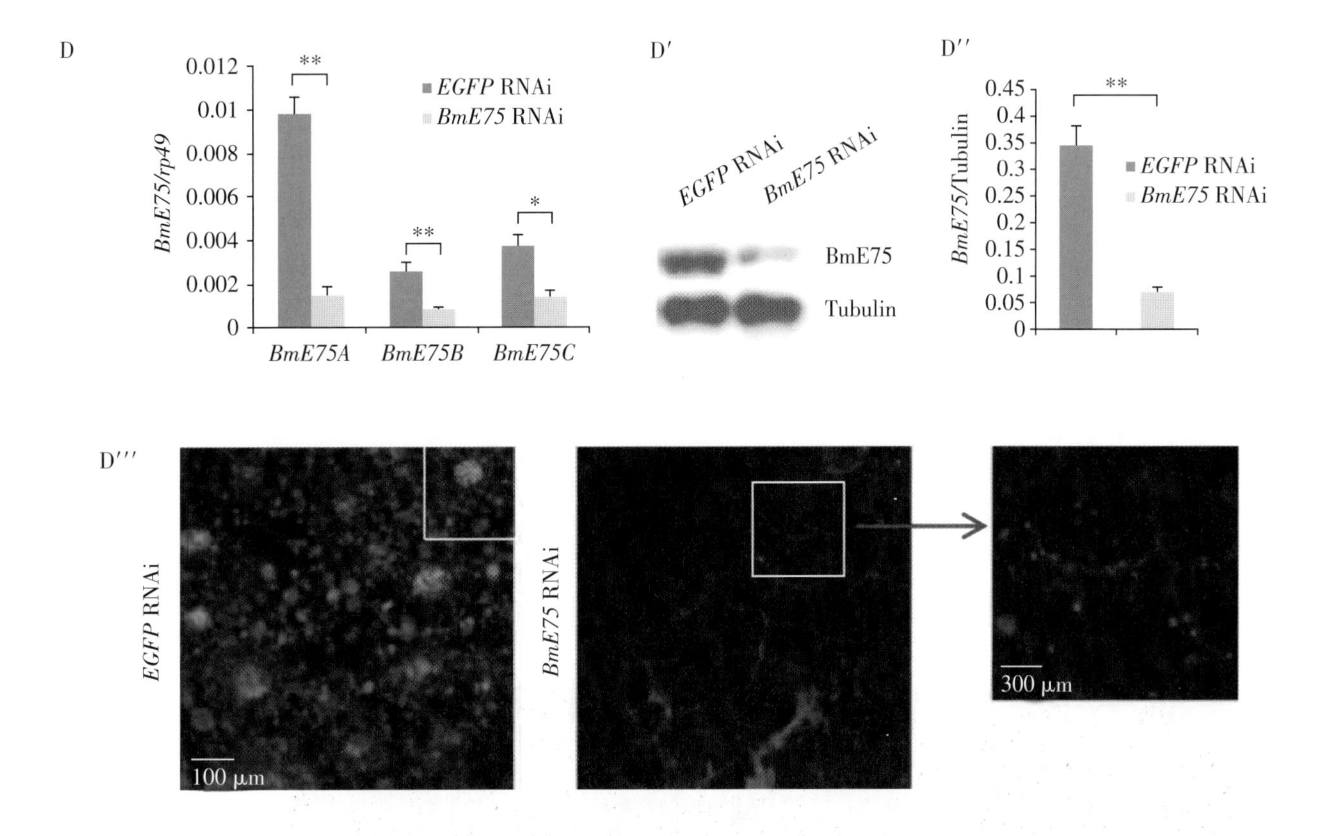

Figure 43. Developmental profiles of *Bm*E75 in fat body. (A-A") Developmental profiles of mRNA levels of *BmE75A* (A), *BmE75B* (A') and *BmE75C* (A") in the fat body from day 2 of the 5th larval instar (L5D2) to day 2 of the prepupal stage (PP2). L, instar; D, day; EW, early wandering; LW, late wandering; PP, prepupal; P, pupa. (B-B') Developmental profiles of protein levels of *Bm*E75A, *Bm*E75B and *Bm*E75C in the fat body from L5D2 to P5. The full-length blots are presented in Supplementary Figure 2. (B') Quantification of *Bm*E75A, *Bm*E75B and *Bm*E75C protein levels in (B). Location of *Bm*E75 proteins in the fat body is detected by immunohistochemistry using the *Bm*E75 antibody from EW to PP2 stage (green). The white box at top-right corner of inset in (L5D2) is merged with DAPI (blue) staining. (D-D"') dsRNA targeting a common region of all three *BmE75* isoforms was injected into *Bombyx* larvae at EW, and the fat body was isolated 48 hours after treatment. *EGFP* dsRNA was used as a control. mRNA levels of *BmE75A*, *BmE75B* and *BmE75C* were detected using isoform-specific primers by qPCR (D). Western blot (D', D") and immunohistochemistry (D"') were performed to detect the small molecular weight *Bm*E75-like protein in the fat body using the BmE75 antibody (green). The white box at top-right corner of inset in (EGFP RNAi) in (D"') is merged with DAPI (blue) staining.（选自：*Bombyx E75 isoforms display stage- and tissue-specific responses to 20-hydroxyecdysone*，参见本书第166页。）

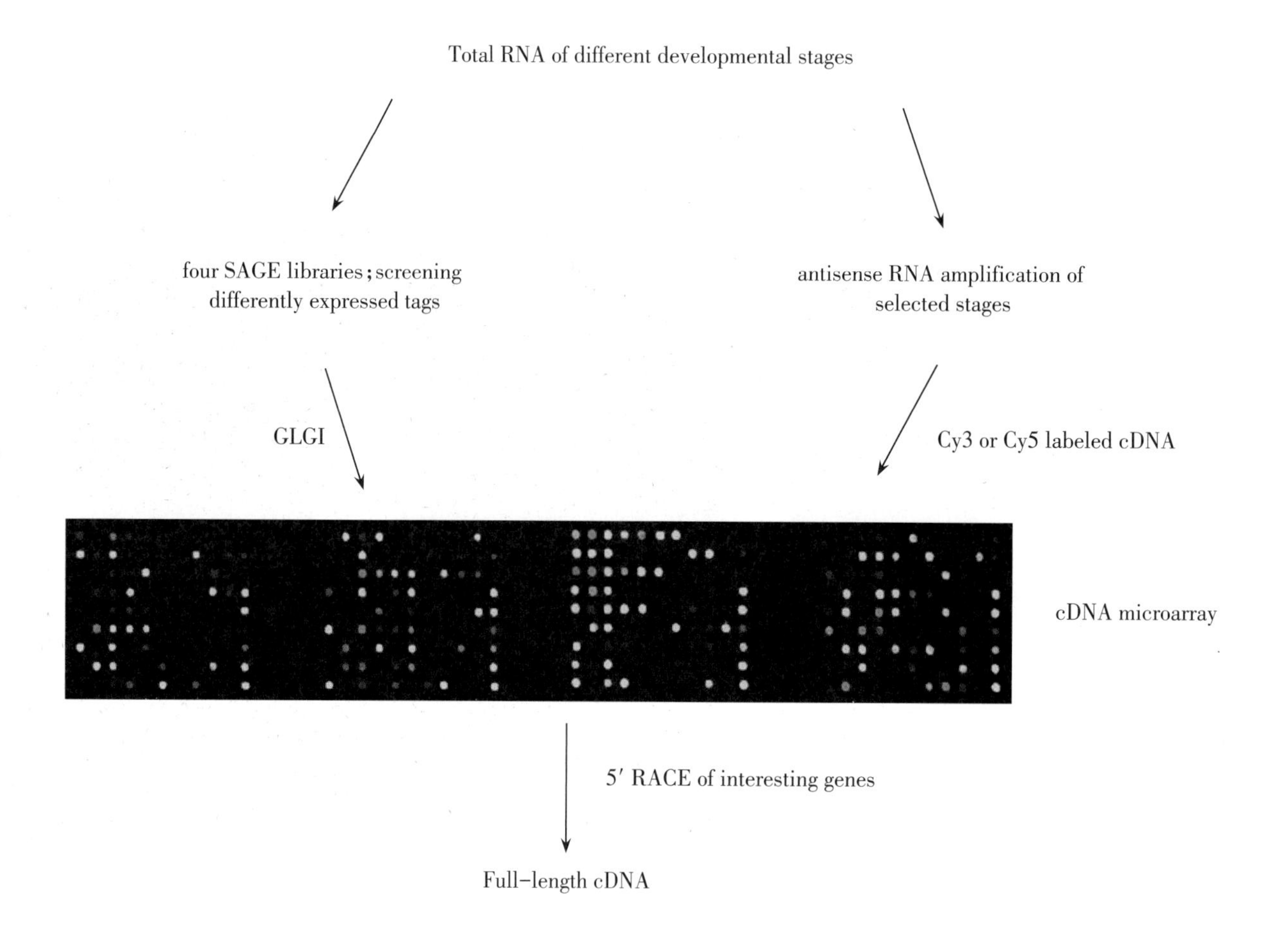

Figure 44. Serial analysis of gene expression (SAGE) tag-based cDNA microarray technique: main steps include total RNA samples extracted from 10 different time points of the silkworm from egg, larval, pupal, and adult developmental stages; construction of a SAGE library for each stage and screening of differently expressed tags; antisense RNA amplification of five larval and five pupal stages; Cy3/Cy5-labeled cDNA; cDNA microarray to confirm gene expression levels of the selected stages; 5' RACE of the genes of interest; cloning of full-length cDNA.（选自：*SAGE tag based cDNA microarray analysis during larval to pupal development and isolation of novel cDNAs in Bombyx mori*，参见本书第170页。）

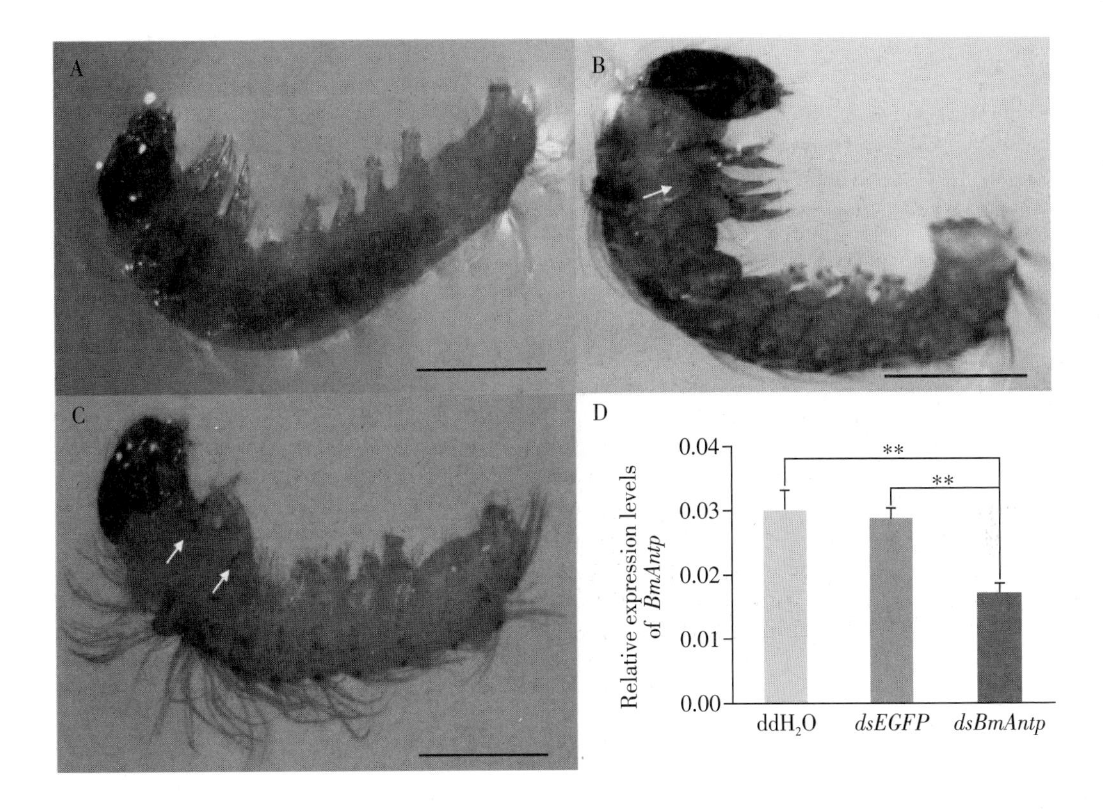

Figure 45. Effect of RNAi on *BmAntp*. (a) Wild type (*Dazao*) embryo injected with doubly distilled water (ddH_2O). Bar =0.5 mm. (b and c) RNAi effects on identities of thoracic segments in embryos. After injection of ds*BmAntp* into the embryo, thoracic segments fused and thoracic legs were reduced partly. Bar = 0.5 mm. (d) Expression levels of *BmAntp* in the embryo injecting ddH_2O, ds*EGFP* and ds*BmAntp*. ddH_2O and ds*EGFP*-injected controls are shown for comparison. The bars indicate mean ±s.d. $^{**}P<0.01$, Student's *t*-test.（选自：*Antennapedia is involved in the development of thoracic legs and segmentation in the silkworm, Bombyx mori*，参见本书第174页。）

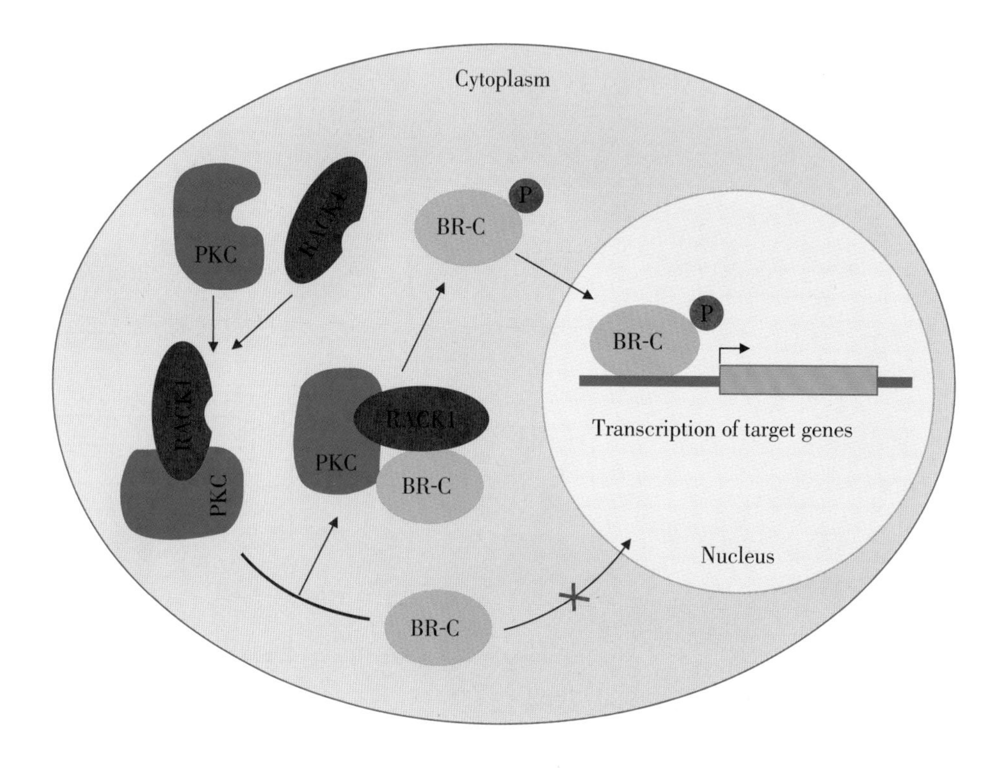

Figure 46. Proposed model for the nuclear import of BR-C upon interaction between BR-C and RACK1. The scaffolding protein RACK1 recruits and activates PKC in the cytoplasm. After being translated in the cytoplasm, BR-C interacts with RACK1 and is phosphorylated by RACK1-anchored PKC at amino acid residues Ser373 and Thr406. Phosphorylated BR-C then translocates into the nucleus and subsequently activates the transcription of its target genes.（选自：*Nuclear import of transcription factor BR-C is mediated by its interaction with RACK1*，参见本书第180页。）

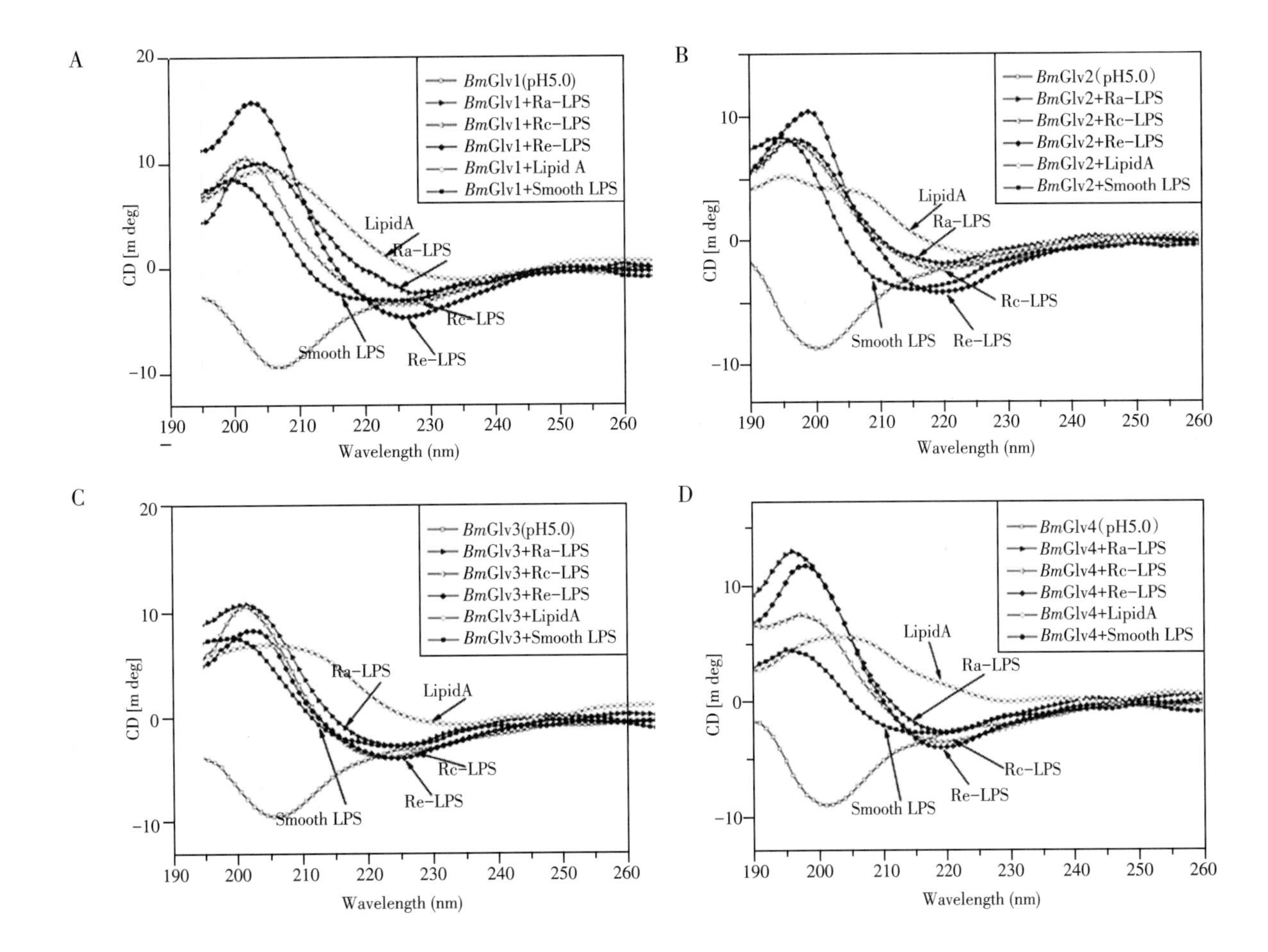

Figure 47. CD spectra of *Bm*Glvs in the presence of smooth LPS, Ra-, Rc- and Re-LPS and lipid A at pH 5.0. Purified *Bm*Glv1-4 were diluted to 0.15 mg/mL in 10 mM phosphate buffer at pH 5.0 in the presence or absence of smooth LPS, Ra-, Rc- and Re-LPS or monophosphoryl lipid A (w/w = 1:1) and CD spectra were recorded on a Jasco-810 spectropolarimeter at 25 °C. Each CD spectrum was obtained after subtracting the signal from protein-free solution.(选自：*Gloverins of the silkworm Bombyx mori: structural and binding properties and activities*，参见本书第220页。)

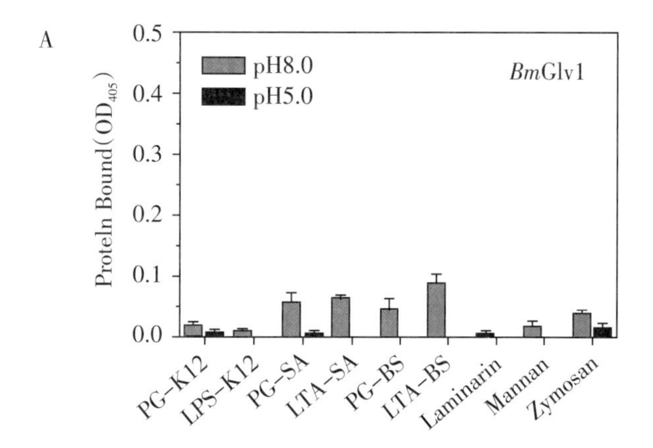
A
Proteln Bound (OD405)
pH8.0
pH5.0
BmGlv1
0.5
0.4
0.3
0.2
0.1
0.0
PG-K12
LPS-K12
PG-SA
LTA-SA
PG-BS
LTA-BS
Laminarin
Mannan
Zymosan

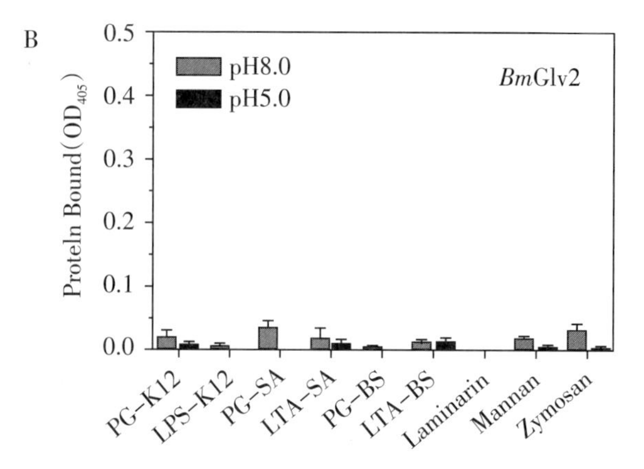
B
Proteln Bound (OD405)
pH8.0
pH5.0
BmGlv2
0.5
0.4
0.3
0.2
0.1
0.0
PG-K12
LPS-K12
PG-SA
LTA-SA
PG-BS
LTA-BS
Laminarin
Mannan
Zymosan

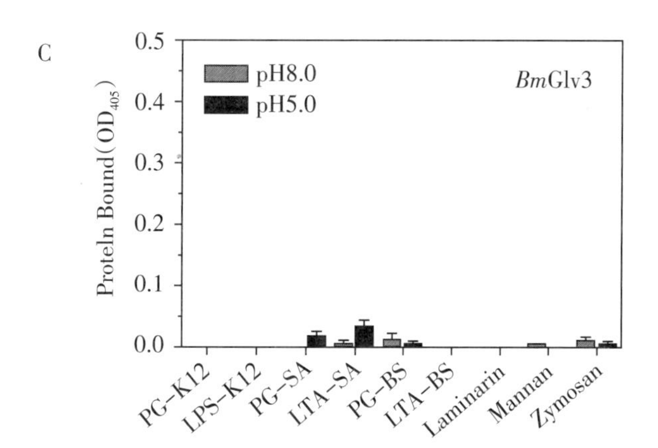
C
Proteln Bound (OD405)
pH8.0
pH5.0
BmGlv3
0.5
0.4
0.3
0.2
0.1
0.0
PG-K12
LPS-K12
PG-SA
LTA-SA
PG-BS
LTA-BS
Laminarin
Mannan
Zymosan

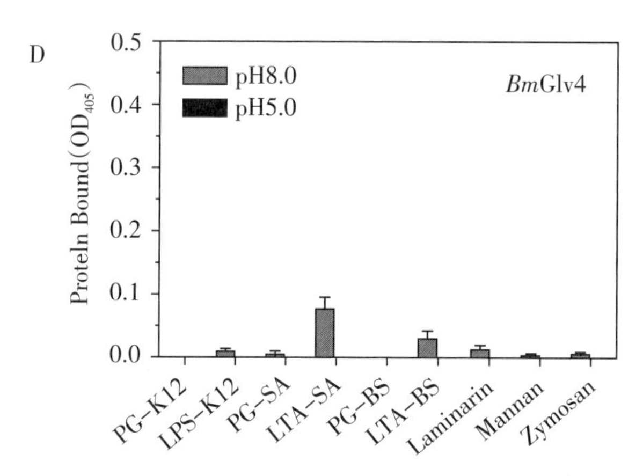
D
Proteln Bound (OD405)
pH8.0
pH5.0
BmGlv4
0.5
0.4
0.3
0.2
0.1
0.0
PG-K12
LPS-K12
PG-SA
LTA-SA
PG-BS
LTA-BS
Laminarin
Mannan
Zymosan

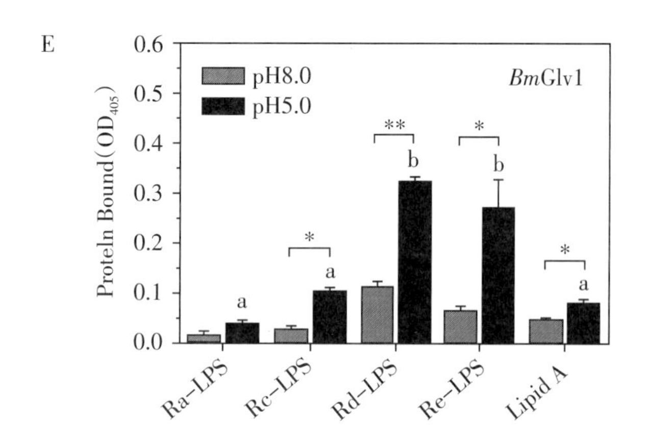
E
Proteln Bound (OD405)
pH8.0
pH5.0
BmGlv1
0.6
0.5
0.4
0.3
0.2
0.1
0.0
**
*
*
*
a
a
b
b
a
Ra-LPS
Rc-LPS
Rd-LPS
Re-LPS
Lipid A

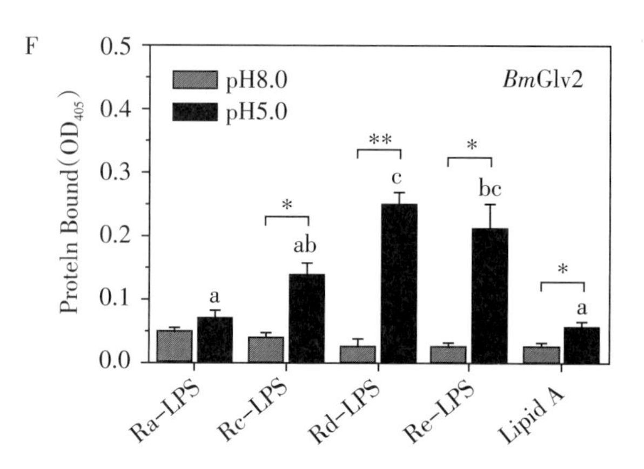
F
Proteln Bound (OD405)
pH8.0
pH5.0
BmGlv2
0.5
0.4
0.3
0.2
0.1
0.0
**
*
*
*
a
ab
c
bc
a
Ra-LPS
Rc-LPS
Rd-LPS
Re-LPS
Lipid A

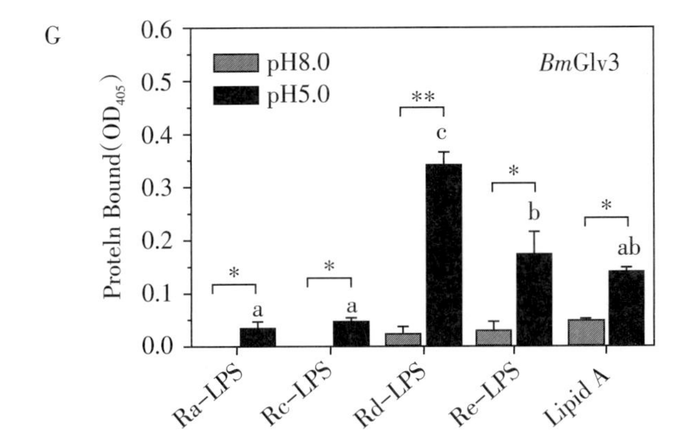

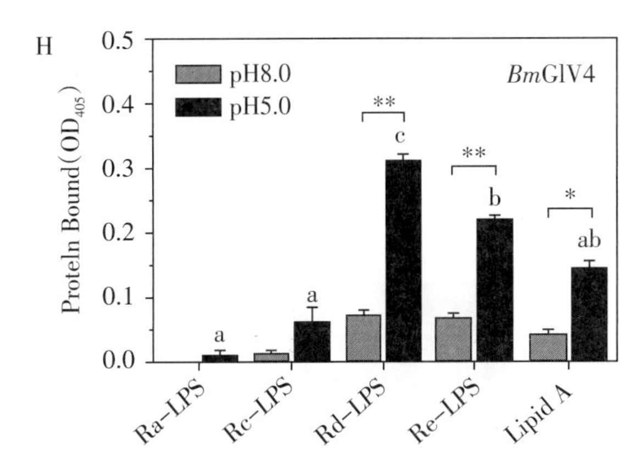

Figure 48. Binding of *Bm*Glvs to microbial cell wall components. Wells of 96-well plates were coated with different microbial components (2 μg/well) and blocked with BSA. Purified *Bm*Glvs and the control CP36 (a recombinant cuticle protein from *M. sexta*) were diluted to 1 μg/mL in 10 mM phosphate, 100 mM NaCl, pH 5.0 or 8.0, and the diluted proteins were added to the coated plates. Binding of proteins to microbial components was determined by plate ELISA assays as described in the materials and methods. The figure showed specific binding of *Bm*Glvs to each microbial component after subtracting the total binding of the control CP36 protein from the total binding of *Bm*Glvs. Each bar represents the mean of at least three individual measurements ± SEM. Comparing binding of each *Bm*Glv to rough LPS and lipid A at pH 5.0 (panels E-H), identical letters are not significant difference ($P > 0.05$), while different letters indicate significant difference ($P < 0.05$) determined by one way ANOVA followed by a Tukey's multiple comparison test. Comparing binding of *Bm*Glv to each rough LPS or lipid A between pH 5.0 and 8.0 (panels E-H), '*' ($P < 0.05$) and '**' ($P < 0.01$) indicate significant differences determined by an unpaired *t*-test.(选自：*Gloverins of the silkworm Bombyx mori: structural and binding properties and activities*，参见本书第220页。)

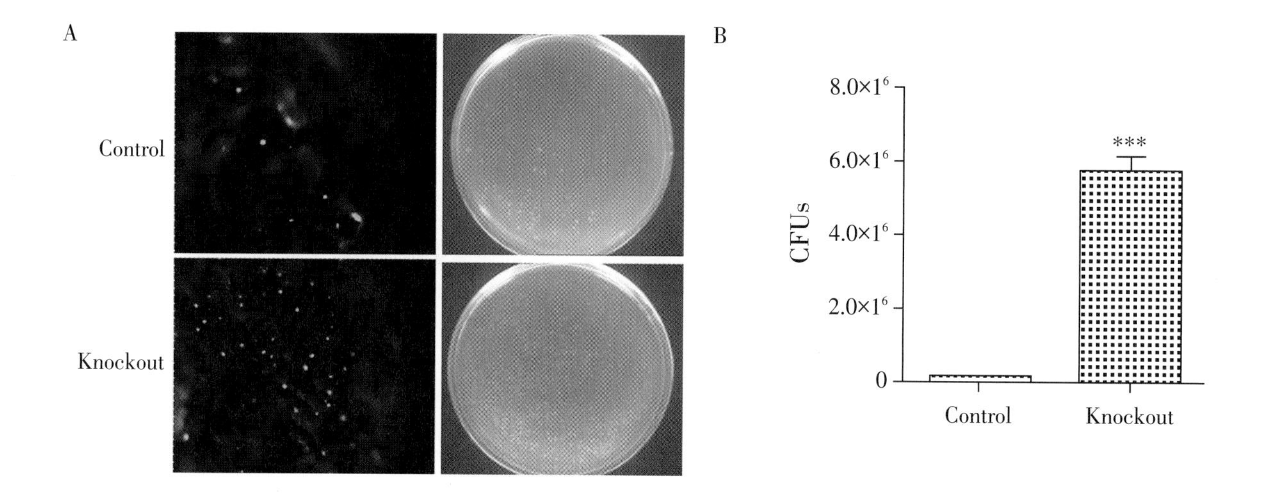

Figure 49. Persistence of *E. coli*-GFP in the peritrophic membrane of *BmDuox* knockout and normal silkworms. (A) Persistence of *E. coli*-GFP (spectinomycin-resistant) in the peritrophic membrane of normal silkworm and representative plates of *E. coli*-GFP in the peritrophic membrane (up); persistence of *E. coli*-GFP (spectinomycin-resistant) in the peritrophic membrane of *BmDuox* knockout silkworm and representative plates of *E. coli*-GFP in the peritrophic membrane (down). (B) Knockdown of *BmDuox* reduces the capacity of larvae to clear *E. coli*-GFP in the midgut. The graph represents relative numbers of *E. coli*-GFP CFUs. Results are expressed as the mean and standard deviation of three different experiments. Statistical differences were evaluated using Student's *t*-test for unpaired samples; $^*P < 0.05$, $^{**}P < 0.01$, $^{***}P < 0.001$.［选自：*Molecular cloning and functional characterization of the dual oxidase (BmDuox) gene from the Silkworm Bombyx mori*，参见本书第226页。］

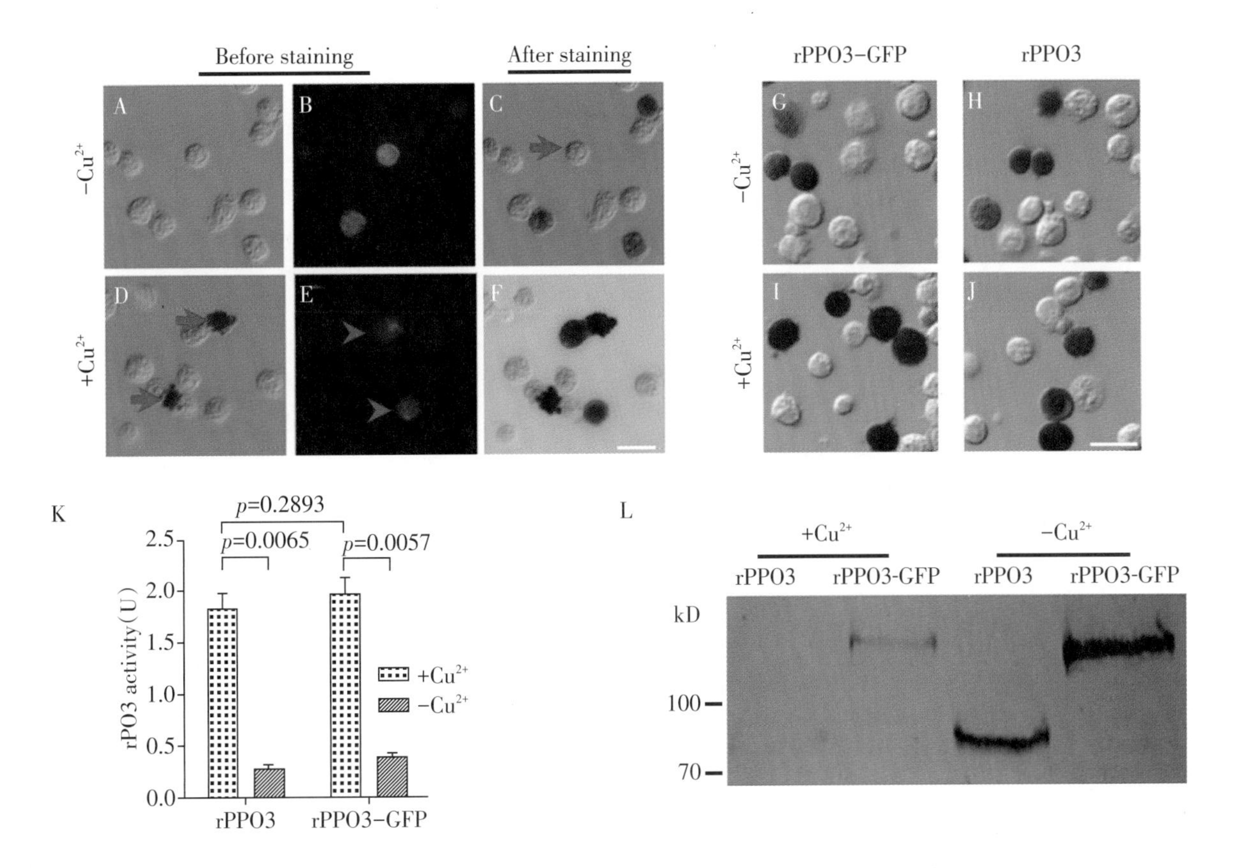

Figure 50. Phenoloxidase in S2 cells over-expressing rPPO1-GFP. DIC and fluorescence images were taken in the absence (A-C; -Cu^{2+}) or presence (D-F; +Cu^{2+}) of Cu^{2+} during cell transfection. In both cases, cells with green fluorescence were detected (A, B, D, E), and in the presence of Cu^{2+} those fluorescent cells displayed PO activity (dark brown) when incubated with dopamine dissolved in 30% ethanol (F). (G-J). Comparison of cells with rPPO1 and rPPO1-GFP expressed. rPPO1 and rPPO1-GFP were over-expressed in S2 cells in the absence (G, H) or presence (I, J) of Cu^{2+} during cell transfection. The cells were then stained for PO activity. Images represent DIC and fluorescence overlays. When Cu^{2+} was not added, no cells stained for PO activity (G, H) even though some cells expressed rPPO1-GFP (green fluorescence). When Cu^{2+} was added, many cells stained for PO activity (I, J; brown pigment), and no fluorescent cells were observed after PO staining due to quenching by melanin. (K) Comparison of rPPO1 and rPPO1-GFP enzyme activities. The amounts of rPPO1 and PPO1-GFP in S2 cell lysates were normalized and determined using purified rPPO1 as a standard by Western blot. Ethanol was used for enzyme activation. When Cu^{2+} was added, rPPO1 had significantly higher enzyme activity than rPPO1-GFP. No enzyme activities were detected if Cu^{2+} was not added, which is in agreement with the cell staining shown in (G-J). Columns represent the mean of individual measurements ± S.E.M (n = 3). Significant differences were calculated using the unpaired t test. (L) A Western blot showed that rPPO1-GFP and rPPO1 protein expression occurred regardless of the presence or absence of Cu^{2+}. Bar: 20 μm. (选自:*Activity of fusion prophenoloxidase-GFP and its potential applications for innate immunity study*,参见本书第228页。)

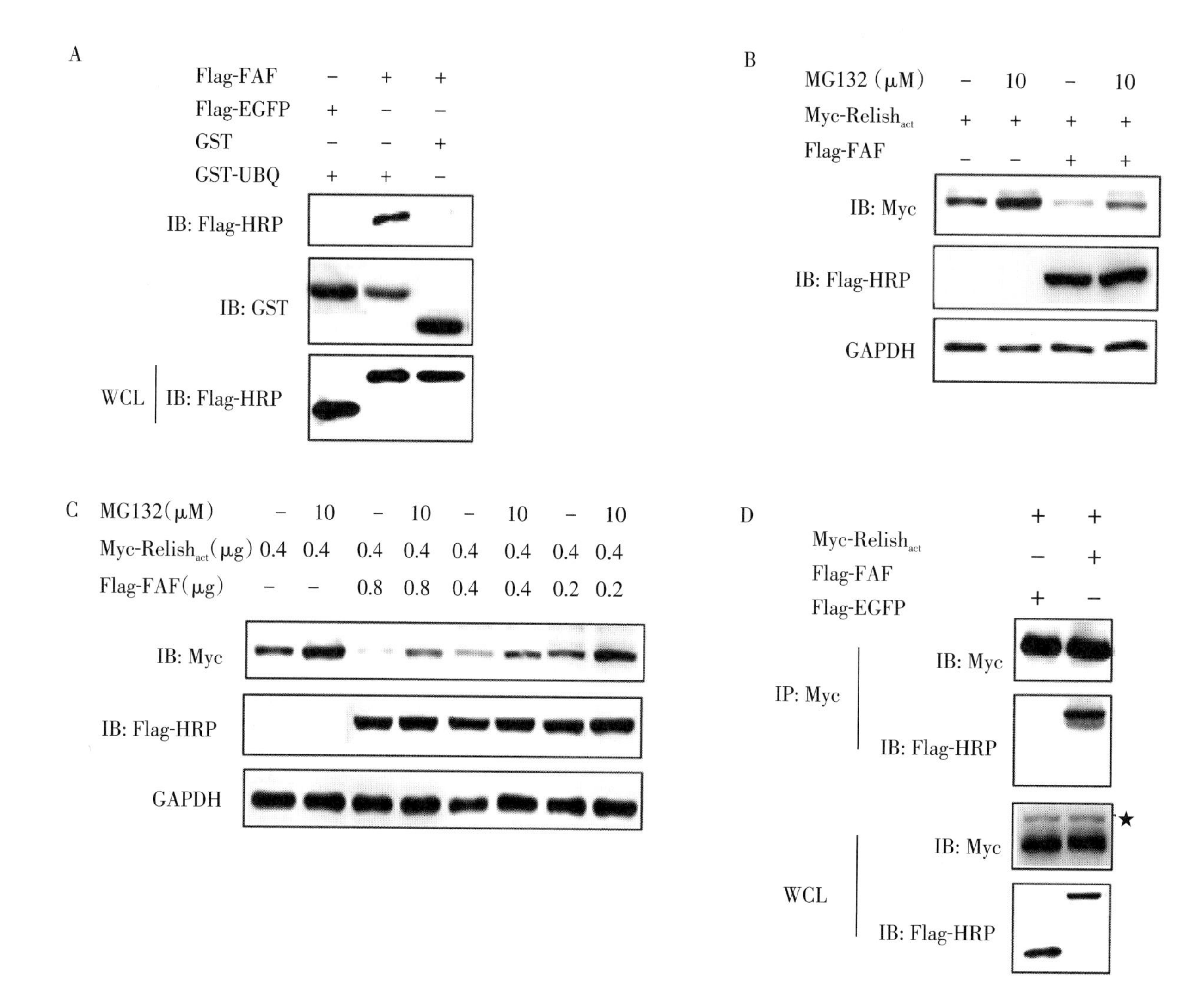

Figure 51. BmFAF promotes Relish proteasomal degradation and binds to Relish. (A) Lysates extracted from BmE cells transfected with Flag-FAF or Flag-EGFP expression plasmid (2 μg) were incubated with immobilized GST or GST-ubiquitin (GST-UBQ). The whole cell lysates (WCL) and bound proteins were subjected to immunoblot with anti-Flag-HRP antibody. (B), (C) *BmE* cells transfected with 0.4 μg Myc-Relishact expression plasmid together with 0.8 μg (B) or indicated amount of Flag-FAF expression plasmid (C) were treated with MG132 (10 μM) for 8 h. The cell lysates were subjected to immunoblot with anti-Myc, anti-Flag-HRP and anti-GAPDH antibodies. (D) *BmE* cells were transfected with 2 μg Myc-Relishact expression plasmid together with 1 μg plasmid encoding Flag-FAF or Flag-EGFP. Cell lysates were subjected to immunoprecipitation with anti-Myc antibody. WCL and precipitated proteins were immunoblotted with anti-Myc and anti-Flag-HRP antibodies. Asterisk indicates nonspecific bands.(选自:*A Fas associated factor negatively regulates anti-bacterial immunity by promoting Relish degradation in Bombyx mori*,参见本书第232页。)

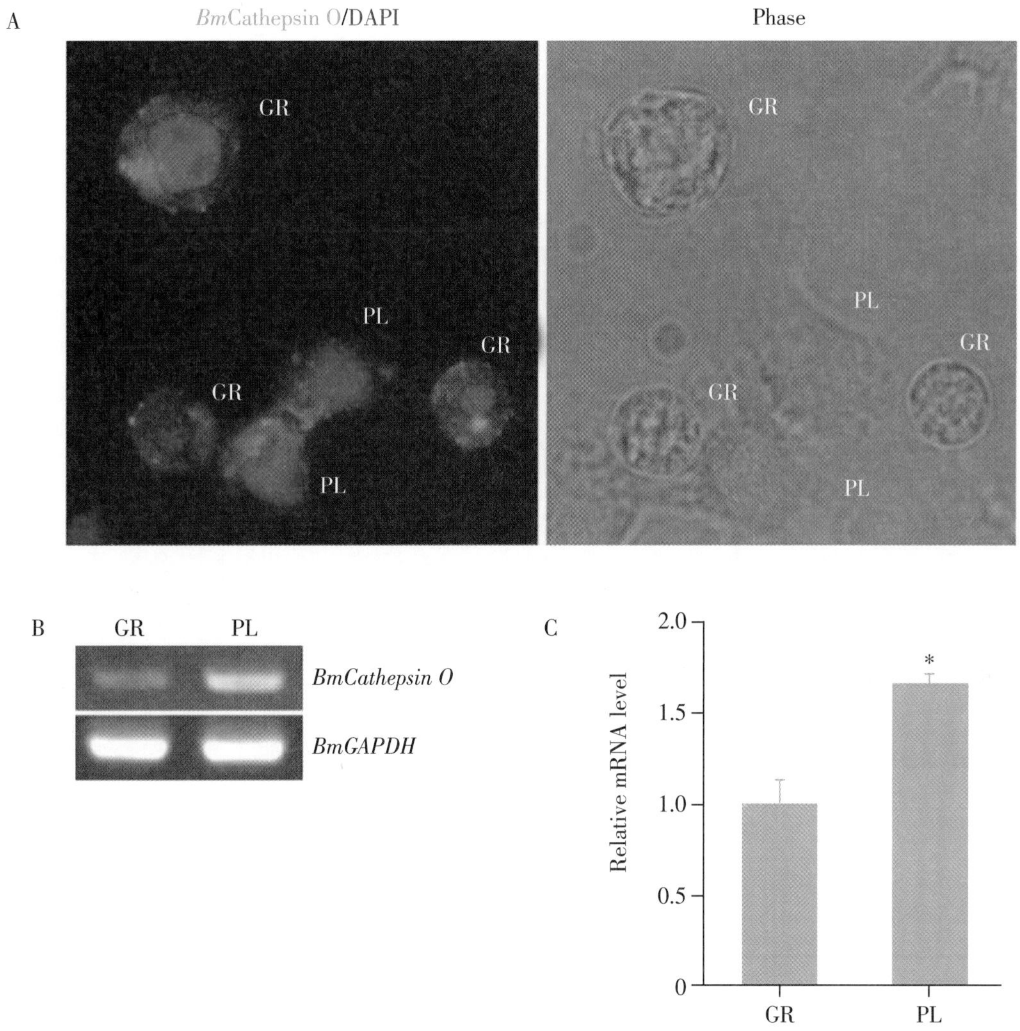

Figure 52. *Bm*Cathepsin O expressed in larval granulocytes and plasmatocytes. (A) distribution of *Bm*Cathepsin O after immunofluorescence in the circulating hemocytes of larval silkworms. GR = granulocytes, and PL = plasmatocytes. Scale bar = 10 μm. B, and C, semi-quantitive PCR (RT-PCR) and qRT-PCR were used to detect the mRNA expression of *BmCathepsin O* in larval granulocytes (GR) and plasmatocytes (PL), respectively. *BmGAPDH* was used as an internal control. The differences between the different groups were analyzed by the Student's *t* test, *$P<0.05$.（选自：*Molecular cloning, characterization and expression analysis of Cathepsin O in silkworm Bombyx mori related to bacterial response*，参见本书第244页。）

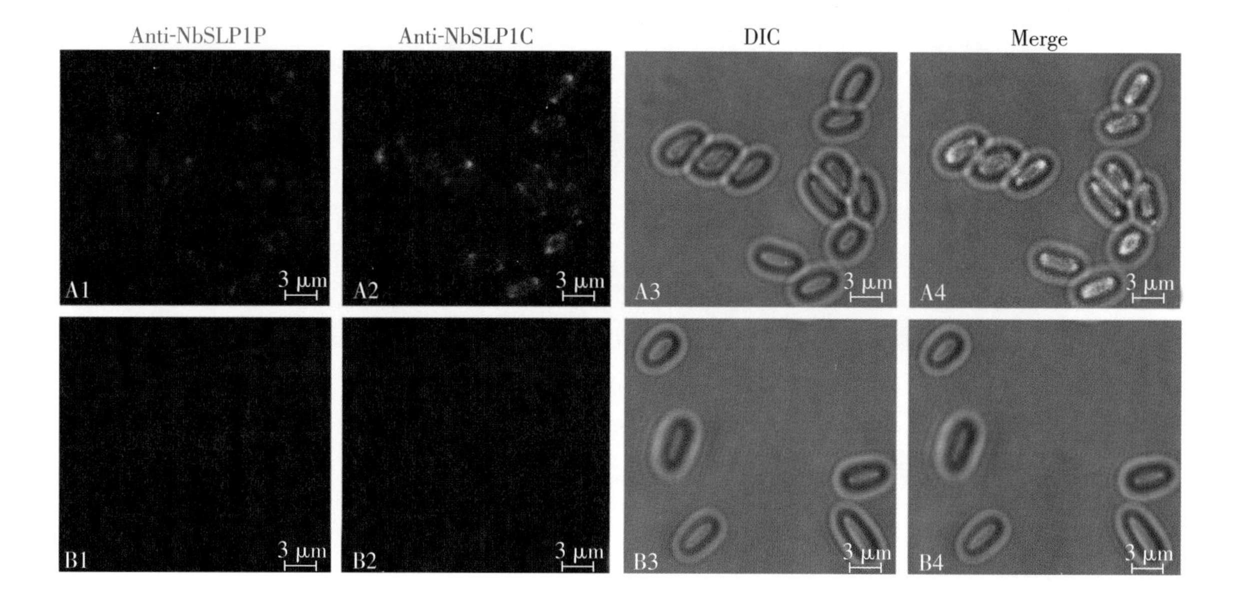

Figure 53. Indirect immunofluorescence assay (IFA) demonstrated the *Nb*SLP1 is localized in two poles of the spore. Antibody against the inhibitor_I9 domain in *Nb*SLP1 (Anti-*Nb*SLP1P, A1, red,) and antibody against the catalytic domain in *Nb*SLP1 (Anti-*Nb*SLP1C, A2, green) detect the surface in purified mature spores treated with Triton; Negative control (B1, B2). A3, B3: Differential interference contrast (DIC) images of the spores. A4 and B4 are the merged images of A1, A2 and A3, B1, B2 and B3, respectively. Anti-*Nb*-SLP1P and anti-*Nb*SLP1C sera were used at a 1:200 dilution. The secondary antibody was Alexa Fluor® 647-R-phycoerythrin goat anti-mouse IgG (Invitrogen) and FITC-conjugated goat anti-rabbit IgG (Sigma) used at a 1:100 dilution. The scale bar represents 3 μm. (For interpretation of the references to colour in this figure legend, the reader is referred to the web version of this article.)(选自：*Characterization of a subtilisin-like protease with apical localization from microsporidian Nosema bombycis*，参见本书第274页。)

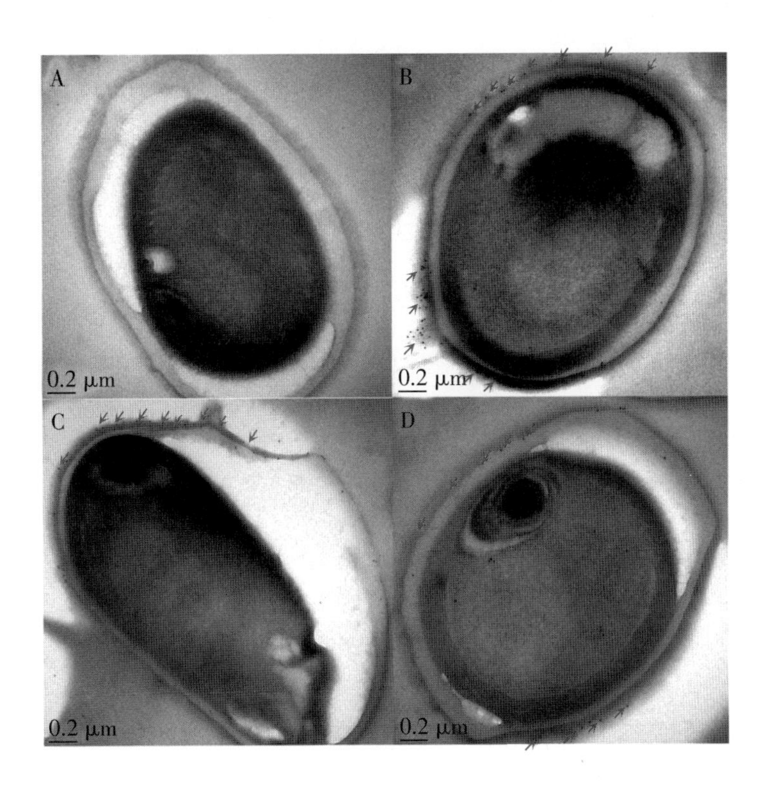

Figure 54. IEM of *Nb*HSWP16 in mature *N. bombycis*. (A) Negative control; (B), (C), (D) spore labelled on the exospore region. Anti-*Nb*SWP16 serum was used at 1:20 dilution. Secondary antibody was goat anti-mouse IgG conjugated with 10 nm colloidal gold particles (sigma). The labelled gold particles were marked with arrowheads. The scale bar represents 0.2 μM. Ex, exospore; En, endospore.［选自：*Characterization of a novel spore wall protein NbSWP16 with proline-rich tandem repeats from Nosema bombycis (microsporidia)*，参见本书第290页。］

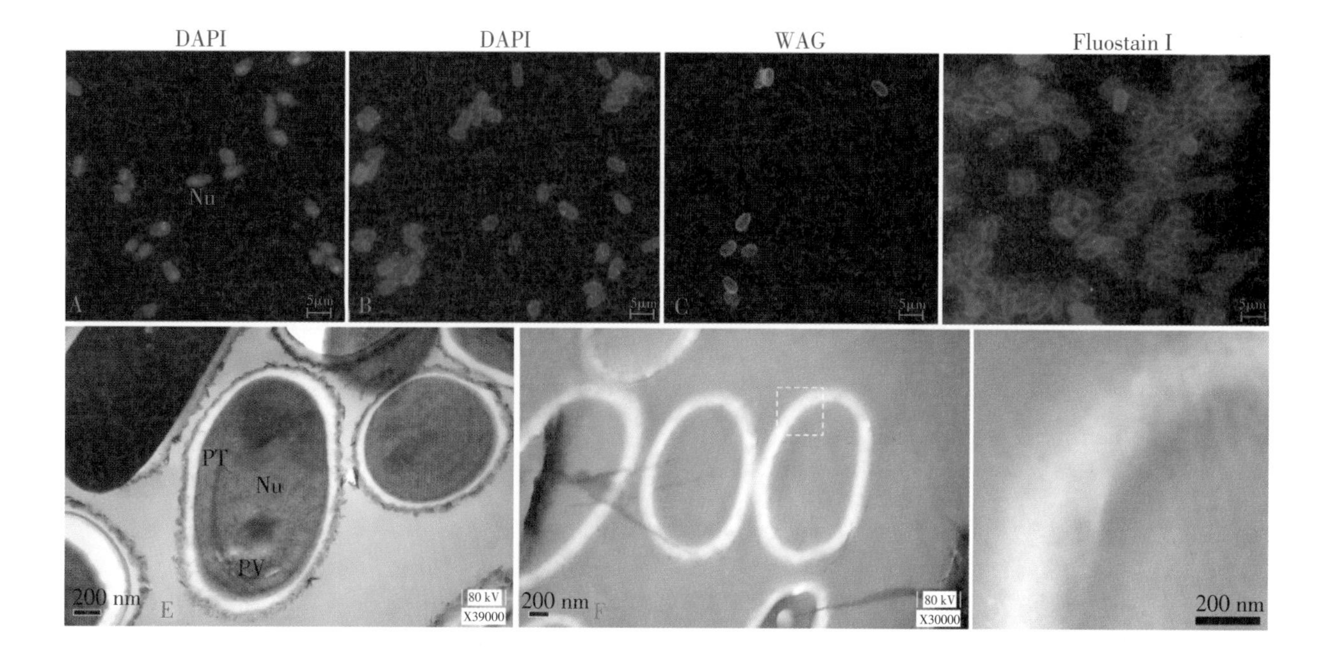

Figure 55. DAPI, WGA and Fluostain I staining and TEM to examine the *N. bombycis* chitin spore coats. (A) The untreated normal mature spores stained with DAPI as control. Nu: nuclei; (B) The DCSCs stained with DAPI; (C) The DCSCs stained with FITC labeled wheat germ agglutinin (WGA) to confirm the chitin component; (D) DCSCs stained with Fluostain I to further confirm the chitin layer; (E) The internal ultrastructure of untreated normal *N. bombycis* to demonstrate the differences between normal spore and DCSCs. PT: polar tube, PV: Posterior Vacuole; (F) The ultrastructure of the *N. bombycis* DCSCs using hot alkali treatment; (G) The enlarged region of (F). The DCSCs is a loose and curled chitin spore coat.［选自：*Development of an approach to analyze the interaction between Nosema bombycis (microsporidia) deproteinated chitin spore coats and spore wall proteins*，参见本书第302页。］

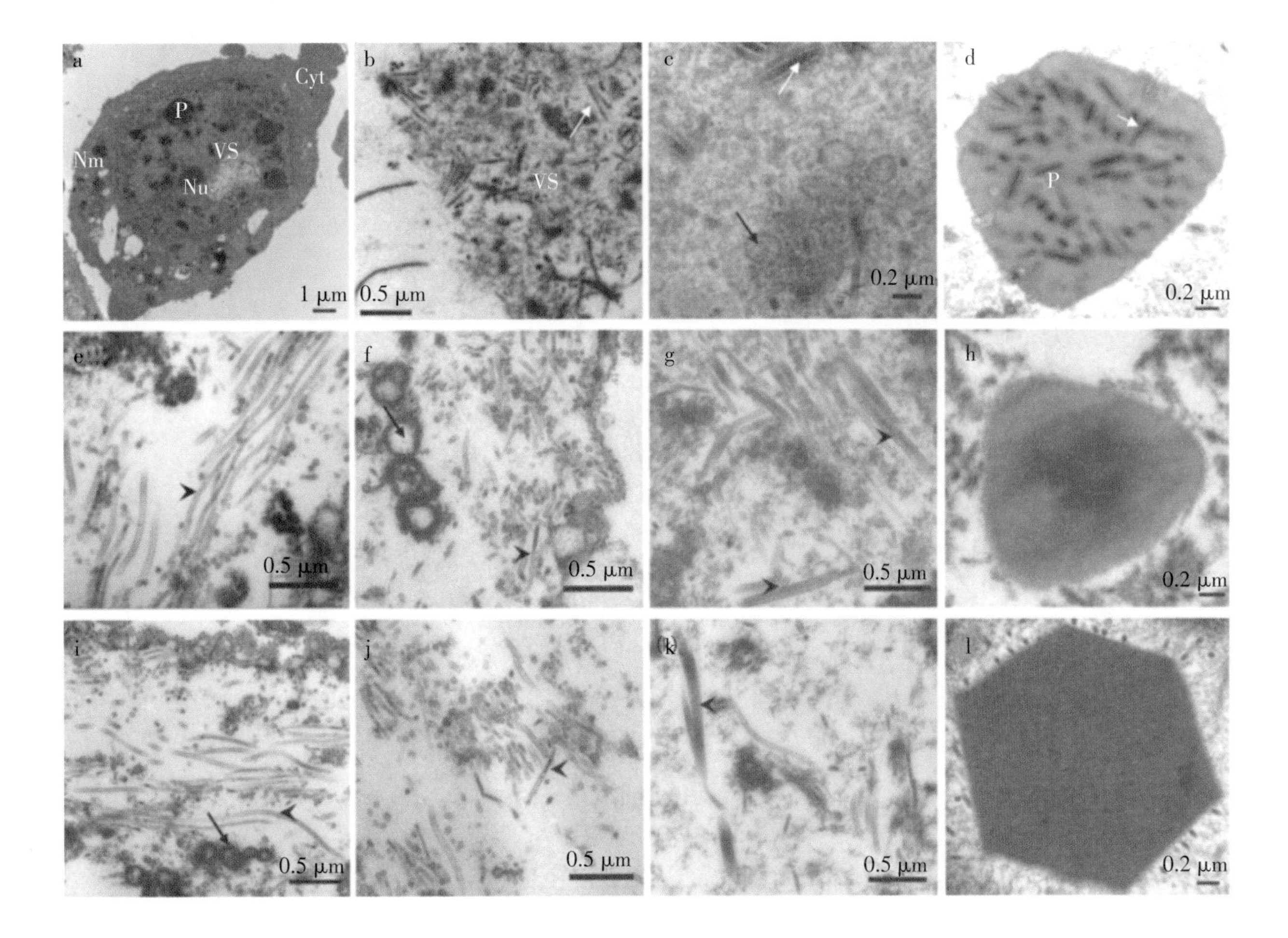

Figure 56. Transmission electron microscopy analysis of *BmN* cells transfected with v*Bm*P95-Re (a-d), v*Bm*P95-De (e-h) and v*Bm*P95-N-Re bacmid (i-l) at 60 h p.t. (a) Image of a whole cell displaying the VS, enlarged nucleus (Nu), nuclear membrane (Nm) and polyhedra (P). Cyt, cytoplasm. (b) Electron-dense nucleocapsids (white arrow) in the VS. (c) Virus-induced intranuclear microvesicles (black arrow), nucleocapsids aligning with the intranuclear vesicles membranes and pre-occluded enveloped virions in the ring zone (white arrow). (d) Normal virions (white arrow) embedded in the polyhedra (P). (e, i) Masses of electron-lucent tubular structures (arrowheads) observed at the VS and inner nuclear membrane. (f, j) Tubular structures with incomplete viral DNA genomes (arrowheads) appeared in the nucleus. In (f) and (i), virus-induced intranuclear microvesicles are indicated by a black arrow. (g, k) Electron-dense and aberrant tubular structures (arrowheads) were seen in the VS. (h, l) Polyhedra (P) devoid of embedded virions.（选自：*Bombyx mori nucleopolyhedrovirus BmP95 plays an essential role in budded virus production and nucleocapsid assembly*，参见本书第320页。）

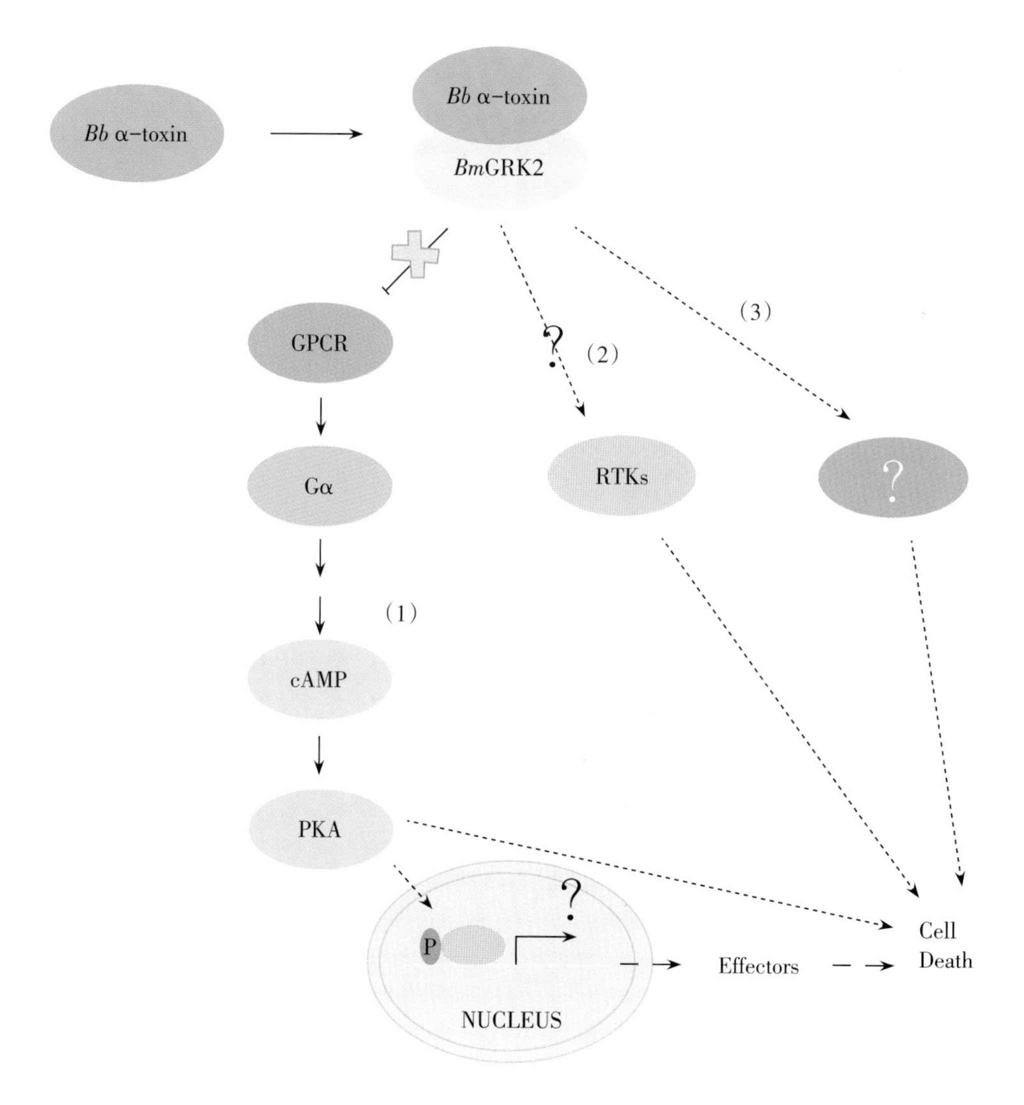

Figure 57. Schematic model for *Bb* α-toxin action. *Bb* α-toxin bound to *Bm*GRK2, a key GPCR regulatory kinase, affecting the modulation of GPCRs signaling pathway, leaded to a continuous up-regulation of downstream activity, which stimulates $G_{\alpha s}$, promotes production of cAMP, and activates PKA. In turn, PKA activation alters effectors that disturb homeostasis and many fundamental biological process that destroy midgut cells, inducing larvae death (pathway 1). Alternatively, *Bb* α-toxin bound to *Bm*GRK2 may also alter the signals downstream of receptor tyrosine kinases (RTKs) that are involved in cell cycle phases, but whether these signaling pathways participate in the *Bb* α-toxin pathogenic mechanism is unknown (pathway 2). Activation of some other signaling pathway is not through pathway 1 or pathway 2 but through other unknown receptors (pathway 3) in *Bb* α-toxin pathogenic mechanism. All of these pathways work together, leading to cell death.（选自：*Bacillus bombysepticus α-toxin binding to G protein-coupled receptor kinase 2 regulates cAMP/PKA signaling pathway to induce host death*，参见本书第342页。）

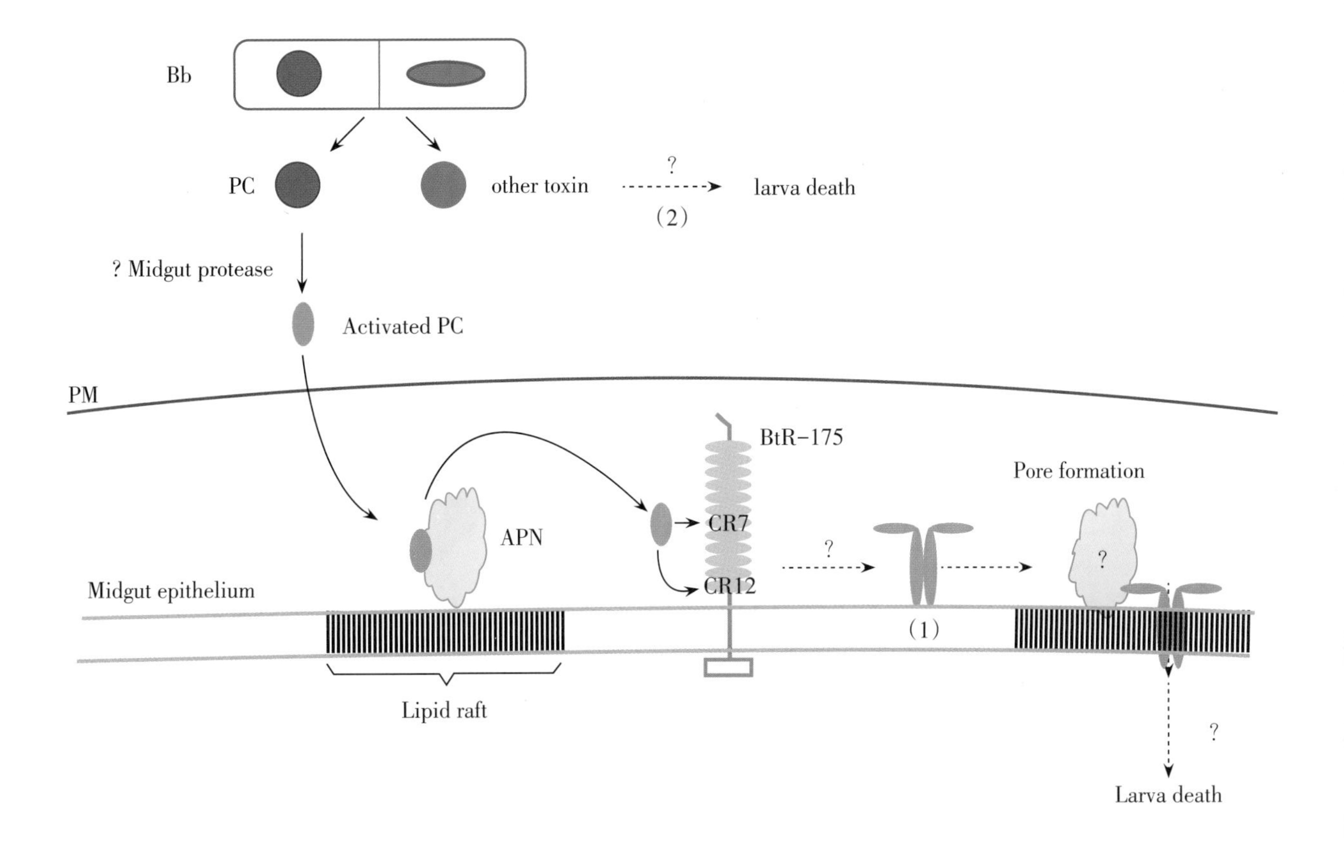

Figure 58. A schematic representation of the model for *B. bombysepticus* or PC causing damage to the silkworm midgut and translocating into the hemolymph. The larvae ingest *B. bombysepticus*, which produces parasporal crystal toxin (PC). PC can be digested by gut proteases. The digested PC pass through the peritrophic membrane (PM) to bind the high-abundance APN receptor, allowing the toxin to be located in close proximity to the membrane. This interaction is followed by high-affinity binding to the BtR-175 receptor by the cadherin fragments CR7 and CR12. Interactions with BtR-175 trigger the oligomerization of toxin that binds to the receptors and leads to pore formation (Pathway 1). However, unknown toxin(s) might also induce larvae death (Pathway 2). All of these toxins and receptors could work together, leading to aberrant gut infiltration and material exchange throughout the insect body, resulting in the death of larvae that is caused by *B. bombysepticus* infection.（选自：*PC, a novel oral insecticidal toxin from Bacillus bombysepticus involved in host lethality via APN and BtR-175*，参见本书第344页。）

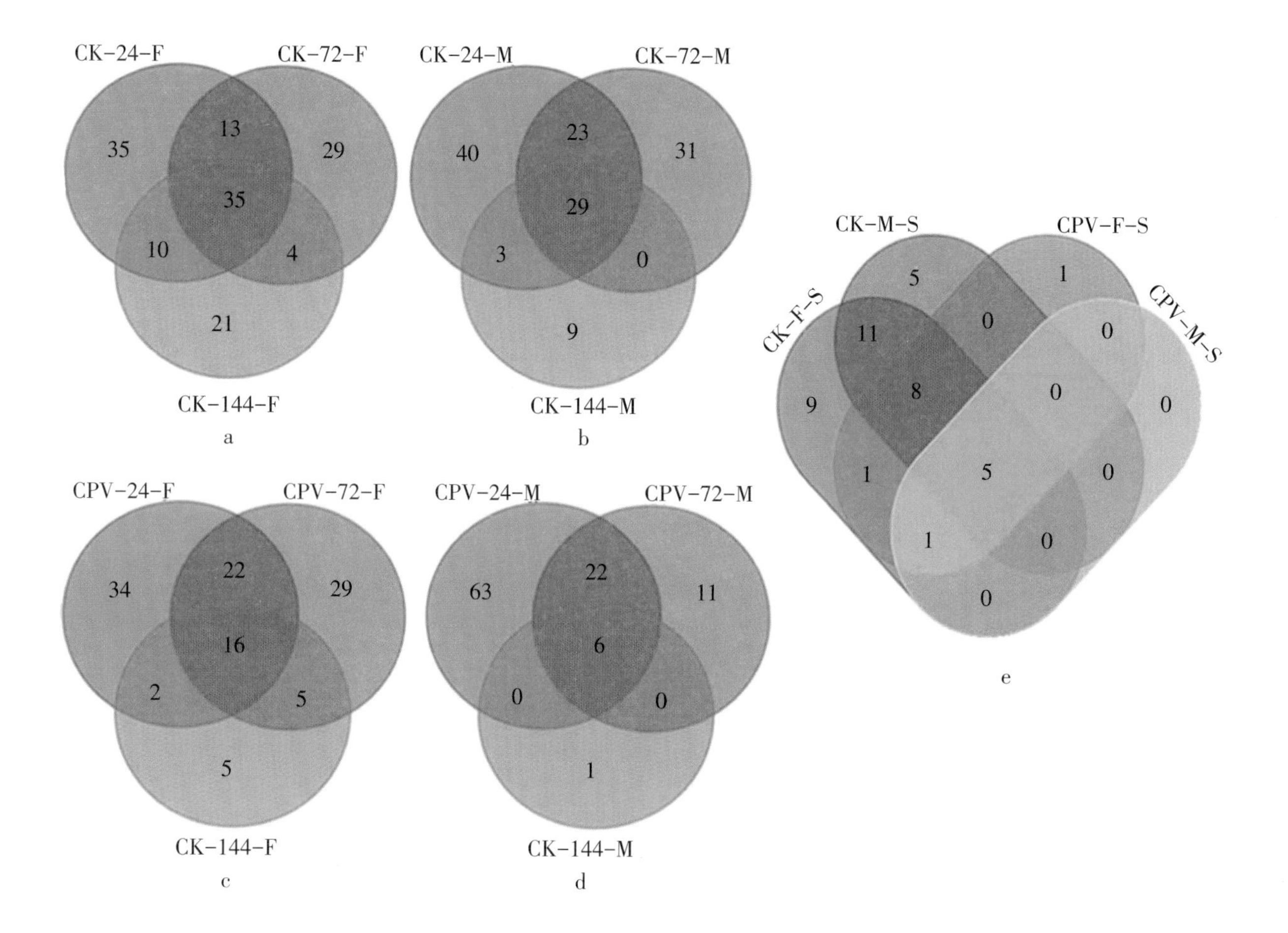

Figure 59. Shared genera analysis of the different samples. Venn diagram showing the unique and shared genera in the different samples. CK (CPV)-24 (72,144)-F (M) are samples mentioned in Table 1. CK-F (M)-S was a general designation of CK-24-F (M), CK-72-F (M) and CK-144-F (M). CPV-F (M)-S was a general designation of CPV-24-F (M), CPV-72-F (M) and CPV-144-F (M). (a) for CK-24-F, CK-72-F and CK-144-F samples; (b) for CK-24-M, CK-72-M and CK-144-M samples; (c) for CPV-24-F, CPV-72-F and CPV-144-F samples; (d) for CPV-24-M, CPV-72-M and CPV-144-M samples; (e) for CK-F-S, CK-M-S, CPV-F-S and CPV-M-S samples.（选自：*Effects of BmCPV infection on silkworm Bombyx mori intestinal bacteria*，参见本书第360页。）

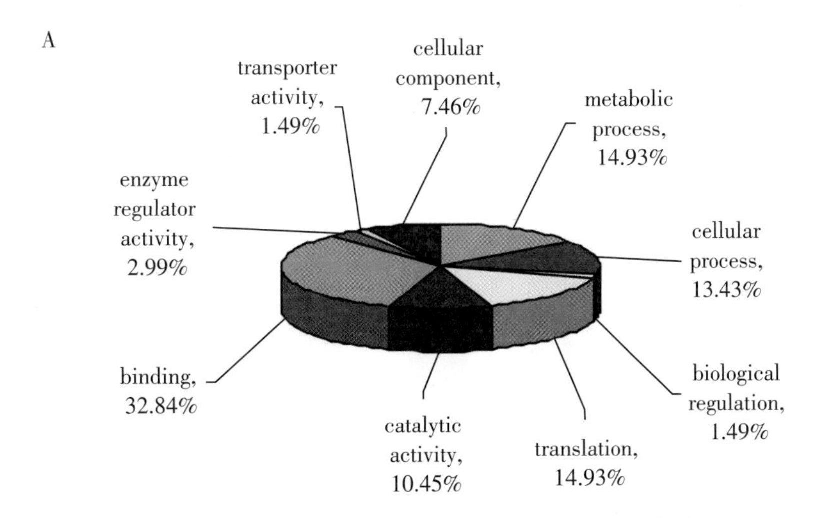

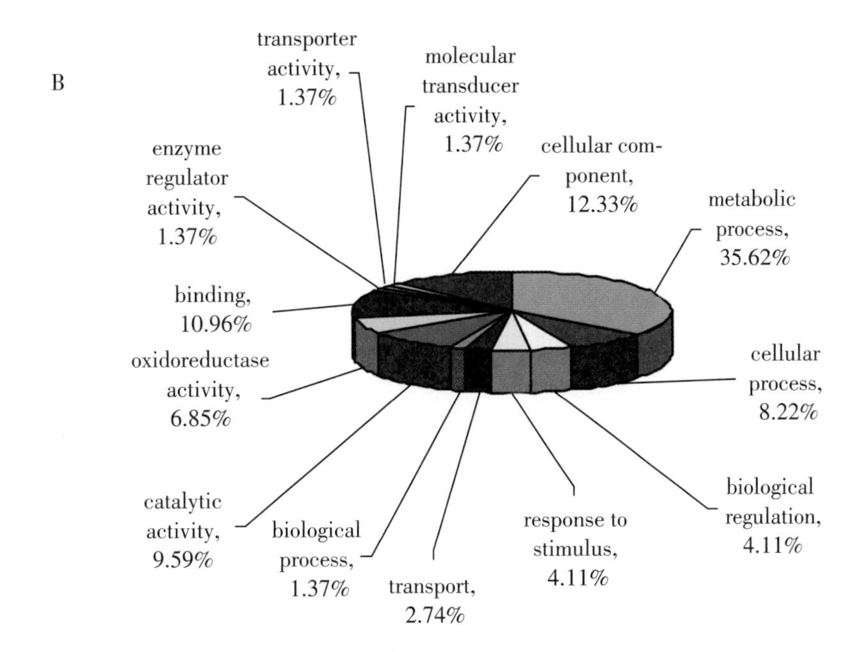

Figure 60. Gene categories of differentially expressed genes at 72 h post-inoculation. According to GO (http://www/geneontology.org/), 67 up-regulated genes were classified into nine categories (a), described as enzyme regulator activity (2.99%), transporter activity (1.49%), binding (32.84%), catalytic activity (10.45%), metabolic process (14.93%), cellular process (13.43%), translation (14.93%), biological regulation (1.49%) and cellular component (7.46%). 73 down-regulated genes were classified into 13 categories (b), described as oxidoreductase activity (6.85%), catalytic activity (9.59%), binding (10.96%), enzyme regulator activity (1.37%), transporter activity (1.37%), molecular transducer activity (1.37%), metabolic process (35.62%), cellular process (8.22%), transport (2.74%), biological regulation (4.11%), response to stimulus (4.11%), biological process (1.37%) and cellular component (12.33%).（选自：*Microarray analysis of the gene expression profile in the midgut of silkworm infected with cytoplasmic polyhedrosis virus*，参见本书第364页。）

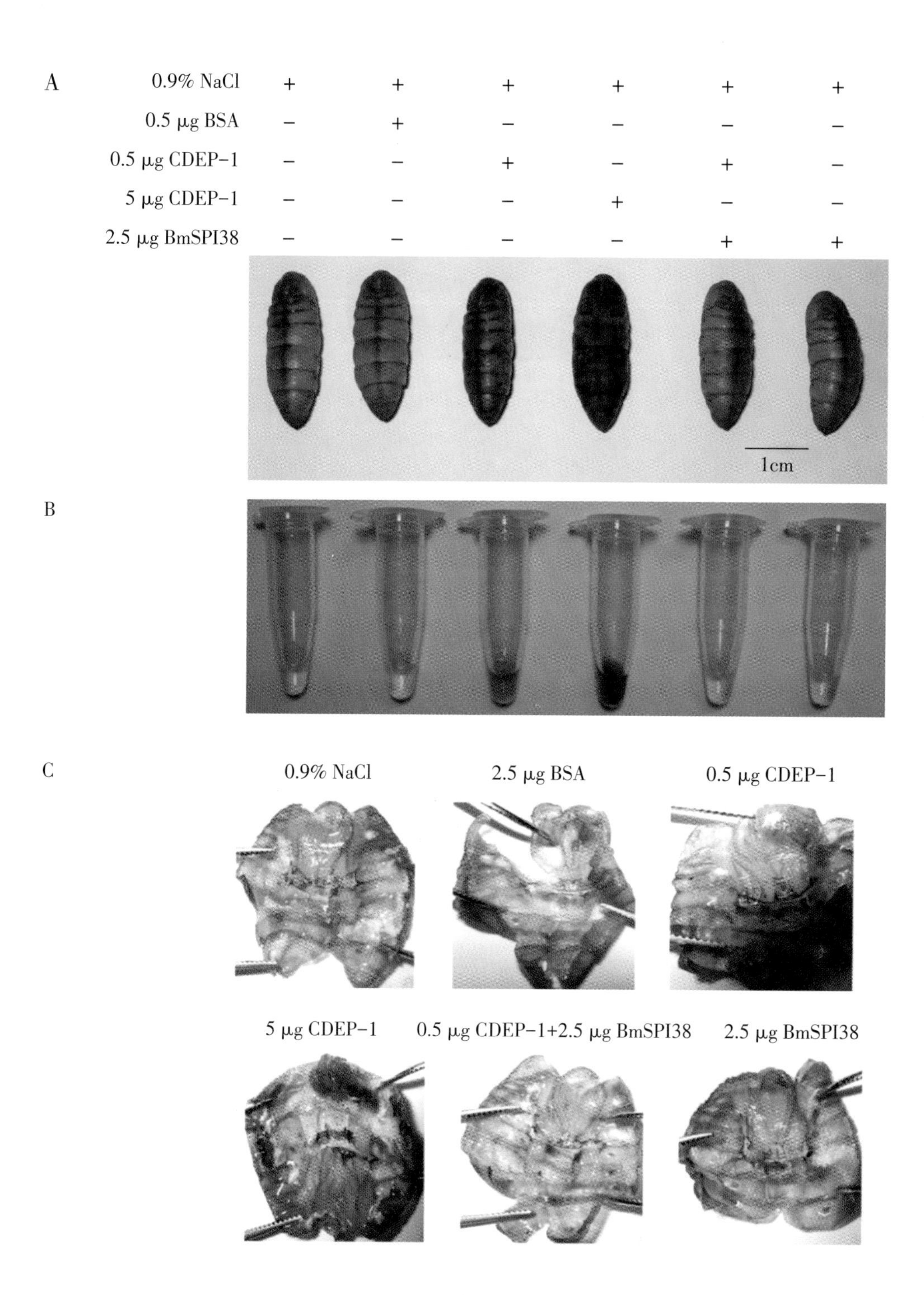

Figure 61. Inhibition of CDEP-l-induced melanization. (A) Induced blackening of silkworms and inhibition of melanization by *Bm*SPI38. (B) The hemolymph and (C) integuments of the injected silkworms. At least eight insects were used in each treatment and each treatment was repeated independently. One insect in each treatment was selected randomly as representative and displayed here. The symptoms were consistent and clear as shown in the figure. The variation in two independent experiments was only in the amount of melanization of the cuticle.（选自：*A novel protease inhibitor in Bombyx mori is involved in defense against Beauveria bassiana*，参见本书第366页。）

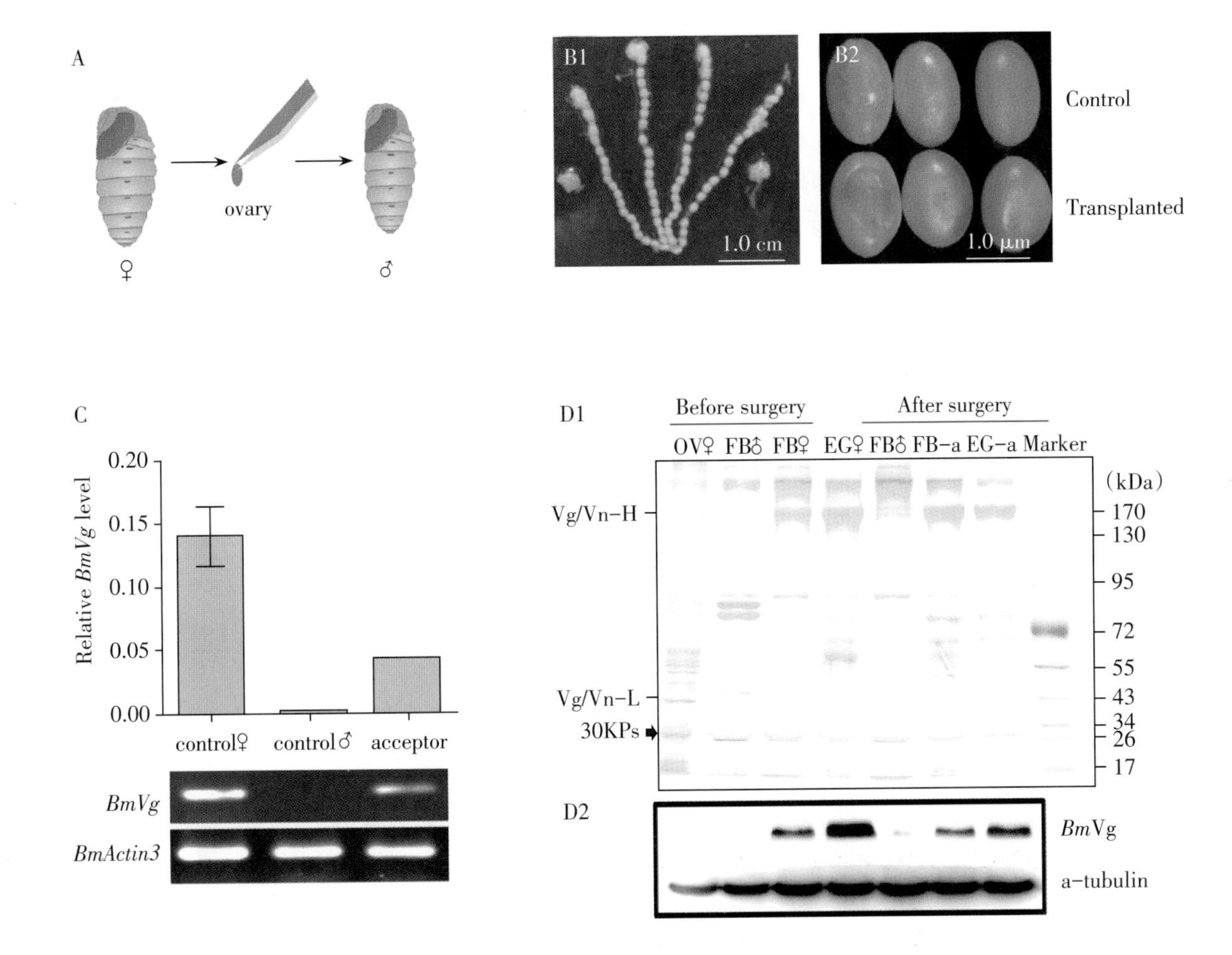

Figure 62. Expression of *BmVg* in male pupae induced by ovary transplanted for egg formation. (A) Day 0 male silkworm pupae with transplanted ovaries. (B) Ovary phenotypes in male pupae. B1, ovary transplanted into a male that developed four complete oviducts with eggs. B2, eggs were slightly whiter than control eggs from female moths. (C) RT-PCR and qPCR for *BmVg* transcription in fat body 2 days after transplant. Acceptor, male pupae transplanted with an ovary; Control♀♂, normal female and male pupae cut and sealed with nail varnish. (D) Analysis of total proteins in silkworm tissues before and after transplant. SDS-PAGE and Western blot for *Bm*Vg (before surgery, day 0 pupa; after surgery, fat body from day 2 pupae and eggs from moths; OV, ovary; FB, fat body; EG, egg; EG-a, FB-a, egg and fat body from the acceptor silkworm; Marker, standard protein marker)（选自：*Female qualities in males: vitellogenin synthesis induced by ovary transplants into the male silkworm, Bombyx mori*，参见本书第376页。）

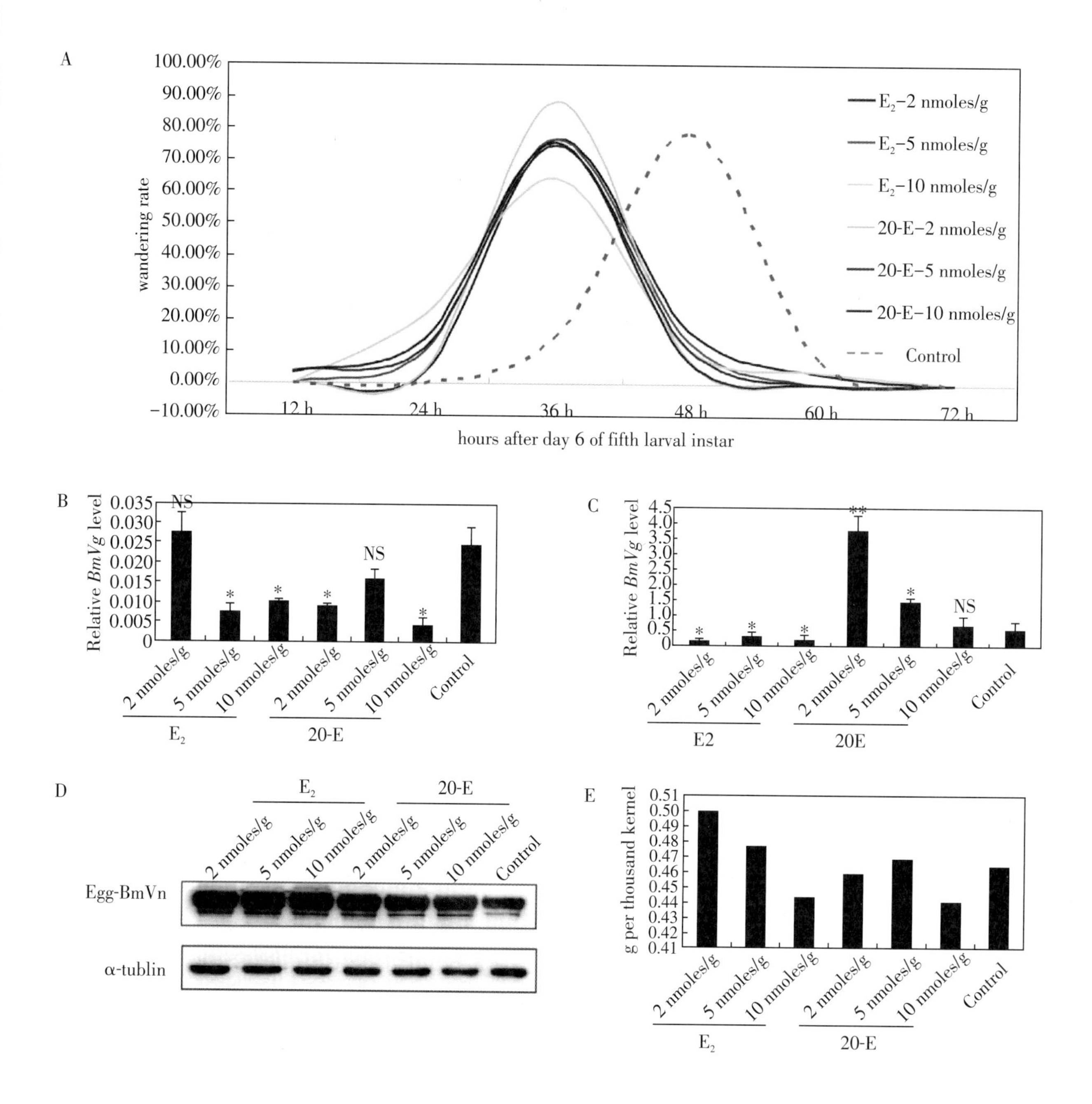

Figure 63. Analysis of female silkworms effects by feeding E_2 or 20-E. (A) Statistical analysis (n = 60) of the wandering rate of female silkworms fed E_2 or 20-E for 4 days at 2 nmol/g, 5 nmol/g or 10 nmol/g. Data were calculated from 12 h to 72 h after the fourth feeding. *BmVg* transcripts by qRT-PCR in female fat bodies treated with E_2 or 20E at 36 h (B) and 72 h (C) after fourth feeding. (D) Western blot for *Bm*Vn in unfertilized eggs from moths fed E_2 or 20-E four times at 2 nmol/g, 5 nmol/g or 10 nmol/g at day 3, 4, 5, and 6 of the fifth instar stage. (E) Weight of eggs (n = 1 000) after hormone treatment. All values are mean ± SD of three biological replicates. *Significance, 0.05; **significance, 0.01 level (t-test) compared with control group; NS, no significant difference at 0.05 level.（选自：*Vertebrate estrogen regulates the development of female characteristics in silkworm, Bombyx mori*，参见本书第378页。）

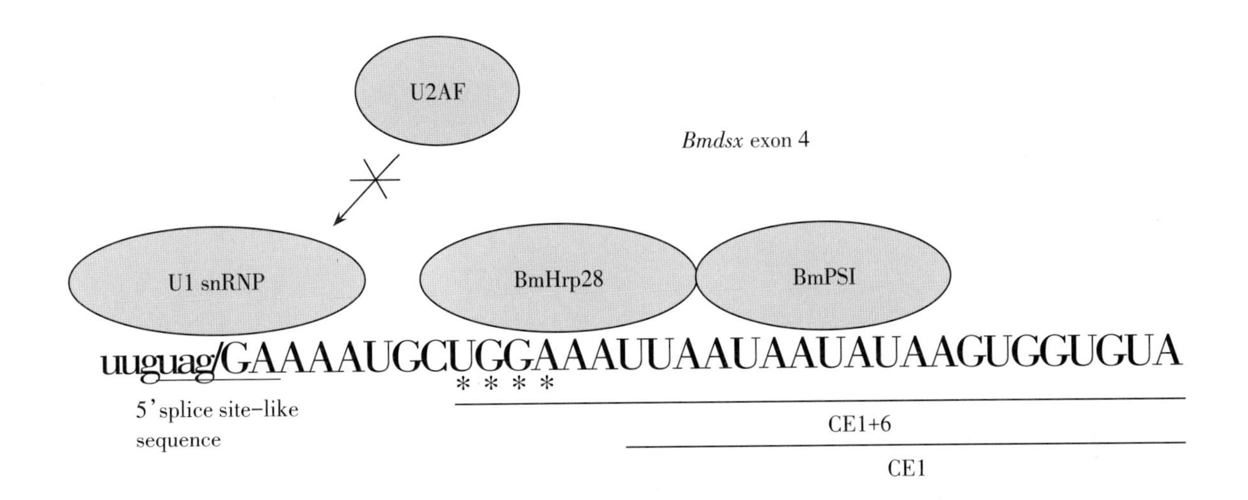

Figure 64. A possible model for the splicing inhibition of *Bmdsx* exon 4. The '/' represents 3' splice site. The sequences of CE1 and CE1+6 are indicated by thick lines. The asterisks indicate the sequence resembling the core sequence UAG (G/A) of hnRNP A1 binding sites. The binding sites of *Bm*PSI and *Bm*Hrp28 on the *Bmdsx* exon 4 are adjacent and near to the 3' splice site of *Bmdsx* intron 3. They act together to recruit the U1 snRNP bind to a 5' splice site-like sequence lied within the 3' splice site, which blocks the binding of general splicing factors, such as U2AF, to the 3' splice site, then preventing the splicing on this site.(选自:*BmHrp28 is a RNA-binding protein that binds to the female-specific exon 4 of Bombyx mori dsx pre-mRNA*,参见本书第384页。)

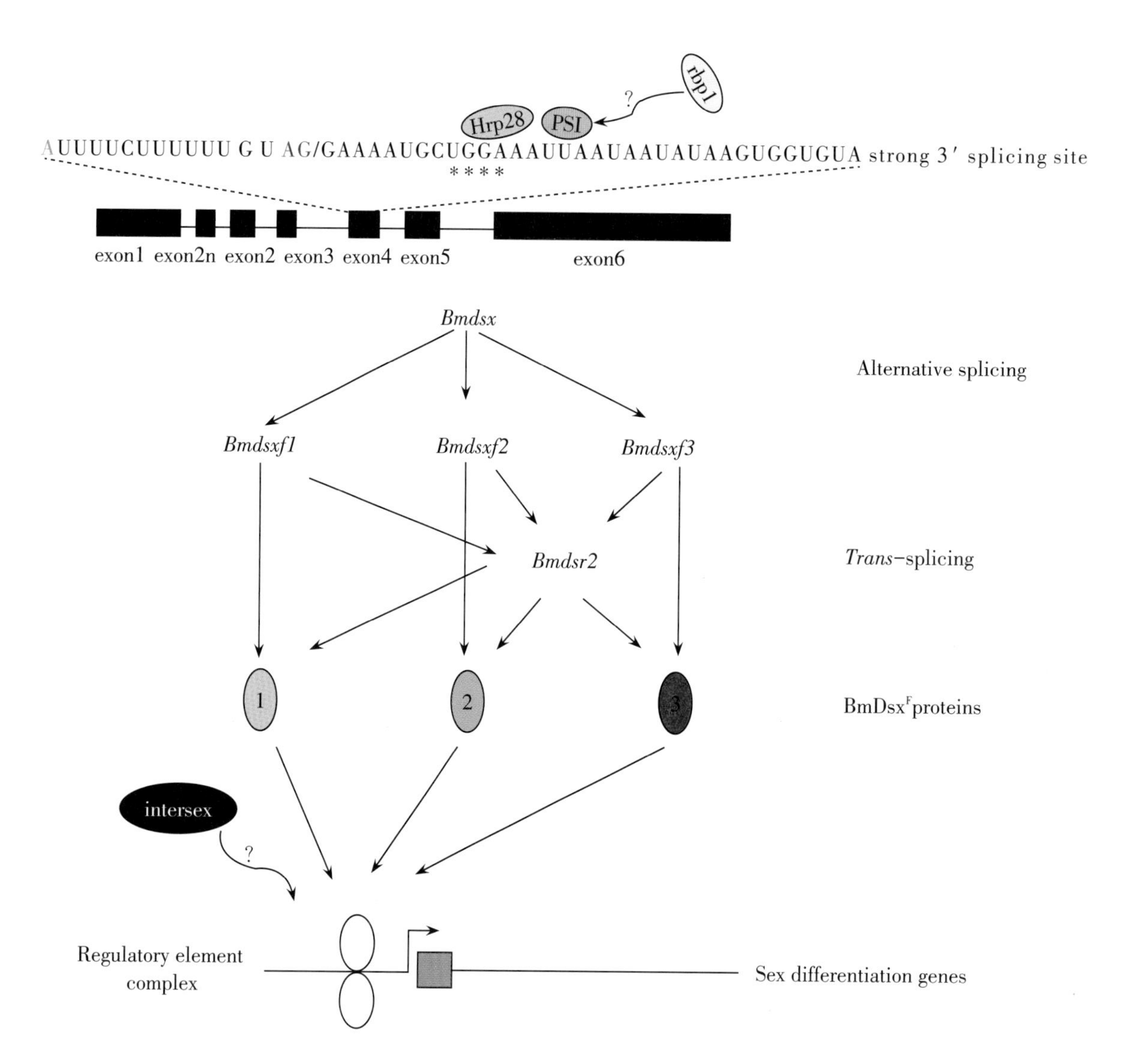

Figure 65. Model explaining the pattern of action of three *Bm*DSX^F proteins and two splicing methods during female sexual differentiation. In females, *Bmdsx* generates three alternative-spliced forms, and the *Bmdsr2* as a 3′ UTR then down-regulates the expression levels of *Bm*DSX^F proteins by the trans-splicing method. Finally, the three proteins further influence the expression of sex-differentiation genes, together with the *Bm*DSXF homodimers or heterodimers, in a female-specific manner.（选自：*Novel female-specific trans-spliced and alternative splice forms of dsx in the silkworm Bombyx mori*，参见本书第388页。）

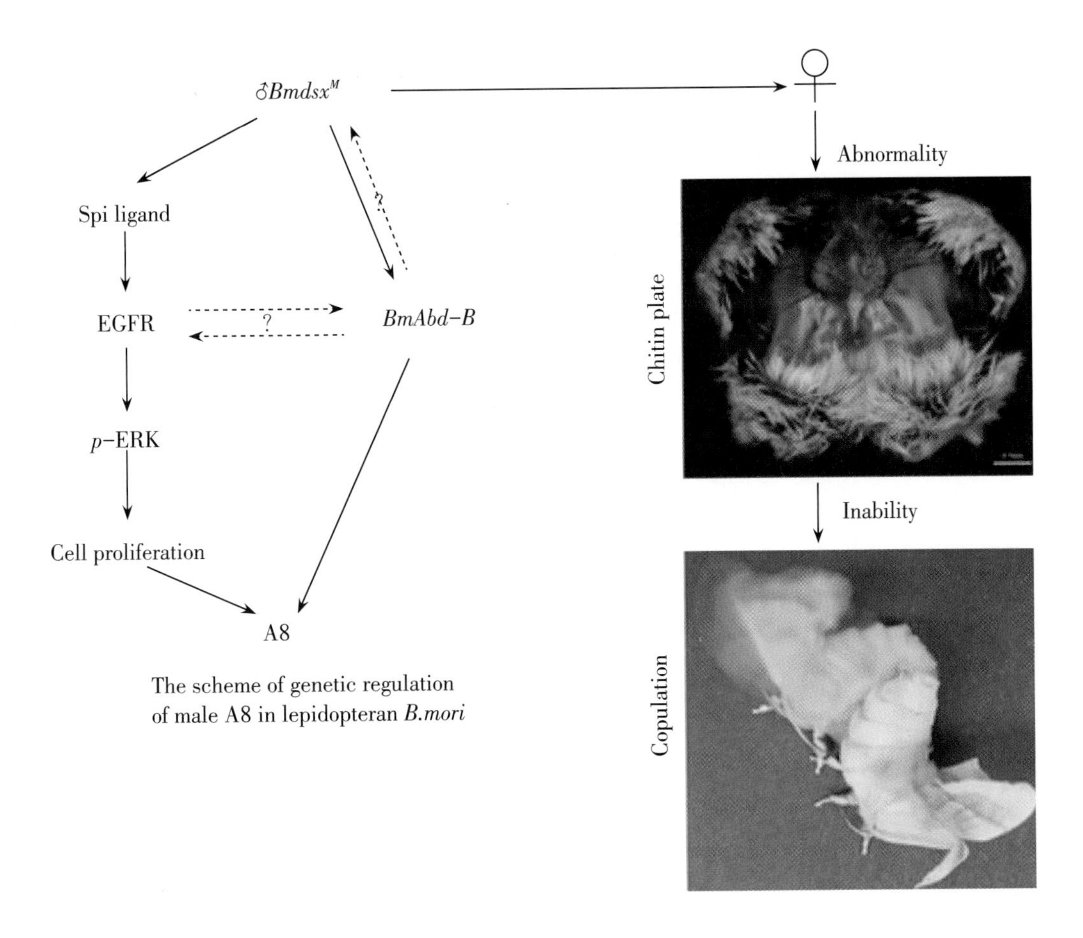

Figure 66. Model explaining the mode of action of *BmAbd-B* and EGFR signaling in the posterior abdomen of wild-type males. The ectopic expression of *$Bmdsx^M$* in females causes formation of an abnormal chitin plate and the inability to copulate. [Color figure can be viewed in the online issue which is available at wileyonlinelibrary.com]（选自：*Ectopic expression of the male BmDSX affects formation of the chitin plate in female Bombyx mori*，参见本书第394页。）

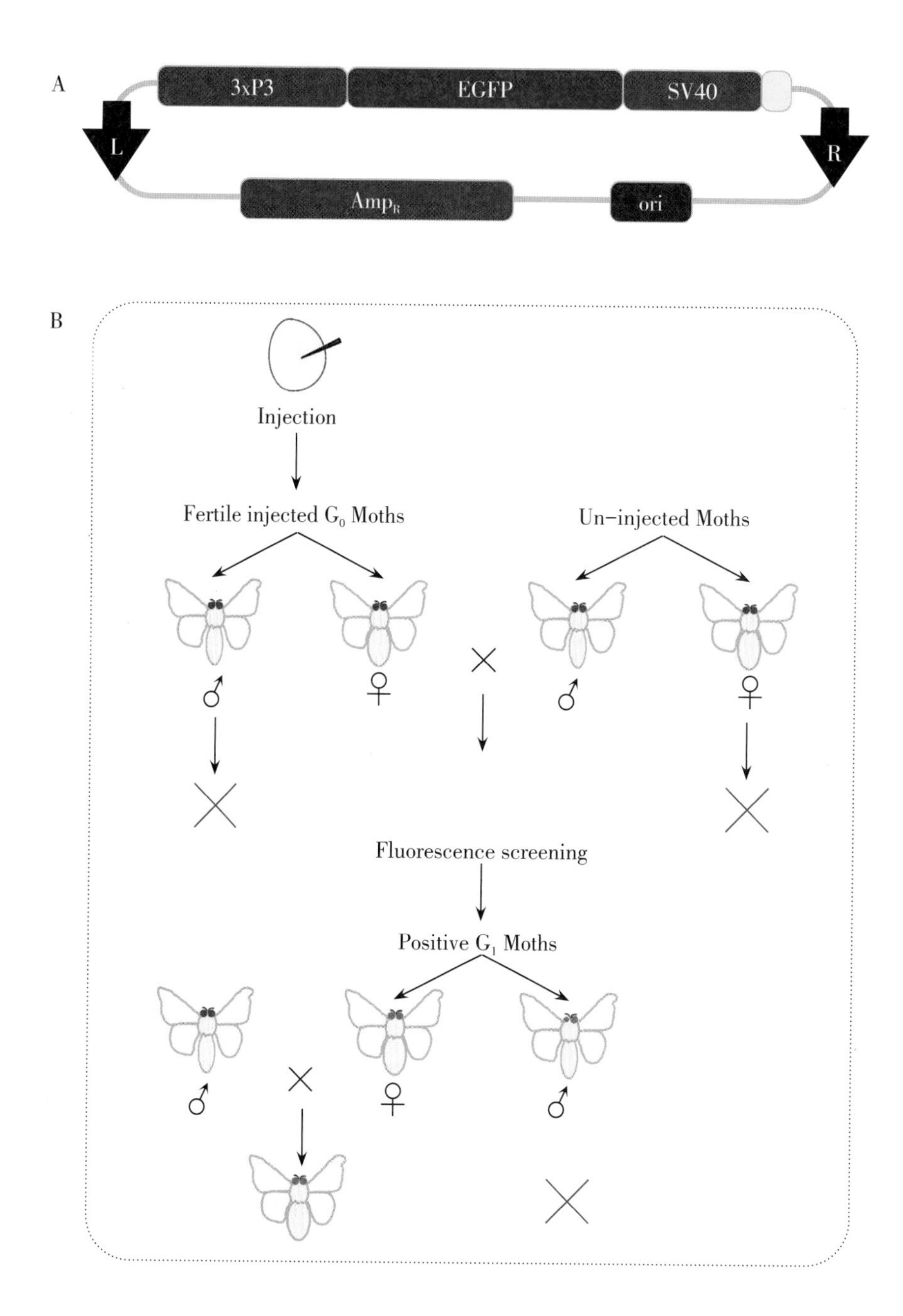

Figure 67. Crossing strategies (a) A representation of transgenic vectors *p*Bac[3×P3-EGFP, af]. 3×P3, an artificial promoter that drives target gene expression in stema and in the nervous system; EGFP, enhanced green fluorescent protein; SV40, transcriptional terminator; L and R, left and right arm of the *piggy*Bac transposon, respectively; Amp$_R$ and ori represent the ampicillin-resistance gene and the replication origin, respectively. (b) A map of crossing strategies used to screen for W chromosome linked transgene.（选自：*Genetic marking of sex using a W chromosome-linked transgene*，参见本书第396页。）

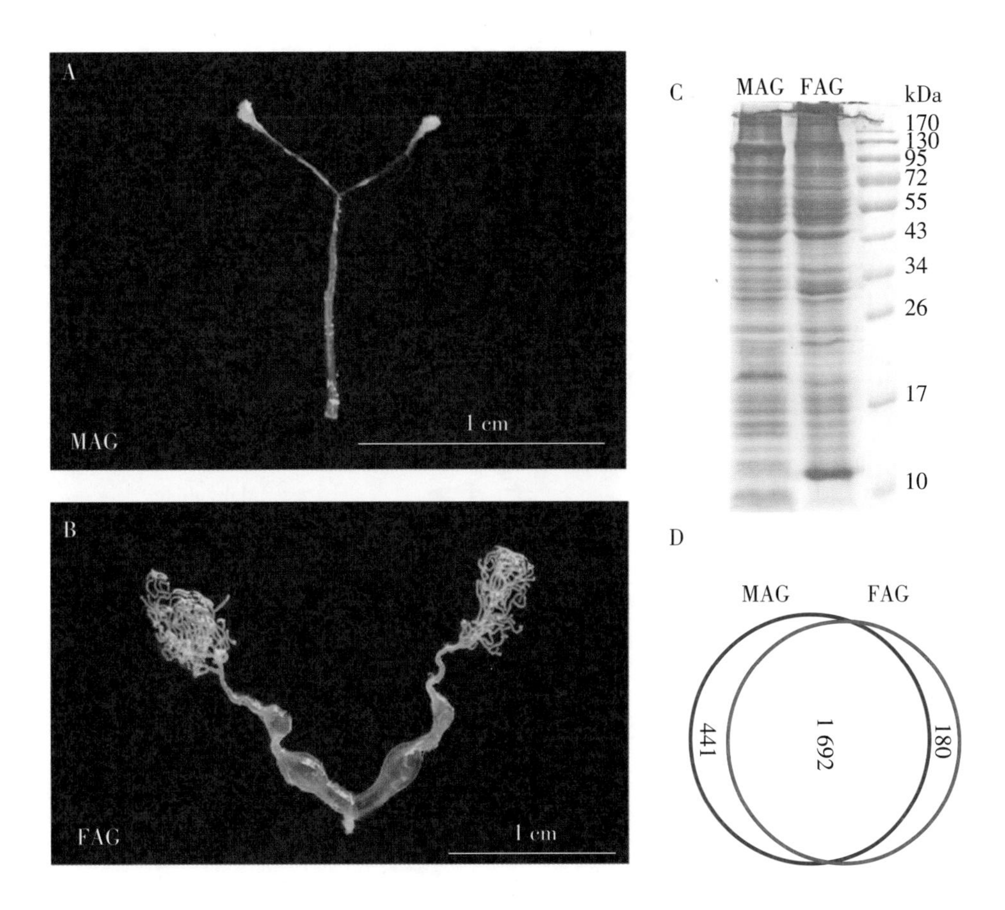

Figure 68. Protein identification in male accessory gland (MAG) and female accessory gland (FAG) of the silkworm. The MAG (a) and FAG (b) were dissected out from unmated moth. Equal amounts of accessory gland proteins were separated on 12% (w/v) polyacrylamide gel and visualized by staining with coomassie brilliant blue (c). The Venn diagram (d) showed that 2 133 proteins and 1 872 proteins were identified in the MAG and FAG (color figure online)（选自：*Proteome profiling reveals tissue-specific protein expression in male and female accessory glands of the silkworm, Bombyx mori*，参见本书第400页。）

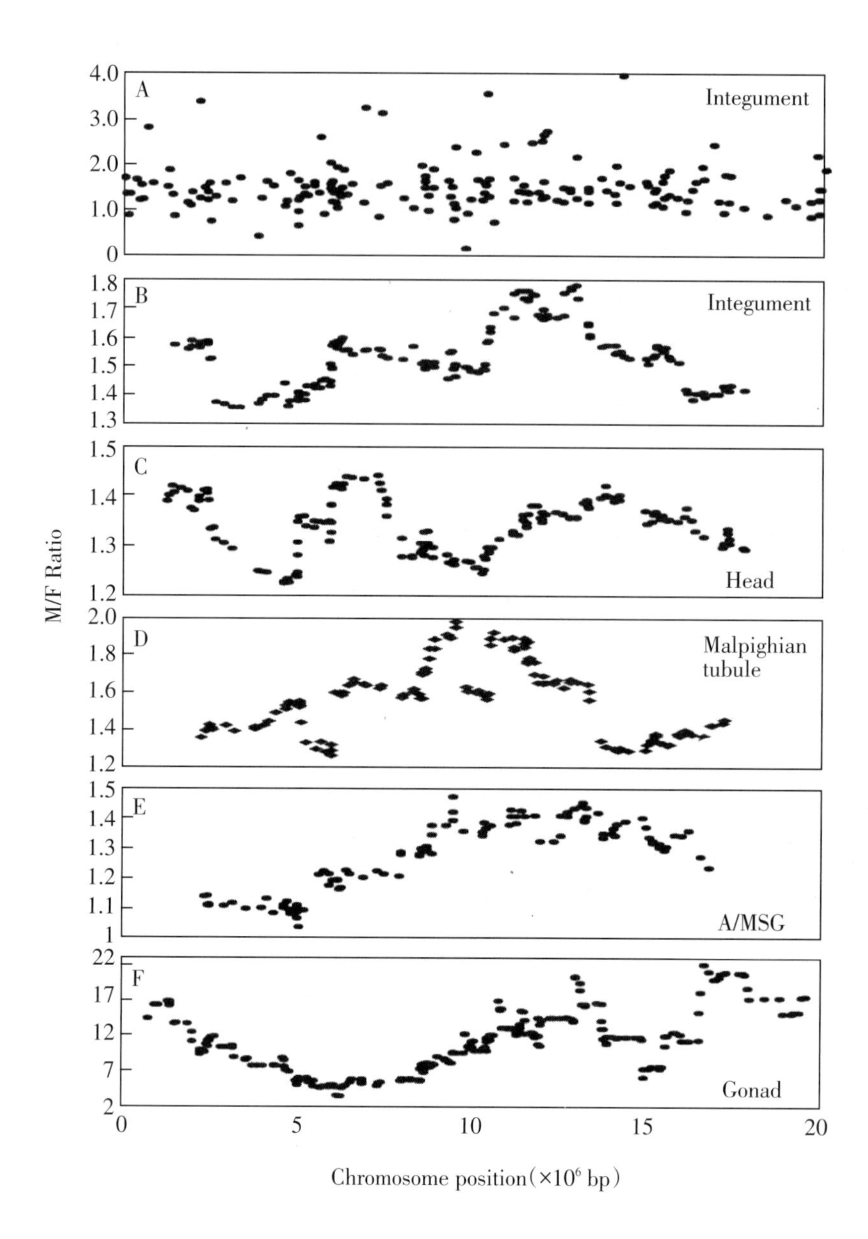

Figure 69. Region distribution of M∶F ratios on Z chromosome. (A) Individual M∶F ratios in the integument, graphed by gene position on the Z chromosome. Two genes (*BGIBMGA002087* and *BGIBMGA000725*) are the outliers with respect to M/F ratio. The former with M∶F expression ratio 5.315 codes macrophage migration inhibitory factor in *Bombyx mori*, and the latter with the ratio 4.010 codes ATP-binding cassette transporter. (B-F) The running average of 30 M∶F ratios is plotted at the median gene position, for integument, head, malpighian tubule, A/MSG (anterior and median silk gland), and gonad.（选自：*Dosage analysis of Z chromosome genes using microarray in silkworm, Bombyx mori*，参见本书第402页。）

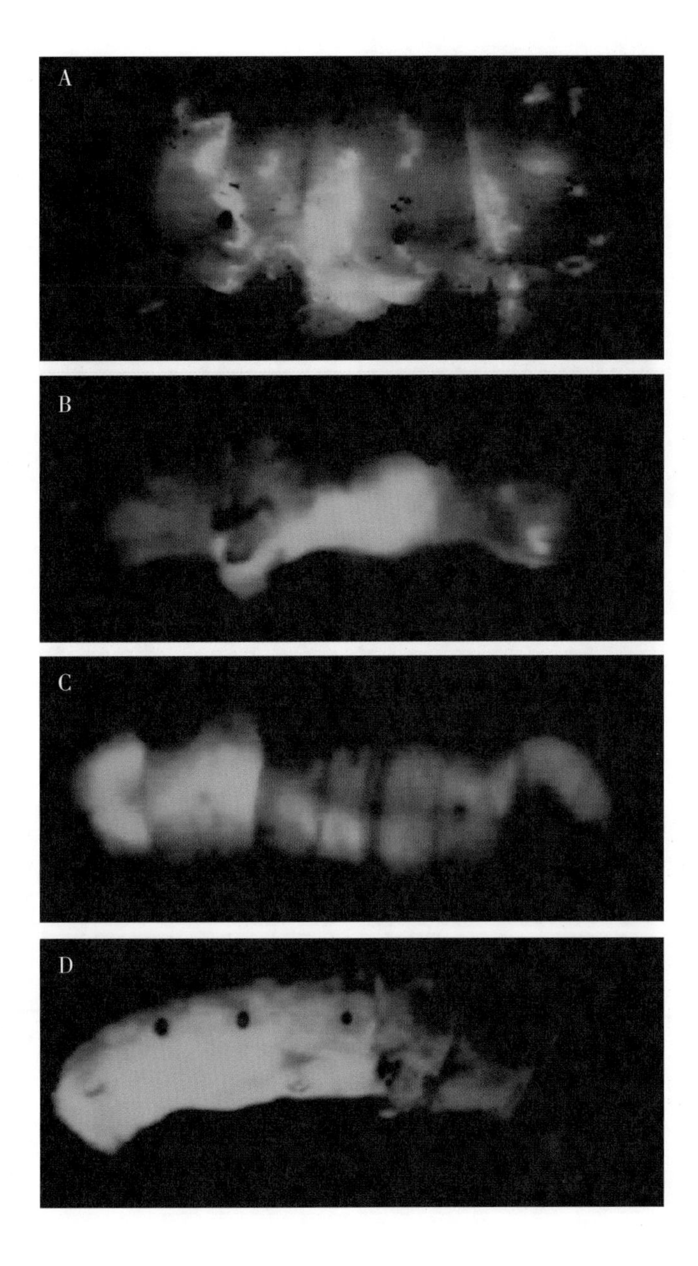

Figure 70. Enhanced green fluorescent protein expression of G_0 transformed individuals of *Nistari* strain of silkworm *Bombyx mori*. The expression patterns appear to be in mosaic displays, and there are diverse types of appearance. (A) Line-like mosaic. (B, C) Short piece-like mosaic. (D) Big block-like mosaic.（选自：*Comparison of transformation efficiency of piggyBac transposon among three different silkworm Bombyx mori strains*，参见本书第414页。）

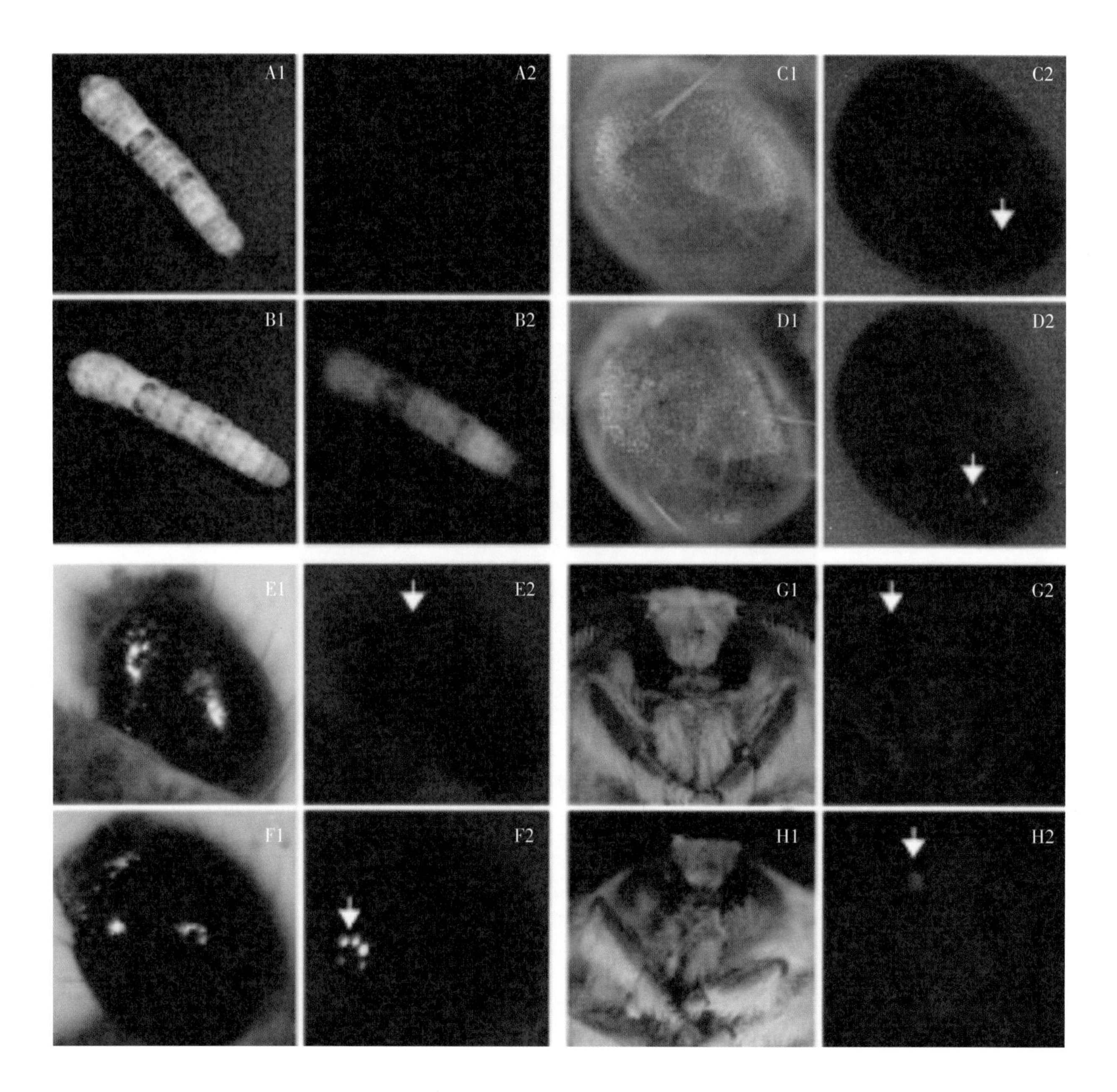

Figure 71. Expression patters in transgenic silkworm. A1 and B1, C1 and D1, E1 and F1, G1 and H1 were in bright field in level, egg, new-hatched silkworm and moth respectively. A2, B2, C2, D2, E2, F2, and G2 were screened for green or red fluorescence. A, C, E, and G series were used as the controls. Arrows indicate the eyes.（选自：*The relationship between internal domain sequences of piggyBac and its transposition efficiency in BmN cells and Bombyx mori*，参见本书第416页。）

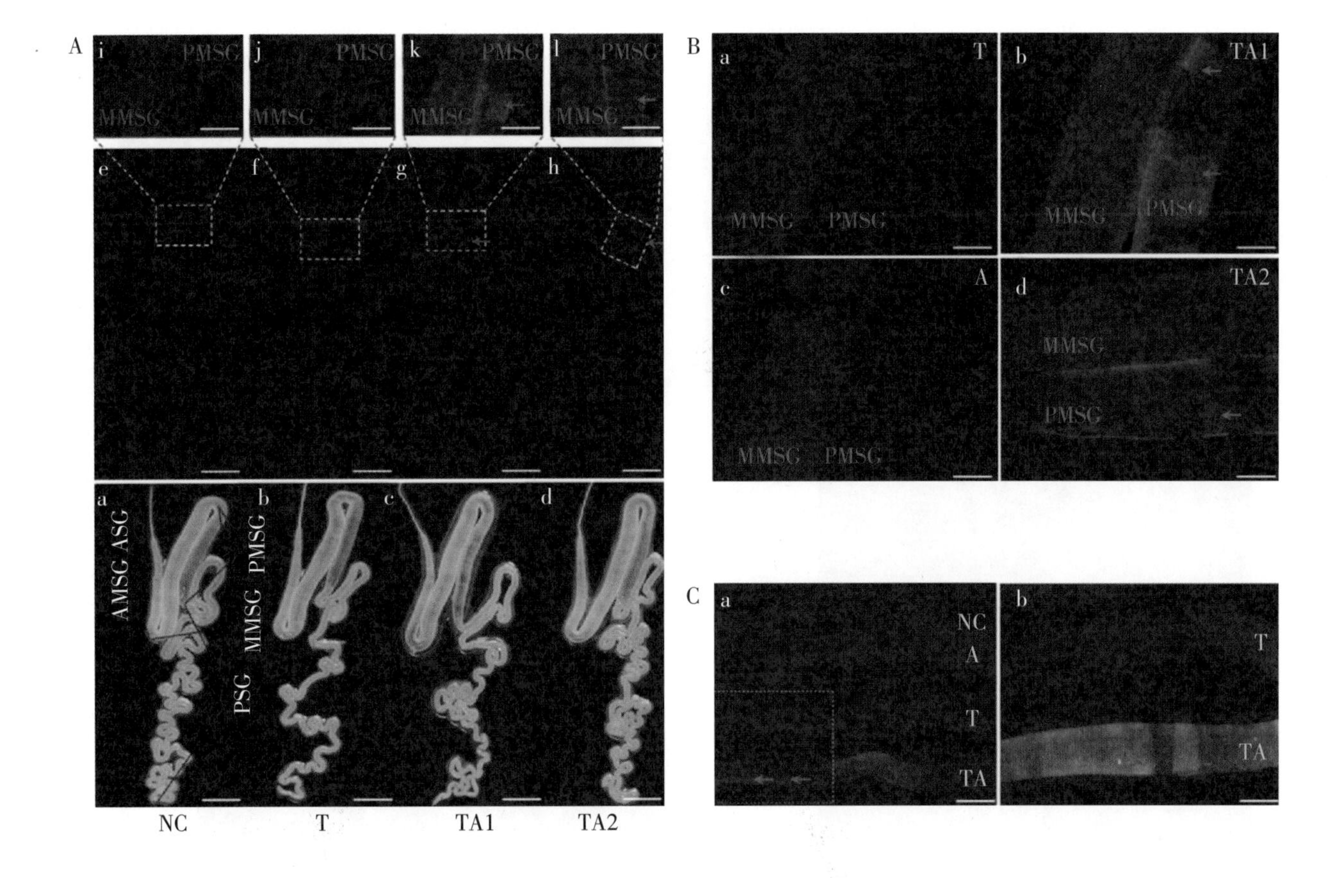

Figure 72. Observation of *Ds*Red fluorescence in the posterior subpart of MSGs. A Silk glands were dissected from one non-transgenic individual (a, e, i), one transgenic T individual (b, f, j), and two double-transgenic offspring TA1 (c, g, k), TA2 (d, h, l) in day 4 fifth instar larvae, and then illuminated under white light (a-d) and *Ds*Red-excitation-wavelength light (e-l). Letters i, j, k and l represent the magnified partial regions of e, f, g and h, respectively. Non-transgenic individual was used as a negative control (NC). Red arrows show the *Ds*Red fluorescent signal. Scale bars a-h 5 mm; i-l 2 mm. Exposure times: a-d 0.5 s; e-l 6 s. B Clear-cut observation of the mosaic expression of RFP in the magnified partial regions of MSGs. PMSGs from single transgenic individuals (T, A) and two double-transgenic individuals (TA1, TA2) were observed. Letters a, b, c, and d represent T, TA1, A, and TA2 individuals, respectively. ASG, AMSG, MMSG, PMSG, and PSG stand for anterior silk gland, anterior subpart of middle silk gland, middle subpart of the middle silk gland, posterior subpart of the middle silk gland, and posterior silk gland, respectively. Scale bar 0.58 mm. Exposure times: a-d 6 s. C Photographs showing the RFP of silk glands from non-transgenic (NC), one transgenic (A, T) and double-transgenic (TA) silkworms in the same field of vision. Letter b represents the magnified partial region of letter a. Scale bar: a 0.16 mm; b 0.38 mm. Exposure time: a 6 s; b 12 s. (Color figure online). [选自：*Cre-mediated targeted gene activation in the middle silk glands of transgenic silkworms (Bombyx mori)*，参见本书第424页。]

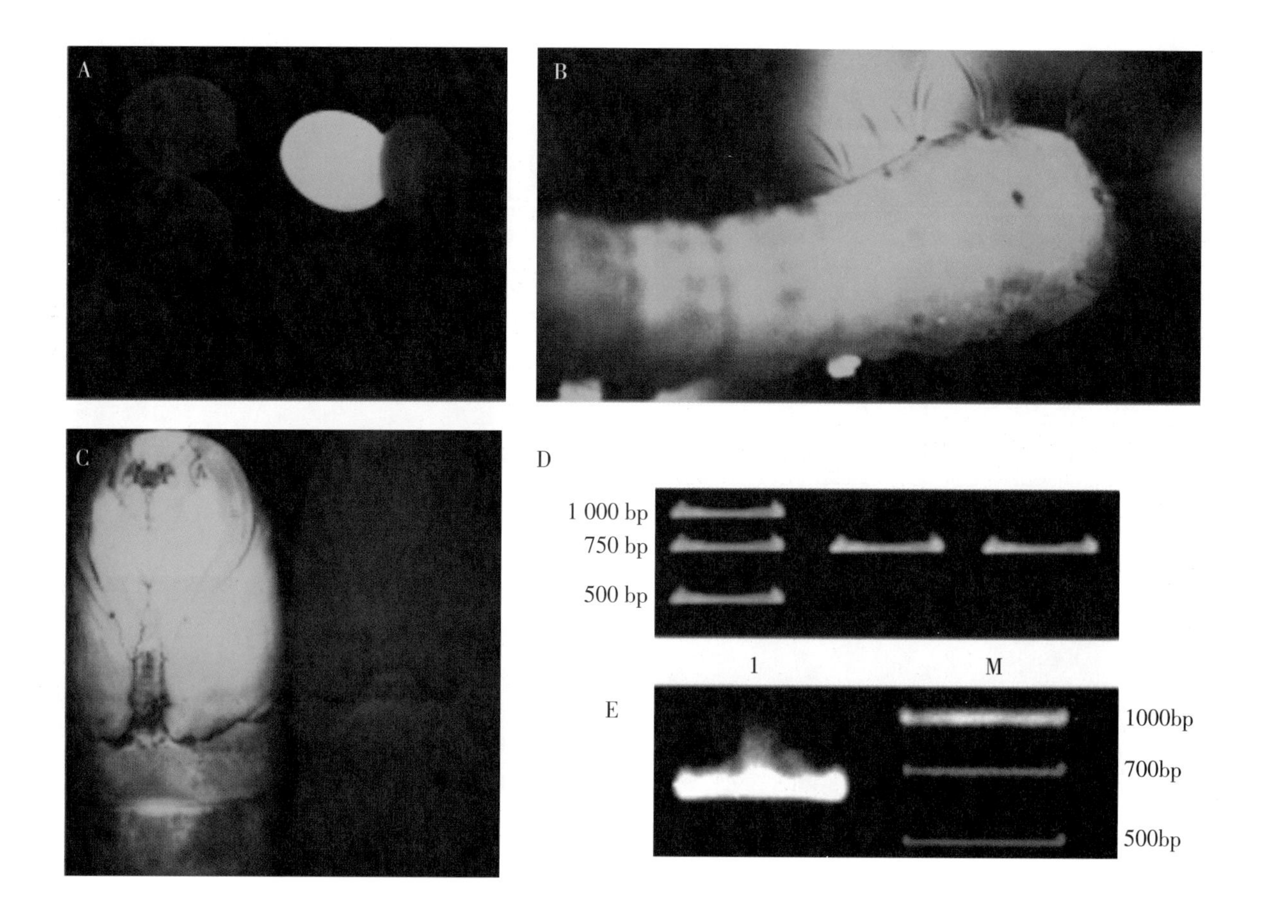

Figure 73. Screening and identification of transgenic silkworms. a Silkworm eggs with green fluorescence. b larva with green fluorescence in G_1 generation; c pupa with green fluorescence in G_1 generation; the left was transgenic, the right was normal; d PCR identification of transgenic silkworm in G_2 generation. M, DNA marker; lanes 1-2, PCR products representing egfp (0.72 kb) amplified from genomic DNA extracted from transgenic silkworm and pSK-attB/Pie1-EGFP/Zeo-PASV40, respectively. e Inverse PCR identification of the transgenic silkworm; M, DNA marker; lane 1, inverse PCR product representing sequence 6.（选自：*Construction of transformed, cultured silkworm cells and transgenic silkworm using the site-specific integrase system from phage φC31*，参见本书第426页。）

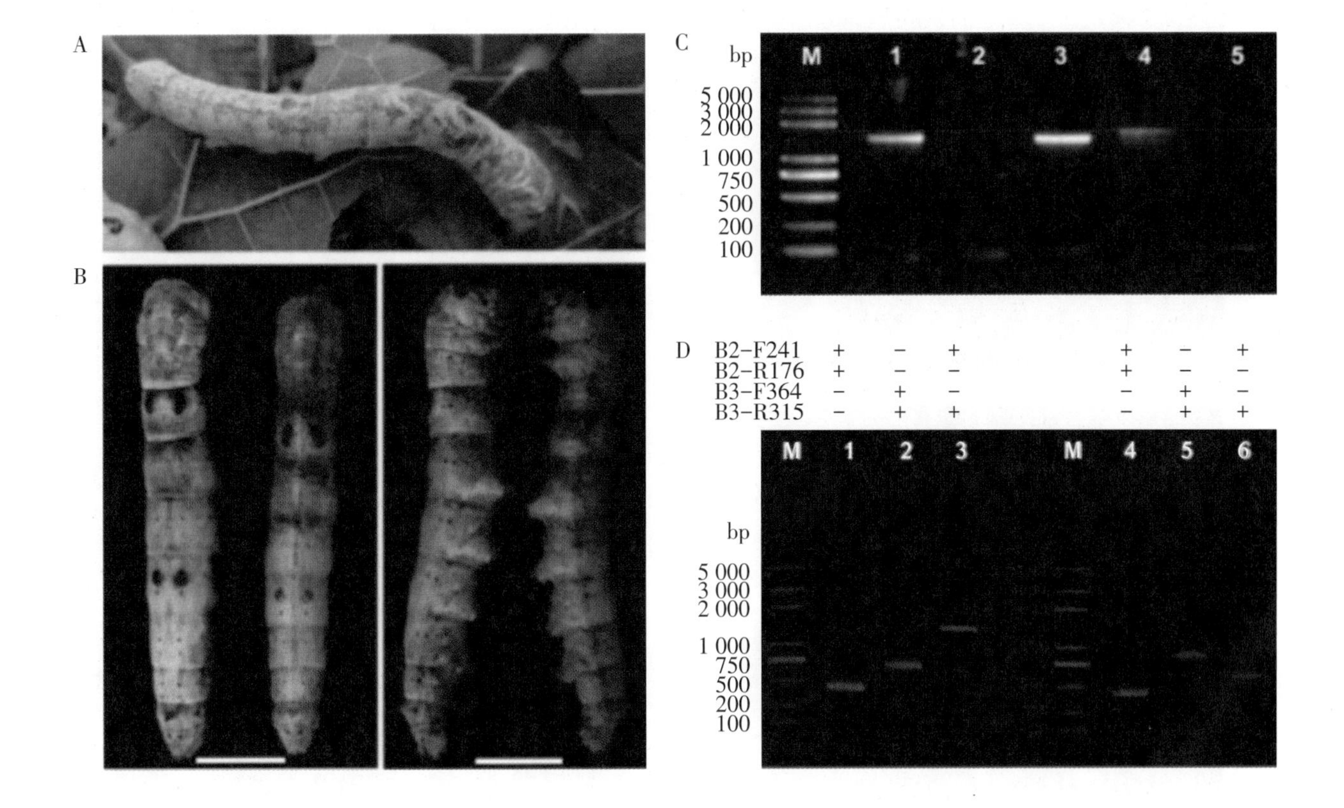

Figure 74. Large chromosomal deletions using a pair of TALENs and ssODN. (A) Mosaic mutations induced by co-injection of TALEN-B3 and ssODN. Silkworms were reared on fresh mulberry leaves. (B) Images of germline mutations from the dorsal side (left panel) and the lateral side (right panel). The worm on the left in each panel is wild type silkworm. Silkworms were reared on artificial diet. The scale bar represents 1 cm. (C) Gel analysis of PCR amplifications using primers B2-F241 and B3-R315 from genomic DNA of 5 G_1 mutant individuals (numbers 1 to 5). M represents the DNA ladders, the letters on the left indicate length of each band. (D) Gel analysis of PCR amplifications using genomic DNA of wild type (panel indicated as 1 to 3) and G_2 mutant silkworm from NO. 4 G_1 mutants (panel indicated as 4 to 6). Primer sets are shown on the top of each panel.（选自：*Multiplex genomic structure variation mediated by TALEN and ssODN*，参见本书第432页。）

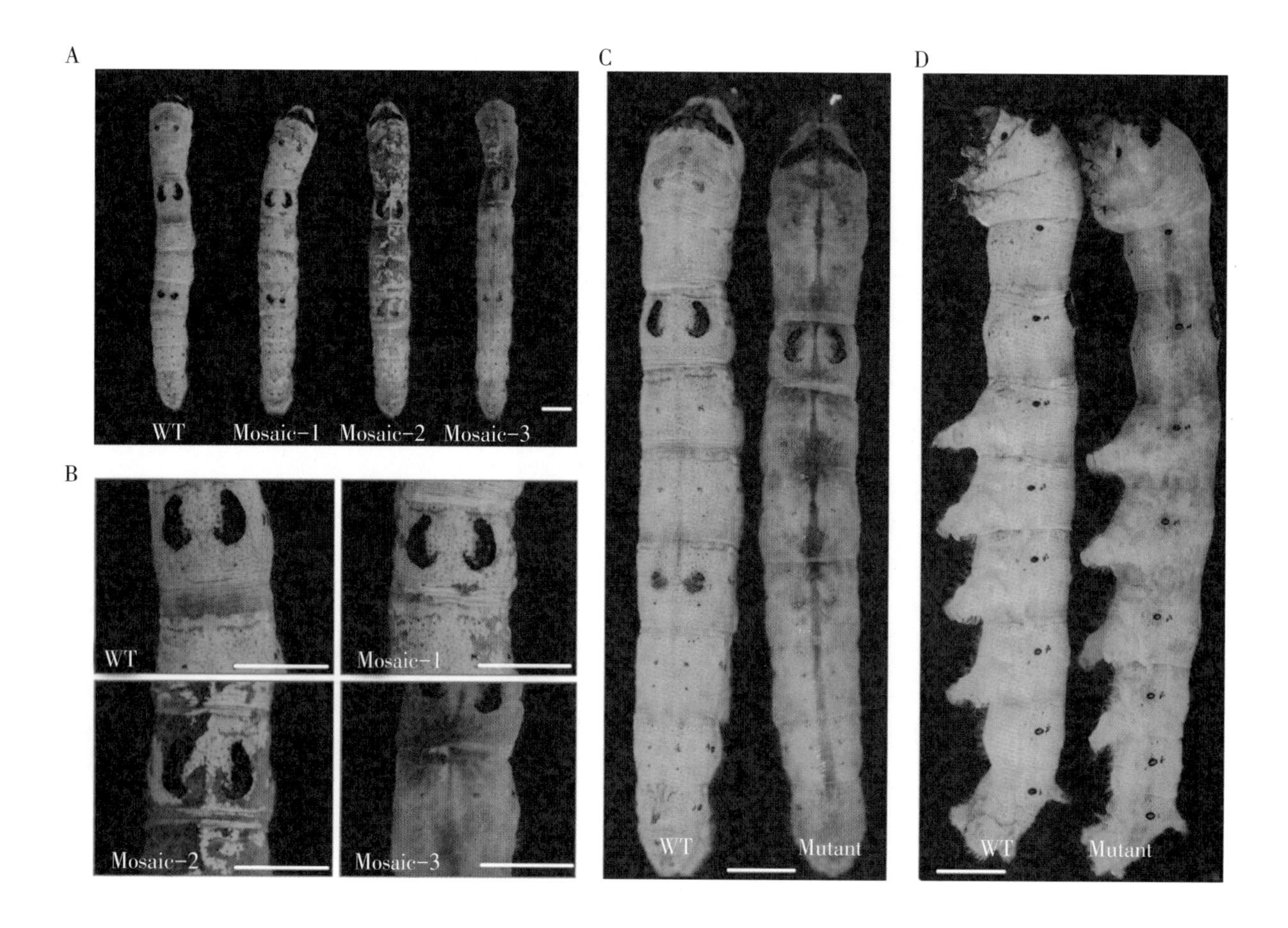

Figure 75. Phenotype of mosaic mutations and germline mutations. (A) Wild type (*Nistari*) and G_0 individuals with mosaic mutations. Mosaic-1, Mosaic-2, and Mosaic-3 represent the three different types of mosaic mutations observed. (B) Mosaic mutation (A) magnified. The epidermal cells were completely opaque in the wild-type silkworms, mostly opaque with translucent spots in G_0 Mosaic-1 mutant silkworms, half opaque and half translucent in G_0 Mosaic-2 mutant silkworms, and mostly translucent with opaque spots in G_0 Mosaic-3 silkworms. (C) Germline mutations from the dorsal side. (D) Germline mutations from the lateral side. The epidermal tissue was completely translucent in the germline mutations. The scale bar represents 5 mm.（选自：*Highly efficient and specific genome editing in silkworm using custom TALENs*，参见本书第434页。）

SILKWORM